Physics for the Health Sciences

Carl R. Nave, Ph.D.

Associate Professor of Physics
Georgia State University
Atlanta, Georgia

Brenda C. Nave, RN

Georgia Baptist Medical Center
Atlanta, Georgia

W. B. Saunders Company

Philadelphia London Toronto
Mexico City Rio de Janeiro Sydney

W. B. Saunders Company: West Washington Square
Philadelphia, PA 19105

1 St. Anne's Road
Eastbourne, East Sussex BN21 2UN, England

1 Goldthorne Avenue
Toronto, Ontario M8Z 5T9, Canada

Apartado 26370—Cedro 512
Mexico 4, D.F., Mexico

Rua Coronel Cabrita, 8
Sao Cristivao Caixa Postal 21176
Rio de Janeiro, Brazil

9 Waltham Street
Artarmon, N.S.W. 2064, Australia

Ichibancho, Central Bldg., 22-1 Ichibancho
Chiyoda-Ku, Tokyo 102, Japan

Library of Congress Cataloging in Publication Data

Nave, Carl R

Physics for the health sciences.

Includes bibliographical references and index.

1. Physics. 2. Biological physics. I. Nave, Brenda C., joint author. II. Title.

QC23.N33 1980 530'.02'461 79-64599

ISBN 0-7216-6666-3

Physics for the Health Sciences ISBN 0-7216-6666-3

Last digit is the print number: 9 8 7 6 5 4

Preface

The writing of this text was prompted by the conviction that students in nursing, the allied health sciences, and other health related fields need a background in physics which is broad in scope but which stresses those applications which will be of importance in their professional work. This is not just another introductory physics text with biological applications tacked on, though it treats most of the topics found in standard texts. The choice of topics, the emphasis given, and the detailed applications are directed toward the needs of students in the health sciences. Every effort has been made to produce an easily readable yet sound development of the physical principles. It is hoped that the numerous applications to biological and physiological problems will not only make these physical principles more useful as professional tools, but also stimulate the student toward a deeper study of the principles.

The order of topical coverage is similar to that in most introductory physics texts, with applications to health related topics integrated into the material. Several applications were considered significant enough to warrant treatment in separate chapters. After the basic treatment of the physics of fluids in Chapters 5 and 6, the applications to the circulatory system are considered in Chapter 7 and further topics concerning the flow of liquids and gases are considered in Chapter 8. After the basic treatment of electricity and magnetism in Chapters 12 and 13, applications are made to electrical safety in Chapter 14, basic instrumentation in Chapter 15, and bioelectricity in Chapter 16.

It is presumed that the student will take the basic science courses early in the curriculum of study. Therefore, the text has been made as nearly self-contained as possible, particularly in regard to the mathematics used. Only a knowledge of elementary algebra is assumed. Since problem solving is fundamental to the study of physics, numerous examples are given in the text to illustrate problem solving techniques. Worked examples are also included among the exercises at the end of each chapter. Instructional objectives are included at the beginning of each chapter as an indication of the level of proficiency sought.

The text has been designed for use in a one semester or one quarter course, but could be extended to a two quarter course with additional emphasis on problem solving and laboratory experience. Some laboratory experience is strongly recommended, and a suggested list of laboratory exercises is included.

We are indebted to many of our colleagues and students who have offered helpful suggestions during the preparation of this text. In particular we wish to express our gratitude to Dr. James E. Purcell for many valuable discussions. We are also grateful to the many users of the first edition who took time to write to us with comments about their experience with the text.

In the second edition the number of problems has been more than doubled, many examples have been added, and the modern physics portion in Chapters 20 and 21 has been significantly extended. In particular, the physical and biological effects of ionizing radiation have been more carefully treated, with more quantitative detail about the assessment of radiation hazards.

The preparation of the second edition was greatly aided by detailed comments from Dr. Richard Dittman (University of Wisconsin-Milwaukee), Dr. Paul Peter Urone (California State University, Sacramento), and Dr. Norman Tepley (Oakland University, Rochester, Michigan). Of course the final product, with any inadequacies which remain, is entirely the responsibility of the authors.

C. R. NAVE
B. C. NAVE

Suggested Laboratory Exercises

1. VELOCITY AND THE ACCELERATION OF GRAVITY

A study of the motion of a falling object to illustrate the relationships between position, velocity, acceleration, and time. The Cenco-Behr free fall apparatus (Cenco #74905) could be used. Alternatively, projectile motion or other types of accelerated motion could be investigated.

2. EQUILIBRIUM

A study involving the balancing of forces and torques to illustrate the concept of equilibrium. A meter stick with a support and a supply of known and unknown masses could be used. A vector force table such as the Cenco #74285 is helpful for demonstrating the balancing of forces.

3. THE GAS LAWS

A Boyle's Law apparatus (Cenco #76365) may be used to study constant temperature processes involving a fixed mass of gas. A constant volume of gas can be used to study the dependence of pressure upon the absolute temperature and to obtain an extrapolated value for absolute zero in the Celsius scale (Cenco #76408).

4. POISEUILLE'S LAW

Volume flow rates through fixed lengths of glass tubing illustrate the dependence of flow rate upon fluid pressure and tubing radius. The exercise shows that the flow rate depends much more strongly upon the tubing radius, but quantitative agreement with Poiseuille's law is difficult to obtain.

5. SPECIFIC HEAT

The measurement of the specific heat of lead shot by using a water-filled calorimeter illustrates the large heat capacity of water and the application of the conservation of energy principle to heat transfer.

6. SERIES AND PARALLEL ELECTRIC CIRCUITS

A simple network using batteries, lightbulbs, resistors, voltmeters, and ammeters can acquaint students with basic electric measurements, circuit fundamentals, and the concepts of voltage, current, and resistance.

7. OSCILLOSCOPE AND ECG MONITORING

Basic exercises in the use of laboratory oscilloscopes can increase the student's understanding of the principles of electricity and the use of oscilloscopes for patient monitoring. ECG simulators could be used for this experiment, or actual ECG measurements may be used if suitable amplifiers are available.

8. ELECTRICAL HEATING

Measurements of electrical power in a small immersion heater can be correlated with the temperature change of a known mass of water.

9. LENS PROPERTIES AND AN EYE MODEL

The lens equation can be investigated with simple convex lenses, a bright object light, and a white card upon which to form images (Cenco #85745). An eye model (Cenco #87660) provides an excellent investigation of the vision process and the correction of defective vision.

10. ATOMIC SPECTRA

The discrete nature of atomic spectra, the quantum nature of light, and the concept of quantization of atomic energy levels can be demonstrated by an investigation of the emission spectrum from a gas-filled spectrum tube. The spectra of hydrogen, helium, and neon provide good examples of diverse spectra. Equipment: spectrum tubes (Cenco #87235, 87215, 87220), spectrometer table (Cenco #87006), a diffraction grating, and a power supply for the spectrum tubes (Cenco #87208). Holograms or other applications of lasers might be demonstrated in conjunction with this experiment.

11. NUCLEAR COUNTING

Various types of safe radioactive samples can be investigated. The penetration of various thicknesses of material and the dependence of the radiation intensity upon the distance from the source could be studied. A general study can be made with a Geiger-Müller counter, but a proportional counter is desirable for more quantitative experiments.

Contents

CHAPTER ONE

Measurement and the Scientific Method

Along with its personal and subjective aspects, modern health care involves the application of the latest scientific and technological advances to the treatment of disease and injury. Building upon decades, even centuries, of research into the nature of the human body and its functioning, a large scientific community is continuously striving to increase our understanding of the unbelievably complex organism. It is clearly impractical for the student in nursing or the allied health sciences to try to survey the whole of this body of scientific knowledge, or even to keep abreast of all the latest developments. However, the understanding of a basic framework of scientific principles is not only practical but extremely important. The many scientific disciplines have a common denominator in the basic sciences of physics and chemistry. Although the boundary between these two sciences if often indistinct, this text is designed to provide a basic understanding of the physical principles most applicable to health care. These principles should not only provide practical information for direct application to health care, but also provide a framework which will be of value in the study of other disciplines such as biology and physiology.

THE ROLE OF THE EXPERIMENT

In addition to presenting the relevant principles of physics and their applications, it is the aim of this text to present the material in such a way that insight is obtained about how scientific investigations are carried out and how the information obtained is systematized for general use.

Science is the study of reproducible phenomena. The final judge of the rightness or wrongness of a concept is a reproducible experiment. A statement about the expected behavior of an object cannot properly be called a "scientific" statement unless it can be backed up by some controlled experiment. For example, the statement, "That heavy rock will fall at the same rate as the small one if you drop them together," is accepted as a statement of scientific fact. It is accepted because, and only because, innumerable experiments have shown that in the absence of significant air friction, all objects accelerate equally in the earth's gravitational field. The knowledge obtained from such experiements is cumulative, and when sufficient information is obtained, physical laws can be formulated, such as the law of gravity. The detailed behavior of objects under the influence of gravity can then be predicted with confidence. All of this proceeds under the basic assumption that nature acts in an orderly and reproducible manner, not capriciously.

But the individual human body is not, strictly speaking, a reproducible phenomenon, so how can medical treatment be really scientific? While the human being as a whole is unique, many of the phenomena taking place within the body can be readily understood from the basic physical laws. Thus, the application of the laws of physics, chemistry, and the other sciences provides a wealth of information which the attending physician or nurse may use to better understand the problems of a particular patient. It is certainly not suggested that the information is complete, nor that the problems are simple. But history has shown that the hours of painstaking experimentation in the laboratory pay off with the vital information needed for medical advances.

The amount which we do not know is vast indeed. But the scientific method provides us with tools with which to chip away at this vast ignorance. This is the area where theories are useful. Often several theories or models may be advanced to explain a given phenomenon. If more than one theory can explain all the observed data, then the theories can be used to predict further, yet unobserved behavior. Then the method of experimentation can be used to test the theories. This method has proved very fruitful in extending our understanding.

MEASUREMENT AND ACCURACY

Experimentation generally implies the measurement of something. At least the observations made must be recorded in such a way that they can be compared with other experiments. The value of the experiments depends upon the accuracy of the measurements made and the clarity with which the measured parameters are defined. In order that people in various parts of the world may compare their experiments directly, a wide system of international standards has been established. Standards for the fundamental quantities such as mass and length are kept for the two main systems of units in current use, the metric system and the British system. The specific units for the British system and the two variants of the metric system, the MKS (meter-kilogram-second) and the CGS (centimeter-gram-second), will be introduced in the text. There is a strong trend toward the metric system in the scientific community, including the medical profession. It is basically more logical and convenient since the subunits are always decimal, and most countries which do not now use it are in active "metrification" programs.

The unit system referred to as the British system evolved over a time span of several centuries, dating back to ancient Rome. Though it was adequate for agriculture and commerce when developed, the complicated relationships between the subunits make it undesirable for scientific use. (12 inches = 1 ft, 3 ft = 1 yard, 1760 yards = 1 mile, etc.) With increasing international commerce, as well as scientific cooperation, there is a pressing need for a worldwide unit system. It is to be hoped that the United States will soon join the rest of the world in a complete conversion to the metric system.

SIGNIFICANT DIGITS

No matter how carefully a measurement is made, it is never exact. A vital part of any good experiment is an evaluation of the degree of uncertainty of the results. The evaluation of the probable range of error in an experimentally determined parameter is often almost as important as the numerical value of the parameter itself. A body temperature reading of 102 degrees Fahrenheit indicates a significant fever if the uncertainty of the thermometer reading is $\pm.2$ degrees. But if a poorly calibrated thermometer with an uncertainty of ±5 degrees is used, then the existence of fever is still in doubt, since the temperature could be anything between 97 and 107 degrees. It might be dangerous to give medication to reduce fever until a more accurate determination of the temperature is made.

One way of keeping track of uncertainties is to record the probable range of uncertainty along with the number, for example 102 ± 5 degrees. It wouldn't make sense to record this temperature as 102.2416 ± 5 degrees because the extra digits don't give any new information. Without the specification of the uncertainty range, these digits might even be misleading because they imply an accuracy of measurement which does not exist. Another way of saying it is to say that the digits .2416 in the above number are not *significant*. The three digits 102 are said to be the significant digits of the measurement. Even though you are not sure about the 2, it is considered to be a significant digit since it specifies the center of the uncertainty range.

Although there is no universal understanding, when a number like 5.41 meters is recorded, it is often assumed that you know the value is between 5.40 and 5.42 meters, or that all three of the digits recorded are significant. This could be misleading, so if the uncertainty in the measurement is ±1 meter, it would be better practice to record the number as simply 5 meters.

In most experiments, the final result will be obtained by combining the results of several measurements, each of which has an associated uncertainty. In laboratory work of any kind, the experimenter should make a habit of assigning an uncertainty to each measurement made. This involves some judgment, and often is an assessment of the upper bound of error which would be made in a given measurement.

For example, if a measuring scale is marked with inches and then subdivided into tenths of inches, you should be able to measure lengths with an error less than one tenth of an inch. In a measurement you could estimate to one hundredth of an inch and might record the length of an object as 5.23 inches. In this case the digit 3 is significant, and you might be confident that the end of the object lies between 5.22 and 5.24, so that the uncertainty is about .01 inch. However, you might make a similar error in lining up the zero end of the scale with the other end of the object, so a conservative record of the measurement would be $5.23 \pm .02$ inches.

The uncertainty of the final result of the experiment must be obtained by considering the uncertainties of the individual measurements. A detailed discussion of the treatment of cumulative errors does not seem warranted here and the interested reader is referred to References 3 and 4. A rough rule of thumb is that the final result should not have more significant digits than the most uncertain individual measurement. This is like judging the strength of a chain by the strength of its weakest link.

For example, suppose the dimensions of a box were measured to be $2.41 \pm .02$ meters, $3.0 \pm .2$ meters, and $.6835 \pm .0005$ meters. The volume calculation would give

$$2.41 \times 3.0 \times .6835 = 4.9417 \text{ meters}^3$$

if all the digits were recorded. But, because the poorest measurement has only two significant digits, it would make more sense to record the volume as 4.9 meters^3 or even just 5 meters^3. If the uncertainties are added as an estimate of the uncertainty of the result, the volume might be recorded as $4.9 \pm .2$ meters^3, dropping the digits beyond the last significant digit in the volume.

An alternate way to assess the uncertainty of a result which combines several measurements is to calculate the percentage of uncertainty of each measurement and add the percentages to obtain an upper bound of uncertainty for the result. This process is often recommended when multiplication or division of measurements is used in obtaining the final result. In the volume calculation above, the dimensions of the box could be written $2.41 \pm .83\%$, $3.0 \pm 6.7\%$ and $.6835 \pm .07\%$ meters. The volume would then be $4.9417 \pm 7.5\%$, and evaluating the actual uncertainty and dropping the non-significant digits would lead to a result of $4.9 \pm .4$ meters^3. Note that this process leads to a larger range of uncertainty than the above, and is therefore a more conservative statement of the experimental result.

Sometimes a more reasonable estimate of the probable error is obtained by using the percentage uncertainty of the least accurate factor involved rather than the sum of the uncertainties, which in this case leads to the intermediate result $4.9 \pm .3$ meters^3. Different experimenters may use different methods for arriving at the "error limits" placed on experimental results, but it is extremely important to have some systematic method for assessing experimental uncertainty. Experiments are often worthless without an assessment of their uncertainty, and scientific journals justifiably refuse to publish measurements without assessments of the possible errors in the reported values.

PROBLEMS

1. A rectangular object has measured dimensions 1.3 cm and 12.5 cm. Calculate the area of the rectangle and express it with the appropriate number of significant digits.

2. A bar of aluminum has measured dimensions $1.2 \pm .2$ cm, $5.7 \pm .2$ cm, and $2.3 \pm .2$ cm. Calculate the volume and express it with the appropriate number of significant digits. Calculate the uncertainty in the volume by

summing the percentage errors of the individual measurements and write the volume with its uncertainty in cm^3.

3. Rough measurements of a cube of material yield a volume of about 25 cm^3. If it were important to know the volume with error limits of ± 1 cm^3, how accurately would you have to measure each side of the cube?

4. If it is difficult to assess the uncertainty of a measurement by other means, it can be estimated by repeating the measurement several times and averaging the results. The maximum deviation from the average is a reasonable estimate of the possible error. (This assumes that the errors are random and not due to faulty measurement instruments or other systematic errors.) In making successive measurements of diastolic blood pressure, the following pressures in millimeters of mercury were obtained; 82, 86, 76, 92, 84, 83. Find the average diastolic pressure with its uncertainty.

5. An object is allowed to fall through a distance of $8.0 \pm .1$ meters and the time elapsed is found to be $1.3 \pm .1$ seconds. If the average speed of fall is the distance divided by the time, calculate this average speed and its uncertainty, using the sum of the percentage errors as an upper limit for the error of the result. Write the speed and uncertainty with the appropriate number of significant digits. Could you gain an extra significant digit by using a stopwatch which read to hundredths of a second, assuming you could measure to an accuracy of $\pm .01$ second?

REFERENCES

1. United States Department of Commerce, National Bureau of Standards. *Brief History of Measurement Systems,* Special Publication 304A, Washington, D.C., 1972.
2. ——— *A Metric America,* Special Publication 345, Washington, D.C., 1971.
3. Pugh, E. M., and Winslow, G. H. *The Analysis of Physical Measurements.* Reading, Mass.: Addison-Wesley, 1966.
4. Bevington, P. R. *Data Reduction and Error Analysis for the Physical Sciences.* New York: McGraw-Hill, 1969.

CHAPTER TWO

The Description of Motion

INSTRUCTIONAL OBJECTIVES

After studying this chapter, the student should be able to:

1. Precisely define the terms velocity and acceleration with appropriate units in the CGS, MKS, and British systems of units.
2. Solve word problems to find the position, velocity, acceleration, or time for motions with constant acceleration, given sufficient data.
3. Plot or interpret a simple graph such as a velocity vs. time graph.
4. Perform unit conversions, given the appropriate numerical conversion factors.

The first topics to be discussed are in the general area of physics known as mechanics. This is the logical first step, since much of the vocabulary of physics is developed in mechanics and many of the units used in other areas are based upon the units of mechanical quantities. The concepts of force, work, energy, power, and so forth which will be developed here are needed in all areas.

VELOCITY AND ACCELERATION

We begin with the question, "What must be specified in order to precisely describe the motion of an object? What would you have to measure and record so that another person who has not seen the motion could reproduce it precisely? For example, consider the description of motion of a ball which is tossed straight up into the air so that it comes back down to the original point. It is clear that the height and the time elapsed must be specified, and the speed of the ball. Further, the direction associated with the speed must be specified, since it is going up part of the time and down part of the time. When one tries to reduce this to precise data, it is evident that many speeds must be indicated since it starts out rapidly, slows down and comes to rest at the top of its trajectory, and then speeds up on the way down. There is actually an infinite number of different speeds during this relatively simple motion, so the three quantities distance, speed, and time are not enough for the description of the motion. The fourth quantity needed is the acceleration. The systematic study of motion using these four quantities is called *kinematics*.

When given precise definitions, the quantities time, position, velocity and acceleration are sufficent to describe any motion of a point if the acceleration is constant. The words speed and velocity in common usage are often taken as synonymous, but in physics they are distinguished; the word *velocity* denotes the speed of an object with its associated direction at a given instant of time. Since the direction is taken to be an intrinsic part of the velocity, then when something is said to have a constant velocity, this means not only that its speed is not changing but also that the direction of motion remains the same. In the special case where the velocity remains constant, the velocity is related to the distance traveled in a given time t by the relationship:

$$v = \frac{s}{t}$$

where v is the velocity and s is the distance. The units of the velocity are those of length divided by time, such as miles/hour or feet/sec. The standard units for the common mechanical quantities in the British, MKS (meter-kilogram-second), and CGS (centimeter-gram-second) systems are summarized in Table 2–1. Not all of these units will be used, but the table is included for reference. The standard units for velocity in the three systems are ft/sec, m/sec and cm/sec, respectively.

TABLE 2–1 The Three Basic Unit Systems

QUANTITY	MKS	CGS	BRITISH
Length	meter	centimeter	foot
Time	second	second	second
Mass	kilogram	gram	slug
Velocity	m/sec	cm/sec	ft/sec
Acceleration	m/sec²	cm/sec²	ft/sec²
Force	newton (kg · m/ sec²)	dyne (gm · cm/ sec²)	pound (slug · ft/ sec²)
Work, energy	joule (nt · m)	erg (dyne · cm)	ft · lb
Power	watt (joule/sec)	erg/sec	ft · lb/sec
Torque	nt · m	dyne · cm	lb · ft
Pressure	nt/m²	dyne/cm²	lb/ft²

A more common type of motion would be that in which the velocity is changing. Unless otherwise noted, the discussion in this chapter will be limited to straight line motion, so the change considered here is one of changing magnitude rather than change of direction. In this case, the *average* velocity obeys the relationship

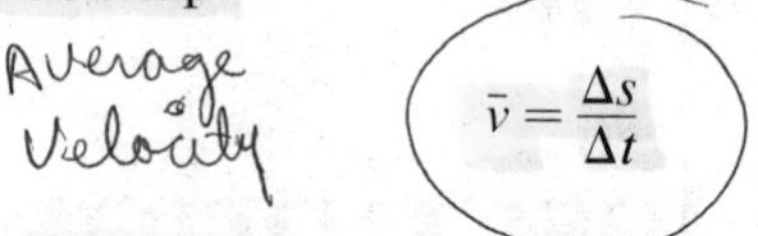

$$\bar{v} = \frac{\Delta s}{\Delta t}$$

where the bar over the v denotes that it is an average rather than an instantaneous velocity and Δs and Δt are the changes in distance and time, respectively. The Greek letter Δ preceding a quantity is used almost universally in scientific literature to indicate a change in the quantity. In general $\Delta s = s_f - s_i$ and $\Delta t = t_f - t_i$ where the subscripts i and f denote the initial and final values. However, it is usually convenient to start with $s_i = 0$ and $t_i = 0$ so that the expression for average velocity can be reduced to simply

$$\bar{v} = \frac{s}{t} \quad \text{or} \quad s = \bar{v}t.$$

In common usage, acceleration is taken to mean the rate of change of speed. When a car accelerates away from a traffic light, the "acceleration" is measured in terms of the number of seconds required to reach a certain speed, for example 60 miles/hour. No distinction is made between acceleration on a straight flat road and acceleration on a curved or hilly road, and the direction of travel is not part of the measurement. As is often the case, to be of the most general use in science, quantities must be given more precise definitions. For use in physics, the acceleration is defined as the rate of change of velocity. That means that an object is said to be accelerated if either the magnitude or direction of its velocity is changed. The logic of this definition will be more apparent in the discussion of circular motion in Chapter 3.

For the special case of motion in a straight line, the average acceleration is related to the change in velocity:

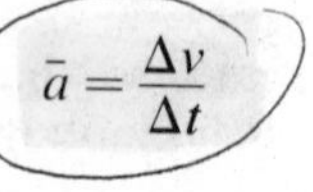

$$\bar{a} = \frac{\Delta v}{\Delta t}$$

Here $\bar{a}$ represents the average acceleration, and Δv and Δt are the changes in velocity and time respectively. The units of acceleration are those of velocity divided by time. If the velocity increases by 10 ft/sec in a time interval of 2 seconds, then the acceleration from the above relationship is 5 ft/sec per second or 5 ft/sec/sec (read five feet per second per second). This is obviously quite cumbersome, so it is usually written 5 ft/sec² and read "five feet per second squared." The corresponding acceleration units in the MKS and CGS systems are m/sec² and cm/sec² respectively.

If the initial velocity is denoted v_0 (for time $t = 0$) and the velocity at some arbitrary time t is denoted v, then the relationship between acceleration and velocity can be rewritten

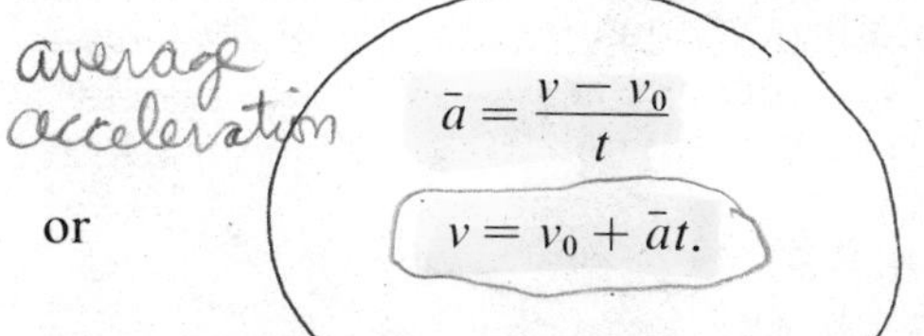

$$\bar{a} = \frac{v - v_0}{t}$$

or

$$v = v_0 + \bar{a}t.$$

For example, suppose a sled is pushed down a smooth incline with an initial velocity of 10 ft/sec and accelerates at the constant rate of 5 ft/sec². After three seconds it will have a velocity

$$v = 10 \text{ ft/sec} + (5 \text{ ft/sec}^2)(3 \text{ sec}) = 25 \text{ ft/sec}.$$

Note that all the quantities in this equation must have the same units of velocity, ft/sec.

The velocity and acceleration are both *vector* quantities; that is, they always have directions associated with them. The acceleration may be directed in a way such that it increases the velocity, or it may be directed opposite to the velocity so that the velocity is decreased (deceleration).

THE ACCELERATION OF GRAVITY

Recalling the example of the ball which is tossed straight up into the air, it can be shown that a constant downward acceleration is sufficient to explain all the features of that motion. Experiments show that an object which is dropped near sea level altitudes on the earth will accelerate downward with a constant acceleration of approximately 32 ft/sec². This is called the acceleration of gravity; the metric equivalents are 9.8 m/sec² and 980 cm/sec². An object dropped from rest will have a downward velocity of 32 ft/sec after one second, 64 ft/sec after two seconds, and so on, increasing 32 ft/sec each second.

Now if a ball is tossed upward with an initial velocity v_0 = 96 ft/sec, it will travel upward for three seconds before reaching its peak. It will then travel downward for three seconds, striking the ground after a total flight time of six seconds with a velocity of 96 ft/sec. All of this information may be obtained from the constant acceleration of gravity with the aid of Figure 2–1. For reference, the upward direction will be denoted the positive direction. Since the acceleration is always downward, it is written g = −32 ft/sec² (g is the symbol generally used to represent the acceleration of gravity). Since the velocity and acceleration are initially in opposite directions, the ball steadily slows down until it reaches

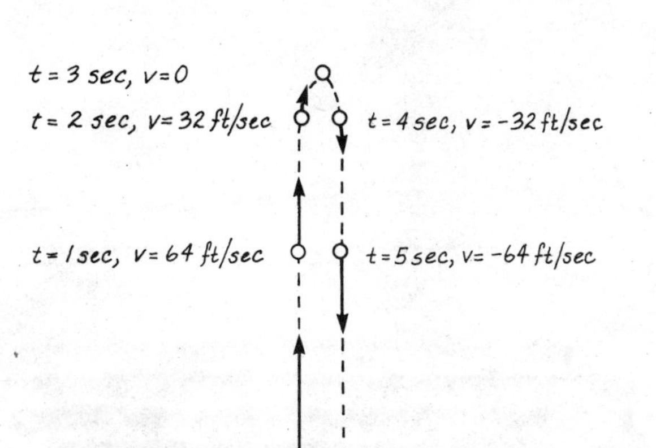

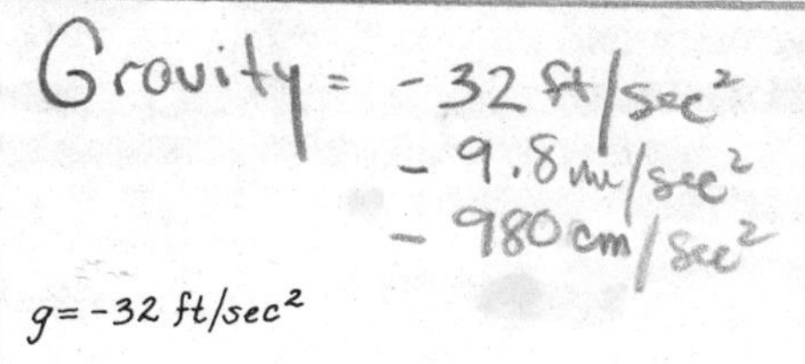

Figure 2–1 Illustration of the acceleration of gravity.

the peak, where its velocity is zero. The velocity at any time during the motion may be calculated from the equation $v = v_0 + at$. Substituting in the appropriate numbers,

$$v = 96 \text{ ft/sec} + (-32 \text{ ft/sec}^2)(t).$$

For example, after five seconds the velocity is

$$v = 96 \text{ ft/sec} - 160 \text{ ft/sec} = -64 \text{ ft/sec}$$

or 64 ft/sec downward.

MOTION WITH CONSTANT ACCELERATION

To find the distance traveled in a given time, the relationship $s = \bar{v}t$ is generally applicable, but in the case where the acceleration is constant, there are some further relationships which are often useful. In this case the average velocity is given by

$$\bar{v} = \frac{v_0 + v}{2}.$$

For example, if an object accelerates from $v_0 = 0$ to a velocity of 60 ft/sec in 2 seconds, the average velocity is 30 ft/sec and the object will have traveled 60 feet in the process. Since $v = v_0 + at$, the distance can be expressed as

$$s = \bar{v}t = \frac{(v_0 + v_0 + at)\,t}{2} = v_0 t + \tfrac{1}{2}at^2.$$

This gives the distance directly in terms of the initial velocity and the acceleration.

Problems involving constant acceleration can usually be solved by direct application of one of the following formulas:

$$s = \bar{v}t \qquad \textbf{2–1}$$

$$v = v_0 + at \qquad \textbf{2–2}$$

$$s = v_0 t + \tfrac{1}{2}at^2 \qquad \textbf{2–3}$$

Example. In order to find the height of a bridge, a rock was dropped off the bridge and timed. If it took three seconds to hit the water, what is the height of the bridge?

Solution. To make sure that directions are used consistently, we will define upward as the positive direction. Applying equation 2–3 with the known data

$$v_0 = 0$$

$$a = g = -32 \text{ ft/sec}^2$$

$$t = 3 \text{ sec}$$

the height is found to be

$$s = \tfrac{1}{2}(-32 \text{ ft/sec}^2)\,(9 \text{ sec}^2) = -144 \text{ ft},$$

where the minus sign indicates that the distance is *below* the starting point.

Example. Suppose the velocity of the blood flowing in an artery was found to be 120 cm/sec at one point and 80 cm/sec at a point 50 cm further along the artery. Assuming constant deceleration, how long would it take the blood to flow from one point to the other?

Solution. Equation 2–1 can be used since the distance is known and the average velocity can be found from the information given.

$$\bar{v} = \frac{v_0 + v}{2} = \frac{120 \text{ cm/sec} + 80 \text{ cm/sec}}{2}$$

$$= 100 \text{ cm/sec}$$

$$t = \frac{s}{\bar{v}} = \frac{50 \text{ cm}}{100 \text{ cm/sec}} = \tfrac{1}{2} \text{ second.}$$

Example. When a diver leaves a 32 foot diving platform, how long does it take for him to reach the water, assuming he has no initial downward velocity?

Solution. Equation 2–3 can be used since v_0, a, and s are known. If we choose upward as the positive direction, the distance is negative (downward), thus

$$-32 \text{ ft} = \tfrac{1}{2}(-32 \text{ ft/sec}^2)\,t^2$$

$$t^2 = 2 \text{ sec}^2$$

$$t = \sqrt{2} = 1.4 \text{ seconds.}$$

GRAPHICAL DESCRIPTION OF MOTION

Since the distance and velocity are explicit functions of the time, it is often useful to plot them on a graph so that the behavior over a given period of time is evident at a glance. In Figure 2–2 a plot of velocity versus time is illustrated. The table at the left gives the data from which the plot was made. From the plot it is evident that the velocity is increasing rap-

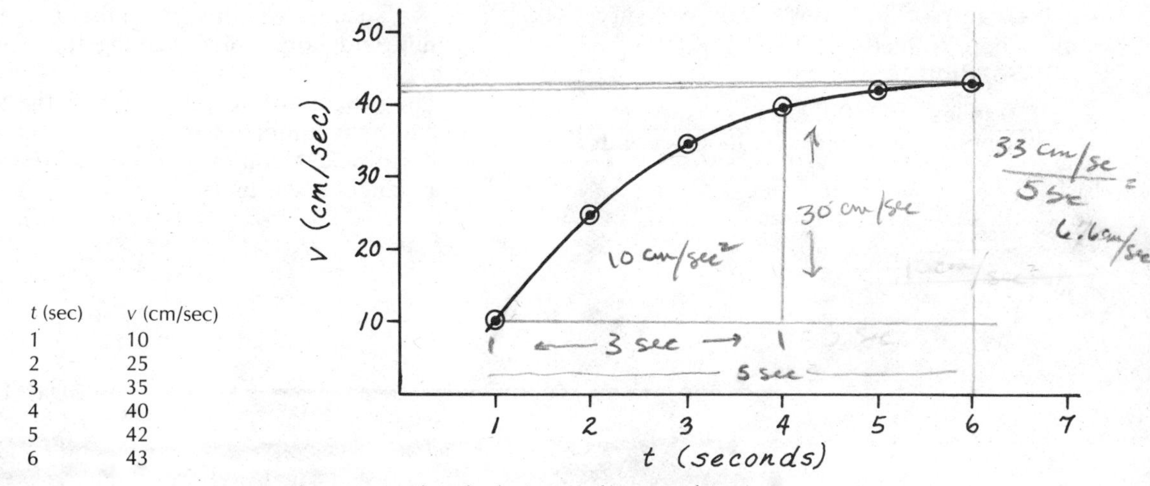

t (sec)	v (cm/sec)
1	10
2	25
3	35
4	40
5	42
6	43

Figure 2–2 Plot of velocity as a function of time.

idly near the beginning of the time interval and very slowly after about 4 seconds. The slope of the velocity vs. time plot gives the acceleration; that is, a steep slope means a large acceleration. A graphical presentation is useful when the motion is too complicated to calculate easily, because it gives a visual display of the general behavior.

THE USE OF POWERS OF TEN

Suppose you had to write down the number of cells in the human body, or the size of a single cell in the usual manner of writing numbers. Either number would be cumbersome, one because it is so large and the other because it is so small. The "powers of ten" notation is quite convenient for writing either very large or very small numbers. The usual practice is to shift the decimal point so that the number is between one and ten and then to multiply by the appropriate power of 10. For example,

$$1{,}000{,}000 = 1 \times 10^6$$

$$0.00000143 = 1.43 \times 10^{-6}$$

$$1740 = 1.74 \times 10^3.$$

For numbers the size of the third example the scheme is of marginal value, but for the extreme cases the text will make use of this notation. The addition, multiplication, and division of numbers expressed in the powers of ten notation can be accomplished with the aid of a few rules. These rules are reviewed and examples given in Appendix B.

Since the metric units are decimal in their divisions, the system of prefixes indicating the divisions is often useful. The commonly used prefixes are listed in Table 2–2.

TABLE 2–2 The Standard Prefixes for Units

nano	$= 10^{-9}$	= 1/1,000,000,000
micro	$= 10^{-6}$	= 1/1,000,000
milli	$= 10^{-3}$	= 1/1,000
centi	$= 10^{-2}$	= 1/100
deci	$= 10^{-1}$	= 1/10
deka	$= 10^{1}$	= 10
hekto	$= 10^{2}$	= 100
kilo	$= 10^{3}$	= 1,000
mega	$= 10^{6}$	= 1,000,000
giga	$= 10^{9}$	= 1,000,000,000

Using these prefixes 1000 grams equal one kilogram, 1/1000 gram equals one milligram, and so forth.

THE CONVERSION OF UNITS

The changing of quantities from one unit system to another is a necessary but often tedious task. A table of conversion factors is included in the Appendix for the reader's convenience (Table T-2). The task of unit conversion can often be made less onerous by carrying out the conversion as a product, with the conversion factors and their associated units treated like numerical fractions.

Example. Convert 12 inches to centimeters.

Solution.

$$(12 \text{ inches})\left(\frac{2.54 \text{ cm}}{1 \text{ inch}}\right) = 30.5 \text{ cm}$$

The inch units cancel, leaving the desired cm.

Example. Convert the velocity 60 miles/hour to feet/sec.

Solution.

$$\left(\frac{60 \text{ miles}}{1 \text{ hour}}\right)\left(\frac{5280 \text{ feet}}{1 \text{ mile}}\right)\left(\frac{1 \text{ hour}}{3600 \text{ seconds}}\right)$$

$$= 88 \text{ ft/sec}$$

The two successive conversion factors cancel the mile and hour units, leaving the desired ft/sec.

The process of keeping track of the units is one to be recommended at all times. Often a mistake in a calculation can be spotted with just a glance at the units.

SUMMARY

The physical quantities distance, velocity, acceleration, and time are required for a precise description of the motion of an object. The basic units involved are those of distance and time, since the velocity is defined as the rate of change of distance and the acceleration is defined as the rate of change of velocity. Three systems of units, the MKS, CGS, and British systems, are in common use. There is a worldwide trend toward the metric systems, the MKS and CGS, since they are more logically developed. In the case of straight line motion with a constant acceleration, the distance *s*, the velocity *v*, and the acceleration *a* can be described by the relationships.

$$s = \bar{v}t \quad \text{where } \bar{v} = \frac{v_0 + v}{2} \qquad \textbf{(1)}$$

$$v = v_0 + at \qquad \textbf{(2)}$$

$$s = v_0t + \tfrac{1}{2}at^2 \qquad \textbf{(3)}$$

With more complicated straight line motion, it is often useful to graph these quantities as a function of time to describe the motion. Distance, velocity, and acceleration are *vector* quantities, requiring both a size and a direction to completely specify them.

REVIEW QUESTIONS

1. What are the advantages of the metric system over the British system of units?

2. What is the difference between a scalar quantity and a vector quantity? Name several examples of each.

3. How is velocity distinguished from speed?

4. Under what conditions would acceleration be different from the "rate of change of speed"?

5. Under what conditions could an object continue to travel at a constant speed and yet be accelerated?

PROBLEMS

Worked Example: (a) Convert 20 ft to inches; (b) convert 20 ft to meters; (c) convert 1 mile to meters.

Solutions: The conversions may be accomplished with the factors in Table T-2, Appendix A.

(a) $20 \text{ ft} = (20 \text{ ft})\left(12 \frac{\text{in}}{\text{ft}}\right) = 240$ inches.

In (b), the conversion factor 3.28 ft/m must be used in division to cancel the units of feet:

(b) $20 \text{ ft} = (20 \text{ ft})\left(\frac{1}{3.28} \frac{\text{m}}{\text{ft}}\right) = 6.1$ meters.

In (c), a two-step conversion may be used since Table T–2 does not contain a single conversion factor from miles to meters:

(c) $1 \text{ mile} = (1 \text{ mile})\left(5280 \frac{\text{ft}}{\text{mile}}\right)\left(\frac{1}{3.28} \frac{\text{m}}{\text{ft}}\right) = 1610$ meters.

1. Convert 30 inches to centimeters. How many centimeters are there in a yard?

2. Find the equivalent of 15 meters in feet.

3. How many square inches are contained in an area of one square foot? How many square feet are contained in an area of one square meter?

4. Automobile speeds in Europe and elsewhere are measured in kilometers per hour. What is the equivalent of 55 miles/hr expressed in km/hr?

5. A passenger in a European sports car glanced at the speedometer and was alarmed to see the needle indicating 110. Then he realized that this number was in kilometers per hour. What was his speed in miles/hr?

Worked Example: The speed of sound in air is about 1100 ft/sec. If lightning strikes 1 mile from you and produces thunder, how long will it take for the sound to reach you?

Solution: The relationship $s = \bar{v}t$ is appropriate.

$$s = 1 \text{ mile} = 5280 \text{ ft}$$

$$\bar{v} = 1100 \text{ ft/sec}$$

$$t = ?$$

$$5280 \text{ ft} = (1100 \text{ ft/sec})t.$$

Dividing both sides of the equation by 1100 ft/sec to isolate the unknown quantity t:

$$t = \frac{5280 \text{ ft}}{1100 \text{ ft/sec}} = 4.8 \text{ sec.}$$

6. An object moving at a constant speed travels 30 meters in 6 seconds. (a) What is the speed of the object? (b) How far does the object move in 10 seconds? (c) How long will it take for the object to move 80 meters?

7. A running time of 10 seconds would be close to the world record for the 100 meter dash. What is the corresponding averge velocity in m/sec? What would be the equivalent in miles/hr?

8. What driving time will be required to travel 600 miles if the average speed

is 50 miles/hr? If the total travel time is 15 hours, what is the average speed?

9. If something falls off the front seat of your car while you are driving at 55 miles/hr and you quickly reach to get it, retrieving it in 0.6 second, how far will your car travel while your eyes are off the road?

10. A typical reaction time to get your foot on the brake in your car is 0.2 second. If you traveling at a speed of 60 miles/hr (88 ft/sec), what distance will your car travel during this reaction time?

11. The pictures of the surface of Mars which were transmitted by the spacecraft Mariner 4 took 12.0 minutes to reach Earth. Given the velocity of light as 186,000 mi/sec (3×10^8m/sec), how far away was Mariner 4 when it transmitted? Express the answer in miles and in meters.

12. A television picture in the United States is composed of 525 "lines" which sweep across the face of the picture tube. These lines represent the path of an electron beam which sweeps out these 525 lines every 1/30 of a second. If the horizontal dimension of the picture is 20 inches, how fast must the beam travel? (The beam must sweep back across the screen to start a new line, but assume that the time for this resweep is negligible compared to the picture sweep.) Express your answer in ft/sec and miles/hr.

13. After being ejected by the "electron gun" in the back of a color television set, the electron may travel in the vacuum of the picture tube at a speed of about 9×10^7m/sec. If it takes 6×10^{-9} seconds for the electron to travel to the face of the picture tube, what is the distance traveled?

Worked Example: If a falling object increases its velocity from 40 m/sec to 60 m/sec in 2 seconds, what is the acceleration? What is the average velocity during that interval? How far will it travel during that time?

Solution: The acceleration, the rate of change of velocity, is given by

$$a = \frac{v_2 - v_1}{t} = \frac{60 \text{ m/sec} - 40 \text{ m/sec}}{2 \text{ sec}} = 10 \text{ m/sec}^2.$$

The average velocity is:

$$\bar{v} = \frac{v_1 + v_2}{2} = \frac{40 \text{ m/sec} + 60 \text{ m/sec}}{2} = 50 \text{ m/sec}.$$

The distance traveled is then

$$s = \bar{v}t = (50 \text{ m/sec})(2 \text{ sec}) = 100 \text{ m}.$$

Note that if the final velocity is less than the initial velocity, the acceleration has a negative sign. A negative acceleration is referred to as a deceleration.

14. If the acceleration of gravity is 9.8 m/sec², how fast will an object be traveling 2 seconds after being dropped? What will be its average speed between times 2 and 3 seconds?

15. The velocity of an object is observed to increase from 30 m/sec to 80 m/sec in a time interval of 5 seconds.

    a. What is the acceleration of the object?
    b. How far does it travel during the 5 seconds?

16. A car traveling 60 mi/hr (88 ft/sec) is brought to a stop in 3 seconds. What is the average acceleration during that time? How far will the car travel before coming to rest, if the acceleration is assumed to be constant?

17. A car traveling at a speed of 50 ft/sec is brought to rest in 2 seconds.

 a. What is the acceleration?
 b. How far will the car travel during the 2 seconds?

18. A car accelerates from rest to 50 mi/hr in 10 seconds. Find the acceleration and the distance traveled in the 10 second interval, assuming that the acceleration is constant.

19. If an object has an initial speed of 5 ft/sec and moves in a straight line for 3 seconds with an acceleration of 2 ft/sec^2, what will be its speed?

20. A bullet traveling at 200 m/sec is stopped in a distance of 6 cm.

 a. Assuming that its deceleration is constant, what is its average speed during the impact process?
 b. What is the acceleration?
 c. How long will it take for the bullet to travel this 6 cm distance?

21. If the velocity of flow of a fluid was found to be 120 cm/sec at one point and 80 cm/sec at a point 50 cm further along the flow path, how long would it take the fluid to flow from one point to the other? What is the acceleration of the fluid?

22. The blood flow in the body slows down as it gets farther from the heart. It is observed to have a speed of 30 cm/sec near the heart and 20 cm/sec at a point 40 cm farther along the arterial pathway. Assume constant acceleration.

 a. What is the average velocity for that 40 cm distance?
 b. How long will it take the blood to flow this distance?
 c. What is the acceleration?

Worked Example: If an object is thrown upward with a velocity of 40 m/sec, what will be its height and velocity after 2 seonds? What will be the height and velocity after 5 seconds? (Assume $g = 10$ m/sec^2.)

Solution: Care must be taken to maintain the correct algebraic sign in this type of problem. Since the object is thrown upward, this direct will be chosen as positive, so that $v_1 = +40$ m/sec. Since the acceleration is downward, $a = -10$ m/sec^2. After 2 seconds,

$$v_2 = v_1 + at = 40 \text{ m/sec} + (-10 \text{ m/sec}^2)(2 \text{ sec}) = 20 \text{ m/sec}.$$

and

$$s = v_1 t + \tfrac{1}{2} a t^2 = (40 \text{ m/sec})(2 \text{ sec}) + (\tfrac{1}{2})(-10 \text{ m/sec}^2)(4 \text{ sec}^2)$$
$$= 80 \text{ m} - 20 \text{ m} = 60 \text{ m} = \text{height at 2 seconds.}$$

After 5 seconds.

$$v_2 = 40 \text{ m/sec} + (-10 \text{ m/sec}^2)(5 \text{ sec}) = -10 \text{ m/sec}.$$

and

$$s = (40 \text{ m/sec})(5 \text{ sec}) + (\tfrac{1}{2})(-10 \text{ m/sec}^2)(25 \text{ sec}^2)$$
$$= 200 \text{ m} - 125 \text{ m} = 75 \text{ m} = \text{height at 5 seconds.}$$

After 2 seconds the object has slowed to 20 m/sec, and after 5 seconds it has passed the peak of the motion and is traveling downward at 10 m/sec.

23. An object is thrown directly upward at a speed of 20 m/sec.

 a. What will be its velocity after 3 seconds?
 b. What will be its average speed between times 1 and 3 seconds?

24. If a stone dropped from the top of a building takes 4 seconds to hit the ground, what is the height of the building? Express the answer in both meters and feet.

25. A stone dropped from a bridge is timed with a stopwatch and is observed to hit the river below 2.3 seconds after being dropped. How high is the bridge?

26. If a baseball is thrown upward with a speed of 120 ft/sec, how high will it go? How long it take to reach the top of its trajectory?

3.0 sec

27. If you drop a stone in a well which is 40 meters deep, how long will it take for the stone to hit the water? If the velocity of sound is 340 m/sec, what total time will elapse before you hear the splash?

28. If you can jump to a height of 3 feet from a flatfooted start, with what speed do you leave the floor? How high could you jump on the moon where the acceleration of gravity is 1/6 as great?

36.4 sec
1456 m

29. A certain jet aircraft is capable of an acceleration of 2.2 m/sec^2 on a runway. how long will it take to attain its take-off speed of 80 m/sec? What runway length will be required?

$D = \bar{V}T$; $\bar{V} = \frac{V_1 + V_2}{2}$

$a = \frac{\Delta V}{T} = \frac{(V_2 - V_1)}{T}$

$S = V_1 t + \frac{1}{2} a t^2$

$V_f = gt$, $t = V_f / g$

$S = \frac{gt^2}{2}$

$V = aT$

$S = \frac{1}{2} a t^2$ (distance) } For a body starting from $0 = V_0$

$S = \frac{at^2}{2}$

$S = V_1 T + \frac{1}{2} a T^2$

CHAPTER THREE

The Causes of Motion

INSTRUCTIONAL OBJECTIVES

After studying this chapter, the student should be able to:

1. State Newton's three laws of motion and give concrete examples of the application of each law.
2. Give practical definitions for force, mass, weight, and density.
3. Work problems involving Newton's second law, $F = ma$, and the resulting motion of objects.
4. Give qualitative examples of the vector addition of forces, applied to traction systems, muscles, or other concrete situations.
5. Explain the conditions for equilibrium and work problems involving the balancing of forces and torques to achieve equilibrium.
6. Explain the concept of "center of gravity" and the relationship of the position of the center of gravity to the stability of an object.
7. Explain why a force is required to move an object in a circular path and calculate the magnitude of the force required, given sufficient data.

It has been pointed out that the motion of an object can be described with the use of the quantities time, distance, velocity, and acceleration. This chapter is devoted to the discussion of forces, the influences which cause changes in motion.

NEWTON'S LAWS

The seventeenth century physicist Isaac Newton (1642–1727) succeeded in formulating three basic laws of mechanics which enabled him to explain the behavior of moving objects with extreme precision. To this day, these laws provide the basis for the study of mechanics. They have had to be modified only for the description of objects moving with velocities near the speed of light, where Einstein's theory of relativity must be used, and for the description of the motion of objects in the submicroscopic world of atoms and nuclei, where quantum mechanics must be used.

Newton's First Law. An object will remain at rest or in uniform motion in a straight line unless some net external force acts upon it.

Which is the most natural state for an object, to be at rest or to be in motion? It is tempting to say, "At rest, of course. You know that you must exert some effort to keep an object moving, and if you stop pushing something, it will just naturally slow down and come to rest." But this line of reasoning quickly leads to some problems. Suppose two rockets pass each other in a remote region of space where there are no planets or nearby stars to serve as reference points (Figure 3–1). If neither was accelerating, then neither would have a sensation of motion, since there certainly wouldn't be any wind whistling by. From all the data either pilot could gather, each would conclude that he was at rest and the other spaceship was moving past him. Which would be correct in assuming he was at rest, and how could you tell? Centuries of experimentation have led to the conclusion that it makes no sense to talk about being "absolutely" at rest; you can only be at rest with respect to some reference point. When we say we are at rest, we usually mean at rest with respect to the earth, but the earth is moving around the sun, the sun is moving with respect to the galaxy, and so forth, so it is only a relative term. Although one pilot is at rest with respect to his own ship, he is moving at a constant speed in a straight line with respect to the other ship. Thus, depending upon your reference frame, being at rest and being in uniform motion in a straight line are actually equivalent, so one is as "natural" as the other.

Of course, such an experiment can only be approximated on the surface of the earth because there is always some kind of friction present to slow down the motion. However, the friction can be greatly reduced, and it can be seen that Newton's first law represents the ideal case. If you slide a hockey puck on a floor and then on ice, it can be seen that it slows down much more gradually on the ice. On an air table where the puck can be supported on a cushion of air, it can be seen that there is almost no deceleration and the puck moves in a straight line at almost constant speed until it bounces off a side. It is easy to take the step to Newton's first law and say that if there were no friction and no barrier of any type, the puck would continue moving forever in a straight line with constant speed.

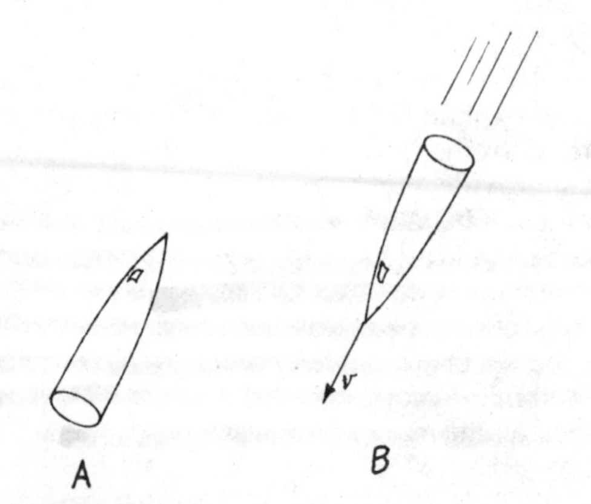

Figure 3–1 At rest or in motion: who can tell?

Newton's Second Law. The acceleration of an object is directly proportional to the net external force exerted upon it and inversely proportional to its mass.

If we let F represent the force and m the mass, then Newton's second law can be written

$$F_{\text{net ext}} = ma \qquad \textbf{3–1}$$

where a is the acceleration. Force and mass are such basic quantities that rigorous word definitions are cumbersome at best, so only operational definitions will be given here with the aid of examples. The mass of an object is an intrinsic property which can be taken as a basic measure of the amount of matter it contains and as a measure of its inertia. The inertia, or resistance to change, is large for a freight car, for example, because it is hard to start in motion when it is at rest and hard to stop when it is in motion. A small cart which has a small inertia (small mass) is easy to either start in motion or stop. The mass is a fundamental constant for a given object, not changing with other physical properties and not changing with a change in its state of motion.*

Once a concept of mass has been obtained, Newton's second law itself could be used, $F_{\text{net ext}} = ma$, to provide a definition of force since acceleration has already been defined. This is not a completely general definition and has exceptions, but it will be useful in gaining some understanding of forces. As a description, a force may be thought of as any influence which tends to change the state of motion of an object.

*The mass of an object does increase when its speed approaches the speed of light, but we will not deal with that extreme here.

The standard units of mass are the kilogram (MKS), the gram (CGS), and the slug (British). These are listed in Table 2–1. International standards for these units are maintained to help insure accuracy and consistency in worldwide use. Once the unit of mass is established, the unit for force can be deduced by applying Newton's second law. The appropriate unit for force in a given unit system must be the unit for mass times the unit for acceleration. In the MKS system, force is then measured in kilograms times meters/sec² or kg-m/sec². It is rather cumbersome to write all this down, so this combination is renamed the "newton." In the CGS system the basic unit is the gm-cm/sec² and is called a dyne. In the British system the basic force unit is the slug-ft/sec², which is called a pound. The slug is a large and cumbersome unit and it is seldom used.

If the mass and force are known, Newton's second law can be used to precisely predict future motion or to describe accurately any motion which has occurred previously. If a man exerts a force of 100 newtons on a crate of mass 5 kilograms, it will, in the absence of friction, acclerate at the rate

$$a = \frac{F_{\text{net ext}}}{m} = \frac{100 \text{ newtons}}{5 \text{ kg}} = 20 \text{ m/sec}^2.$$

The relationships developed in Chapter 2 can then be used to predict its future motion. If the crate was initially at rest, then after 3 seconds its speed and distance can be predicted:

$$v = at = 20 \text{ m/sec}^2 \times 3 \text{ sec} = 60 \text{ m/sec}$$

$$s = \frac{1}{2} at^2 = \frac{1}{2} \times 20 \text{ m/sec}^2 \times 9 \text{ sec}^2$$

$$= 90 \text{ meters.}$$

If the same force of 100 newtons is exerted on a large crate with mass 40 kg, the acceleration will be only 2.5 m/sec² and the speed and distance after three seconds will be only 7.5 m/sec and 11.25 meters, respectively (Figure 3–2).

The assumption of no friction in this problem is obviously an unrealistic one. In practice, if you exert a 100 newton force on a 40 kg crate, it may not move at all, because frictional forces may exert 100 newtons of force against you. In the application of Newton's second law, it must be kept in mind that the force F in the equation $F = ma$ is the *net* force. More realistically, suppose a force of 100 newtons is exerted on the 40 kg mass and is opposed by a 20 newton frictional force. The net force is then 80 newtons and the acceleration will be 2 m/sec² rather than the 2.5 m/sec² obtained above.

Newton's Third Law. When one body exerts a force on a second body, the second body exerts an equal but oppositely directed reaction force on the first body.

This law is sometimes stated, "For every action there is an equal and opposite reaction." This statement of the law is misleading in that it does not clearly specify that two objects are required. A force cannot be exerted by one body in isolation; forces always occur in pairs. For example, when one fires a shotgun, a large forward force is exerted on the shot to move it out of the gun barrel (Figure 3–3). This force does not occur in isolation, as the one who pulls the trigger knows all too well, for he feels the equally large reaction force backward on his shoulder. The huge rockets which carry satellites into orbit operate according to Newton's third law. They thrust burning fuel out their exhausts at extremely high velocities. The reaction force, just like the shotgun kick, propels the rocket in the opposite direction. These rockets are

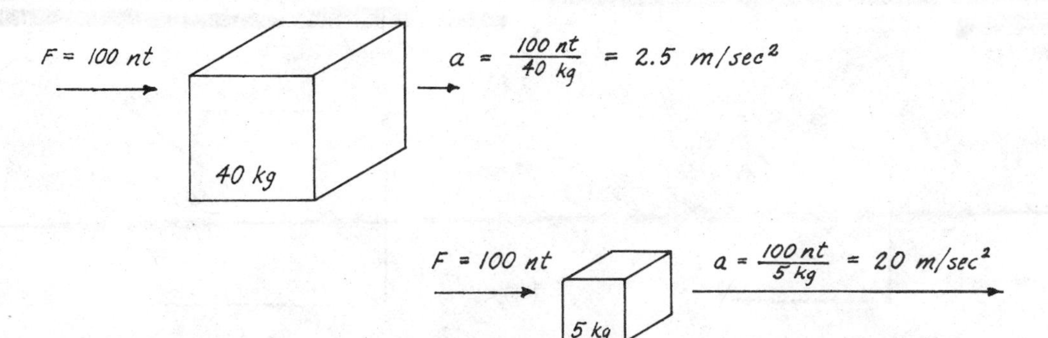

Figure 3–2 Illustration of Newton's second law. (No friction is present here.)

Figure 3–3 Newton's third law: Forces occur in pairs.

particularly useful in space, because they don't have to push on anything else in order to move. Even in the void of space the reaction force pushes them forward when they push the exhaust out backward.

Related to Newton's third law and Newton's first law is the principle of conservation of momentum. This principle states that the momentum of an isolated object or collection of objects cannot change. The momentum is defined as the mass times the velocity, mv. Consider the example of a boy sitting in a canoe on a lake. The momentum of the collection consisting of canoe plus boy is zero since neither is moving, and thus it must remain zero. If the boy dives out of the canoe forward, then the canoe will move backward with a velocity such that its momentum, mv, backward just cancels out the boy's forward momentum. Suppose the mass of the canoe is twice the mass of the boy. If the boy dives forward with a velocity of 10 ft/sec, then the canoe will move backward with a velocity of 5 ft/sec such that these momenta cancel, leaving the total equal to zero according to the principle of conservation of momentum. (Conservation of momentum means momentum kept constant.)

FORCES AS VECTORS

It has been pointed out that velocity and acceleration are vector quantities; both a magnitude and a direction are required to specify them completely. By contrast, mass is called a scalar quantity since it is completely specified by a number and a unit. It has no direction associated with it. Force is a vector quantity, and there are often several forces acting in different directions on a single object. The force contained in Newton's second law, $F = ma$, is the *net* external force, the sum of all the forces acting on an object. Thus it is necessary to take the sum of all the forces with their directions taken into account. This is called a vector sum.

If you had two 50 pound forces directed in opposite directions, their vector sum would be zero since they would cancel each other out. If they were directed in the same direction their vector sum would be 100 pounds since they would add directly, just like scalar quantities. If they are oriented at arbitrary angles, their vector sum can take any value between 0 and 100 pounds.

To find the resultant force when two forces act at right angles, the Pythagorean theorem can be used. As shown in Figure 3–4, the resultant has the magnitude of the length of the hypotenuse of the right triangle whose sides are the two forces. The hypotenuse c is given by the relationship $c = \sqrt{a^2 + b^2}$ where a and b are the sides of the triangle. If a thirty pound force and a forty pound force act at right angles, this is equivalent to a force:

$$F = \sqrt{(30 \text{ lb})^2 + (40 \text{ lb})^2} = \sqrt{2500 \text{ lb}^2} = 50 \text{ lb}$$

acting in the direction indicated in Figure 3–4. The process may be accomplished entirely graphically by drawing the forces to scale from the same point and then completing the rectangle. The diagonal of the rectangle drawn from the starting point gives the direction of

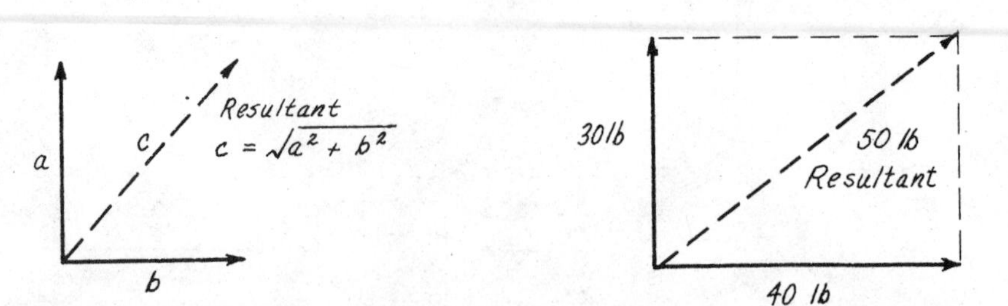

Figure 3–4 Finding the resultant of two vectors which act at right angles to each other.

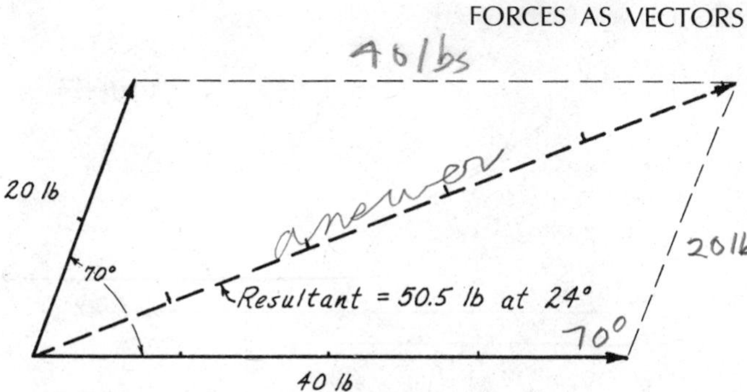

Figure 3–5 Finding the resultant of two forces which act at angles other than 90 degrees.

the resultant force, and measuring its length to scale will give the size of the force.

If the forces are not at right angles, a similar graphical method can be used to find the resultant. If the two forces are drawn to scale from a single point as shown in Figure 3–5, the resultant may be found by completing the parallelogram. The direction of the resultant force will be the direction of the diagonal of the parallelogram drawn from the original point. The angle can be measured with a protractor, and measuring the length of this diagonal to scale gives the magnitude of the force.

An alternate method for finding vector sums has the advantage that it will work for any number of forces. Consider the five forces drawn to scale in Figure 3–6(a). They are all acting on a single object, so the resultant must be found before any prediction of its motion can be made. The vectors can be drawn successively, tail to head, *in any order,* as shown in Figure 3–6(b). The vector drawn from the tail of the first to the head of the last will give the direction and magnitude of the resultant force.

Although the numerical details will not be

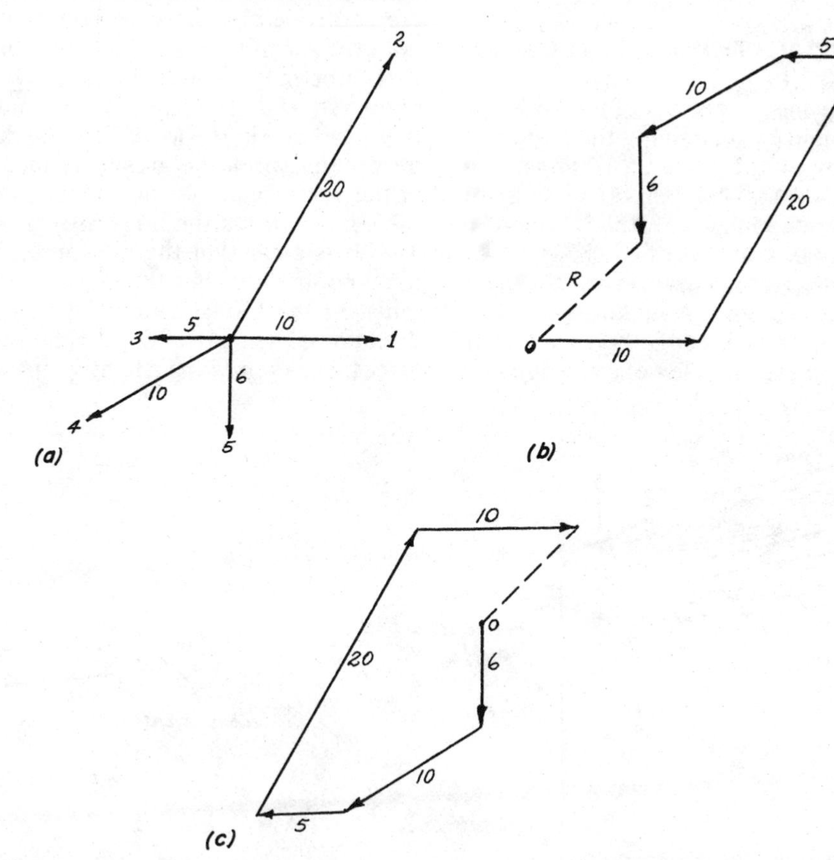

Figure 3–6 The graphical method for vector addition. (a) Forces acting at a point. (b) Illustration of graphical methods for vector addition. (c) Vector addition in the reverse order. The vector sum is independent of the order in which the vectors are added.

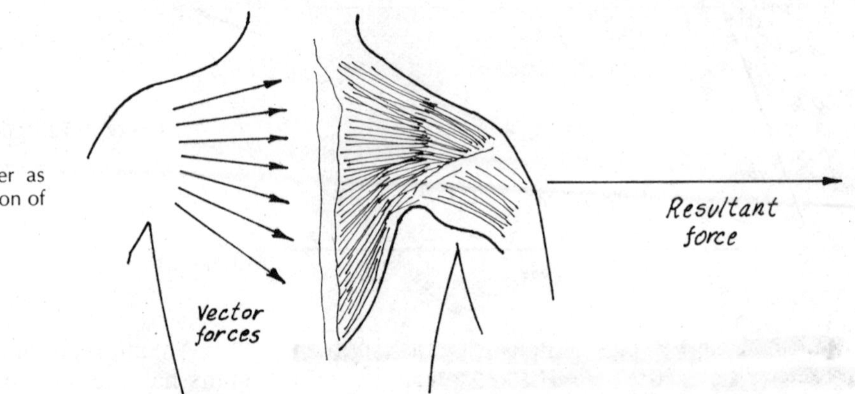

Figure 3–7 Force on shoulder as an example of the vector addition of forces.

emphasized here, a qualitative understanding of the process of adding vectors can give considerable insight into the behavior of objects on which several forces are acting. The numerically preferable methods for adding a number of vectors involve the use of trigonometry and can be found in one of the general physics references listed at the end of this chapter.

APPLICATIONS OF VECTORS

Some of the best examples of the addition of forces are found by examining the action of various muscles of the body. The hand can exert a force in virtually any direction. The net force exerted by the hand is actually the vector sum of a number of forces exerted by the muscles of the hand, arm, and shoulder. To move the entire arm and shoulder, a number of muscles in the back act together, as shown in Figure 3–7. These muscles exert a number of forces in different directions to achieve the net force in the desired direction.

In the everyday activity of walking, the muscles of the hips and legs are called upon to exert forces in a number of different directions in a short interval of time. The flexibility to exert these various forces exists because a large number of individual muscles exert forces in different directions to produce the desired vector sum. Since all of these muscles are controlled by the central nervous system, both the magnitude and direction of the resultant force may be varied at will, within limits, of course. If there were only one muscle in the thigh and one in the lower leg, the flexibility of movement would be greatly reduced. Consider the vector addition of forces required to lift the leg in front of the body as shown in Figure 3–8. It is clear that the net force on the knee must be upward, but the thigh muscles which must do the lifting cannot be pointed in that direction. The contracting thigh muscles exert forces backward and slightly upward on the

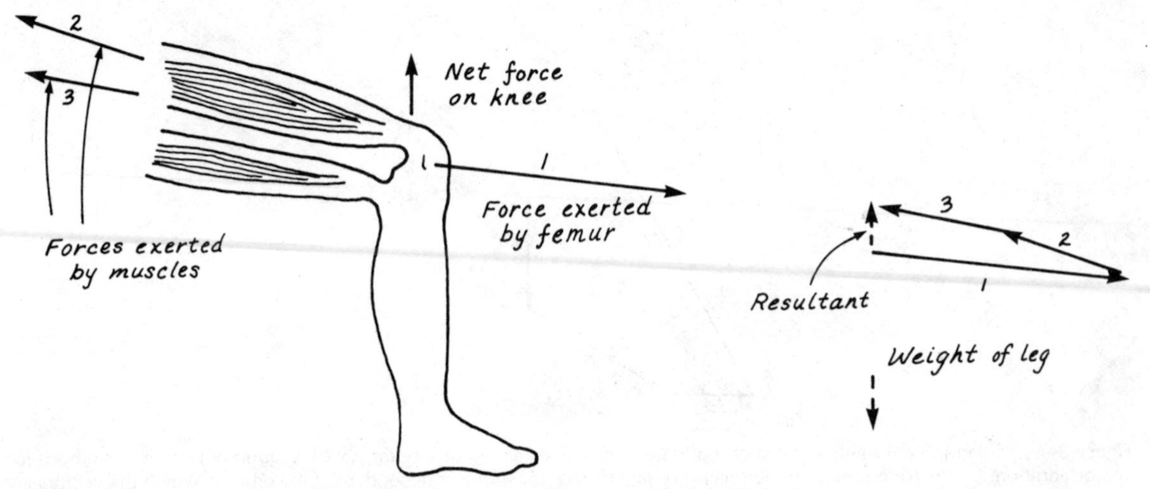

Figure 3–8 Vector sum of forces required to lift the knee.

knee. Acting against this contraction, the femur exerts a forward reaction force against the knee, canceling out the horizontal components of the muscle forces. The resultant is a relatively small upward force on the knee, as shown in the figure. Since large forces must be exerted to achieve a small resultant force, this is a rather inefficient process. This conclusion will no doubt be supported by the reader's experiences in lifting a leg in this position.

In the care of orthopedic patients, a traction apparatus is often used to exert a force on one of the patient's extremities. It is often convenient to use several forces to achieve the desired net force. In the traction apparatus shown in Figure 3–9, a combination of pulleys is used with a weight to exert three forces in different directions. Forces exerted by strings (or ropes, cables, etc.) can only be in the direction of the strings and not perpendicular to them. Such forces are commonly called tensions. When the traction apparatus is at rest, the tension at all points in the rope is presumed to be the same, and equal to the weight which is suspended from the end of the rope. The resultant force on the patient's femur can be obtained by the method previously discussed for adding several vectors. When the vectors representing three equal forces acting in different directions on the patient are drawn successively as shown in the figure, the resultant force can be seen to be directed along the femur.

THE FORCE OF GRAVITY

The humorous stories about the apple falling on Newton's head as he slept beneath the tree serve to illustrate one aspect of gravity, namely, "The earth attracts all objects to itself." This aspect of gravity is also illustrated by the statement, "What goes up must come down." But the law of gravity as discovered and generalized by Newton reaches far beyond these statements. Experiments and observations indicate that every object exerts a gravitational force on every other object which is proportional to the product of the masses of the two objects and inversely proportional to the square of the distance between them. Stated in symbolic form, the attractive force between masses m_1 and m_2 is given by

$$F_{\text{gravity}} = \frac{G \times m_1 \times m_2}{r^2} \qquad \textbf{3–2}$$

where r is the distance between the masses and G is a numerical proportionality constant.

$$(G = 6.67 \times 10^{-11}\ \text{nt} \cdot \text{m}^2/\text{kg}^2)$$

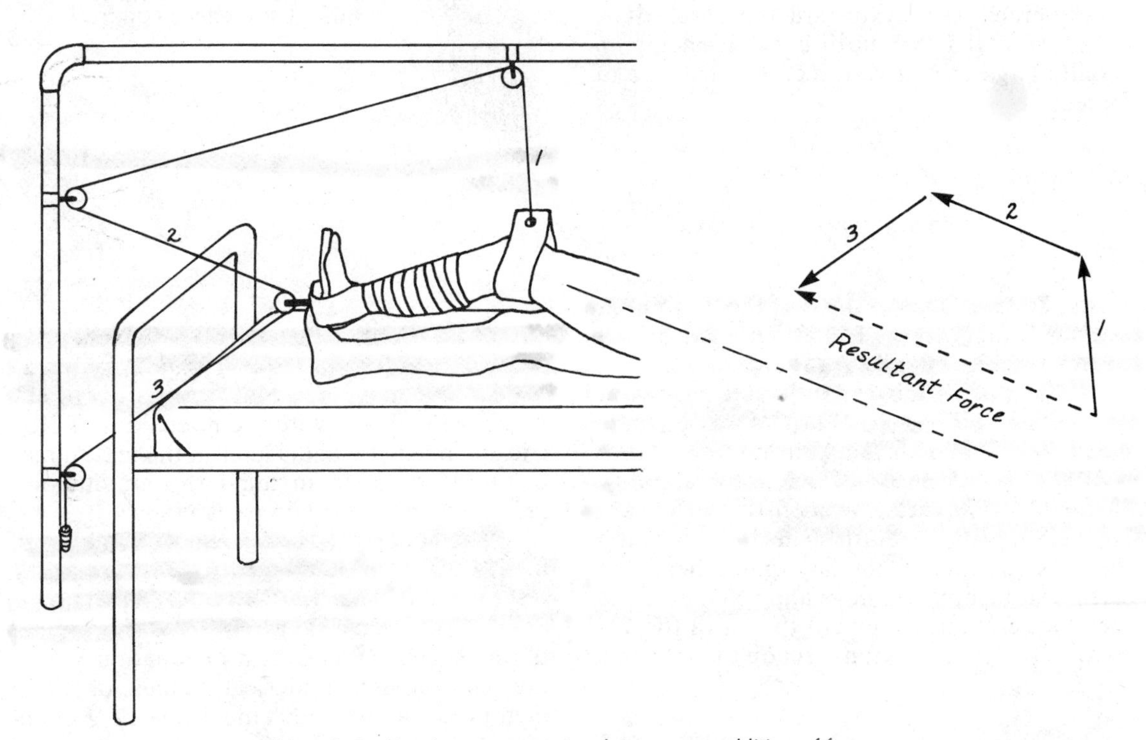

Figure 3–9 Traction apparatus involving vector addition of forces.

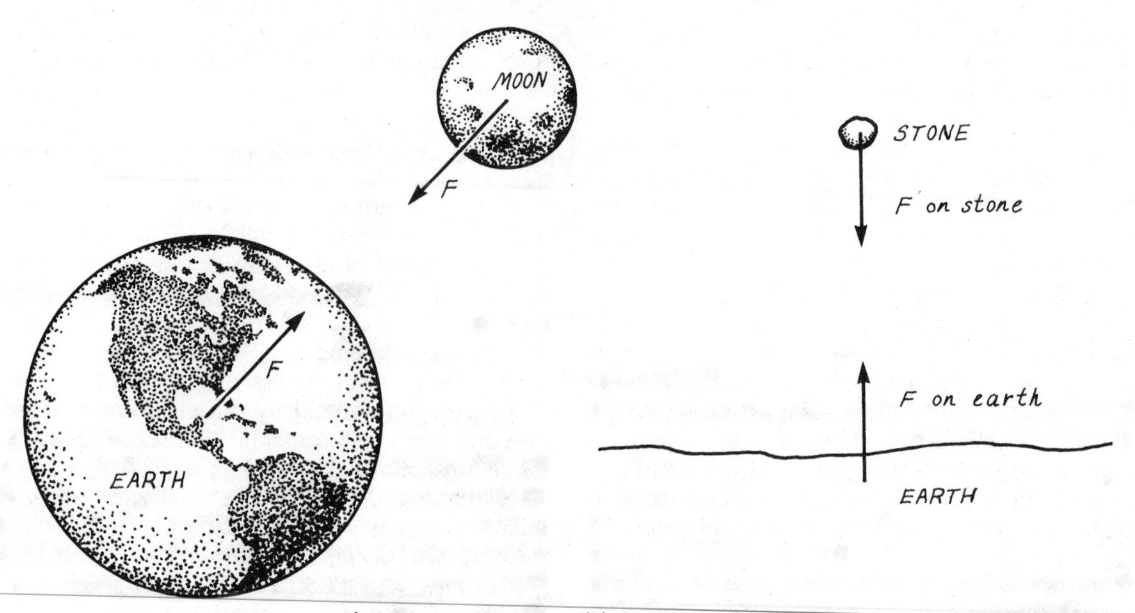

Figure 3–10 Gravitational forces occur in pairs.

As illustrated in Figure 3–10, the forces of attraction are equal and opposite, in agreement with Newton's third law. Forces always occur in pairs. This pairing of equal forces occurs regardless of the relative sizes of the two masses. When a stone is released near the surface of the earth, it is obvious that the stone is experiencing an attractive force, since it accelerates rapidly toward the earth. It is not obvious that the earth is experiencing an equally large attractive force. The balance of forces

$$F_{\text{earth}} = m_{\text{earth}} \times a_{\text{earth}} = m_{\text{stone}} \times a_{\text{stone}} = F_{\text{stone}} \qquad \textbf{3–3}$$

exists, but because the mass of the earth is so enormous (about 6×10^{24} kilograms), its acceleration is essentially zero.

Since the earth is approximately a sphere, all objects on the earth's surface are approximately the same distance from the center. This implies that the earth will exert approximately the same force on an object, regardless of where it is on the earth's surface. Thus, if a stone is dropped from any point near the earth's surface, it will have approximately the same accleration, *g*, toward the earth (downward). Applying Newton's second law to this force,

$$F_{\text{gravity}} = m_{\text{stone}} \times a_{\text{stone}} = mg. \qquad \textbf{3–4}$$

The acceleration of gravity, g, does not depend upon the mass of the stone, m. This can be seen by applying equation 3–2, letting $m_1 = m$ (stone) and m_2 = mass of the earth. The acceleration of gravity, g, can be solved for by substitution:

$$g = G \times \frac{\text{(mass of the earth)}}{\text{(radius of the earth squared)}} = \frac{GM_{\text{earth}}}{r^2}. \qquad \textbf{3–5}$$

If the acceleration of gravity g is put in the form

$$g = \frac{F_{\text{gravity}}}{m},$$

it is seen to be the *force per unit mass* for the gravitational field and thus g can be interpreted as the intensity of the gravitational field. Since we will deal with the intensities of electric and magnetic fields later in the text, it may be useful to keep in mind the gravitational field intensity g for comparison.

The acceleration of gravity (gravitational field intensity) thus depends only upon the mass of the earth, the radius of the earth if you are on its surface, and the universal gravitation constant, G as shown in equation 3–5. If the appropriate numerical values of these factors are substituted, the value of g is obtained:

$$g = 32 \text{ ft/sec}^2 = 9.8 \text{ m/sec}^2 = 980 \text{ cm/sec}^2,$$

or equivalently, to emphasize that it is the gravitational field intensity,

$$g = 32 \text{ lb/slug} = 9.8 \text{ nt/kg} = 980 \text{ dyne/gm}.$$

These numerical constants are listed in Table T–3 in Appendix A.

The main point here is that the acceleration is independent of the mass of the object dropped; a large stone and a small stone will accelerate downward at the same rate. This was the result of Galileo's famous experiment in which he dropped two stones from the leaning tower of Pisa. (It is doubtful that he actually performed this experiment, but it is popularly attributed to him.) When dropped from the same height, the stones hit the ground at the same time and with the same velocity. Of course, if a stone and a feather are dropped simultaneously, the feather will fall more slowly because of the resisting force of air friction. If the feather and the rock were dropped in a vacuum, they would fall at the same rate. This was actually demonstrated by the Apollo astronauts, who dropped a hammer and a falcon feather simultaneously in the vacuum on the moon's surface.

The acceleration of gravity does show slight variations since the radius of the earth is not exactly constant. It will be less on top of a high mountain because that point is farther from the center of the earth. These variations are small and usually insignificant.

MASS, WEIGHT, AND DENSITY

The *weight* of an object is defined as the force of gravity acting upon it. It is proportional to the mass, but it is distinct from it and is measured in different units. The mass is a fundamental measure of an object's inertia and the amount of matter present; it does not change with the position on the earth. Since the force of gravity varies with the altitude above sea level, your weight will also vary. It has been pointed out that these variations are usually insignificant, but this serves to illustrate that the mass is the more fundamental measurement. The relationship between mass and weight can be seen from the law of gravitation

$$F_{\text{gravity}} = (\text{mass}) \times \left(\frac{GM_{\text{earth}}}{r^2}\right)$$

$$= (\text{mass}) \times (\text{intensity of gravity}).$$

Therefore the weight W is

$$W = mg. \qquad \textbf{3–6}$$

This can also be seen from Newton's second law $F_{\text{net ext}} = ma$ applied to a freely falling body. The external force is the weight and the acceleration is g.

The most difficult problem here is that of keeping the units straight. Keeping in mind that the weight is a force will perhaps help with that problem.

Example. What is the weight of a 10 kilogram object?

Solution. Since the intensity of gravity in the MKS system is 9.8 newton/kg, the weight is

$$W = mg = (10 \text{ kg}) \times (9.8 \text{ newton/kg})$$
$$= 98 \text{ newtons}.$$

Example. If a person has a mass of 5 slugs, what is his weight?

Solution. Using the value $g = 32$ pounds/slug for the British system:

$$W = (5 \text{ slugs}) \times (32 \text{ pounds/slug}) = 160 \text{ pounds}.$$

The appropriate unit for weight in the British system is the pound, and in the metric system (MKS) the appropriate weight unit is the newton. The confusion of weight and mass units occurs largely because the *weight* in pounds is the most common measure of quantity in the British system, whereas the *mass* in kilograms is the most common measure of quantity in the metric system. (See Table 2–1.) The British mass unit, the slug, is too cumbersome for normal use, so the grocery shopper buys items by their weight. One pound of meat may be just about right, but one slug of meat (32 pounds) would be too heavy to carry home, to say nothing of costing a fortune. The corresponding European shopper might ask for a kilogram of meat (which weighs 2.2 pounds) since the newton is an inconveniently small quantity. Convenience is sufficient justification for mixing units in such cases, but when Newton's laws are used for the calculation of motion, the masses and forces must be expressed with appropriate units.

As a further illustration of the distinction between mass and weight, consider an object which is carried to the moon. Since the mass is an intrinsic characteristic of the object, it will not change. The weight, however, will change rather drastically, since the moon's intensity of gravity is only about a sixth as strong as the earth's gravity. The fact that things weigh only one-sixth as much on the moon enabled the Apollo astronauts to handle massive pieces of equipment with relative ease, even in their cumbersome space suits. In deploying the experimental instrument packages on the moon, loads which would have weighed approximately 400 pounds on the earth were carried. In the moon's gravity they weighed less than seventy pounds.

Density is defined as the mass per unit volume and is most often expressed in CGS units. The density will be represented by the symbol d:

$$d = \frac{m}{V} \qquad \textbf{3–7}$$

where m is the mass and V the volume. For example, if we measure the volume of a 100 gram block of aluminum and find that it occupies a volume of 37 cm³, we can calculate the density:

$$d = \frac{m}{V} = \frac{100 \text{ grams}}{37 \text{ cm}^3}$$

$$= 2.7 \text{ gm/cm}^3.$$

If a similar experiment is done to measure the density of water, it is found to have a density of 1.0 gm/cm³. (Actually it is exactly 1.0 gm/cm³ at a temperature of 4°C.) This is, of course, not just a coincidence; the density of water at a specific temperature was used as a standard in setting up the metric system of units. In fact, for practical purposes a gram could be defined as the mass of a cubic centimeter of water. The densities of other substances are characteristic of those materials and can be used in a limited way to help identify those materials. Some typical densities are listed in Table 3–1, showing the range from the light metal, magnesium, to the heaviest element, osmium. A more extensive listing of densities may be found in Table T–4, Appendix A.

TABLE 3–1 Typical Densities (gm/cm³)

air	1.3×10^{-3}	iron	7.8
ice	0.92	lead	11.
water	1.0	mercury	13.6
magnesium	1.7	gold	19.
aluminum	2.7	osmium	22.5

When the British system is used, it is sometimes useful to define another quantity, the *weight density*. The weight density is the weight per unit volume.

$$d_w = \frac{\text{weight}}{\text{volume}} = \frac{mg}{V} = dg \qquad \textbf{3–8}$$

For example, the weight density of water is 62.4 pounds/ft³.

The specific gravity of a material is defined as its density divided by the density of water. A specific gravity of two means that the substance is twice as dense as water. Since the density of water in the CGS system is 1 gm/cm³, the specific gravity of any substance is numerically equal to its density in gm/cm³. Density and specific gravity will be discussed further in Chapter 5 along with Archimedes' principle.

THE PHENOMENON OF WEIGHTLESSNESS

Since the weight of an object is defined as the force of gravity on it, there is no way to be really weightless while remaining close to the earth. The earth's gravitational field extends out into space and penetrates all matter; there is no known way to construct a "shield" against it. However, there are circumstances under which the body feels weightless.

A person's sensation of weight comes not so much from the direct gravitational force as from the support forces exerted by the floor, a chair, a bed, and so forth. If these support forces are suddenly removed, we feel "weightless" and have the sensation of falling. If you are in an elevator and its support cable breaks, you have the sensation of weightlessness and may float around relative to the elevator. This might be exhilarating if it weren't for the anticipation of the quick stop at the bottom.

It must be emphasized that the "weightlessness" experienced while orbiting the earth in an artifical satellite is not the absence of a gravitational force, but the absence of a supporting force. Consider the case of a satellite in orbit 100 miles above the earth's surface. An astronaut would actually weight only about five per cent less at that altitude than he weighs on the surface of the earth. A person who weighs 160 pounds at sea level would still actually weight about 152 pounds in that orbit, but he would feel weightless because of the lack of support. The phenomenon of weightlessness occurs because both the astronaut and the satellite are in a state of "free-fall" in the earth's gravitational field. This state of free-fall does not cause the satellite to hit the earth because of its high speed (over 17,000 miles per hour) around the earth. This motion will be discussed further in the section on circular motion.

EQUILIBRIUM AND TORQUES

When you place an object on a table and expect it to remain there, you are making use of the concept of *equilibrium*. Equilibrium is related to the terms "stable" and "balanced," as will be explained. An object is said to be in *equilibrium* when the sum of the forces acting upon it is zero ($F_{net\ ext} = 0$) and the sum of the torques (turning influences) exerted upon it is zero. It is possible to be in equilibrium while moving, but the remarks here will be limited to the case of *static equilibrium*. Any object which is at rest and remaining in a fixed position is said to be in static equilibrium.

An object which is remaining at rest may have many external forces acting upon it, but its observed equilibrium implies that the *net* force is zero; i.e., the vector sum of all the external forces is zero. If you place a box on a scale and determine its weight to be 40 pounds, you know that gravity exerts a 40 pound downward force on it. From the fact that the box is at equilibrium you can imply that the scale exerts a 40 pound upward force on the box to keep it at equilibrium.

As a useful application of this first condition for equilibrium, consider the problem of weighing a patient who cannot be moved from the bed. The whole bed could not be put on most scales, but it might be possible to weigh one end of the bed at a time. Could the total weight be obtained in this manner? Suppose the scale registers 170 pounds when one end of the bed is weighed and 215 pounds when the other end is put on the scale (see Figure 3–11). This means that the floor is exerting a total upward force of 170 lb + 215 lb = 385 pounds to support the bed and patient. Since the bed is normally at equilibrium, the forces up must equal the forces down, so the total downward force of gravity (the weight of the patient and bed) must be 385 pounds. If the bed is known to weigh 200 pounds, this gives the patient's weight as 185 pounds.

Conditions for Static Equilibrium

1. Vector sum of forces = 0:
   sum of forces up = sum of forces down
   sum of forces right = sum of forces left
2. Sum of torques about any point = 0:
   sum of clockwise torques = sum of counterclockwise torques

A zero net external force is necessary for equilibrium, but that condition alone is not enough to guarantee equilibrium. If you had a stick lying on a table and you exerted a 10 lb force to the left on one end of it and a 10 lb force to the right on the other end, the net force would be zero but the object would clearly not be at equilibrium. The stick would rotate because the forces exert a net *torque* on it. The second condition for equilibrium is that the net torque must be zero.

The effectiveness of a force in producing rotation depends not only upon the size of the force but also upon its effective *lever arm* with respect to the axis about which the rotation takes place. The torque exerted by a force F about an axis A can be expressed as

$$\tau_A = \ell_\perp F \qquad \textbf{3–9}$$

where the lever arm $\ell_\perp$ is the perpendicular distance from the axis to the line of action of the force (see Figure 3–12). For the process of turning a bolt with a wrench, the torque is produced by exerting a force at the end of the wrench, the axis is along the center of the bolt and the lever arm is the *effective* length of the wrench. In Figure 3–12(a) the force is exerted perpendicular to the handle so the lever arm is two feet, the full length of the wrench handle. In 3–12(b) the force is exerted on the handle at an angle, which renders it less effective in producing torque. The effective length of the wrench is now only 1.2 feet. In 3–12(c) the force is exerted directly along the handle and no torque is produced. There is no tendency to rotate in this case since the line of action of the force passes directly through the axis, and the lever arm is then zero.

For rotation about a given axis, the condition of zero net torque can be expressed by saying that the sum of the clockwise torques must be equal to the sum of the counterclockwise torques about that axis:

$$\tau_{CW} = \tau_{CCW}.$$

Figure 3–13 shows some examples of this

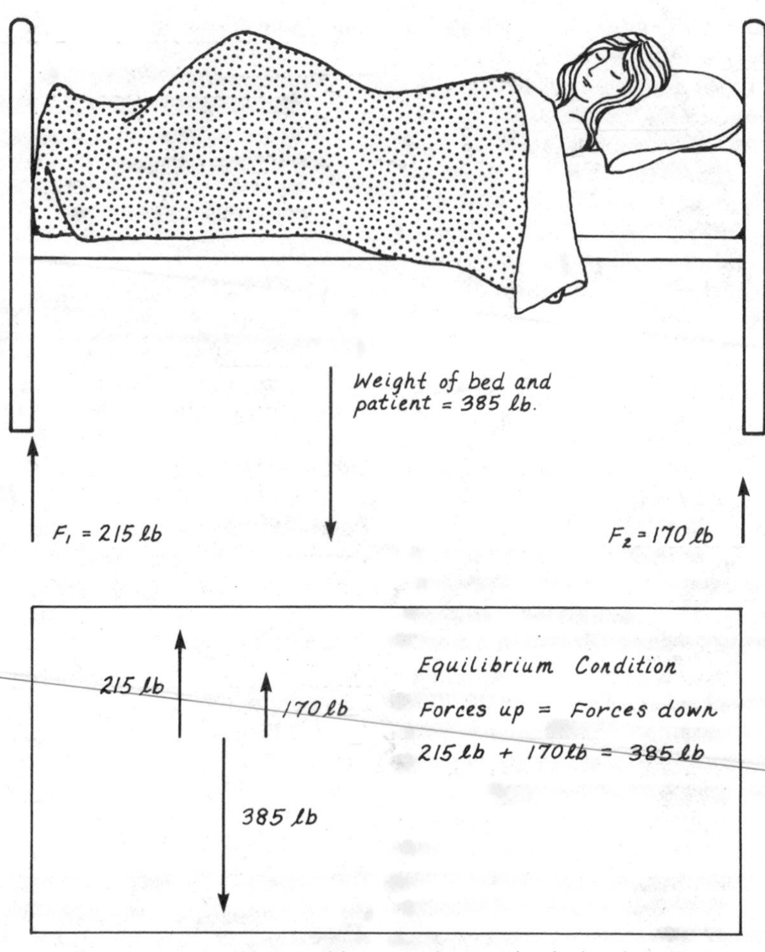

Figure 3–11 Equilibrium of forces used to weigh a bed and patient.

equilibrium condition. In Figure 3–13(a) one end of a uniform board is being supported by a person while the other end rests on a table. The end on the table is chosen as the axis for the torque calculation. (Any line perpendicular to the plane of rotation can be chosen as the axis for the torque calculations; a line through the end of the board is chosen for

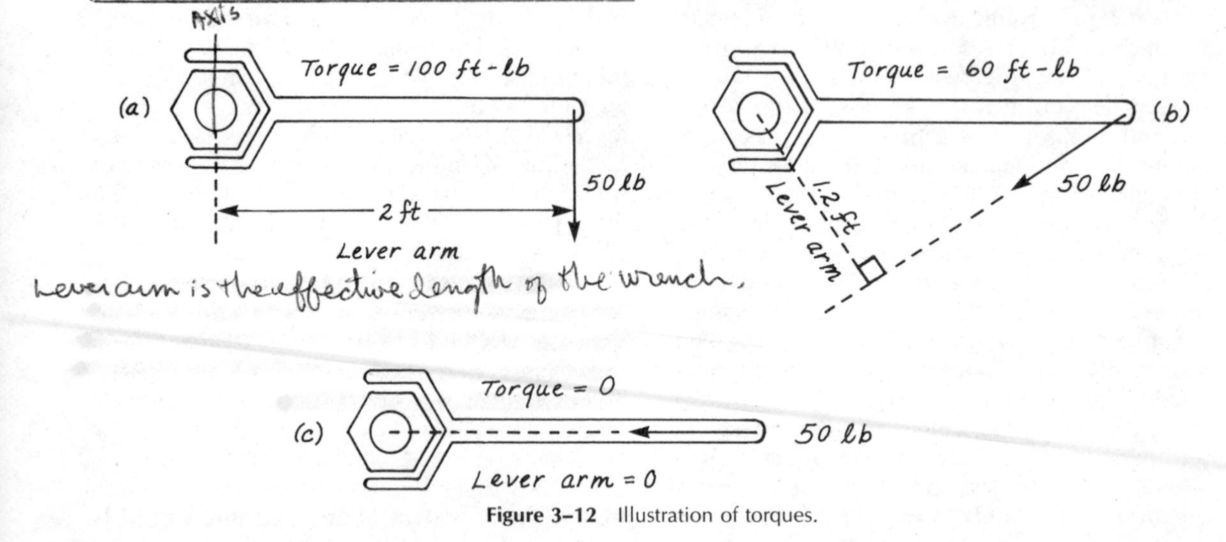

Figure 3–12 Illustration of torques.

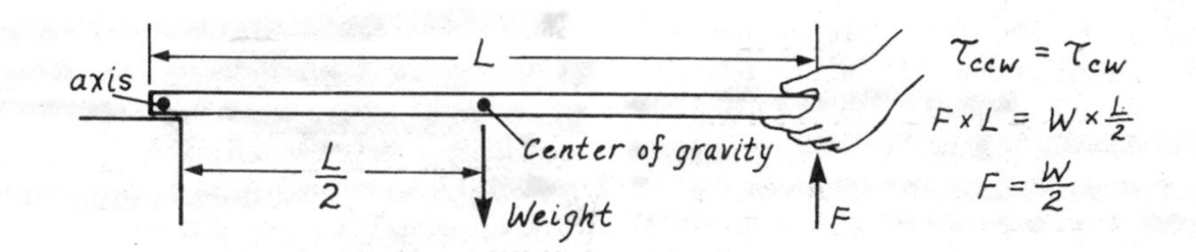

a. The weight of an extended object can be considered to act at its center of gravity for calculating torques

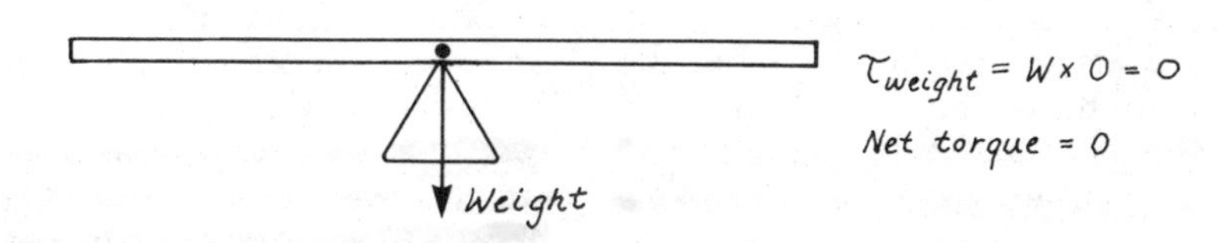

b. A uniform board supported at its halfway point will balance since the effective lever arm associated with the weight is zero.

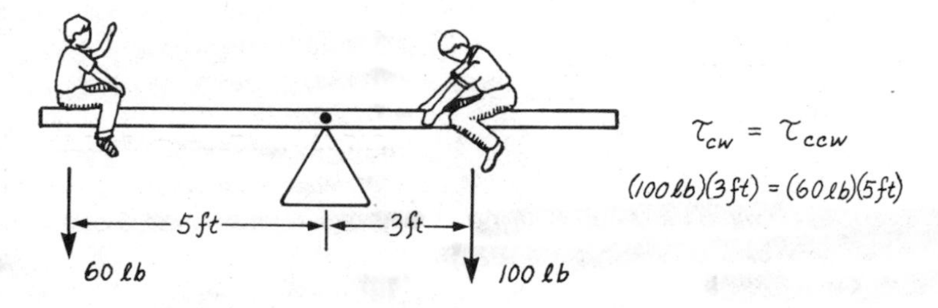

c. Children's see-saw with balanced torques

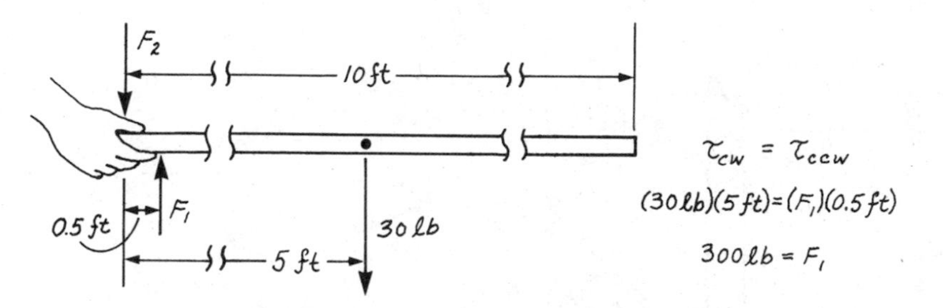

d. Extreme example: lifting a board by one end

Figure 3–13 Examples of torques and equilibrium.

convenience.) The force exerted by the person would tend to cause the board to rotate counterclockwise about the chosen axis, so we say that he exerts a counterclockwise torque. Since the force is perpendicular to the board, the lever arm is the full length of the board.

Since gravity acts downward along the entire length of the board in Figure 3–13(a), the torque exerted by gravity requires special consideration. The weight would cause the board to rotate clockwise with respect to the chosen axis, and the force involved is the weight of the board, but what about the lever arm? If the board were divided into a large number of equal segments and the weight of each of these segments was treated as an individual force, then the lever arm associated with each segment would be its distance from the axis of rotation. Therefore, the lever arm of the first segment at the axis end would be essentially zero and the lever arm of the final segment at the other end would be equal to the length of the board L. The sum of the torques which would be obtained by multiplying the weights of all of these individual segments times their respective lever arms would be the same as that obtained by multiplying the *entire weight of the board* times the *average lever arm* of the segments since all the segments have the same weight. This average lever arm would be half the length, L/2. This is equivalent to saying that the entire weight of the board could be considered to be acting at the center of the board for the purpose of calculating torques. This point at the center of the uniform board is said to be the *center of gravity* of the board. The equilibrium condition applied to the system in Figure 3–13(a) would then be

$$\tau_{CW} = \tau_{CCW}.$$

$$(W)\left(\frac{L}{2}\right) = (F)(L),$$

$$\frac{W}{2} = F.$$

Therefore the force F required to hold up one end of the uniform board would be equal to half the weight of the board.

This example has been used to introduce the concept of *center of gravity*. Every object has a point at which the entire weight of the object can be considered to act for the purpose of calculating torques. If the object is supported under this center of gravity point it can be *balanced*, or placed in equilibrium by this single supporting force, as illustrated in Figure 3–13(b). Since all the weight can be considered to be acting at this point, it has a zero lever arm and exerts no torque. For a nonuniform board the center of gravity would not be at the geometrical center. For objects made of different kinds of materials of differing densities, it is more complicated to calculate the location of the center of gravity, but it is learned by experience that there is always a point at which the object can be lifted such that it will "balance" and not rotate.

The position of the center of gravity and the nature of the support determine whether an object will be in stable equilibrium. This is illustrated in Figure 3–14. The structure in Figure 3–14(a) is in stable equilibrium since the line of action of the gravitational force acting as its center of gravity falls well inside the base. Even if the structure were tipped to the right, it would tend to tip back because its weight would exert a counterclockwise torque to restore it to its upright position. The structure in Figure 3–14(b) is in unstable equilibrium since its weight acts directly through its right edge and thus could exert no counterclockwise torque to stabilize it. If it were

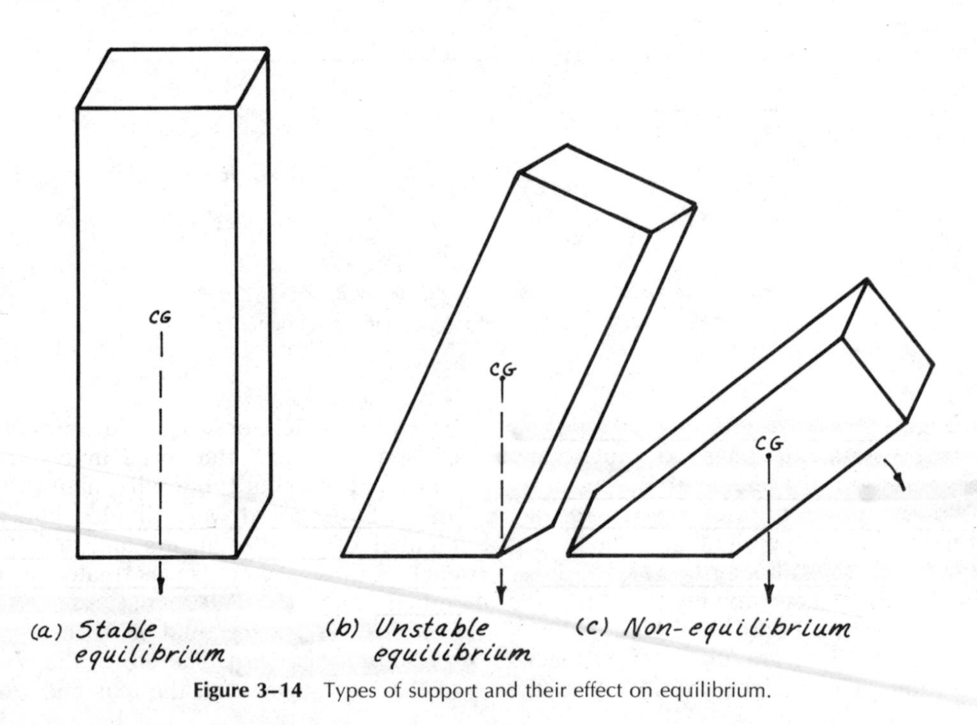

Figure 3–14 Types of support and their effect on equilibrium.

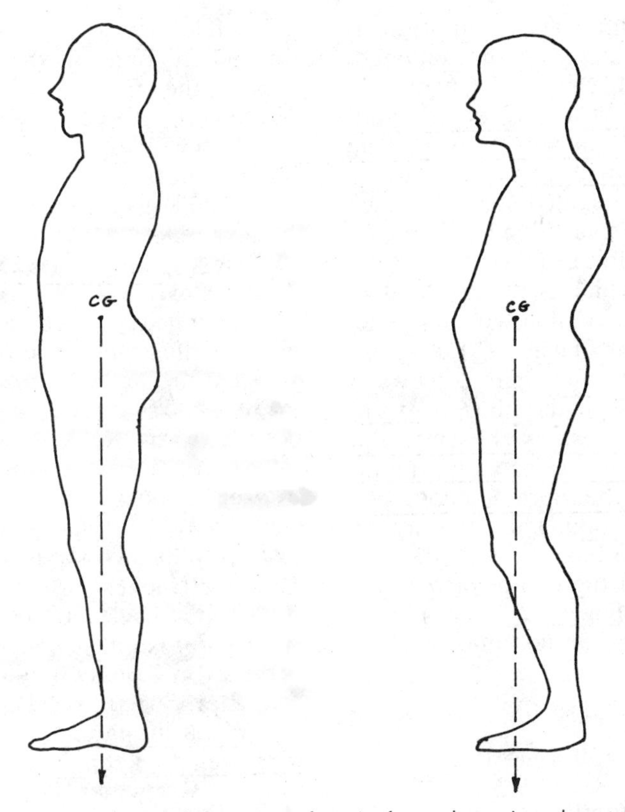

Figure 3–15 A slumping posture tends to shift the center of gravity forward, causing a less stable equilibrium and abnormal muscle strain.

tipped the slightest amount to the right, it would fall over. The structure in Figure 3–14(c) is unstable and cannot be in equilibrium. The line of action of its weight falls outside the base and thus exerts an unbalanced clockwise torque, so it will tip over.

The concept of stable equilibrium is important in relation to good posture and to the proper methods for lifting and carrying heavy objects. As illustrated in Figure 3–15, good posture places the center of gravity of the body solidly above the feet so that balance is easily maintained. A slumping posture shifts the center of gravity forward, producing a less stable balance. The balls of the feet must now exert torques to maintain the balance, and more muscle tension is required, particularly in the back muscles, to keep the torques balanced.

As another example of the second condition for equilibrium, consider a children's see-saw composed of a 10 foot uniform board weighting 50 pounds and supported under its center of gravity. This is illustrated in Figure 3–13(c). The board alone would balance at the center since (1) the support would exert an upward force of 50 pounds, and (2) the force of gravity acting down on it would produce no torque since it passes directly through the support point (axis) and thus has a zero lever arm. Now if a 60 pound child sat on the left end of the see-saw, this would produce a counterclockwise torque of (60 pounds) × (5 foot lever arm) = 300 ft-lb. A 100 pound child wishes to sit on the other end so that they will balance. He cannot sit on the end of the board since that would produce (100 lb) × (5 ft) = 500 ft-lb of clockwise torque. Applying the second condition for equilibrium,

$$\tau_{CCW} = \tau_{CW}$$

$$(60 \text{ lb}) \times (5 \text{ ft}) = (100 \text{ lb}) \times (x \text{ ft}).$$

Now, dividing both sides of the equation by 100, we find that the 100 pound child must sit at a distance $x = 3$ ft from the rotation axis (the support) and thus 2 ft from the end of the board. To satisfy the first condition for equilibrium, the support must now exert a total upward force of 210 pounds to support the board plus the children.

This second condition for equilibrium is the basic principle underlying the action of levers, which will be discussed in Chapter 4.

If you had to carry a long board, your common sense would tell you to pick it up under its center of gravity. As an extreme example, suppose you decided instead to pick up a board by one end, using one hand. Figure 3–13(d) illustrates the forces involved in picking up a 30 lb board which is uniform and 10 feet long. It is presumed that you exert an upward force with your fingers (F_1) at a distance 0.5 feet from the end and downward force at the end with your thumb (F_2). If the end where your thumb rests is chosen as the axis of rotation, the lever arm of the thumb force F_2 is zero and therefore F_2 does not enter the torque calculation. The lever arm of the 30 lb weight of the board is 5 ft (the distance from the axis to the center of gravity) and the force of your fingers, F_1, has a lever arm 0.5 ft. The equilibrium condition is then

$$\tau_{CCW} = \tau_{CW}$$

$$(F_1)\ (0.5\ \text{ft}) = (30\ \text{lb})\ (5\ \text{ft})$$

$$F_1 = 300\ \text{lb}.$$

The fingers must exert 300 lb of force to balance a 30 lb board because they are acting through such a short lever arm! The thumb gets off easier; it has to exert only a 270 lb downward force on the end of the board to satisfy the first condition for equilibrium (forces up = forces down).

You wouldn't be so foolish as to pick up a board as described in the above example, even if you had strong enough hands. But the extreme stresses we sometimes place on our back muscles by putting them in comparable situations are also foolish. If you bend over to pick up a heavy object which is far from your body as illustrated in Figure 3–16, the torque exerted on your body by the weight must be counteracted by torques exerted by your back muscles, which act through a very short lever arm with respect to a pivot axis in the pelvic region. Suppose you pick up a 10 lb weight in such a way that it has a 20 inch lever arm with respect to the pivot axis, producing a torque of 10 lb × 20 inches = 200 lb · inches. If the lever arm for the back muscles with respect to this pivot point is only one inch, they will have to exert a force of 200 pounds to counteract the torque produced by the load!

When lifting and carrying heavy objects, it is well to keep in mind the balancing of torques. When a load is held close to the trunk with the body erect, then the process is mainly one of balancing the forces (first condition for equilibrium). When a load is held out from the body it exerts a considerable amount of torque; this requires tension in the back mus-

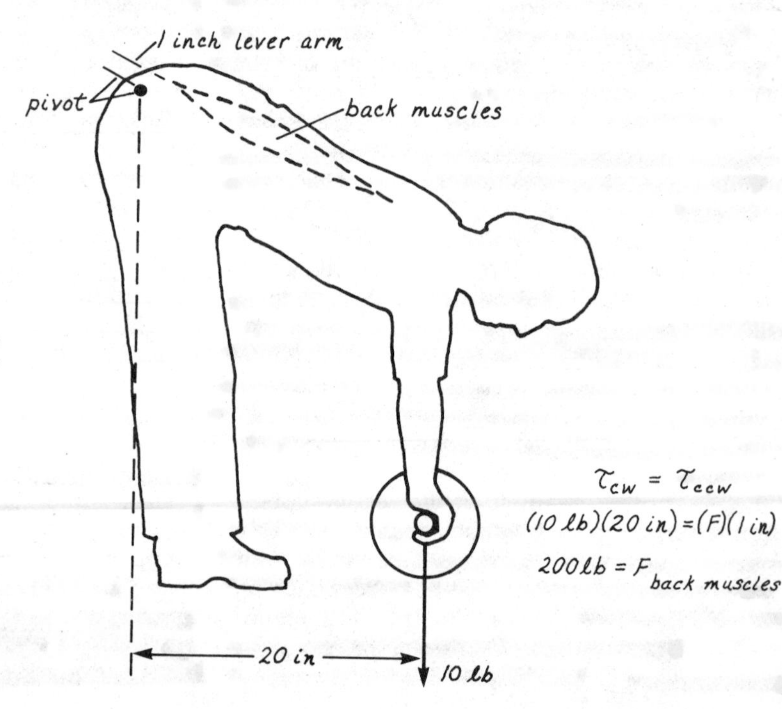

Figure 3–16 Stress on back muscles from lifting objects which are far away from the body.

cles and a backward leaning posture to produce a compensating torque.

CLINICAL APPLICATIONS OF GRAVITY

Under the influence of the constant downward pull of gravity, all things tend to move to the lowest possible position. There are numerous ways to take advantage of this force, particularly in the handling of fluids. Gravity drainage may be used, for example, to remove fluid from the lungs without the need for a suction apparatus. The gravitational force provides the pressure for administration of intravenous fluids, transfusions, and so on. These procedures will be discussed further in Chapters 5 and 6.

The force of gravity has an ever-present effect upon the circulation of fluids in the body. In the standing position the pressure in the blood vessels of the lower extremities is in general higher than that in the head and neck because of the gravitational pull. People who spend long hours on their feet have a tendency toward edema of the feet and legs. The gravity effect can be seen in the distended veins of the arms and hands if the arms are left hanging by the sides for long periods. Changing the body posture will result in changes in the blood pressure distribution and changes in the amount of work the heart has to do to maintain normal pressures.

Various treatments make use of the effects of gravity. Premature infants are sometimes placed on electrical rocking beds to improve circulation to and from the extremities. Post-operative patients are often placed in special positions; this is partially to improve gravity drainage. Brain surgery may be performed with the patient in a sitting position to lessen the danger of hemorrhage. The head may then be elevated after surgery to facilitate drainage of blood from the brain.

The "weightlessness" associated with manned orbital flight and space travel may produce circulatory difficulties if maintained for long periods of time. Since the circulatory system is adapted for operation under the influence of the earth's strong gravitational field, prolonged periods of "weightlessness" must represent a major adjustment. Long term orbiting space platforms are designed with "artificial gravity" to help with this problem. This artificial gravity effect can be produced by rotation, as discussed in the next section.

CIRCULAR MOTION

Acceleration was defined in Chapter 2 as the rate of change of velocity. Since the velocity is a vector quantity, with both a magnitude and a direction, an object is said to be accelerated if either the size or the direction of its velocity changes. Thus, motion in a circle is always accelerated, since its direction is constantly changing. This is true with any curved motion. The discussion of circular motion was deferred to this point because Newton's second law and the concept of force are useful in examining it.

Driving a car provides a great deal of experimental evidence about circular motion. It is easy to perceive that rounding a curve at constant speed is different from traveling along a straight road at a constant speed. While rounding the curve, the road must exert a force on the tires toward the inside of the curve to keep the car on the road. If that force is not exerted, as in the case of an ice-covered road, then the car will not turn but keep going straight, with perhaps disastrous results. It is also observed that the greater the speed of the car, the greater the force which the road must exert. Further, the sharper the curve, the greater the force required to keep the car on the road.

This readily available experimental evidence contains most of the important aspects of circular or curved motion. The fact that a road force directed inward toward the center is required to keep a car on a circular track is evidence that the circular motion is indeed an accelerated motion, since by $F_{\text{net ext}} = ma$, if there is no acceleration, no force is required. Further, this is evidence that the acceleration is toward the center of the circle. If the force were removed, the car would continue in a straight line, off the track, according to Newton's first law. More detailed experiments indicate that the force required is proportional to the square of the velocity and inversely proportional to the radius of the circle. This required force is called the *centripetal* (center-seeking) force. For a given mass, m, the centripetal force is given by the relationship

$$F = \frac{mv^2}{r} \qquad \textbf{3–10}$$

where v is the velocity and r is the radius of the curve. This is the net force required to keep an object moving in a circular path and it acts in a direction perpendicular to the velocity and toward the center of the circle. Actually, any part of a curved path could be extended to make a circle of some radius, r, so it is a general expression for the centripetal force required at any point on a curved path. The tendency for an object to leave the circular path and travel in a straight line according to Newton's first law is often characterized in terms of an effective *centrifugal* force (center-fleeing force). Since nothing exerts this force, it must be looked upon as a fictitious or "effective" force; it merely expresses the tendency of objects to travel in a straight line. Nevertheless, it is often convenient to describe the behavior of objects flying off a rapidly whirling wheel, objects sliding across your car seat on a curve, etc.

Returning to the example of driving a car around a curve, the fact that the necessary centripetal force depends upon the velocity squared is of great practical significance. Suppose it required 500 pounds of center-directed force to round a certain curve at 30 miles/hour. The force required at 60 miles/hour would be 2000 pounds, or four times as much, since it is proportional to the square of the velocity. The road could probably not supply this much force, particularly if it were wet, so a skid might result.

Example. A 64 pound weight is to be swung in a horizontal circle of radius 2 feet with successive speeds of 10 ft/sec, 20 ft/sec, and then 30 ft/sec. If this is done by attaching a rope to it, what must the tension in the rope be in the three cases?

Solution.

Case 1: $v = 10$ ft/sec

The mass of the object must first be found. If its weight is $W = 64$ lb, then the mass is given by

$$W = mg;\ m = \frac{W}{g} = \frac{64 \text{ lb}}{32 \text{ ft/sec}^2} = 2 \text{ slugs}.$$

Then from equation 3–10, the centripetal force is given by

$$F = \frac{mv^2}{r} = \frac{(2 \text{ slugs})(10 \text{ ft/sec})^2}{(2 \text{ feet})},$$

$$F = 100 \frac{\text{slug-ft}}{\text{sec}^2} = 100 \text{ pounds}.$$

Case 2: $v = 20$ ft/sec

The force required is

$$F = \frac{(2 \text{ slugs})(20 \text{ ft/sec})^2}{(2 \text{ feet})} = 400 \text{ pounds},$$

so when the speed is doubled, the force increases by a factor of four. This result could have been obtained by noting that the ratio of the forces is equal to the ratio of the square of the velocities:

$$F_2 = \frac{(20)^2}{(10)^2} \times F_1 = 2^2 \times 100 \text{ lb} = 400 \text{ lb}.$$

Case 3: $v = 30$ ft/sec

$$F = \frac{(2 \text{ slugs})(30 \text{ ft/sec})^2}{(2 \text{ feet})} = 900 \text{ pounds}$$

or

$$F = (3)^2 \times 100 \text{ lb} = 900 \text{ pounds}.$$

The fact that the force required goes up so rapidly with speed puts a very large strength requirement on ultracentrifuges and high-speed dental drills. High-speed rotating equipment should be checked often for cracks since the outer edges of the rotors must withstand enormous forces.

In the case of the circular orbital motion of satellites, the earth's gravity exerts the required centripetal force to keep the satellite from traveling in a straight line off into space. The same is true of the moon's orbit, and the sun's gravity provides the centripetal force to keep the planets in orbit around it.

CLINICAL APPLICATIONS OF CIRCULAR MOTION

The laboratory centrifuge makes use of the principles described above to separate materials according to their densities. Since the force required to keep a particle moving in a circle is proportional to its mass, it follows that the heavier particles in a centrifuge like the one illustrated in Figure 3–17 will move to the outside. Less force will be required to keep them moving in a circle at that point since the radius is larger (the radius means the distance from the rotation axis in this case). Besides its use in separating materials of differing densities, the centrifuge can be used to measure densities. This can be accomplished by centrifugation of the unknown with several known substances with graduated densities. By the final position of the unknown, its density can be bracketed between two known densities. Centrifuge microscopes have rotating centrifuge heads so that cells and small organ-

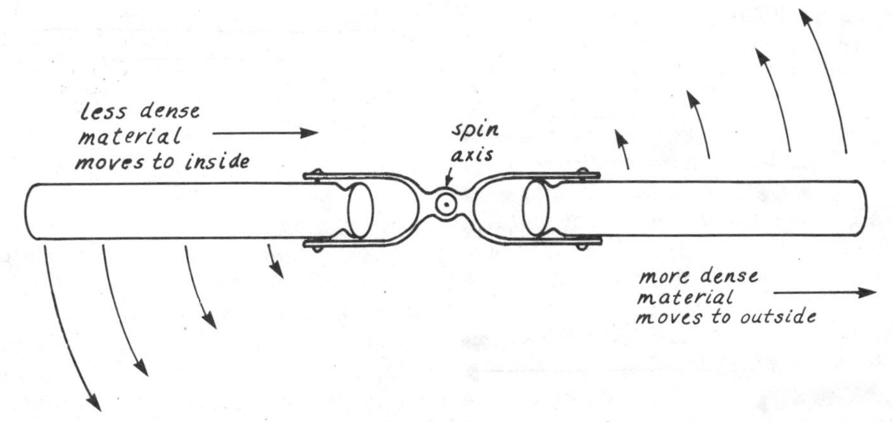

Figure 3–17 Simple laboratory centrifuge.

isms can be observed while rotating rapidly. This makes possible the immediate analysis of the relative densities of different parts of the cells.

If the centripetal acceleration of the sample is increased, the centrifuge's selectivity is improved. That is, it can separate substances with smaller density differences. Ultracentrifuges which rotate at rates over 100,000 revolutions per minute and produce accelerations 30,000 times that of gravity are important tools in microbiology research. They have been used for such difficult tasks as separating different types of proteins from cellular material.

Since different types of microorganisms may have different densities, a high speed centrifuge might be used to separate different types of microorganisms. If the concentration of a certain type of microorganism is small in a fluid or sputum sample, the centrifuge can be used to concentrate them in a small portion of the sample.

FRICTIONAL FORCES

When there is relative motion between two surfaces which are in contact, there will be a force which resists this motion. The size of this frictional force will depend upon the nature of the surfaces, but it is always present. Even when surfaces appear to be perfectly smooth, there will be irregularities when the surfaces are examined microscopically, as illustrated in Figure 3–18. There may also be adhesion between the two surfaces as a contributing factor. Besides the nature of the surfaces, the frictional forces depend mainly upon the amount of force pressing the two surfaces together.

Frictional forces are sometimes harmful and sometimes useful. The harmful effects include the following: (1) it increases the effort necessary to operate any kind of mechanical device and therefore reduces the efficiency of machines, (2) it produces heat by converting mechanical energy into heat energy, and (3) it wears away or otherwise damages the surfaces which are rubbing against each other. On the other hand, if there were no friction, we could not walk. Cars could not move because there would be no traction, and if put in motion they could not stop because braking depends entirely upon friction. Friction must be maximized when it is useful, and methods must be found to reduce it when it is harmful.

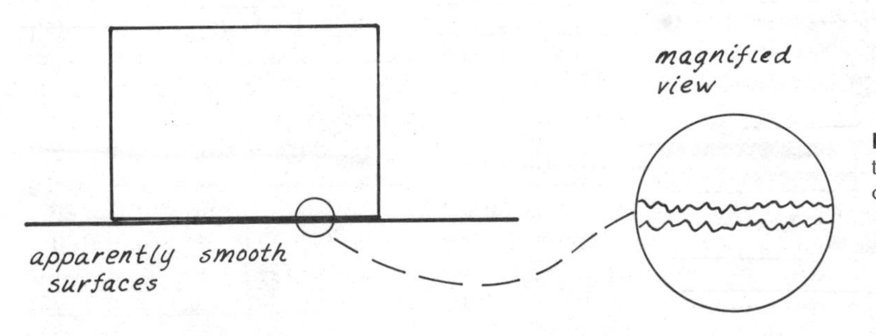

Figure 3–18 Microscopic irregularities on even highly polished surfaces contribute to friction.

In clinical applications such as the insertion of gastric tubes, catheters, and rectal tubes, the friction between these tubes and the membranes over which they must pass may present a severe problem. Measures must be taken to reduce the burning and irritation caused by such tubes. Modern plastic tubing has been found to be superior to rubber in this regard; a considerable amount of research has gone into the development of tubing with minimal friction drug.

Since the forces caused by fluid friction are in general much smaller than the friction forces between solids, the use of lubricants is usually advisable. Lubricating oils or salves separate the two surfaces and the friction is limited to the much smaller fluid friction. Special antiseptic, water-soluble lubricants are now in general use for such applications. In the absence of such lubricants, water can be used as a lubricant in some cases. That is the function of the water taken with a capsule or tablet.

When frictional forces are treated quantitatively, it is found that the force opposing the relative motion of two surfaces is proportional to the force pressing the surfaces together, which is called the normal force and is designated N. The frictional force is found to be nearly independent of the area of the surfaces, unless of course one surface is so small that it cuts into the other. For low velocities, frictional forces do not change very much with changes in the speed of relative motion. Under these conditions, the frictional force F can be written

$$F = \mu N$$

where μ is called the coefficient of friction. One immediate observation about friction is that it takes a considerably greater force to start the motion of a heavy object than to keep it in motion at a constant velocity. The surface irregularities which are responsible for friction become interlocked when the surfaces are at rest and a larger force is required to start the motion. When the surfaces are moving, the surface irregularities of one surface may bounce along on the other surface and provide less resistance to movement. Because of this distinction, it is convenient to define two coefficients of friction, μ_s and μ_k, representing the coefficient of static friction and the coefficient of kinetic friction, respectively. The coefficient of static friction, μ_s, is used in the calculation of the maximum resistance force an object can exert before it starts to move:

$$F_{max} = \mu_s N = \text{maximum static frictional force.} \quad \textbf{3–11}$$

For example, if an 80 lb crate was characterized by a coefficient of static friction $\mu_s = .4$, it would take a force F = (.4)(80 lb)= 32 lb to start the crate in motion since the normal force in this case is the weight of the crate. If a 25 lb horizontal force was exerted on the crate, it would not move and the frictional resistance force would be just 25 lb, in accordance with Newton's third law. Therefore, the coefficient μ_s represents the threshold of motion condition, and is used only for that condition. On the other hand, the coefficient of kinetic friction, μ_k, can be used to calculate the frictional resistance force:

$$F = \mu_k N = \text{kinetic friction force} \quad \textbf{3–12}$$

under any conditions of motion so long as our assumptions about the velocity independence of friction are valid. If the kinetic friction coefficient were $\mu_k = 0.3$ for the crate described above, a force

$$F = (0.3)(80 \text{ lb}) = 24 \text{ lb}$$

would be required to keep the crate moving with a constant velocity once it is started.

As a practical application of the distinction between static and kinetic friction, consider the process of inserting a naso-gastric tube. Each time the tube is started in motion, static friction must be overcome. The insertion can be accomplished with less effort and less resistance if it is inserted in one continuous motion than if it is inserted with a large number of stops and starts.

Other applications of these concepts may be made in the case of rolling objects. For example, when a car tire is rolling, the part of the tire in contact with the road is instantaneously at rest and static frictional forces exist. However, if the tire is sliding on the surface, the lesser kinetic friction forces apply. It is common experience that more traction is achieved on an icy road if the tires can be kept from spinning, since static friction forces are at work when the tires are rolling on the surface without slipping. It is also observed that a car can decelerate faster if the wheels are turning, rather than locked and skidding. Modern computer-controlled braking systems for aircraft and large vehicles obtain optimum braking by releasing and reapplying the brakes when slipping starts, so that the wheels continue to roll and take advantage of the larger braking forces of static friction.

This distinction between rolling and sliding friction while braking is not so important for automobiles when you are on dry flat road surfaces since the static and kinetic friction coefficients for modern tire designs have been shown to be about the same. In fact, a lot of engineering effort has been invested toward that end, to reduce the seriousness of a skid. Tests have shown that the friction coefficients average about 0.7 for new tires under optimum conditions whether the tires were sliding or still rolling (8). But the presence of water or snow and ice on the roadway should again produce an advantage in braking for the tire which continues to roll rather than slide.

SUMMARY

Forces and torques are the influences which cause motion, or more accurately stated, they are the influences which cause changes in the motion of an object. The relation of force to motion can be summarized by Newton's three laws. Newton's second law, $F_{\text{net ext}} = ma$, relates the net force on an object to the acceleration of the object and its inertia, as measured by its mass m. Newton's first law is a special case of the second law dealing with the case $F = 0$. It states the physical fact that an object will remain at rest or in a uniform motion unless acted upon by an external net force. The third law states that forces occur in pairs, with an equal but oppositely directed reaction force accompanying every force. Torques, the rotational analogs of forces, tend to cause rotation. The torque associated with a given force is equal to the size of the force times the perpendicular distance to the axis of rotation. An object which remains at rest is said to be in static equilibrium. The conditions for equilibrium are (1) zero net force and (2) zero net torque. In determining the net force, it must be noted that forces are vector quantities and must be added by vector methods to find the resultant force. In determining the net torque, it is often convenient to define the *center of gravity*, the point at which all the weight could be considered as acting for the purpose of calculating torques. The center of gravity is also useful for determining the stability of equlibrium of large objects.

According to Newton's first law, a force is required to change either the speed or the direction of the motion of an object. Therefore, all curved motion is accelerated motion. The force required to make an object follow a curved trajectory is called the centripetal force, which is given by the relationship $F(\text{centripetal}) = mv^2/r$ where r is the radius of curvature of the path.

In the absence of a net force, an object would continue to move at constant speed in a straight line forever, but we don't observe such motion in practice because of frictional forces. The frictional resistance forces which oppose the relative motion of two solid surfaces are proportional to the force pressing the surfaces together and to the coefficient of friction, which indicates the relative roughness of the surface. The force needed to overcome static friction and start something in motion is generally greater than that needed to overcome kinetic friction to keep it moving. Fluid frictional forces contribute to viscosity and wall friction and oppose the motion of fluids.

REVIEW QUESTIONS

1. What conditions are necessary for an object to move with constant velocity?

2. What conditions are necessary for an object to move with constant acceleration?

3. According to Newton's third law, every force is accompanied by an equally large reaction force. Why does a gun not recoil backward as fast as the bullet goes forward?

4. Why is it necessary to lean forward when pushing a heavy cart? Explain with the use of vector forces.

5. A force of 40 lb and a force of 30 lb act on the same object. What is the maximum total force they can exert? What is the minimum force?

6. A given muscle can exert a force only along its length. How can you account for the fact that a combination of several muscles in the arm can exert a force in virtually any direction on the hand?

7. What is the difference between weight and mass?

8. Why does lengthening the handle of a crank reduce the force necessary to turn it?

9. Can an object be in equilibrium if there are forces acting on it?

10. What conditions are necessary for an object to be in equilibrium? Can a moving object be in equilibrium?

11. What would happen to the weight of an object if it were taken to the moon? What would happen to its mass?

12. If an object is moving at a constant speed in a circle, is it accelerated? Explain.

13. When you whirl a ball on the end of a string, it pulls outward on your hand. If you release the string, the ball moves in a straight line. Explain these effects in terms of Newton's laws.

14. What is meant by an acceleration of 1 *g*? How can a centrifuge produce accelerations of hundreds of *g*'s?

PROBLEMS

Worked Example: If a 2 kg mass is acted upon by a force of 10 nt, what is the acceleration? If it is intially at rest, what will be its velocity after 3 seconds?

Solution: From Newton's second law, $F_{\text{net ext}} = ma$, the acceleration is

$$a = \frac{F_{\text{net ext}}}{m} = \frac{10\ nt}{2\ \text{kg}} = 5\ \text{m/sec}^2.$$

Using the equation $v = at$, the velocity is found to be

$$v = at = (5\ \text{m/sec}^2)(3\ \text{sec}) = 15\ \text{m/sec}.$$

Worked Example: A crate weighing 160 pounds is acted upon by a 100 lb horizontal force. The friction between the box and the floor produces a 40 lb opposing horizontal force. Find the acceleration and the distance moved in 5 seconds.

Solution: The mass must be found before Newton's second law can be applied. Since the weight is defined as $w = mg$, the mass is

$$m = \frac{w}{g} = \frac{160 \text{ lb}}{32 \text{ ft/sec}^2} = 5 \text{ slugs}.$$

In $F = ma$, the force F is the net external force

$$F_{\text{net external}} = 100 \text{ lb} - 40 \text{ lb} = 60 \text{ lb}.$$

The acceleration is then

$$a = \frac{F_{\text{net external}}}{m} = \frac{60 \text{ lb}}{5 \text{ slugs}} = 12 \text{ ft/sec}^2.$$

The distance moved in 5 seconds is

$$s = v_0 t + \tfrac{1}{2}at^2 = \tfrac{1}{2}(12 \text{ ft/sec}^2)(5 \text{ sec})^2$$
$$= (6 \text{ ft/sec}^2)(25 \text{ sec}^2) = 150 \text{ ft}.$$

Alternatively, the distance may be found by noting that

$$v = at = (12 \text{ ft/sec}^2)(5 \text{ sec}) = 60 \text{ ft/sec}$$

after 5 seconds and that the average velocity is then

$$\bar{v} = \frac{v_0 + v}{2} = \frac{0 + 60 \text{ ft/sec}}{2} = 30 \text{ ft/sec}.$$

The distance is

$$s = \bar{v}t = (30 \text{ ft/sec})(5 \text{ sec}) = 150 \text{ ft}$$

as obtained above.

1. What force in newtons is required to accelerate a 2 kg object at a rate of 8 m/sec^2?

2. If an object weighing 32 lb is acted upon by a net force of 32 lb, what will be its acceleration?

3. If a 2 kg object is observed to accelerate from rest to 15 m/sec in 5 seconds, what was the net external force acting upon it (assuming that the force was constant)?

4. A 64 lb box is pushed along a flat surface by a net external force of 20 lb. If it starts from rest, what will be its speed after 2 seconds?

5. A 10 newton net external force is exerted upon a 2 kg object.

   a. What is the acceleration?
   b. How far will it travel in 4 seconds if it starts from rest?

6. A 3200 lb car accelerates at the rate a = 5 ft/sec^2.

   a. What net external force must the road exert forward on the car to produce this acceleration?
   b. What will be the velocity of the car 10 seconds after it starts?

7. A 20 lb block rests on top of a 12 lb block which rests on a scale. Find the force exerted

   a. by the 20 lb block on the 12 lb block,
   b. by the 12 lb block on the 20 lb block,
   c. by the 12 lb block on the scale,
   d. by the scale on the 12 lb block.

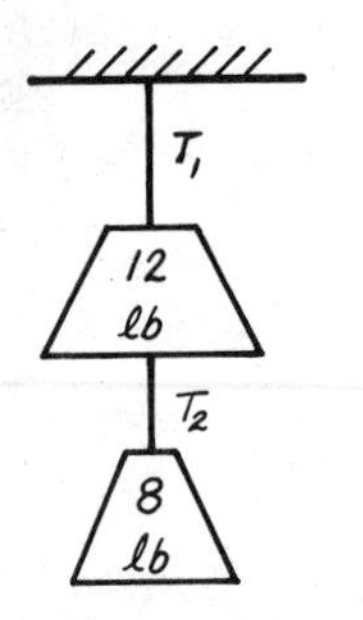

Figure 3–19

e. by gravity on the 12 lb block.
f. Find the net force on the 12 lb block.

8. Find the tensions, T_1 and T_2 in Figure 3–19.

9. If two ropes are tied to an object, one exerting a 15 lb force horizontally and one exerting a 20 lb force vertically, what is the resultant force? Draw a sketch to show the direction of the resultant force.

10. A post has three ropes attached to it. One exerts a 100 nt force east and another exerts a 50 nt force north. It is found that the third force must be directed at an angle 30 degrees south of west to keep the post vertical (to produce equilibrium). What force must be exerted by this third rope?

11. What is the weight of a 20 kg block on the surface of the earth? What is the mass of a 224 pound object?

12. The acceleration of gravity on the moon is $g = 1.63$ m/sec^2. What would be the weight of a 5 kg mass on the earth and on the moon? What would be the weight of a 160 lb man on the moon?

13. The Saturn V rocket used in the Apollo moon missions weighed 6.1×10^6 lb before takeoff. During the initial takeoff thrust, the rocket engines produced a thrust of 7.5×10^6 lb.

    a. Neglecting air resistance, what would be the acceleration of the rocket at takeoff?
    b. Assuming constant acceleration, what would the speed be after one minute?

14. A flea jumps by releasing its spring-like legs and can reach great heights, compared to its size. Certain types of fleas have been observed to reach an upward velocity of 130 cm/sec in about 1 millisecond by thrusting their legs about 0.07 cm. The resting height of the flea is in the neighborhood of 0.08 cm.

    a. Calculate the average acceleration of the flea during its takeoff thrust.
    b. Calculate the acceleration in g's and compare to the maximum take-off acceleration of the Apollo astronauts on their way to the moon, 15 g's.
    c. Calculate the maximum height of the flea's jump, and express it as a multiple of the flea's rest height. How high could you jump if you could jump the same multiple of your height?

15. How far will a 2 kg mass be moved in 4 seconds by a 5 nt force if it starts from rest?

16. What force must be exerted upon a 3200 lb car to accelerate it to 60 miles/hr in 10 seconds? (60 miles/hr = 88 ft/sec).

17. A child is pushing another child in a wagon. The combined weight of the second child and the wagon is 96 lb. The child doing the pushing can exert a 40 lb force, but he is resisted by a 10 lb frictional resistance force.

    a. What is the acceleration?
    b. How far must he push the wagon to achieve a speed of 8 ft/sec?

$A = \frac{\Delta V}{T}$; $T = V/a$

18. A 3200 lb car traveling at 44 ft/sec (30 mi/hr) strikes another car in the rear and comes to rest after pushing the car 10 feet.

    a. Assuming uniform acceleration, what force was exerted on the car to bring it to rest in 10 ft?

b. What force was exerted on the 160 lb driver by his seatbelts to bring him to rest in 10 ft?

19. A motorcyclist traveling at 60 mi/hr (88 ft/sec) catches a 0.5 ounce bug in the center of his facemask. A realistic time for this collision is 0.0005 sec, and during that time you may assume that the bug's speed is increased from zero to 88 ft/sec. Calculate the force of impact of the bug, assuming the acceleration to be constant.

20. A private aircraft approaching a landing at a speed of 60 m/sec collides with a 100 gram bird. The time involved in the collison is 0.001 seconds, during which the bird is assumed to be accelerated from v = 0 to 60 m/sec. Assuming constant acceleration, calculate the force of impact.

Worked Example: When you are being accelerated upward by an elevator, you feel "heavier" than normal, and if you were standing on a scale at the time, the scale would give you a "weight" measurement higher than normal. This larger apparent weight arises from the fact that the support force exerted upon you by the elevator is larger than your weight because it is producing a net external force to accelerate you upward. If a person who weighs 120 lb is accelerated upward by an elevator at the rate a = 4 ft/sec², what is his apparent weight?

Solution: The support force must overcome gravity and provide the upward net external force to give the acceleration.

$$F_{\text{net ext}} = ma$$

$$F_{\text{support}} - mg = ma = \left(\frac{120 \text{ lb}}{32 \text{ lb/slug}}\right)(4 \text{ ft/sec}^2)$$

$$F_{\text{support}} = 120 \text{ lb} + (3.75 \text{ slugs})(4 \text{ ft/sec}^2)$$

$$F_{\text{support}} = 135 \text{ lb} = \text{apparent weight}$$

21. A 160 lb person is being accelerated upward by an elevator at the rate 8 ft/sec².

  a. What force must the elevator exert upon him to produce this upward acceleration?
  b. What would be his apparent weight if the elevator were accelerating *downward* at 8 ft/sec²?
  c. If the elevator cable broke so that he was accelerating downward at a = 32 ft/sec², what would be his apparent weight?

22. A 120 lb person is being lifted aloft by a rocket which accelerates uniformly to an upward speed of 300 ft/sec in 2 seconds.

  a. What is the upward acceleration in multiples of g?
  b. What is the apparent weight of the person during the acceleration?

23. A 5 kg mass is given an upward acceleration of 15 m/sec² from the earth by a rope. What is the tension in the rope?

24. A metal object of mass 500 grams is found to have a volume of 185 cm³. What is its density? If it is assumed to be a pure metal, what is the metal?

25. What would be the mass of a rectangular bar of iron with dimensions 4 cm × 6 cm × 12 cm?

Worked Example: Three children weighing 40 lb, 50 lb, and 80 lb wish to balance on a see-saw. If the 80 lb child sits 5 ft from the pivot on the left end and the 50 lb child sits 5 ft from the pivot on the right end, where must the 40 lb child sit to achieve balance?

Solution: The 40 lb child clearly must sit on the end with the 50 lb child, and if his distance from the pivot is represented by x:

$$(50 \text{ lb})(5 \text{ ft}) + (40 \text{ lb})(x) = (80 \text{ lb})(5 \text{ ft})$$

$$250 \text{ ft-lb} + (40 \text{ lb})x = 400 \text{ ft-lb}$$

$$x = \frac{400 \text{ ft-lb} - 250 \text{ ft-lb}}{40 \text{ lb}}$$

$$x = 3.75 \text{ ft.}$$

26. A 60 lb child sits 4 feet from the pivot point on a see-saw. If the center of gravity of the see-saw is just above the pivot point how far from the pivot must a 40 lb child sit in order to balance?

27. A meter stick is balanced at its center and three masses (20 gm, 60 gm, and 100 gm) are suspended from it. If the 100 gm mass is placed at the 80 cm mark on the stick and the 60 gm mass is placed at 10 cm, where must the 20 gm mass be suspended to achieve equilibrium?

Worked Example: A 50 kg load is placed on a 10 kg uniform stretcher to be carried by two persons. The stretcher is 2 meters long and the load is placed 0.6 meters from the left end. How much force must each person exert to carry the stretcher?

Solution: The first step in a torque-equilibrium problem is to choose an axis for the torque calculation. It may be chosen at any point, but the calculation is simplified by choosing it at the point of action of one of the unknown forces so that the lever arm for that force is zero, eliminating one unknown from the calculation. The left and right support forces will be labeled F_L and F_R, and the axis will be chosen at the left end to eliminate F_L from the torque calculation. With $g = 9.8$ nt/kg, the weights of the load and stretcher are 490 nt and 98 nt respectively.

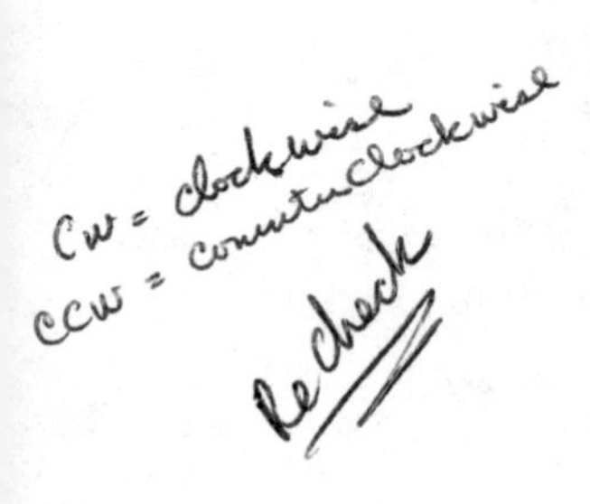

$$\tau_{CW} = \tau_{CCW}$$

$$(490 \text{ nt})(.6 \text{ m}) + (98 \text{ nt})(1 \text{ m}) = (F_R)(2 \text{ m})$$

$$F_R = \frac{392 \text{ nt} \cdot \text{m}}{2 \text{ m}} = 196 \text{ nt.}$$

The total support $F_L + F_R = 490 \text{ nt} + 98 \text{ nt} = 588$ nt,

so $F_L = 392$ nt.

29. Two men are carrying a 150 lb weight on a uniform 5 ft board which weighs 20 lb. The weight is 2 ft from one end of the board. Draw a diagram of all the forces acting on the board and calculate the force which each man must exert to carry the load.

30. If a uniform 7 ft bed weighs 120 lbs, what force will be required to lift one end of it?

31. A uniform 2 meter stretcher has a mass of 5 kg. It is used by two persons to carry three masses of 40 kg, 10 kg, and 20 kg, which are located at 0.2 m, 0.5 m, and 1.5 m respectively from the left end of the stretcher. How much support force must each person exert to carry the stretcher?

32. One end of a 6 ft board is supported by a table, and you are supporting the other end. If the uniform board weighs 30 lb and there is a 60 lb additional load at a point 2 ft from the table end, how much force will you have to exert to hold up your end?

33. Assume that a person's arm weighs 8 lb and has a center of gravity 12 inches from the effective pivot point of the shoulder when the arm is stretched out horizontally. Further assume that the effective lever arm for the muscles which must support the arm in this position is one inch.

    a. How much muscle force is required to support the arm in this outstretched horizontal position?
    b. What total muscle force would be required to support a 5 lb weight in the hand, 24 inches from the pivot, in addition to supporting the arm?

34. A person picks up a 20 ft uniform pole which weighs 15 lb by grasping one end with his left hand and placing his right hand at a point 1 ft from the end. How much force must each hand exert to hold the pole in a horizontal position?

Worked Example: On a light rod of length 2 meters and negligible mass, three concentrated masses 2 kg, 6 kg, and 8 kg are placed at 0 m, 1 m, and 2 meters respectively. Where would you place a support under this system so that it would balance (i.e., where is the center of gravity)?

Solution: Finding the center of gravity corresponds to finding the point at which all the weight could be concentrated to give the same torque about any axis. The end of the rod at 0 meters is chosen as an axis for the calculation and the position of the center of gravity is represented by $\overline{x}$. The torque about the axis is.

$$\tau = (2\text{ kg})(g)(0\text{ m}) + (6\text{ kg})(g)(1\text{ m}) + (8\text{ kg})(g)(2\text{ m})$$

where the masses are multiplied by g to get the weights.

If all the weights were concentrated at the center of gravity $\overline{x}$, the torque would be the same:

$$\tau = (16\text{ kg})(g)\bar{x}.$$

Therefore

$$\overline{x} = \frac{(6\text{ kg}\cdot\text{m} + 16\text{ kg}\cdot\text{m})(g)}{(16\text{ kg})(g)} = 1.38\text{ m}.$$

Note that g cancels out in this expression, so that for practical purposes, either the mass or the weight could be used in this calculation. If supported at $x = 1.38$ m, the system would balance.

35. A 10 lb ball and a 2 lb ball are placed on a rigid bar such that their centers are 2 ft apart. Neglecting the weight of the bar, where is the center of gravity of this combination? (Hint: Where would you support the bar to balance it so that the weights exert no net torque?)

36. A uniform bar of length 50 cm is observed to balance when supported at the center of the bar. When the support is moved 8 cm away from the center, it is found that a 150 gm mass hung 12 cm from the new pivot point will again balance the rod. What is the mass of the rod?

37. An irregularly shaped object is too large to be weighed directly on one scale, so the two ends of the object are placed on separate scales. The scales read 40 lb and 120 lb. If the length of the object is 8 ft, find the position of the center of gravity of the object and its weight (i.e., find the horizontal distance from one of the scales to the center of gravity).

38. Two masses, 3 kg and 8 kg, are attached by a rigid, uniform bar of mass 2 kg such that their centers are 50 cm apart. Find the location of the center

of gravity of this combination. (Assume that the masses may be treated as point masses on the ends of a 50 cm rod, and that the mass of the bar itself can be treated as a point mass at the center of the bar.)

39. a. What force would be required to swing a 0.5 kg mass in a circle of radius 2 meters with a velocity of 8 m/sec?
    b. What force would be required to swing a 7.3×10^{22} kg moon in a circle of radius 3.8×10^{8} m with a velocity of 1000 m/sec?

40. If a force of 500 lb toward the center must be exerted by the road to keep your car moving around a circular track at 30 miles/hr, what would be the required road force at 60 miles/hr? at 90 miles/hr?

41. A highway curve with a radius of curvature of 1000 ft is unbanked, so that friction between your tires and the road must supply all the necessary centripetal force to keep your car on the road.

    a. If your car weighs 3200 lb, what frictional force toward the center of the curve would be required to keep your car on the road if you rounded the curve at 60 mi/hr (88 ft/sec)?
    b. What coefficient of friction between you and the road would be required in (a)?
    c. If the coefficient of friction were 0.1 because of snow on the road, what would be the maximum speed you could drive around the curve without slipping?

42. A 160 lb person rides "over the hump" on a roller coaster and feels lighter in the process. The radius of curvature of the top of the hump is 40 ft and his speed at the top is 20 ft/sec.

    a. What is the centripetal acceleration experienced at the top of the ride?
    b. What is his apparent weight as a result of this downward acceleration at the top?

43. An airplane which is pulling out of a dive is traveling at 200 m/sec at the bottom of the curve where the radius of curvature is 1000 meters.

    a. What is the upward centripetal acceleration of the airplane at this point?
    b. Calculate the actual and apparent weight of the 70 kg pilot at this point.

Worked Example: A 50 lb box is to be moved across a rough floor. The coefficient of static friction is 0.5 and the coefficient of kinetic friction is 0.3. How much force is required to start the box in motion and how much force is required to keep it moving with a constant speed?

Solution: To move the box, static friction must be overcome. This requires a force

$$F = \mu_s N = (0.5)(50 \text{ lb}) = 25 \text{ lb}.$$

However, to keep it moving, the required force is only

$$F = \mu_k N = (0.3)(50 \text{ lb}) = 15 \text{ lb},$$

since the kinetic friction resistance is less than the static friction resistance.

44. The friction between a crate and the floor is characterized by a static friction coefficient $\mu_s = 0.4$ and a kinetic friction coefficient $\mu_k = 0.2$. How much force will be required to start the 140 lb crate in motion and

how much force will be required to keep it in motion at a constant speed?

45. A typical coefficient of friction for the braking of a car with good tires on a dry surface is 0.7 (Ref. 8). What distance would be required to stop a 2500 lb car traveling at 60 mi/hr? What distance would be required to stop a 4500 lb car traveling at 60 mi/hr? What does this tell you about the relation of the braking distance to a car's weight?

46. A 3200 lb car is traveling at a speed of 50 ft/sec.

    a. Assuming a coefficient of friction 0.7, what distance would be required to stop the car after the brakes are applied?
    b. If this same car were on the moon with the same speed and the same coefficient of friction, but where the weight of the car is one sixth as great, how far would the car travel while braking?

47. A 10 kg sled traveling at an initial speed of 5 m/sec on level snow came to rest after traveling 20 meters. What was the force of friction opposing the sled's motion and what was the coefficient of friction?

48. A 5 kg block is initially moving with a speed of 8 m/sec on a horizontal surface where the coefficient of kinetic friction is 0.25 and the coefficient of static friction is 0.40.

    a. What is the force of friction acting against the moving block?
    b. What is its acceleration?
    c. How far will it move before coming to rest?
    d. How much force would be required to start the block in motion again after it stops?

REFERENCES

1. Blackwood, O. H., Kelly, W. C., and Bell, R. M. *General Physics,* 4th ed. New York: Wiley, 1973.
2. Krauskopf, K. B., and Beiser, Arthur. *The Physical Universe,* 3rd ed. New York: McGraw-Hill, 1973.
3. Flitter, H. H. *An Introduction to Physics in Nursing,* 7th ed. St. Louis: C. V. Mosby Company, 1976.
4. Sackheim, G. I. *Practical Physics for Nurses,* 2nd ed. Philadelphia: W. B. Saunders Company, 1962.
5. Greenwood, M. E. *An Illustrated Approach to Medical Physics,* 2nd ed. Philadelphia: F. A. Davis Company, 1966.
6. Jensen, J. T. *Introduction to Medical Physics.* Philadelphia: J. B. Lippincott Company, 1960.
7. Stearns, H. O. *Elementary Medical Physics.* New York: The Macmillan Company, 1947.
8. Bartels, R. A., *Braking Distance Versus Mass for Automobiles,* Am. J. Phys. Vol. 45, p. 398, 1977.

CHAPTER FOUR

Work, Energy, and Machines

INSTRUCTIONAL OBJECTIVES

After studying this chapter, the student should be able to:

1. Define precisely the terms work, power, and energy and give the appropriate units in the metric and British systems of units.
2. State and explain the principle of conservation of energy.
3. Give examples of kinetic energy and potential energy.
4. Work problems involving work, power, and energy.
5. Explain the concepts of mechanical advantage, ideal mechanical advantage, and efficiency for machines.
6. Work word problems involving the simple machines, particularly lever systems.

The terms work, energy and power arise in almost every area of science. In everyday use these words have many shades of meaning, but for use in physics or the other sciences they must be given rather precise and limited meanings. In this chapter these terms will be defined and used to explain the physical principles associated with machines.

WORK

If a person exerts a force, F, on a box and slides it a distance, s, across the floor, he has done work on the box. The work done on the object is defined as the force exerted on it times the distance moved *in the direction of the force*. If we let $\mathscr{W}$ represent the work, then this definition can be expressed as

$$\mathscr{W} = Fs \tag{4–1}$$

where it is understood that s is the distance moved in the direction of the force. In the physical sense, no work is done by the force F if the object doesn't move, and no work is done by the force if the object moves in a direction perpendicular to the direction of the force.

1 Nt-Meter = Joule

The units of work are the units of force times distance, as shown in Table 2–1 and in Table T–1 in Appendix A. In the British system the units are foot-pounds (ft-lb). In the MKS system the unit of work is the newton-meter (nt-m), which is called a *joule*. It is common practice to change the name of the unit combination just about every time another unit is added to it. This will undoubtedly be a source of confusion until the units become familiar, so the use of Table T–1 in Appendix A is recommended to see how the various units are built up from the basic mass, length, and time units. After some familiarity is gained, it is more convenient to use "joule" than to use the equivalent combination of mass, length, and time units (kg-m²/sec²).

As illustrated in Figure 4–1, if a 50 pound force is exerted on a crate and it moves 10 feet across the floor in the direction of the force, then the work done is 50 pounds times 10 feet, or 500 ft-lb. On the other hand, suppose a person supports a 50 pound box on his shoulder while walking 10 feet at a constant speed. In the physical sense, no work is done on the box since there is no motion in the direction of the force. Although this may seem illogical, it must be kept in mind that the work referred to is the work done on the box. The 50 pound upward force exerted by the person just balances the downward force of gravity and there is no net force on the box. Since it is moving along at a constant velocity, it is in fact at equilibrium. Of course, the person will feel that he has done some work, in the everyday sense of the term, and there will be some physical work done inside his body. In the process of walking with the load, there will be internal forces and movements of muscles. So the person must do work internally, but in the physical sense the work done on the box is zero. If his shoulder is 5 feet high, he did an amount of work 50 pounds times 5 feet (250 ft-lb) in the process of lifting the box to his shoulder.

When the motion of an object is not in the same direction as the force, then the distance, s, in equation 4–1 is not the total distance but the component of the distance in the direction of the force. Alternatively, the component of the total force which is in the direction of the motion can be found and multiplied by the total distance; these two approaches give equivalent results. For example, consider the process of pulling a load in a wagon as illustrated in Figure 4–2. The force exerted on the handle is 50 pounds, but since it is directed at an angle of 53 degrees upward, not all of this force contributes to the work done on the cart. This force is equivalent to a 40 pound force acting upward and a 30 pound force directed horizontally. Only the horizontal component of the force contributes to the work. When the cart is pulled 10 feet along a horizontal floor, the work done is (30 lb) (10 ft) = 300 ft-lb. The 40 pound upward component tends to lift the front end of the cart off the floor, but does not contribute to the useful work done. It would be more efficient, in terms of the force required, to exert the total force directly horizontally, but this would require an awkward posture.

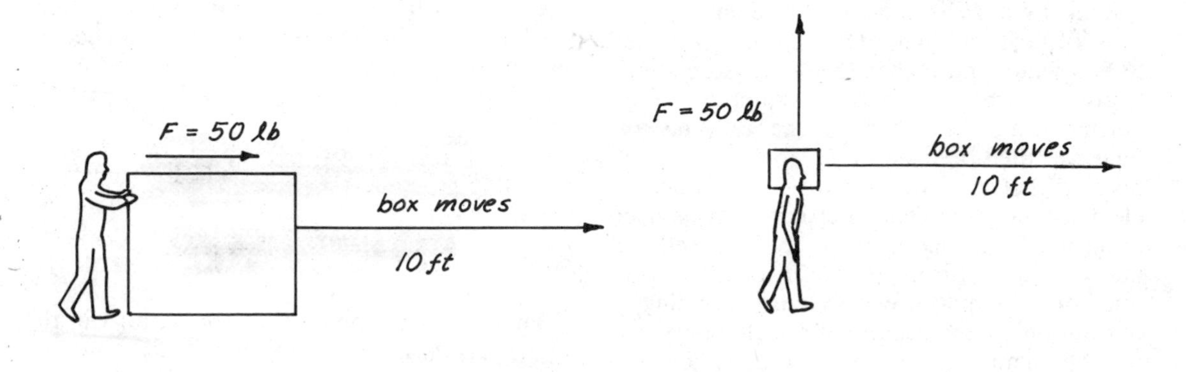

Work = F × s = 50 lb × 10 ft

= 500 ft-lb

Work = F × s = 50 lb × 0 ft

= 0

Figure 4–1 Illustrations of work.

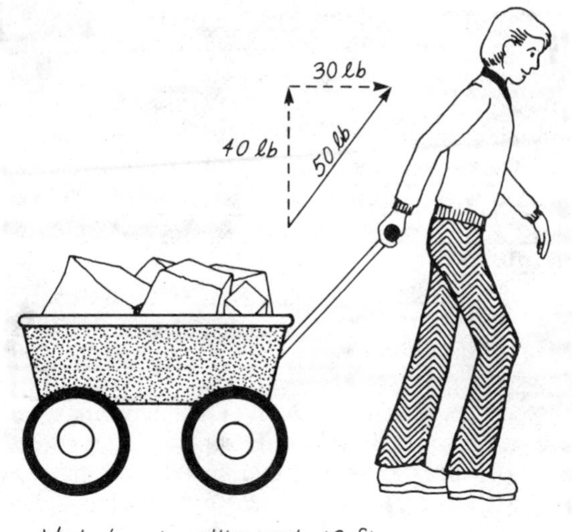

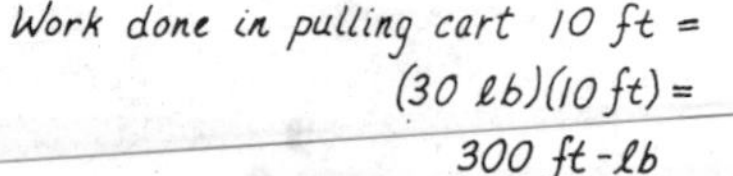

Figure 4–2 Work done when the force and the motion are not parallel.

ENERGY

Energy can be loosely defined as the capacity for doing work. Thus, the quantities work and energy are intimately related and are in fact measured in the same units (joules in the MKS system and ft-lb in the British system). Energy can take many forms, such as heat, electrical energy, nuclear energy, or mechanical energy. Many of the tasks accomplished by machines can be classified as changing energy from one form to another. For example, when coal is burned, the stored chemical energy is converted to energy in the form of heat and light. The heat may be used to boil water and provide high pressure steam to turn a turbine. This is an example of converting heat to mechanical energy. The turbine can then be used to generate electricity, involving a transformation from mechanical to electrical energy. That electricity can be used to produce mechanical energy hundreds of miles away by turning a motor, or it can produce other forms of energy corresponding to our innumerable uses of electrical energy.

Mechanical energy will be considered in more detail here, and other forms of energy will be discussed along with their appropriate applications. Mechanical energy is usually divided into two classes: kinetic energy and potential energy.

Kinetic energy is the energy associated with motion. An object may have kinetic energy because it is moving as a whole, or because it is rotating, or both. The main part of the discussion here will be limited to the energy associated with the motion of the object as a whole (translational motion as opposed to rotational motion). Kinetic energy will then mean translational kinetic energy unless otherwise noted.

The kinetic energy of an object is found to be proportional to its mass and proportional to the square of its velocity. The precise relationship is:

$$\text{K.E.} = \tfrac{1}{2}mv^2. \qquad \textbf{4–2}$$

The relationship to the mass should not be surprising, since one would expect a more massive object to have more energy. The relationship between the velocity and the energy is not quite so obvious. If the velocity is doubled, the kinetic energy is multiplied not by a factor of two, but by a factor of *four,* since it depends upon the square of the velocity.

The expression for kinetic energy in equation 4–2 can be made more plausible by showing its relationship to the work done on an object. When something is accelerated by the application of a net external force which does work on it, its velocity and therefore its kinetic energy will increase. Suppose a mass m is accelerated from rest by the action of a constant force $F_{\text{net ext}}$. The work done by this force is

$$\mathscr{W} = F_{\text{net ext}}s = mas.$$

But this work can be expressed in terms of the velocity with the use of the motion equations 2–1 and 2–2. For this case of constant acceleration from rest these equations can be put in the form:

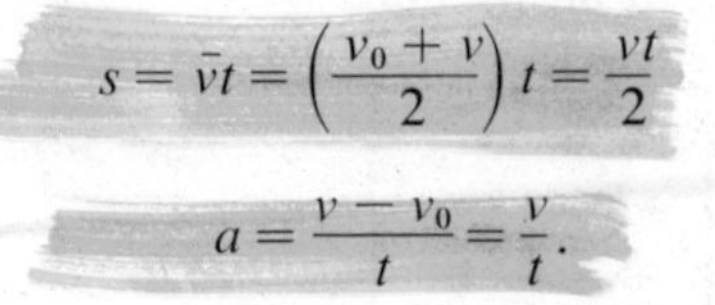

$$s = \bar{v}t = \left(\frac{v_0 + v}{2}\right)t = \frac{vt}{2}$$

$$a = \frac{v - v_0}{t} = \frac{v}{t}.$$

Making these substitutions for a and s in the work expression,

$$\mathscr{W} = m\left(\frac{v}{t}\right)\left(\frac{vt}{2}\right) = \tfrac{1}{2}\,mv^2.$$

we see that the work done by the net external force is equal to the kinetic energy $\tfrac{1}{2}\,mv^2$.

The result obtained here is far more general than this specific case; it is an example of a fundamental physical principle, the *work-energy principle*.

Work-Energy Principle: The change in the kinetic energy of an object is equal to the work done on it by the net external force acting upon it.

This principle can be expressed in the form

$$(\text{K.E.})_f - (\text{K.E.})_i = \mathscr{W}_{\text{net ext}} \qquad \textbf{4–3}$$

where the subscripts i and f refer to the initial and final states.

When work is done on an object by any force, energy is transferred either to the object or away from it; work is an energy transfer process. If you exert a force on a cart by pushing it, you are doing work on it and transferring energy to it. The work done on the cart by any frictional resistance forces will be negative and will take energy away from the cart. This is perhaps best illustrated by a series of examples.

Example. First an ideal case with no frictional resistance will be considered. A cart of mass 20 kg is pushed a distance of 20 meters by a force of 50 newtons. If the cart starts from rest and all the energy is transferred into the kinetic energy of the cart, what is the final velocity of the cart?

Solution. As outlined above, the work done on the cart is

$$\mathscr{W} = Fs = (50 \text{ nt})\ (20\ m) = 1000 \text{ joules.}$$

With the assumption made, the kinetic energy is

$$\text{K.E.} = 1000 \text{ j} = \tfrac{1}{2}mv^2.$$
$$1000 \text{ j} = \tfrac{1}{2}(20 \text{ kg})(v^2)$$
$$100 \text{ m}^2/\text{sec}^2 = v^2$$
$$10 \text{ m/sec} = v.$$

So the final velocity of the cart will be 10 m/sec. The same result could have been obtained by the use of Newton's laws, but the process is more complicated.

Example. A 50 newton force pushes a 20 kg cart a distance of 20 meters, but must overcome a 20 newton frictional resistance force. If the cart starts from rest, what will be its velocity at the end of the 20 meter distance?

Solution. Both the 50 nt push and the 20 nt resistance force do work on the cart. Choosing the direction of the cart's motion as the positive direction, the work done by the 50 nt force is

$$\mathscr{W}_1 = (50 \text{ nt})(20 \text{ m}) = 1000 \text{ joules.}$$

The frictional force acts in the opposite direction and is therefore negative. The work done on the cart by the frictional force is

$$\mathscr{W}_2 = (-20 \text{ nt})(20 \text{ m}) = -400 \text{ joules.}$$

The negative work indicates that energy is taken from the system. The net work done on the system gives it energy in the form of kinetic energy. Using equation 4–2:

$$\mathscr{W}_1 + \mathscr{W}_2 = 1000 \text{ j} - 400 \text{ j} = 600 \text{ j} = \tfrac{1}{2}mv^2$$

$$v^2 = \frac{(2)(600 \text{ j})}{(20 \text{ kg})} = 60 \text{ m}^2/\text{sec}^2$$

$$v = 7.8 \text{ m/sec.}$$

The work done by the individual forces was calculated separately here to emphasize that the work done by each force represents an energy transfer. The process is simplified by making use of the *work-energy principle* discussed above. Since the change in kinetic energy is equal to the work done by the net external force, we could just calculate the net external force and the work done by it:

$$F_{\text{net ext}} = 50 \text{ nt} - 20 \text{ nt} = 30 \text{ nt}$$

$$\mathscr{W}_{\text{net ext}} = (30 \text{ nt})(20 \text{ m}) = 600 \text{ j} = \text{K.E. gained}$$

This yields the same result as above with fewer steps.

Example. The 20 kg cart has been given a velocity of 10 m/sec and is then released to roll in a straight line. It is acted upon by a 20 nt frictional resistance force. How far will the cart roll before coming to rest?

Solution. The work-energy principle may be used to advantage here. The net external force is the frictional resistance force which does negative work on the cart, reducing its kinetic energy. The cart will come to rest (zero kinetic energy) when the work done by friction is equal to the negative of the initial kinetic energy.

$$(K.E.)_f - (K.E.)_i = \mathscr{W}_{net\ ext}$$

$$0 - \tfrac{1}{2}(20\ kg)(10\ m/sec)^2 = (-20\ nt)(x)$$

$$\frac{-1000\ joules}{-20\ nt} = x$$

$$50\ m = x$$

Potential energy is the energy associated with position or configuration. If a brick is lifted over your head, it has the capacity to do work if it is released, so it is said to have potential energy. When it is released, that energy is converted into kinetic energy, and it has the capacity to do work on your toe. This type of potential energy which an object possesses because of its elevation above the earth is referred to as gravitational potential energy. The fact that the gravitational potential energy is directly proportional to the height can be seen from the following example.

Example. A 25 pound weight is lifted to a height of 6 feet above the ground. What is its potential energy at that point?

Solution. To lift the weight, a 25 pound force must be exerted in the upward direction. If the weight is lifted a distance of 6 feet, the work done by the lifting force is

$$\text{work} = (25\ \text{pounds})\ (6\ \text{feet}) = 150\ \text{ft-lb}.$$

Since we are assuming that the weight is brought to rest at that height, it has no kinetic energy and all the work went into potential energy. The potential energy is 150 ft-lb. which is just the weight times the height.

This example can be generalized to show that the gravitational potential energy of an object with respect to the surface of the earth is equal to its weight times its height above the surface. To lift any object you must exert a force equal to or greater than its weight, mg. To lift it straight up to a height h will require work mgh. It can be shown that regardless of the path taken to reach a state of rest at height h, the work done by the lifting force against gravity is mgh. Thus the energy given to the object in the form of gravitational potential energy is

$$P.E._{gravitational} = mgh. \qquad \textbf{4–4}$$

Another type of potential energy is that associated with a spring. If a spring is compressed, it has the capacity to spring back and do some work. Therefore, in its compressed configuration it has some potential energy. Likewise, a stretched rubber band has potential energy because of its configuration.

CONSERVATION OF ENERGY

If asked the question, "What are the most fundamental relationships in physics?" a physicist around the beginning of the twentieth century might well have responded by quoting some of the specific laws, like Newton's laws. Soon after that, however, the foundations of physics were shaken by Einstein's development of the theory of relativity and by the discovery that a whole new formulation of physics called quantum mechanics was required to explain the behavior of nature on the atomic and nuclear level. It was necessary to search for unifying principles which were valid not only for ordinary phenomena, but also at speeds approaching the speed of light where Einstein's theories applied and in the submicroscopic world of the atom. Most physicists today would agree that the most fundamental relationships are those expressed as "conservation principles," that is, statements of physical quantities which always remain the same under the specified conditions. One of the most important of these principles is the principle of conservation of energy:

Energy, including mass energy, can neither be created nor destroyed, only changed from one form to another.

All scientific principles are based upon experimental evidence, and if an experimental counter-example to the above could be found, the principle would be set aside or at least amended. The sum total of experimental evidence thus far suggests that the total energy remains the same, or is "conserved," in all physical processes.

This principle takes into account the fact that energy is always changing form, from mechanical to heat, to electricity, to mechanical, to chemical, in many processes. But if there were 100 joules of energy to begin with, there will be 100 joules of energy in some form at the

Mass can be changed into energy.

end of the process. One of Einstein's discoveries was that mass could be changed into energy (and vice versa) according to his famous equation $E = mc^2$, where c = velocity of light = 3×10^8 m/sec. Since c is squared in the equation, and since c is such a large number, the yield in energy from even a tiny mass is enormous. This conversion is responsible for the enormous energy of nuclear bombs and other nuclear devices. Einstein's equation will be discussed further in the chapter on nuclear processes. It is mentioned here to explain why the phrase "including mass energy" must be included in the statement of the conservation of energy principle.

In the ideal mechanical case in which there is no friction, the energy of a body is composed of kinetic energy and potential energy. The implication of the conservation of energy principle is that the sum of the potential energy and the kinetic energy is a constant:

$$\text{mechanical energy} = \text{K.E.} + \text{P.E.} = \text{constant.}$$

However, the energy can be transformed back and forth between the two types of mechanical energy. A good example of this is a mass bobbing up and down on a spring. When a mass is attached to a spring, it rests at an equilibrium point where the force of the stretched spring exactly balances the gravitational pull on the mass. In this position it could be said that both the kinetic energy and the potential energy are zero. If a force is then exerted on the mass to pull it down further, this force does work on it and gives it energy in the form of potential energy. If the mass is then released, it will not just return to its equilibrium point but will bob up and down about its equilibrium point. If the amount of work done on the mass is represented by E, then the potential energy where it is released is E (see Figure 4–3). When it returns to the equilibrium point (Figure 4–3(c)), there is no potential energy and the energy, E, has been transformed into kinetic energy. As it moves to the top of its travel and comes to rest, the energy E is transformed back into potential energy (Figure 4–3(d)). As it repeatedly bobs up and down, the energy is transferred back and forth between kinetic and potential energy. It would keep doing so forever if it were not for the fact that in any real experiment there is some frictional loss of energy. As the energy is gradually changed to other forms by frictional heating, etc., the mass slowly comes to rest.

A note is in order here about the potential energy for the mass on the spring. There are two types of potential energy involved, gravitational potential energy and the elastic potential energy of the stretched spring. The potential energy referred to above is the result of both the gravitational and spring forces, and the equilibrium point is the point at which these two forces cancel. At this equilibrium point there is no capacity to do work (energy) from the combination of these two forces, so we say that the potential energy is zero. When the mass is either above or below the equilibri-

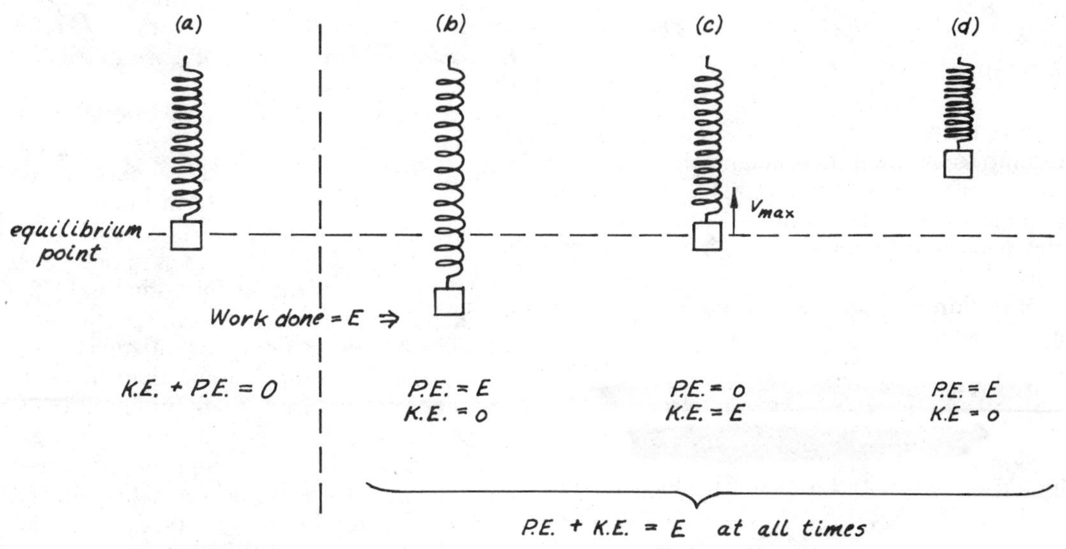

Figure 4–3 Mass on a spring as an illustration of conservation of energy.

um point there is a net force which tends to bring the mass back to equilibrium.

The conservation of energy is a powerful tool for solving certain types of problems. For a given ideal mechanical process, the conservation of energy could be expressed as

$$(\text{K.E.})_i + (\text{P.E.})_i = (\text{K.E.})_f + (\text{P.E.})_f$$

where the subscripts i and f refer to the initial and final states.

Example. A boy dives straight down from a 16 ft diving board. Use conservation of energy to find his speed when he enters the water.

Solution. Use conservation of energy in the form

$$(\text{K.E.} + \text{P.E.})_{\text{diving board}} = (\text{K.E.} + \text{P.E.})_{\text{entering water}}.$$

The initial K.E. is zero, and the initial P.E. is the boy's weight times the 16 ft height. This gravitational potential energy can be expressed as mgh, where m is the boy's mass and g is the acceleration of gravity (weight $= mg$). The final P.E. is zero, and the final K.E. $= \frac{1}{2}mv^2$. The equation for conservation of energy takes the form

$$0 + mgh = \tfrac{1}{2}mv^2 + 0.$$

Dividing the equation by m eliminates the mass, showing that the final velocity is independent of the mass. The equation becomes

$$gh = \tfrac{1}{2}v^2$$

and substituting the numbers,

$$v^2 = 2gh = (2)(32 \text{ ft/sec}^2)(16 \text{ ft}) = 32^2 \, ft^2/\text{sec}^2,$$

$$v = 32 \text{ ft/sec}.$$

Example. A skier is poised at the top of the ramp to ski jump at a height of 40 meters. The take-off point of the jump is at a height of 20 meters. If friction on the ski jump is negligible, what will be his speed as he leaves the jump?

Solution. Using conservation of energy

$$(\text{K.E.})_i + (\text{P.E.})_i = (\text{K.E.})_f + (\text{P.E.})_f$$

$$0 + mgh_i = \tfrac{1}{2}mv^2 + mgh_f$$

$$v^2 = 2g(h_i - h_f) = (2)(9.8 \text{ m/sec}^2)(40 \text{ m} - 20 \text{ m})$$

$$v^2 = 392 \text{ m}^2/\text{sec}^2$$

$$v = 19.8 \text{ m/sec}.$$

Note that the mass of the skier cancels out and does not affect the velocity. This implies that for a nearly frictionless ski jump, no skier would have an advantage because of his mass.

This example further demonstrates the power of the conservation of energy principle as a problem solving tool. Note that we did not have to take into account the details of the contour of the ski jump since the gravitational potential energy change depends only upon the difference in height between the initial and final points.

The conservation of energy has been strongly emphasized here because of its numerous applications in the other areas of physics which will be discussed. A good understanding of its meaning in mechanical terms will aid in the understanding of its application in other areas.

In light of the energy crisis which is a continuing part of our lives, some further reflection on energy sources and the conservation of energy principle seems appropriate. Since we cannot create energy, we must use the energy "currency" which we have on the earth wisely and efficiently. The stored energy reserves in the form of fossil fuels and minerals are not replaceable, and the direct use of the incoming energy from the sun presents many problems.

When the energy stored in a ton of coal is used by burning the coal, the energy is not destroyed but it is changed from a highly ordered, compactly stored state (the coal) to a highly disordered, broadly distributed state (random molecular motion). This energy which is distributed by heating the environment is not practically recoverable, so the use of the coal reduces our reserve of usable energy.

Some perspective on the energy in joules available from common sources may be obtained from Figure 4–4. The energies are expressed in the "powers of ten" notation to get a wide spread of energies. The number for the energy use of one U.S. citizen in one year, 3.6×10^{11} joules, is found by taking a 1973 estimate of the entire U.S. energy budget and dividing by the number of citizens. It seems very large, but keep in mind that this includes the energy required to manufacture all the products we use, as well as the energy for heating and lighting our houses. The numbers in Figure 4–4 are based on data collected in the energy textbook by Romer (2). Though outside the scope of this text, the projections of our energy needs and the prospects for sup-

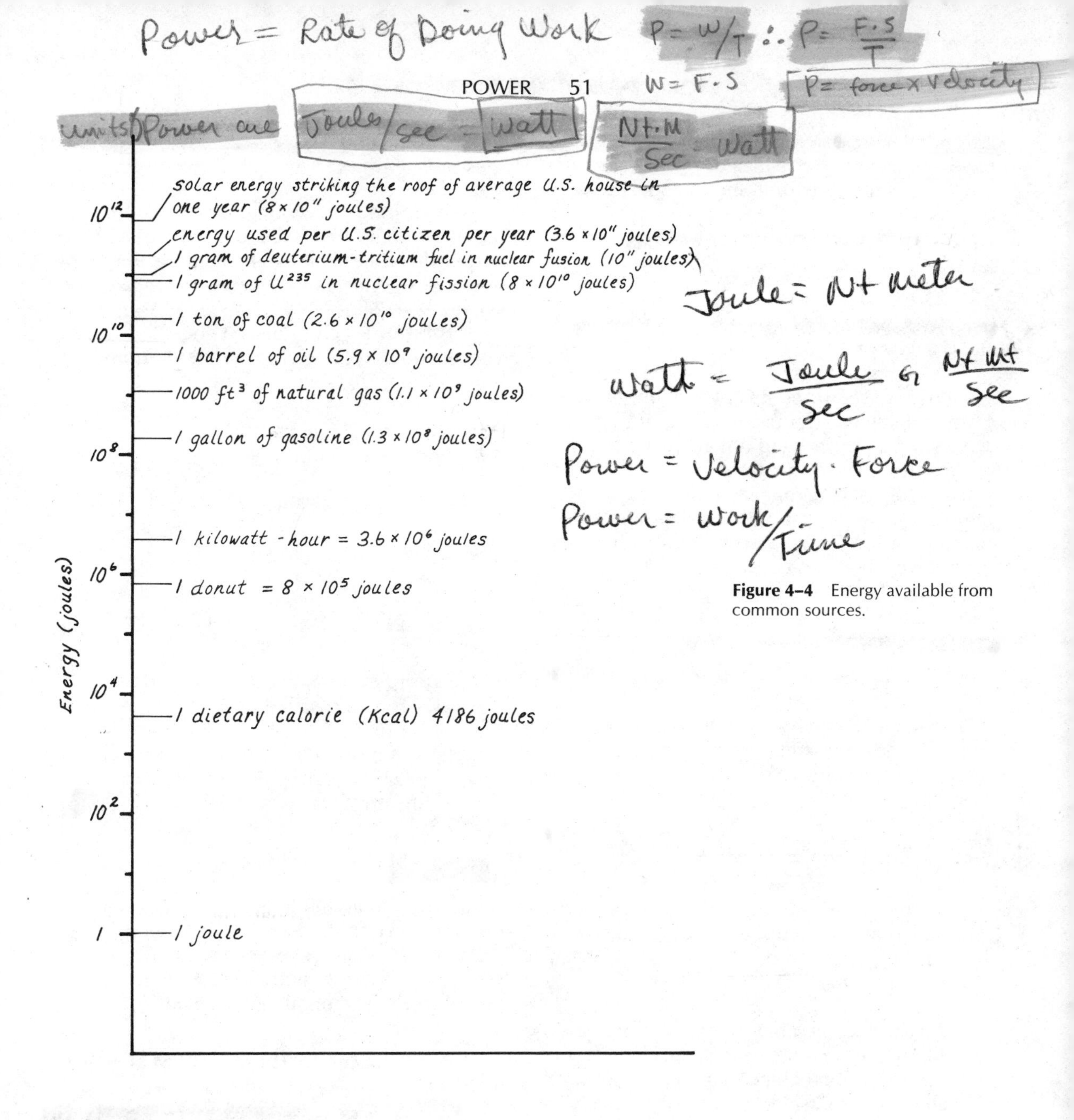

Figure 4–4 Energy available from common sources.

plying them are topics of importance of the first magnitude. References 2 to 4 are recommended for further discussion of these ideas.

POWER

Power is defined as the rate of doing work. It is thus distinct from work and energy, with which the term is often confused. Power bears the same relation to work as velocity bears to distance. The average power is defined by the relationship

$$\text{average power} = \frac{\text{work}}{\text{time}}$$

or

$$P_{\text{avg}} = \frac{\mathscr{W}}{t}. \qquad \textbf{4–5}$$

Since the work is given by $\mathscr{W} = Fs$, then this could be expressed as

$$P_{\text{avg}} = \frac{Fs}{t} = Fv \qquad \textbf{4–6}$$

or

$$\text{power} = \text{force} \times \text{velocity}.$$

The units of power are the units of work divided by time, which are joules/sec in the MKS system. Consistent with the practice of using a new name when a unit is added, this unit is called a watt. The kilowatt, 1000 watts, is also commonly used. The basic power unit in the British system is the ft-lb/sec. The horsepower is derived from this unit and is defined by

$$1 \text{ horsepower} = 550 \text{ ft-lb/sec} = 33{,}000 \text{ ft-lb/hr.}$$

One horsepower is equivalent to 746 watts or 0.746 kilowatt.

A common misuse of the word "power" is illustrated by comments about buying "power" such as electrical power. It is energy which is purchased, and not power; the utility companies charge the same whether the energy is all used in one day (high power) or spread out over a month (low power). The energy used may be obtained by multiplying the power by the time during which it is used:

$$\text{energy (joules)} = \text{power (watts)} \times \text{time (seconds).}$$

The commonly used unit of electrical energy is the kilowatt-hour. The equivalent energy in joules may be obtained from the above relationship:

$$1 \text{ kilowatt-hour} = (1000 \text{ joules/sec})(3600 \text{ sec}) = 3{,}600{,}000 \text{ joules.}$$

The economy of electrical energy is illustrated by the fact that a kilowatt-hour of electrical energy costs only a few cents.

Example. A man exerts a force of 50 pounds on a heavy cart and pushes it 60 feet in 100 seconds. A machine can exert a force of 250 pounds and push it the same distance in 20 seconds. What is the power in each case?

Solution. For the man the power is

$$P = \frac{Fs}{t} = \frac{(50 \text{ pounds})(60 \text{ feet})}{(100 \text{ sec})} = 30 \text{ ft-lb/sec} = 0.054 \text{ horsepower.}$$

The machine does more work in less time:

$$P = \frac{(250 \text{ pounds})(60 \text{ feet})}{(20 \text{ sec})} = 750 \frac{\text{ft-lb}}{\text{sec}} = 1.4 \text{ horsepower.}$$

The power depends not only upon the amount of work done but also upon the amount of time required to do it.

Example. A 70 kg person runs up three flights of stairs in 9 seconds. He stops at the end of this time, having changed his height by 10 meters. What is his average output power for the 9 seconds?

Solution. The work he does against gravity is equal to his change in gravitational potential energy. Using equation 4–4:

$$\text{Work done} = mg\Delta h = (70 \text{ kg})(9.8 \text{ nt/kg})(10 \text{ m}) = 6860 \text{ joules.}$$

The power is then

$$P_{avg} = \frac{\mathscr{W}}{t} = \frac{6860 \text{ j}}{9 \text{ sec}} = 762 \text{ watts.}$$

Example. To push a box across the floor at a constant speed of 0.5 m/sec requires the application of a 200 newton force in the direction of motion to overcome the frictional resistance. What is the power required?

Solution. The power is given by equation 4–6,

$$P = Fv = (200 \text{ nt})(0.5 \text{ m/sec}) = 100 \text{ watts.}$$

This means that 100 joules of energy per second are given to the box, but its kinetic energy remains constant. The energy is required to overcome friction and is dissipated (transferred to the surroundings) by heat.

THE PRINCIPLES OF MACHINES

Machines in general may be looked upon as devices for transforming energy. The applications of simple machines such as hand tools are often for the purpose of multiplying force, that is, for obtaining a larger output force than one exerts on the tool. (Examples will be given in the discussion of the simple machines.) However, a machine cannot supply more output energy than it is given as input energy; i.e., the output energy is always less than or equal to the input energy. This follows from the conservation of energy principle; energy cannot be created by the machine.

Three of the terms most often used to describe machines are (1) Ideal Mechanical Advantage (IMA), (2) Actual Mechanical Advantage (AMA), and (3) efficiency.

ACTUAL MECHANICAL ADVANTAGE. As mentioned above, one of the reasons for using a machine is to overcome a large resistance by exerting a relatively small effort; i.e., the machine is used to multiply force. The actual mechanical advantage is equal to the output force of the machine divided by the input force, and as such gives the factor by which the machine multiplies the input force. It may be written

$$\text{AMA} = \frac{\text{force overcome (resistance)}}{\text{force applied (effort)}} \qquad \textbf{4–7}$$

$$\text{or AMA} = \frac{\text{force output of the machine}}{\text{force input to the machine}}.$$

In the case of a simple machine, the actual mechanical advantage is often the most important factor.

EFFICIENCY. Although the machine can multiply force, in no case can it multiply work or energy. This would violate the conservation of energy principle. In fact, in all real machines, the output energy will be somewhat less than the input energy because of losses to frictional heating, etc. The efficiency as a percentage is defined by

$$\text{efficiency (\%)} = \frac{\text{work output}}{\text{work input}} \times 100. \qquad \textbf{4–8}$$

The theoretical maximum of 100% efficiency is set by the conservation of energy principle. Although some simple machines, such as levers, approach this 100% efficiency, most machines operate at much less than 100% efficiency. An automobile engine may have an efficiency as low as 12 to 18%, even when it is running properly.

IDEAL MECHANICAL ADVANTAGE. Since all real machines dissipate some energy because of frictional heat loss and other factors, the relationship between input and output work can be written

$$\text{work input} = \text{work output} + \text{energy dissipated in friction, etc.}$$

In equation 4–1 work is defined by $\mathcal{W} = Fs$. The above relationship can then be written

$$F_e \times s_e = F_r \times s_r + \text{energy lost to other forms} \qquad \textbf{4–9}$$

where F_e is the effort force applied to the machine, s_e is the distance moved by the effort force, and F_r and s_r are the corresponding quantities associated with the resistance. If the efficiency is 100% so that the energy losses are zero, then the relationship becomes

$$F_e s_e = F_r s_r \qquad \textbf{4–9(a)}$$

or equivalently,

$$\frac{F_r}{F_e} = \frac{s_e}{s_r}.$$

The quantity F_r/F_e (the resistance force divided by the effort force) is the number of times the machine multiplies the force and is thus a mechanical advantage. Since it was obtained under the assumption of 100% efficiency, this is called the *Ideal* Mechanical Advantage. It can be defined as

$$\text{IMA} = \frac{s_e}{s_r}. \qquad \textbf{4–10}$$

This relationship illustrates the fact that, in order to obtain a mechanical advantage greater than one, the effort distance must be greater than the resistance distance. For example, if you want to lift a heavy boulder six inches by using a long lever, you must move the long end of the lever much farther than six inches.

Although the actual mechanical advantage is the most important in applying a machine, the ideal mechanical advantage is useful in designing machines and in classifying the different types of machines. It is sometimes convenient to express the efficiency in terms of the ratio of the ideal to the actual mechanical advantage:

$$\text{efficiency (\%)} = \frac{\text{AMA}}{\text{IMA}} \times 100. \qquad \textbf{4–11}$$

This is equivalent to equation 4–8.

THE SIMPLE MACHINES

Though machines vary greatly in type and complexity, most of them can be described in terms of combinations of a relatively small number of basic "simple machines." These machines fall into three broad classes: (1) levers, (2) inclined planes, and (3) hydraulic presses. The hydraulic press will be discussed in the next chapter. The basic derivations of the lever and the inclined plane will be discussed here. The wheel and axle and the pulley can be looked upon as modified levers, and the wedge and the screw are adaptations of the inclined plane. Some introduction to these basic machines can give a better understanding of the action of the muscles in the body, and can give some insight into how to use machines most advantageously.

Lever Systems

The basic lever consists of any bar or rod (bent or straight) arranged in such a way that it can pivot about some definite point, so that it can overcome some resistance force F_r by the application of an effort force F_e. The simplest form of lever is illustrated in Figure 4–5. The relation between the effort force F_e and the resistance force F_r can be determined by the second condition for equilibrium discussed in Chapter 3. If the effort force in Figure 4–5(a) is balancing the resistance force, then the clockwise torque associated with F_e must be equal to the counterclockwise torque associated with F_r. The lever arms associated with the effort and resistance are l_e and l_r, respectively. Equating the torques gives

$$F_e l_e = F_r l_r. \qquad \textbf{4–12}$$

Note that this is a torque equation and is not equivalent to the work equation 4–9(a). However, the ideal mechanical advantage of the lever is just the effort arm l_e divided by the resistance arm l_r since l_e is proportional to s_e and l_r is proportional to s_r. For most practical purposes the rigid lever can be assumed to be 100% efficient, so the ratio of the lever arms can be taken as the actual mechanical advantage.

Example. A seven foot rigid bar is available for use as a lever. If the pivot point is placed one foot from the end, what force will be required to lift a 300 pound load? If the pivot is moved to a point 6 inches from the end, what effort force will be required then? What is the mechanical advantage in each case?

Solution. Using equation 4–12 with $F_r = 300$ lb, $l_r = 1$ foot, and $l_e = 6$ ft, we find

$$(F_e)(6 \text{ ft}) = (300 \text{ lb})(1 \text{ ft})$$

$$F_e = \frac{300}{6} \text{ lb} = 50 \text{ lb}.$$

Now with the pivot shifted to 6 inches from the resistance, $l_r = 0.5$ ft and $l_e = 6.5$ ft. This gives

$$F_e = (300 \text{ lb}) \frac{0.5}{6.5} = 23.1 \text{ lb}.$$

The ideal mechanical advantages in each case are

$$(1)\ \text{IMA} = \frac{l_e}{l_r} = \frac{6 \text{ ft}}{1 \text{ ft}} = 6,$$

$$(2)\ \text{IMA} = \frac{6.5 \text{ ft}}{0.5 \text{ ft}} = 13.$$

If the pivot is placed closer to the resistance, the mechanical advantage is greater and less effort is required to lift the load. A disadvantage of the closer pivot is that the effort force must be exerted through a greater distance.

Levers can be divided into three classes which are referred to as first, second, and third class levers. The *first class lever* is one such that the pivot point is somewhere between the effort and resistance forces, as in Figure 4–5. Pliers, hemostats, scissors, and other types of hinged instruments are examples of first class levers. Scissors (Figure 4–6) are composed of two first class levers with a common pivot point. It is a readily observable fact that moving the material to be cut closer to the pivot point makes possible a greater cutting force with a given effort on the handles. This is an illustration of the fact that the mechanical advantage of the lever is equal to the effort arm divided by the resistance arm; if the resistance arm is decreased, the mechanical advantage is increased. If a large mechanical advantage is required, the first class lever is usually employed.

Numerous examples of lever systems are found in the cooperative action of the muscles, ligaments, and bones of the body. The classification of these levers as first, second, or third class is sometimes difficult because of ambiguities about the exact locations of the effort and resistance forces. However, the classification is possible in some cases.

The triceps brachii muscle acts as part of a first class lever system. The effort force of this muscle acts behind the elbow joint, which acts as the pivot point. The resistance force consists of the weight of the arm and hand plus whatever load is being supported by the hand. If the palm is upward and supporting a load, a contraction of the triceps causes the hand to move downward. Of course, the triceps does not act alone in this process, and the action of

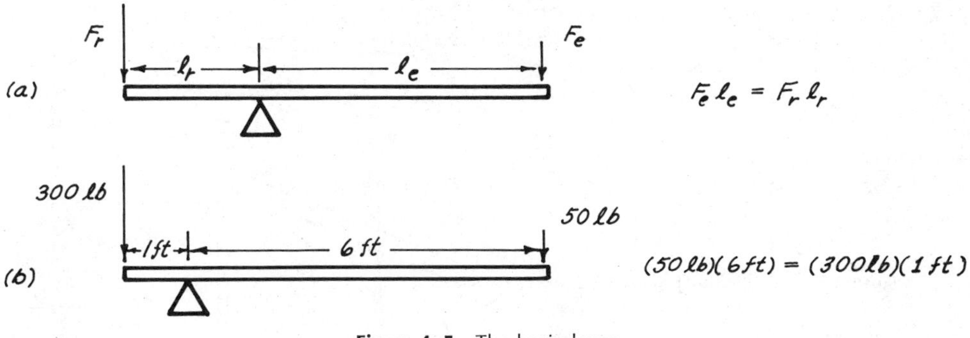

Figure 4–5 The basic lever.

the biceps will be mentioned below in the section on third class levers.

A lever is classified as a *second class lever* if the resistance force is exerted between the pivot point and the effort force. The wheelbarrow is the classic example. Hand trucks and carts for carrying compressed gas bottles are also second class lever systems. The processes of lifting one end of a bed, turning a mattress, or lifting one end of any extended object are examples of second class levers. The ideal mechanical advantage of the second class lever is found in the same way as for the first class lever, by dividing the lever arm associated with the effort force, l_e, by the resistance lever arm, l_r. The mechanical advantage obtained depends upon the distance from the pivot point to the center of gravity of the load (see Figure 4–7).

Example. What force will be required to lift one end of a fifty pound uniform mattress, and what is the mechanical advantage of this second class lever system?

Solution. The pivot point will be the other end of the mattress, and since the mattress is uniform, the center of gravity will be the geometrical center of the mattress. Equating the torques, with the mattress length represented by L:

$$\text{resistance torque} = \text{effort torque}$$

$$(50 \text{ pounds})\left(\frac{L}{2}\text{ ft}\right) = F_e \times L \text{ ft}$$

$$F_e = 25 \text{ pounds.}$$

Since the resistance force of 50 pounds was overcome by an effort force of 25 pounds, the mechanical advantage is two.

The effort force required to pick up one end of any uniform extended object will be one-half of its weight, as in the above example. This fact is sometimes used to obtain the weight of very large objects.

The mechanical advantage for second class levers may be considerably greater than

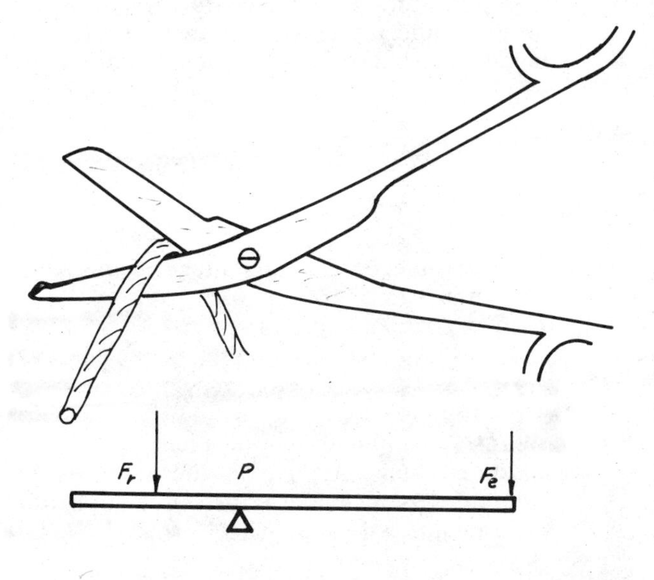

Figure 4–6 First class lever.

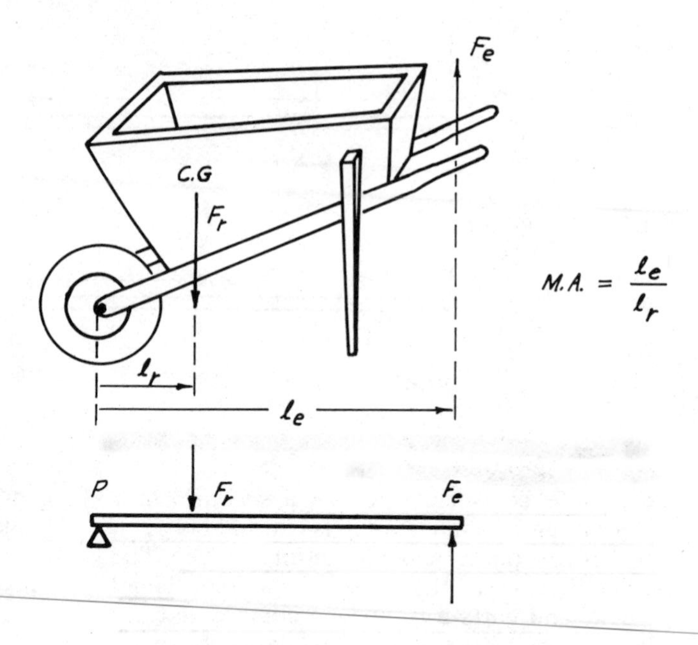

Figure 4–7 The second class lever.

two when the load is concentrated closer to the pivot point, but generally speaking, the mechanical advantages obtainable with second class levers are not as high as those obtainable with first class levers.

Third class levers are those for which the effort force is exerted between the resistance force and the pivot. Since the effort arm l_e is necessarily less than the resistance arm l_r, the mechanical advantage is always less than one for this type of lever. Such levers are often useful, even though the effort force which must be exerted is greater than the resistance to be overcome.

In the body, the action of the biceps in lifting the forearm is an example of a third class lever (Figure 4–8). Forceps and tweezers are examples of third class levers. Even though they exert less force at the ends than the effort forces on the sides, they are useful tools because of their ability to hold small objects and to probe into small areas.

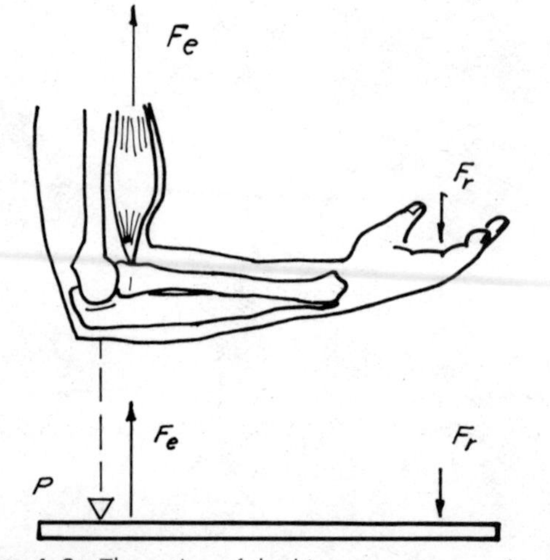

Figure 4–8 The action of the biceps as an example of a third class lever.

Example. A person lifts a five pound tray by balancing it on the palm of one hand so that the center of gravity of the tray is 15 inches from the elbow pivot point. What force must be exerted by the biceps to support this load if the effort force is assumed to be directed vertically and acts at a point three inches away from the pivot point?

Solution. Using the effort lever arm $l_e =$ 3 inches and the resistance lever arm $l_r = 15$ inches, the balancing of torques requires

$$F_e l_e = F_r l_r$$

$$F_e \times (3 \text{ in}) = (5 \text{ lb})(15 \text{ in})$$

$$F_e = 25 \text{ pounds.}$$

The ideal mechanical advantage of this lever system is 3/15 or IMA = 0.20.

One of the main limitations of the simple lever is that it can move only through a limited arc. The *wheel and axle* as a machine uses the lever principle and is capable of continuous rotation. As illustrated in Figure 4–9(b), a large resistance applied at the axle can be overcome by a smaller effort force applied at the outside of the wheel. Equating the

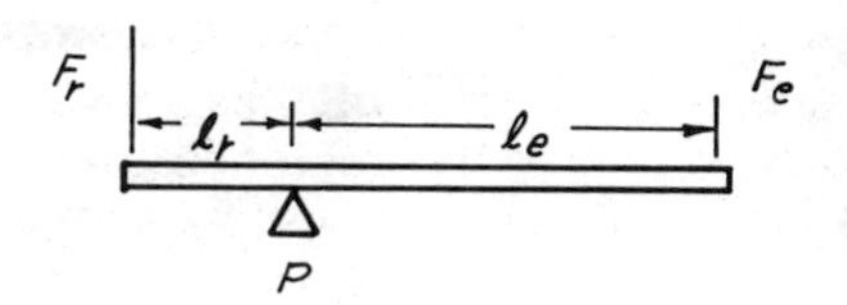

$$IMA = \frac{l_e}{l_r}$$

(a) simple lever

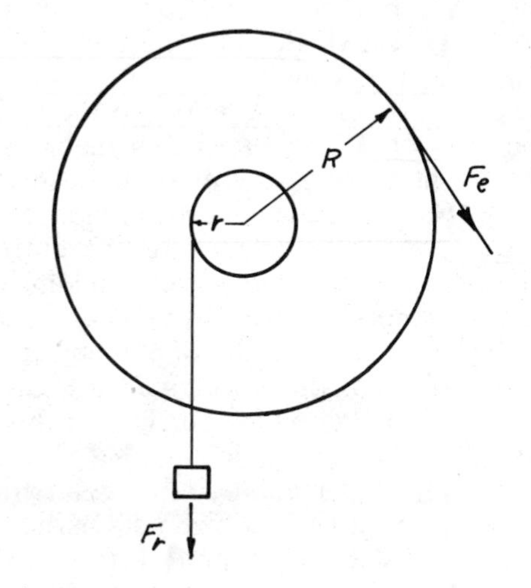

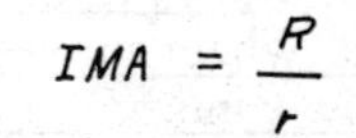

(b) wheel and axle

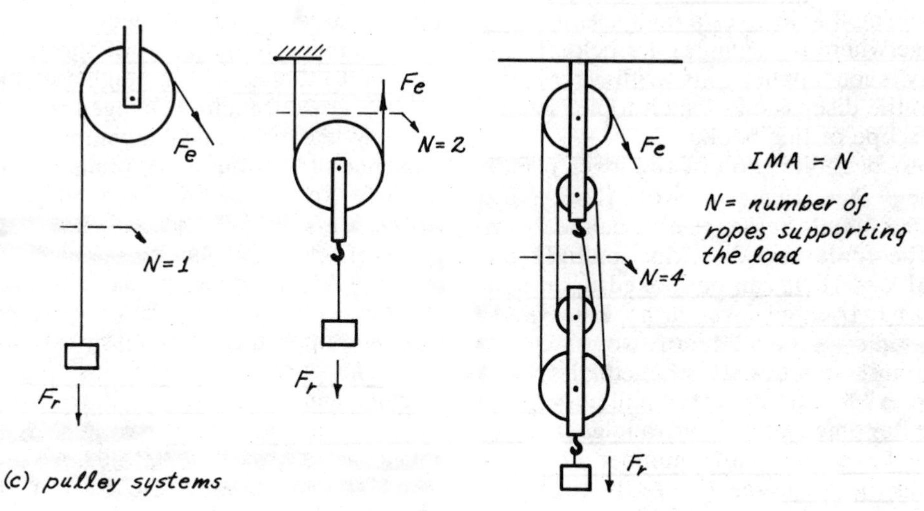

Figure 4–9 Lever systems.

torques, $F_r r = F_e R$, gives the expression for the ideal mechanical advantage

$$\frac{F_r}{F_e} = \frac{R}{r} = \text{IMA}.$$

If a wheel has a 12 inch radius and an axle of 1 inch radius, then it will have a mechanical advantage of 12. Assuming that dissipative forces are negligible, a ten pound force acting on the outside of the wheel could overcome a 120 pound resistance at the axle.

It is not necessary to elaborate upon the applications of the wheel and axle. It might be pointed out that any device which uses a large diameter handle to turn a smaller diameter shaft is an application of the wheel and axle principle. A significant mechanical advantage can be obtained by the use of screwdrivers, faucet handles, and other machines which use this principle.

It has been pointed out that a user of a machine "pays" for the multiplication of force achieved by exerting the effort force over a longer distance than would have been necessary without the machine. In the example above, a force multiplication of 12 was achieved. The user of the machine pays for this by having to move the outside of the wheel 12 inches to get the outside of the axle to move 1 inch. In terms of speed, if the outside of the wheel moves 12 in/sec, the axle moves only 1 in/sec, so a reduction in speed results from this machine. Many uses of the wheel and axle work in the opposite way. That is, a large force is exerted on the axle in order to get a multiplication of speed on the outside of the wheel. The automobile tire is an example of such an application. The engine, working through the transmission, drive shaft, and differential, exerts a large force on the rear axle to achieve a highspeed motion on the outside of the tires. Another example is a laboratory centrifuge. A relatively large force exerted on a small axle gives a high speed at the outer edge where the samples are held. There are many such applications with gears and cogs, but the discussion of such topics is outside the scope of this book.

Another modification of the lever principle is the *pulley* (Figure 4–9(c)). The single fixed pulley merely serves to change the direction of the force and does not multiply the force (IMA = 1). It can be looked upon as a lever with two equal lever arms. The single movable pulley has an IMA of two, since the fixed support supplies a force equal to the effort force. More complicated pulley systems give greater mechanical advantages. (The IMA will be equal to the number of ropes supporting the load; see Figure 4–9(c).) The frictional forces increase with the number of pulleys, however, so the actual mechanical advantage is often considerably less than the IMA.

Single fixed pulleys and sometimes movable pulleys are used in traction setups. These pulleys are mainly for the purpose of redirecting the forces. They allow a hanging weight to exert a force in almost any direction, and make possible the exertion of equal forces in two or more directions with a single rope (see Figure 3–9). Since the tension is the same at all points in the rope, the same force is exerted at each contact point. This simplifies the process of adding the vector forces to determine the direction of the net force on the patient.

Inclined Plane Systems

It requires less force to push a heavy object up an inclined plane than to lift the object directly. The output force, or resistance force, is that which lifts the object from one level to another and is equal to the weight of the object. Since the incline achieves an effective multiplication of force, it may properly be called a machine. Common examples of this type of machine are the entrance ramps to hospitals and other buildings, parking lot ramps, and so forth.

The weight of an object on an incline, such as the wheelchair in Figure 4–10(a), acts directly downward. Part of this weight is supported by the incline. To push the wheelchair up the incline, only the component of the weight force which acts parallel to the incline must be overcome. On a steeper incline, a larger force would be required since it is more nearly parallel to the direction of the gravitational force. The ideal mechanical advantage of an incline is the ratio of the length of the incline to the change in height achieved by it. Thus, if a 20 ft incline changed the height by 2 ft, the ideal mechanical advantage is 10. In the absence of friction, a 500 pound load could be pushed up the incline by a 50 pound effort force. This is not very realistic, however, because friction is usually a major factor with inclines. The actual mechanical advantage is usually only a small fraction of the IMA, depending upon the roughness of the surfaces.

The general fact remains that a gradual incline will have a larger mechanical advantage than a steep incline. A very large mechanical advantage can be obtained by winding a very gradual incline around a shaft to form a *screw*. Although frictional losses are quite high, a machine like the screw jack illustrated in Figure 4–10(c) can exert enormous lifting forces with a reasonably small effort force. Usually a screw is used in conjunction with a lever, as in the case of the screw jack. The ideal mechanical advantage quoted in Figure 4–10(c) includes the mechanical advantage of the lever. When the handle is moved through a distance $2\pi L$ to turn the screw a full circle, the screw moves up one thread (p = pitch of thread). Screws are used to lift and support

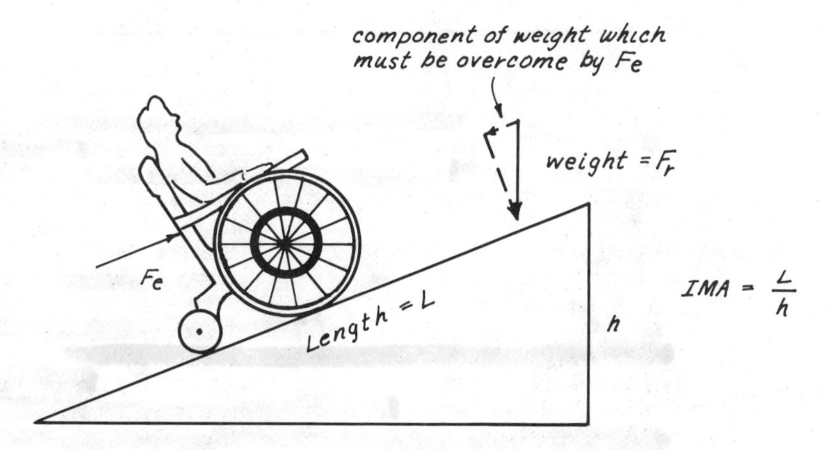

(a) simple inclined plane

(b) wedge

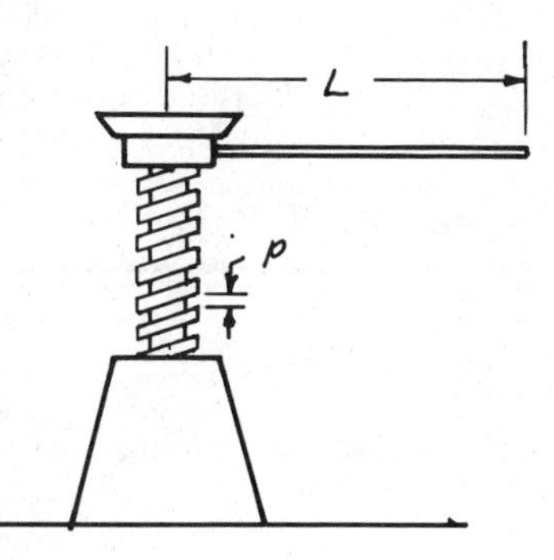

(c) screw jack

Figure 4–10 Inclined plane systems.

heavy machinery; in this application the low efficiency can be tolerated to get the large mechanical advantage. The holding power of ordinary fastening screws is attributable to this large mechanical advantage.

The *wedge* is another common simple machine; it consists of a double inclined plane. The ideal mechanical advantage depends upon the angle between the surfaces, as illustrated in Figure 4–10(b). A sharper wedge will have a larger mechanical advantage and will therefore exert a larger separating force with a given effort force. Scalpels and other surgical cutting tools utilize this principle.

SUMMARY

The physical concepts of work, energy, and power are important in all areas of science. The work done on an object may be defined as the net force times the distance moved in the direction of the force. Power is defined as the rate of doing work. Energy has the same units as work and may be generally defined as the ability or capacity to do work. One of the most basic concepts of physics is the principle of conservation of energy: energy, including mass energy, can neither be created nor destroyed. Mechanical energy, one of the many forms of energy, can be divided into kinetic energy and potential energy. The kinetic energy of an object is the energy due to its motion and is given by the expression $\frac{1}{2}mv^2$. An object is said to have potential energy if it has the ability to do work because of its position (such as gravitational potential energy due to its height above the ground) or its configuration (such as a stretched spring). If an object has only mechanical energy, then the conservation of energy principle implies that the sum of the kinetic energy and potential energy is a constant. Therefore, the only energy transformations taking place are from kinetic to potential energy and vice versa. In general, other types of energy transformations will occur, such as the production of heat energy owing to friction.

Machines may be defined as devices for transforming energy. Machines cannot provide an energy output which is greater than the input energy, since this would violate the conservation of energy principle, but they can provide an output force which is larger than the input force. The multiplication of force achieved is called the actual mechanical advantage of the machine and may be compared to the ideal mechanical advantage of an ideal machine of that type. The efficiency of a machine is defined as a percentage by

$$\text{efficiency} = \frac{\text{output work}}{\text{input work}} \times 100$$

and is always less than 100% because of frictional and other losses. Complex machines can be described in terms of combinations of simple machines such as levers, inclined planes, and hydraulic presses. Wheel and axle systems and pulleys can be described as modified lever systems, and the screw and the wedge can be seen as applications of the inclined plane.

REVIEW QUESTIONS

1. Define work, power, and energy.

2. Explain how it is possible for a force to act for a long time and yet do no work. How is it possible for a force to move and yet do no work?

3. What is meant by "conservation of energy"?

4. When a moving cart slows down and comes to rest, does this violate the conservation of energy principle? If not, where did the energy go?

5. Can a machine achieve a multiplication of energy? Explain.

6. Can a machine achieve a multiplication of force? Explain.

7. Explain what is meant by actual mechanical advantage, ideal mechanical advantage, and efficiency.

8. Scissors may be made with long blades and short handles, but bolt cutters are made with short blades and long handles. Explain the difference by using the principles of machines.

9. What is the difference between first, second, and third class levers? Which would give the greatest mechanical advantage?

PROBLEMS

Worked Example: A 60 newton force moved an 8 kg object through a distance of 24 meters in 4 seconds, starting from rest. The frictional force opposing the motion was 36 nt.

(a) How much work was done by the 60 nt force? What was the power?
(b) How much energy was used to overcome friction?
(c) What is the velocity of the object at the end of 4 seconds?

Solution: (a) The work was done by the 60 nt force is

$$\mathscr{W} = Fs = (60 \text{ nt})(24 \text{ m}) = 1440 \text{ joules}.$$

The power is then

$$P = \frac{\mathscr{W}}{t} = \frac{1440 \text{ joules}}{4 \text{ sec}} = 360 \text{ watts}.$$

(b) A force of 36 nt was required to overcome friction; therefore,

$$\mathscr{W}_{\text{against friction}} = (36 \text{ nt})(24 \text{ m}) = 864 \text{ joules}$$

represents the energy lost in the form of heat owing to friction.
(c) The remaining energy supplied to the object appears in the form of kinetic energy:

$$1440 \text{ j} - 864 \text{ j} = 576 \text{ j} = \text{kinetic energy}$$

$$\tfrac{1}{2}mv^2 = 576 \text{ j}$$

$$v^2 = \frac{2(576 \text{ j})}{8 \text{ kg}} = 144 \text{ m}^2/\text{sec}^2$$

$$v = 12 \text{ m/sec}.$$

1. A force of 20 nt is required to move an object of mass 5 kg at a constant velocity of 10 m/sec parallel to the force.

   a. How much work is required to move the object 50 m?
   b. What is the power?
   c. What is happening to the energy supplied to the object?

2. If a force of 50 nt is exerted to push a 25 kg cart down a 40 m hall, how much work is done? If there were no frictional resistance, what would be the velocity of the cart at the end of the 40 m? If the time required to travel 40 m was 6.3 seconds, what is the power supplied in watts?

3. A 5 kg object which is initially at rest is acted upon by a 25 newton net external force in the same direction for 4 seconds.

   a. What is the speed of the object after 4 seconds?
   b. What is the distance traveled in 4 seconds?
   c. How much work is done on the object?
   d. What is the average power?
   e. Show that the work done is equal to the change in kinetic energy (work-energy principle).

4. A 100 lb force moves a 64 lb object through a distance of 40 ft in the direction of the force. The object was initially at rest.
   a. How much work is done by the 100 lb force?
   b. Assuming all energy is in the form of kinetic energy, how much kinetic energy does the object have after the work is finished?
   c. What is the acceleration of the object, if 100 lb is the net external force?
   d. If this object is now to be brought to rest in a distance of one foot, how much force will be required?

Worked Example: A skier starts at the top of a ski jump, which is 100 ft above his landing point on the slope below. The ski jump is assumed to be frictionless, and his takeoff point is 60 ft above the slope. With what velocity will the skier touch down on the slope?

Solution: This type of problem illustrates the usefulness of the conservation of energy principle as a problem solving tool. If Newton's laws were used, the detailed contour of the ski jump would have to be known, and even then the problem would be prohibitively difficult. Assuming no frictional loss, we know that the skier's mechanical energy, consisting of potential energy plus kinetic energy, is the same at all points of the motion. Therefore,

$$(\text{P.E.} + \text{K.E.})_{\text{top}} = (\text{P.E.} + \text{K.E.})_{\text{bottom}}.$$

Since P.E. $= mgh$ and K.E. $= \frac{1}{2}mv^2$, we can substitute

$$(mgh + 0)_{\text{top}} = (0 + \tfrac{1}{2}mv^2)_{\text{bottom}}.$$

The mass, m, may be canceled, indicating that in the absence of friction, no skier has an advantage because of his weight.

$$gh = \tfrac{1}{2}v^2$$

$$v^2 = 2gh = 2(32 \text{ ft/sec}^2)(100 \text{ ft}) = 6400 \text{ ft}^2/\text{sec}^2$$

$$v = 80 \text{ ft/sec} = 54.5 \text{ mi/hr}.$$

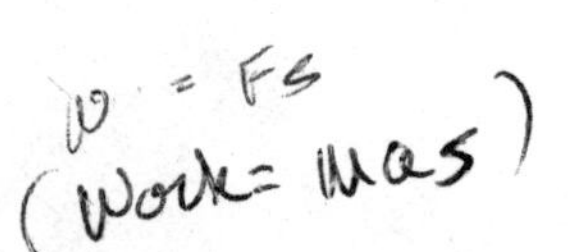

5. How much work is required to lift a 40 kg mass to a height of 20 meters? With what velocity would the object strike the ground if dropped from that height? (Hint: Use conservation of energy.)

6. An object is projected vertically with an initial speed of 40 m/sec. Use conservation of energy to calculate the maximum height of the object.

7. A 1200 kg car traveling at a speed of 20 m/sec (about 45 mi/hr) collides with a barricade and comes to rest in a distance of 0.5 meter. How much energy is taken from the car while it is being stopped (by work done on the car by the barricade)? What net external force is exerted on the car during the collison (assuming it is constant)? Convert the force to pounds.

8. One dietary calorie is equivalent to about 4186 joules of energy. If a person uses 2500 dietary calories of food per day in metabolism, how much energy in joules is used? What is the average metabolic rate, or average power, in watts?

9. The amount of oxygen used by the body in metabolism is proportional to the energy yield, with 1 liter of oxygen combining with foods to produce about 20,000 joules. If the oxygen consumed in one minute is 1.5 liters during vigorous exercise, what is the metabolic rate in watts (i.e., what is the average power during this minute)?

10. Assuming an efficiency of 25% for the muscle system in the process of converting food energy into mechanical work, how much energy would be used by a person of mass 75 kg (weight = 165 lb) in the process of climbing three flights of stairs for a total height of 12 meters? Find the answer in joules and convert to dietary calories (1 dietary calorie = 4186 joules).

11. If a 165 lb man runs up a 12 ft flight of stairs in 3 seconds, what is his power output in horsepower?

12. An energy of 500 joules is put into a machine, but 100 joules of that energy is lost to friction. What is the efficiency of the machine?

13. What is the output power of a car's engine and drive mechanism if it keeps a car traveling at a constant speed of 20 m/sec against a resistance force of 2000 nt?

14. If a 10 gram bullet traveling at a speed of 300 m/sec was stopped after penetrating to a depth of 5 cm in a solid material, what was the average force acting against the bullet to stop it?

15. A block of mass 10 kg is to be pushed up a frictionless incline of length 10 meters to reach a vertical height of 2 meters.

    a. How much force would have to be exerted to lift the block in the absence of the incline?
    b. What minimum amount of work would have to be done to lift the block directly upward 2 meters?
    c. What minimum amount of work would have to be done to push the block up the incline to a height of 2 meters?
    d. What minimum average force would be required to push the block up the incline?

16. If an object is pushed 5 m along a certain incline, its height above the ground increases by 2 m. If the incline were frictionless it would require a force of 196 nt exerted parallel to the incline to push a 50 kg crate up the incline at a constant speed.

    a. How much work would be done in pushing the crate 5 m up the incline?
    b. If the "output work" of the machine is to lift the crate 2 meters vertically, how much work is accomplished?
    c. What is the mechanical advantage of this incline as a machine?

17. A 20 kg block slides 5 meters along a rough horizontal surface where the coefficient of friction is 0.4. What is the work done by the frictional drag force on the block? What is the work done by the gravitational force on the block?

18. A person pushes a 10 kg cart a distance of 20 meters by exerting a 60 newton horizontal force. The frictional resistance force is 50 newtons. How much work is done by each force acting on the cart? How much kinetic energy does the cart have at the end of the 20 meters if it started from rest?

19. What would be the kinetic energy of a 2-ton truck traveling down a highway at 55 mi/hr? How high would you have to lift the truck to give it an equivalent amount of gravitational potential energy?

20. A 12 ft rod is to be used as a lever to lift a 300 lb rock. If a pivot is placed under the rod at a distance of 2 ft from one end, how much downward force must be exerted on the other end to lift the rock? What is the

mechanical advantage of the lever? Assuming that the bar is perfectly rigid, how far will the long end of the lever have to be moved to lift the rock 1 ft?

21. Assume that the biceps is attached to the forearm at a point 3 inches from the elbow pivot point. If a 10 lb object is to be held in the hand such that its distance from the elbow is 15 in, how much force upward must be exerted by the biceps? What is the mechanical advantage of this system?

22. A lever has a total length of 6 meters, with a pivot point placed 1 meter from the end.

    a. If this lever is used to lift a 100 kg load through a distance of one meter vertically, what is the output work?
    b. How much force would you have to exert on the long end of the lever to lift the load?
    c. How far would you have to move your end vertically to lift the load one meter (Assume rigid lever, 100% efficiency)?

23. One end of a stretcher is to be lifted from the floor, so that the other end may be considered to be the pivot point. The stretcher is 8 ft long and the combined weight of stretcher plus patient is 160 lb.

    a. How much force would you have to exert to lift your end if the center of gravity were 2 feet from the end which is on the floor?
    b. What is the mechanical advantage of this lever system?
    c. How much weight does the floor support?
    d. How much work would you do if you lifted your end 2 ft?

24. Use the rough estimate 4×10^{11} joules for the total annual energy consumption per U.S. citizen and calculate the amount of each type of energy source which would be used if it were the sole source of energy for one citizen. Use the energy yields from Figure 2–4: a, Coal; b, Gasoline, c, Natural gas; d, Oil; e, Uranium metal, considering that only .72% of it is the fissionable isotope U^{235}.

REFERENCES

1. Harris, N. C. and Hemmerling, E. M. *Introductory Applied Physics*, 3rd ed. New York: McGraw-Hill, 1972.
2. Romer, R. H. *Energy, An Introduction to Physics*. San Francisco: W. H. Freeman, 1976.
3. Rouse, R. S. and Smith, R. O. *Energy, Resource Slave, Pollutant*. New York: Macmillan, 1975.
4. Fong, P. *Physical Science, Energy, and Our Environment*. New York: Macmillan, 1976.

CHAPTER FIVE

The Properties of Liquids

INSTRUCTIONAL OBJECTIVES

After studying this chapter, the student should be able to:

1. Define pressure and work problems involving the pressure in stationary liquids.

2. State Pascal's principle and give examples of its application in the hospital and within the human body.

3. Work problems involving the hydraulic press, such as finding the force required to lift a specified load, given the dimensions of the cylinders of the hydraulic press.

4. Calculate the buoyant force on a submerged object, given its mass and volume and given the density of the liquid in which it is submerged.

5. State the physical variables which determine the flow rate of a liquid through a tube (Poiseuille's law) and calculate the change in the flow rate associated with a specified change in any of these variables.

6. State the conditions of fluid flow for which Poiseuille's law is valid.

THE LIQUID STATE

The three normal states of matter are the solid, liquid, and gaseous states. Most solids have a microscopic order associated with their atoms, as illustrated in the simplified sketch in Figure 5–1(a). There are many different patterns, or arrays, in solids but there is a regular framework for the atoms. It is sometimes helpful to visualize a solid as a number of atoms suspended on a system of springs which hold them in approximately fixed positions. This model of the solid state will be developed later, but it is mentioned here to point out that the solid state is characterized by internal order and more or less fixed atomic positions. The solid thus keeps its shape and does not readily take the shape of its surroundings.

In the liquid state, as in the solid state, the atoms or molecules have very strong cohesive

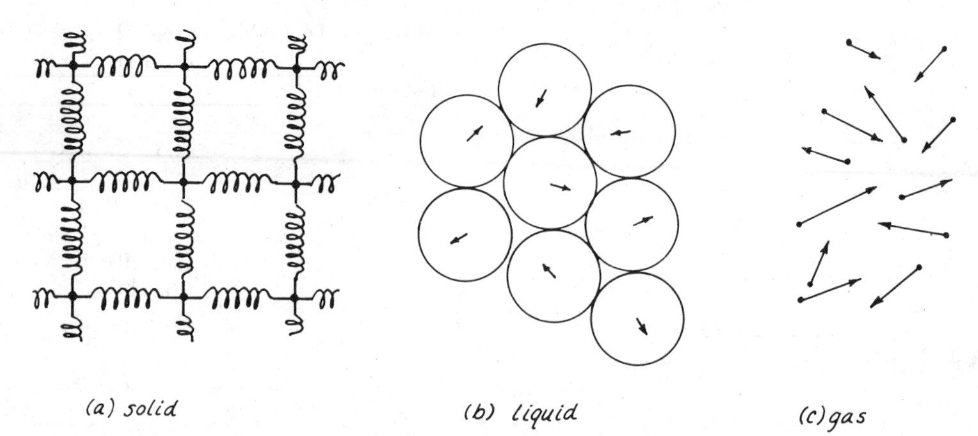

Figure 5–1 Simplified models of the three states of matter.

forces which hold them together. (Cohesion refers to the attraction of like molecules for each other, whereas adhesion refers to the attraction between unlike molecules.) However, the attractive forces between liquid molecules do not have preferred directions like the "spring" type forces of the solid. Thus the molecules are free to move relative to one another. The behavior of liquid molecules is somewhat analogous to a group of sticky balls, as illustrated in Figure 5–1(b). The freedom for relative motion between liquid molecules accounts for the fact that they readily take the shape of any container. Liquids flow readily, but it takes a considerable force to separate the molecules. The strong cohesive forces account for such phenomena as surface tension and viscosity, which will be discussed in Chapter 9.

The atoms or molecules of the liquid state are pictured in Figure 5–1(b) as much larger than in the solid, whereas of course they are actually the same size. The scale was chosen to suggest the fact that the distances between the centers of atoms in a liquid are of the same order of magnitude as the interatomic distances in a solid. Both solids and liquids are compact and not easily compressed. Even though a liquid takes the shape of a container, it is said to be "incompressible" and cannot be made to occupy a smaller volume without the exertion of enormous pressures.

In the gas phase, there are no longer appreciable attractive forces between molecules, and the only forces they exert upon each other are due to collisions. The interatomic distance scale is actually much larger than indicated in Figure 5–1(c), but that diagram illustrates the random motion and lack of cohesion. The general term *fluids* ("things which flow") will be used to include both liquids and gases. The discussion of gases will be deferred to Chapter 6.

Water is, of course, the most abundant liquid on the earth, covering three-fourths of the earth's surface and composing more than 60 per cent of our body weight. The only other natural liquid which exists on the earth in large quantities is petroleum. Many other naturally occuring liquids, such as blood and the sap of trees and plants, are mostly water. Most of the other liquids which occur in the world today are man-made, such as gasoline, alcohols, glycerine, vegetable oils, and numerous liquid organic chemicals.

THE DEFINITION OF PRESSURE

Since the words pressure and force are often confused in common usage, it is important to make a clear distinction between them. The definition of force is found in Chapter 3 along with a discussion of applications. Pressure is defined as the force per unit area:

$$\text{pressure} = \frac{\text{force}}{\text{area}} \quad \text{or} \quad P = \frac{F}{A}. \qquad \textbf{5–1}$$

As an example of the distinction between force and pressure, consider a woman wearing low, flat heels and then the sometimes-stylish high spike heels. Since the weight supported by the two types of heel is the same, there is no difference in the force exerted on the ground by the heels. However, since their areas are different, the *pressure* on the area of ground contacted is rather drastically different.

Example. A woman who weighs 120 pounds is supported by flat-heeled shoes. Each heel has an area of 5 in^2 and supports a weight of 40 lb (the sole of the shoes support the remaining 40 lb). What is the pressure exerted upon the ground by the heel? What would the pressure be if she were wearing spike heels which have an area of 0.25 in^2, assuming that they also supported 40 pounds?

Solution. Applying equation 5–1, the pressure exerted by the flat heel would be

$$P = \frac{F}{A} = \frac{40 \text{ lb}}{5 \text{ in}^2} = 8 \text{ lb/in}^2.$$

With the same force exerted, the spike heel would exert a pressure

$$P = \frac{40 \text{ lb}}{0.25 \text{ in}^2} = 160 \text{ lb/in}^2.$$

Because of the smaller area, the spike heel exerts a much larger pressure with the same force. This would be particularly noticeable when walking across soft ground because the penetrating ability of an object depends largely upon the pressure rather than just the force.

The hypodermic needle makes use of this fact to penetrate the skin with ease. Since the tip area is extremely small, a small force will result in a rather large penetrating pressure. A sharper needle will require less force to produce the required penetrating pressure. This is the whole point of the sharpening of nails, screws, chisels and other penetrating tools. An effective multiplication of the pressure is achieved by the sharpening. After the initial penetration is achieved, the wedge principle discussed in Chapter 4 is used to widen the opening for the larger part of the tool. Figure 5–2 further illustrates the different pressures which can be obtained with a given force.

The units of pressure are those of force divided by area, and the basic units in the three unit systems are listed in Table T–1, Appendix A. There are variations of these units, depending upon the application. For example, the basic British unit for pressure is lb/ft^2, but lb/in^2 is also commonly used. The conversion between these units can be obtained by applying the conversion factor 1 ft^2 = 144 in^2. Thus, a pressure of 1 lb/in^2 is equivalent to a pressure of 144 lb/ft^2. In the MKS and CGS systems the basic units are nt/m^2 and $dynes/cm^2$. (See Table T–2, Appendix A.)

PRESSURE IN LIQUIDS

The pressure caused by a liquid which is at rest is proportional to the depth of the liquid and its density. When a person dives into water, he quickly becomes aware of an additional pressure on his body. (A person is usually unaware of the atmospheric pressure.) The liquid pressure is particularly felt by the ears and becomes painful if one dives too deep. Submarines do not normally travel near the bottom of the ocean because the pressures

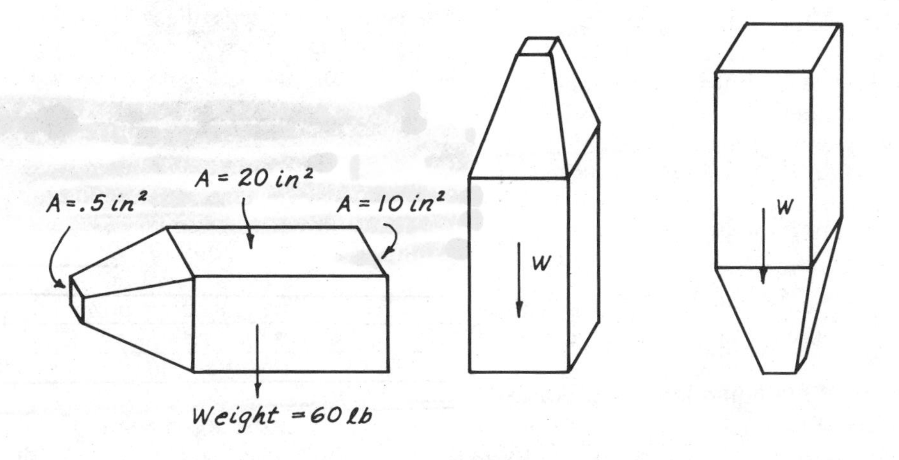

Figure 5–2 Illustration of the different pressures obtainable from a given force.

are enormous there and would crush the vessel. Depths of about a hundred feet are more common for submarine travel.

The pressure of a liquid at rest is called the hydrostatic pressure. This pressure may be determined from the depth and weight density of the liquid. The weight density of water is 62.4 lb/ft^3. A cubical vessel of water that measured one foot on each side would have a bottom area of 1 ft^2 and a depth of one foot. The pressure would be 62.4 lb/ft^2, using equation 5–1. If the bottom area were kept the same and the depth of the water were increased to 2 ft, keeping the sides of the vessel vertical, there would be 2 ft^3 of water which would weigh 124.8 lb. The pressure on the bottom would then be 124.8 lb/ft^2. This process could be extended to establish the basic relationship

hydrostatic pressure = depth × weight density

$$P_h = hd_w = hdg \qquad \textbf{5–2}$$

The symbol h is chosen to represent the depth of the liquid; it stands for the "head" of the liquid. The term "head" is commonly used to specify the height of the surface of the liquid above the point where the pressure is to be measured. The density d and the weight density d_w are defined in Chapter 3 (equations 3–7 and 3–8). When working in the British system of units d_w is usually specified, but in the metric system the mass density d is almost universally used, so the form $P_h = hdg$ is more convenient. Applied to the example above, this relationship gives the pressure

$$P = hd_w = (2 \text{ ft})(62.4 \text{ lb/ft}^3) = 124.8 \text{ lb/ft}^2.$$

To find the pressure in lb/in^2 at any depth in pure water, the numerical relationship is

$$P = (62.4h) \text{ lb/ft}^2 = (0.433h) \text{ lb/in}^2$$

where h is the depth in feet.

Example. The pressure required for the infusion of intravenous fluids into a vein is obtained by lifting the fluid bottle above the point of infusion. If the liquid level in a bottle is 18 inches above the point where the needle enters the vein and the liquid is assumed to have the same density as water, what is the liquid pressure at the needle?

Solution. The liquid pressure is

$$\begin{aligned} P_h = hd_w &= (1.5 \text{ ft})\,(62.4 \text{ lb/ft}^3) \\ &= 93.6 \text{ lb/ft}^2 \\ &= (93.6 \text{ lb/ft}^2)\left(\frac{1 \text{ ft}^2}{144 \text{ in}^2}\right) = .65 \text{ lb/in}^2 \end{aligned}$$

It is more common in the British system to express pressures in lb/in^2. Note that no details about the liquid bottle, tubing, etc. are necessary for the solution; the pressure is determined by the density and depth.

Example. An irregularly shaped pool has slanted walls so that the volume of water in it is difficult to determine. Its depth is measured to be 3 meters. How could you calculate the hydrostatic pressure at the bottom of the pool?

Solution. The liquid pressure at the bottom is determined solely by the depth and density; the total volume, total weight and shape of the pool are not directly relevant. The density of water is

$$d = 1 \text{ gm/cm}^3 = 1000 \text{ kg/m}^3.$$

Therefore

$$\begin{aligned} P_h = hdg &= (3 \text{ m})(1000 \text{ kg/m}^3)(9.8 \text{ m/sec}^2) \\ &= 29{,}400 \text{ nt/m}^2 = 29{,}400 \text{ Pascals}. \end{aligned}$$

The pressure unit nt/m^2 is called a *Pascal* in S.I. (Système Internationale) units.

It is important to note that the pressure in a container of liquid is dependent upon the depth of the liquid only and not upon the shape of the container. As illustrated in Figure 5–3, the pressure at a depth h below the surface is the same, regardless of the shape of the vessel holding the water. Actually the vessel sketched in Figure 5–3 is designed to show that when a liquid is free to flow, it will seek a common level, regardless of the shape of the container. The height in all commonly connected vessels will be the same unless the vessels are so small that capillary action is significant (see Chapter 9). As illustrated in Figure 5–4, the liquid level in an opaque container can be measured by connecting it to a small, transparent vertical tube. The water level in the tube will be the same as that inside the container.

Since the pressure at the bottom of a column of liquid varies in direct proportion to the height of the column, it is often convenient to use the height of a liquid column as a pressure measuring instrument. The simplest of such

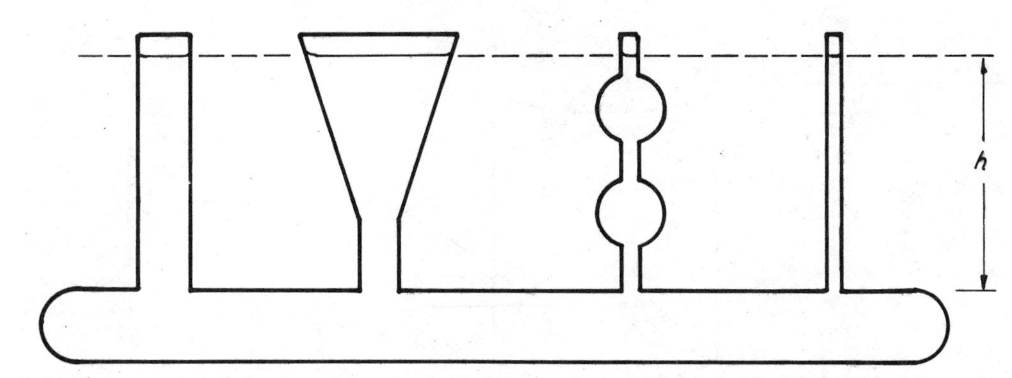

Figure 5–3 The pressure is dependent only upon the height *h* and not upon the shape of the vessel or the total volume of water.

devices is the manometer, which consists of a single vertical tube of liquid, usually mercury or water. If a pressure at the bottom of the tube is sufficient to push mercury into the tube to a height of 20 centimeters, then the pressure is said to be 20 cm Hg or 200 mm Hg. Likewise, a pressure sufficient to raise the level of water 20 cm in a manometer could be said to be a pressure of "20 cm of water" or "20 cm H_2O." Since the density of mercury is 13.6 times as great as that of water, it takes 13.6 times as much pressure to push a column of mercury as to push a column of water to the same height. The pressure required to push mercury 20 cm high would push a column of water 13.6×20 cm = 272 cm high. Thus, as a measure of pressure,

$$20 \text{ cm Hg} = 272 \text{ cm } H_2O$$

or

$$1 \text{ cm Hg} = 13.6 \text{ cm } H_2O.$$

Such units of pressure prove to be convenient for the measurement of blood pressure, atmospheric pressure, the pressure of suction machines, and so forth. (See Table T–2, Appendix A.)

DISTRIBUTION OF PRESSURE IN A STATIC LIQUID

The distribution of pressure in a static liquid (hydrostatic pressure) makes possible the use of fluids to transmit pressure. It has been stated that the pressure at a depth *h* below the surface of a liquid is proportional to that depth and to the density of the liquid (equation 5–2). If the pressure at a given depth is 15 lb/in^2, then on a square inch of liquid at that depth there is a downward force of fifteen pounds. If there is no movement, then there must be an equal upward force acting on that square inch, so that the net force is zero. Such arguments can be extended to show that the *pressure is exerted equally in all directions in a static liquid*. If a tiny droplet of the liquid at that depth could be examined, it would be found that the droplet is in equilibrium but that it has forces exerted upon it from all directions, as illustrated in Figure 5–5. These forces must be exerted in pairs such that the net force is zero.

These observations about the distribution of pressure can be used in conjunction with equation 5–2 to find the pressure on the wall of a liquid container. At a depth of 10 feet in water the pressure is 624 lb/ft^2. Since this pressure acts equally in all directions at that depth, the pressure sideways against the wall at a depth of 10 ft is 624 lb/ft^2. Since the pressure downward increases with depth, the horizon-

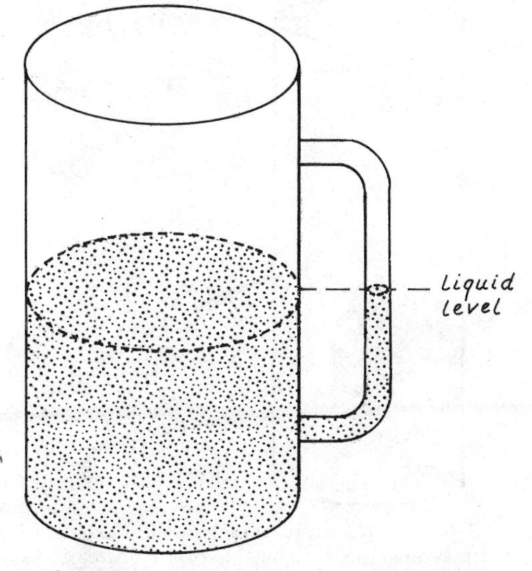

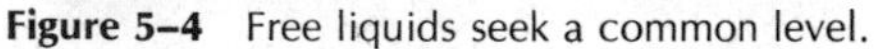
Figure 5–4 Free liquids seek a common level.

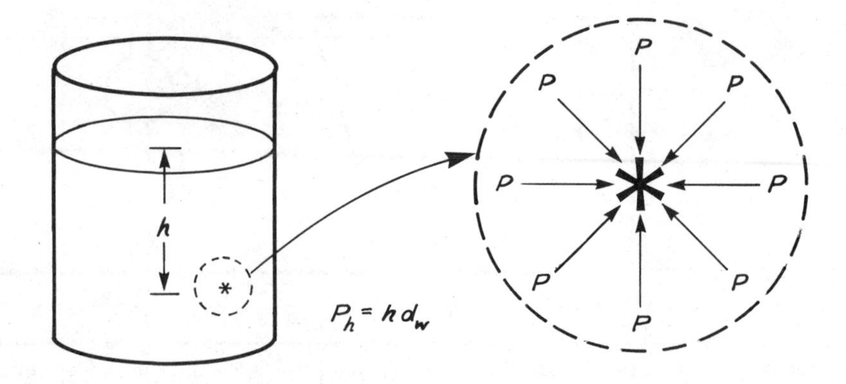

Figure 5–5 The same pressure is exerted in all directions at a point in a static liquid.

tal pressure on the walls also increases. If holes are punched in the side of a vessel as illustrated in Figure 5–6, the increase in horizontal pressure with the increase in depth can be readily demonstrated.

The fact that the pressure acts in all directions makes possible the transmission of pressure through tubing such as that used for administering intravenous fluids. In the case of the I.V. apparatus illustrated in Figure 5–7, the pressure is determined by the height of the liquid surface above the point of input to the patient. The pressure is transmitted through the tubing, even if it has turns or coils. (Of course, it can be obstructed by kinks or constrictions.) Since the hydrostatic pressure is determined by the height from the point to the open liquid surface, the pressure at the patient under static (no flow) conditions will be the same in configurations A and B of the tubing in Figure 5–7. When the fluid is flowing, there will be some drop in pressure compared to the static case because of fluid friction, as will be discussed in a later section. If the length of the tubing is the same, the frictional effects should be about the same and configurations A and B should be about equivalent under slow flow conditions. Under these conditions the pressure at the patient depends on the height of the bottle and not on the path taken to reach the patient.

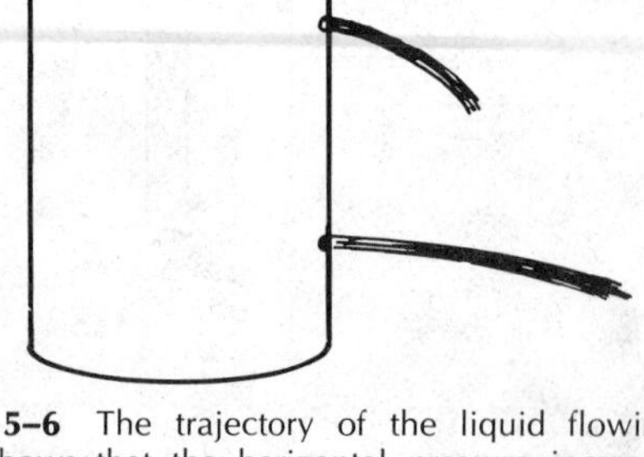

Figure 5–6 The trajectory of the liquid flowing from holes shows that the horizontal pressure increases with depth.

TRANSMISSION OF PRESSURE: PASCAL'S PRINCIPLE

The static pressure in an open liquid caused by its weight is equal to the weight density times the depth at the point being considered, and that pressure acts equally in all directions. Any external pressure exerted on an enclosed liquid must be added to this pressure to get the total pressure. For example, the atmosphere exerts a pressure of almost 15 lb/in^2 on every exposed object at sea level. Thus the pressure at a depth h below the surface of a liquid is actually given by

$$\text{pressure} = hd_w + 15\ lb/in^2.$$

This transmission of the 15 lb/in^2 pressure to all parts of the liquid is an example of *Pascal's Principle:*

Any change of pressure in an enclosed fluid is transmitted undiminished to all parts of the fluid.

The Hydraulic Press

In Chapter 4 it was stated that the simple machines could be broadly divided into three classes: levers, inclined planes, and hydraulic

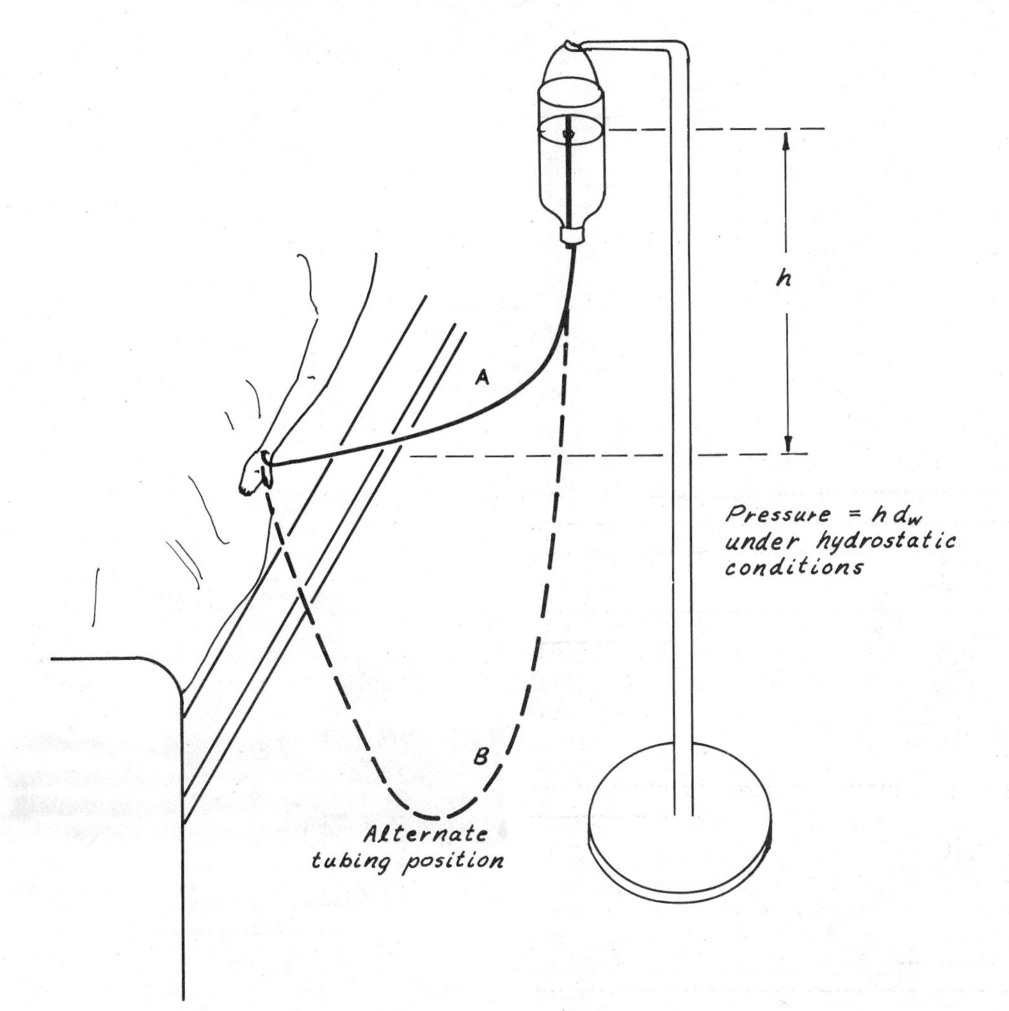

Figure 5–7 I.V. apparatus as an example of the transmission of pressure. Under hydrostatic (no flow) conditions, the pressure at the patient's arm would be the same in configurations *A* and *B*.

presses. The hydraulic press is an application of Pascal's principle and has thus been deferred until this point.

Since pressure is transmitted undiminished in an enclosed liquid, there will be the same number of pounds per square inch exerted at each point in the liquid as a result of any externally applied pressure. (This is, of course, in addition to the pressure due to the liquid's weight.) This fact can be used to obtain a multiplication of force. This is one of the basic functions of a machine, to overcome a large resistance force by exerting a small effort force. Figure 5–8 illustrates how this is done with the hydraulic press. A small effort force of 10 lb is exerted on a piston in the right-hand cylinder, which has a cross-sectional area of 1 in^2. This pressure of 10 lb/in^2 is transmitted to all parts of the liquid and acts on the bottom of the large piston, which has an area of 100 in^2. With a pressure of 10 lb/in^2 on this piston, the upward force can be calculated.

$$P_1 = P_2$$

$$\frac{F_1}{A_1} = \frac{F_2}{A_2}$$

$$\frac{10 \text{ lb}}{1 \text{ in}^2} = \frac{F_2}{100 \text{ in}^2}$$

$$F_2 = (10 \text{ lb/in}^2)(100 \text{ in}^2) = 1000 \text{ lb}$$

Thus a mechanical advantage of F_r/F_e = 1000/10 = 100 is achieved. This mechanical advantage is also equal to the ratio of the areas of the two pistons.

The machine known as the hydraulic press has many applications. For example, the hydraulic lifts used to raise patients are hy-

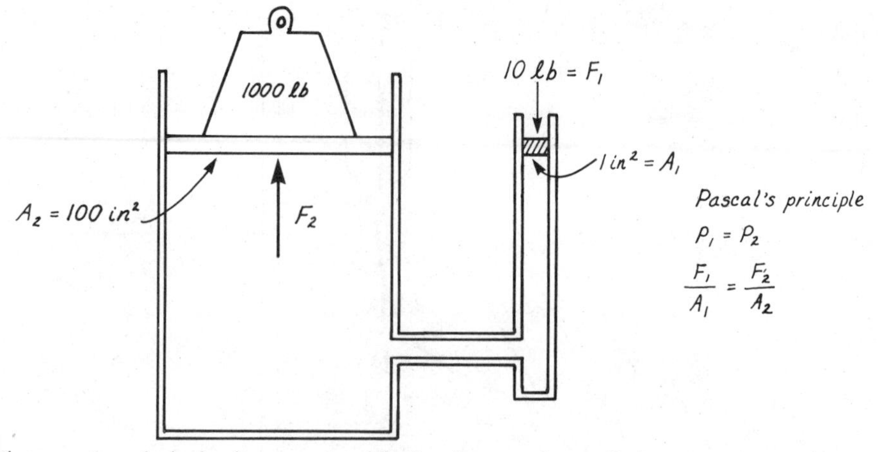

Figure 5–8 The operation of a hydraulic press. A small effort force on the small piston overcomes a large resistance force on the large piston.

draulic presses. In Figure 5–9 the patient is lifted by a small effort force exerted on the handle of the hydraulic lift. This handle acts as a lever which exerts a force on a small piston filled with oil. The pressure is transmitted through the oil to exert a larger upward force on a large piston which raises the patient.

Hydraulic lifts are often used to raise and lower operating tables and sometimes a simpler hydraulic lift may be included as part of a bed. Dentists' chairs are usually raised and lowered by means of a hydraulic mechanism. In these cases a small electric motor is often used to pump the oil into the small cylinder to supply the effort force. Similar hydraulic devices are used to raise cars in a garage. A small pump forces oil into a small cylinder to supply the effort force. The pressure is then transmitted to a large cylinder to provide the multiplied force necessary to raise the car.

The automobile brake system (power or manual brakes) is a further application of the

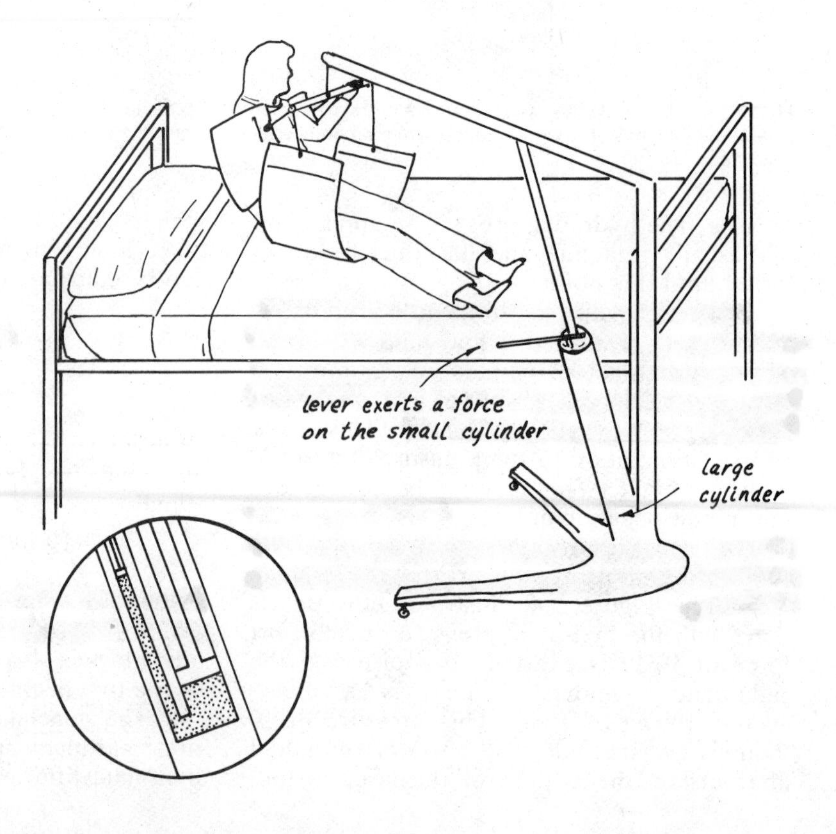

Figure 5–9 Example of a hydraulic lift.

foot
fluid
heels;
linder
n the
exert a
means
anical
ic ma-
heavy
has the
e pres-
nce by
smitted

ATIONS
NCIPLE

ndency to
es) when
r a length
an air or
formation
lication of
to develop
which have
the largest pressures exerted upon them by an ordinary mattress. Since the air and water mattresses constitute closed fluid systems, the pressure is evenly distributed in them. When the patient lies on the mattress, the same pressure is exerted against every part of the body which is in contact with the mattress. To obtain the full advantage, the entire body must be supported by the enclosed fluid. The mattress must be sufficiently inflated or filled with liquid so that no part of the body touches the bottom of the mattress.

Pascal's law applies to any enclosed fluid. There are several examples of enclosed or nearly enclosed fluids in the body. One interesting case is that of the cerebrospinal fluid. This fluid circulates around the spinal cord and up into the subarachnoid space around the lower part of the brain. Though it is not a completely closed liquid system, it is a close approximation to one. An increase in pressure at any point in the fluid will increase the pressure in all parts of the fluid. Thus, a brain tumor or any abnormal growth which protrudes into the space normally occupied by the fluid may cause a measurable increase in pressure in all parts of the fluid. A measurement of the pressure of the fluid at any convenient location can detect the increase in pressure. The measurement of the fluid pressure is normally made between the third and fourth lumbar vertebrae by means of a spinal tap. A water manometer is normally used for this measurement, and the pressure is then recorded in cm of water or mm of water.

The Queckenstedt test for obstructions in the cerebrospinal fluid is based on Pascal's principle. If the flow of blood from the venous sinuses into the internal jugular veins is shut off temporarily by squeezing the jugular veins, the intracranial pressure rises. If there is no obstruction in the cerebrospinal fluid, this pressure increase will be transmitted to all parts of the fluid and will be indicated by a rise in the water level in the spinal tap manometer. If the manometer pressure is unaffected by the pressure on the jugular veins, an obstruction is indicated.

The unborn fetus is protected from external forces by the fluid in the amniotic sac surrounding it. This fluid tends to distribute the effect of a force exerted on the abdominal area, but since it is an enclosed fluid, Pascal's principle applies. Any pressure applied to the abdominal wall is transmitted to all parts of the fluid and will be exerted upon the fetus. The pre-natal patient is usually cautioned against wearing tight clothing since there is a possibility that continuous pressure might produce fetal deformity.

Since the eye contains an enclosed fluid, any blow to the front of the eye will transmit pressure to the back of the eye. The blood vessels, retina, and optic nerve are delicate structures and may be injured by excess pressure. It is possible that a blow to the eye may not produce visible injury to the front of the eye but may damage the optic nerve because of the transmitted pressure.

Pascal's law may be applied to explain the effects of collected fluid in the pericardial and pleural cavities. Normally there is not enough fluid there to constitute a sizeable enclosed fluid. When fluid collects abnormally, then the pressures transmitted by these fluids may lead to abnormal pressures on the heart and lungs.

BUOYANT FORCE AND ARCHIMEDES' PRINCIPLE

When something is immersed in water or another fluid, it seems to weigh less than it does in air. This effect is called *buoyancy,* and it provides the lifting force required to float a

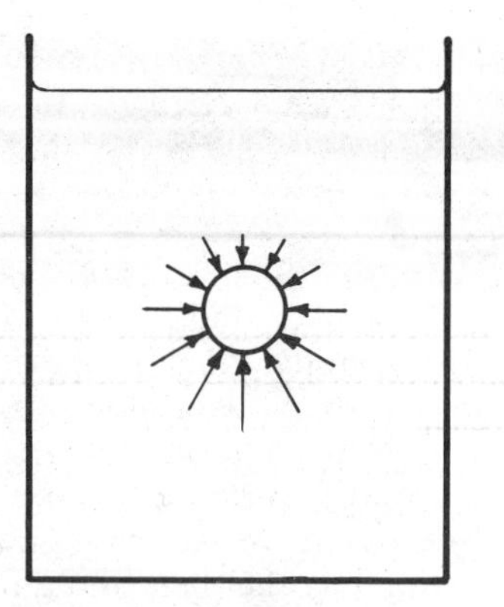

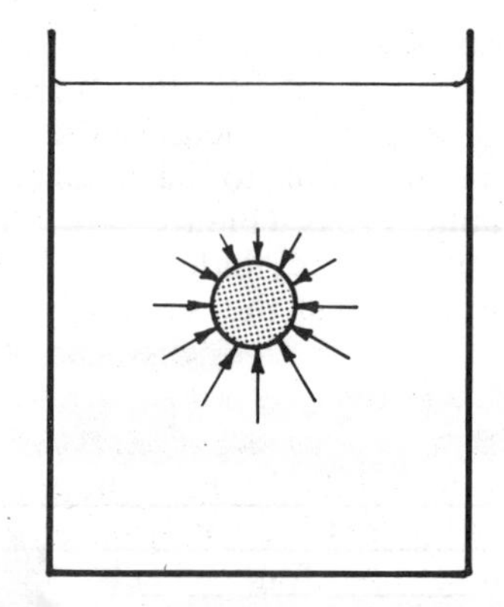

Figure 5–10 Origin of buoyant force.

ship on water or a balloon in the air. If the upward buoyant force on a submerged object is greater than its weight, it will rise. If the buoyant force is equal to the weight it will float; if the force is less than the weight it will sink.

Every part of a stationary vessel of liquid will be at rest and can be said to be at equilibrium. This requires that the vector sum of the forces acting on any small volume of the liquid is exactly zero. Consider a small volume of liquid in a larger container of liquid, as illustrated in Figure 5–10. There is a gravitational force acting downward on the liquid volume equal to

$$Vd_w = \text{weight}$$

where V = volume considered and d_w = weight density. Therefore, there must be an upward buoyant force equal to the weight of this volume of liquid. This buoyant force arises from the fact that the pressure exerted on the bottom of the volume V is slightly greater than the pressure at the top of the volume. Thus, the buoyant force arises from the difference in liquid pressure above and below the object. If a solid object with the same volume V is submerged in the liquid (Figure 5–10(b), the liquid pressure distribution remains the same and, therefore, the buoyant force on that object is equal to the weight of the water displaced.

The generalization of this idea is known as Archimedes' principle.

Archimedes' principle: The buoyant force on a submerged object is equal to the weight of the fluid displaced by the object.

$$F_{\text{buoyant}} = Vd_w = Vdg \qquad \textbf{5–3}$$

where V is the volume of the fluid displaced and therefore equal to the volume of the submerged object. The weight density d_w refers to the weight density of the fluid. Note that this applies in general to *fluids,* either liquids or gases.

If the average density of the submerged object is less than that of water, then it will displace a weight of water which is greater than its own weight. Thus, the buoyant force will be larger than the weight, and the net upward force will move the object to the surface of the water. For example, if an object which weighs 10 pounds and has a specific gravity of 0.5 is completely submerged, it will displace 20 pounds of water since the water is twice as dense. The resulting 20 pound buoyant force overcomes the weight and moves the object upward to the surface with a net force of 10 pounds. It will be at equilibrium when one half of its volume is submerged, since it will then displace precisely 10 pounds of water. Similarly, a 10,000 ton ocean freighter will be at equilibrium when it displaces 10,000 tons of water. If it is loaded with 1000 tons of cargo, it will settle deeper until it displaces 11,000 tons of water.

To measure the specific gravity of a liquid, a glass tube of standard specific gravity can be floated in the liquid as illustrated in Figure 5–11. Such a device is called a hydrometer. If the specific gravity of the liquid is increased, the hydrometer will float higher because a smaller liquid volume must be displaced to displace its weight in liquid. A specialized hydrometer called a urinometer is used to measure the specific gravity of urine.

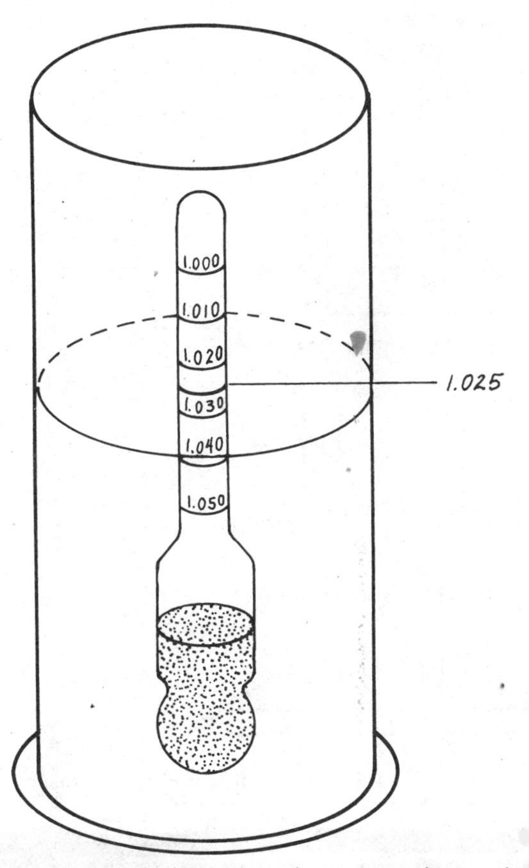

Figure 5–11 A hydrometer used to measure the specific gravity of a urine specimen. The narrow tubing is calibrated so that the specific gravity 1.025 can be read from the tube.

This specific gravity ranges from 1.001 to 1.035 and is an important physiological indicator. Some diseases alter the composition of the urine and change its specific gravity.

Archimedes discovered that objects apparently weigh less when weighed under water, and he used this fact to measure densities. Archimedes' principle is used to advantage in widely used types of therapy for persons with motor difficulty. If the patient is exercised while partially submerged in water, the buoyant force aids the patient's muscles. This aid makes possible greater movement of the extremities and helps prevent disuse atrophy.

PRESSURE IN FLOWING FLUIDS

The pressure in a flowing liquid depends on the details of the flow process, in contrast to the case of the static liquid, where the pressure depends only upon the depth and density of the fluid and the externally applied pressure. When a liquid flows through a tube, there will be a pressure drop; the pressure at the exit point of a length of uniform tubing will be lower than the pressure at the entrance to the tube. The pressure drop in a flowing liquid is illustrated in Figure 5–12. In Figure 5–12(a) the pressure at all points in the horizontal tube is the same since there is no flow and the pressure depends only upon the depth of the water or the "head," h. The pressure is indicated by the height of the liquid in the vertical tubes.

In Figure 5–12(b), the liquid is allowed to flow out the end of the tube and there is then a pressure drop along the tube as indicated by the liquid height in the vertical tubes. It will be noted that the amount of pressure drop is the same between successive equally spaced vertical tubes. That is, the drop in pressure along each successive unit length in a flowing fluid will be the same. It is convenient to discuss pressure changes in terms of the *pressure gradient,* which is defined as the pressure drop per unit length:

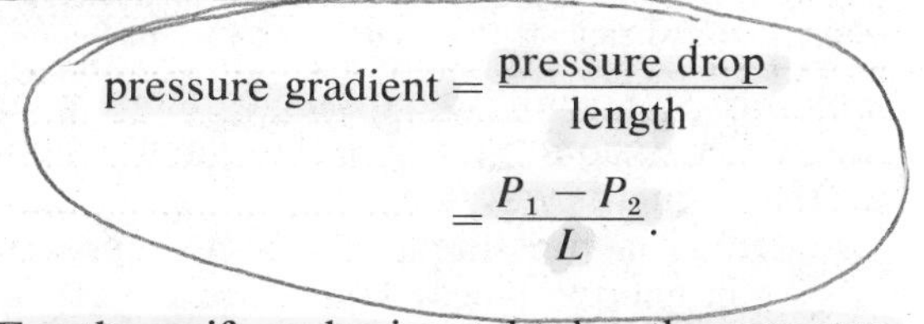

$$\text{pressure gradient} = \frac{\text{pressure drop}}{\text{length}} = \frac{P_1 - P_2}{L}.$$

For the uniform horizontal tube, the pressure gradient will be the same at all points in the tube if a perfectly uniform flow pattern is maintained.

Several factors affect the rate of flow through a length of tubing. For the smooth flow of an ideal fluid, the factors can be summarized in the relationship

$$\mathscr{F} = \frac{P_1 - P_2}{R} \qquad \textbf{5–4}$$

where P_1 and P_2 are the pressures at the upstream and downstream ends of the tube as shown in Figure 5–13, $\mathscr{F}$ is the volume flow rate (volume flow rate = volume/time) and R is the effective resistance to flow. This relationship is sometimes referred to as "Ohm's law for fluid flow" because of its similarity to the law for electric current flow discussed in Chapter 13. Under ideal conditions of flow, the resistance depends upon the diameter and length of the tube and upon the viscosity of the fluid. The factors affecting the resistance to

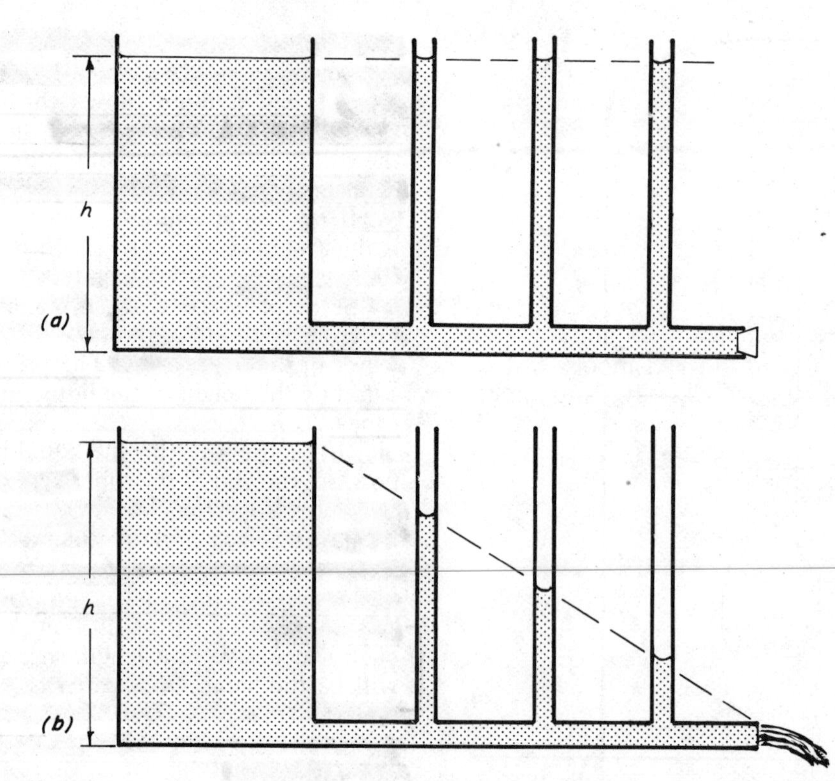

Figure 5–12 (a) The pressure is the same at all points along the horizontal tube when there is no flow. (b) A uniform pressure drop occurs when there is smooth flow through a uniform tube.

flow are sufficiently important to merit a detailed discussion because of their applications to the circulatory system.

Drops in pressure during flow represent losses in energy. These losses are largely attributable to frictional effects. One type of friction is that between the fluid and the walls of the tubing. This friction increases the fluid resistance and thus impedes the flow. Plastic tubing material is found to be superior to rubber tubing for many applications because of its smaller wall friction effects. Frictional forces also exist within the fluid. These forces which oppose the flow are referred to as *viscosity*. Viscosity is a molecular phenomenon which will be discussed further in Chapter 9. It is clear that viscosity is a very important factor in determining the fluid resistance and hence

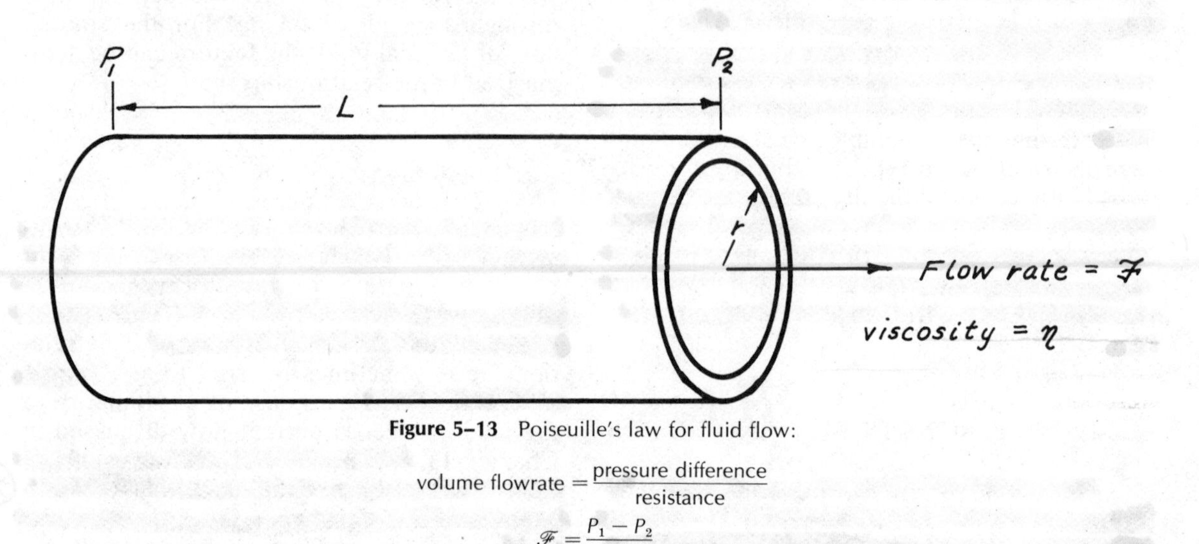

Figure 5–13 Poiseuille's law for fluid flow:

$$\text{volume flowrate} = \frac{\text{pressure difference}}{\text{resistance}}$$

$$\mathscr{F} = \frac{P_1 - P_2}{R}$$

the flow. Saline solutions have a considerably lower viscosity than blood or plasma and thus tend to flow more readily. A higher pressure at the beginning of the tube will be required to produce the same flow of blood or plasma. With a given pressure gradient, the volume flow rate is inversely proportional to the viscosity.

The flow pattern of the liquid will also affect the pressure gradient. The two types of flow are referred to as *laminar* and *turbulent* flow. In a straight tube with smooth walls, a fluid tends to flow in smooth layers. The layer nearest the wall will be essentially at rest. The speed gradually increases to a maximum on the central axis of the vessel. The average flow speed is then about half the maximum speed found at the center. This type of flow is known as laminar flow, and it represents the minimum energy loss. If the flow is speeded up past a certain critical speed, or if there is an obstruction in the flow path, eddies will start to form and the laminar flow may break down into turbulent flow. The pressure gradient for a given flow rate will be lower if laminar flow can be maintained. The maintenance of laminar flow is important for efficient fluid circulation, whether it is in an intravenous fluid apparatus or the circulatory system of the body. To prevent turbulent flow, it is important to make smooth transitions to different tubing sizes. Right angle bends and kinks can generate turbulent flow. Flow regulators which constrict the size of fluid tubing should do so in a manner such that smooth contours are maintained inside the tubing. Deposits or constrictions in arteries or veins can generate a turbulent flow pattern. To maintain the normal flow rate, the pressure must then be raised, so an additional burden is placed on the heart. A sudden change in the diameter of a blood vessel owing to weakening and dilation of the wall of the vessel will produce turbulent flow and have a similar effect.

The flow of liquids through very narrow cylindrical tubes was investigated in detail by Poiseuille, who was primarily interested in the flow of blood through veins and arteries. Under laminar flow conditions in tubes where wall friction was not a significant factor, Poiseuille found that pure liquids approximate ideal flow. The resistance to flow under these conditions is

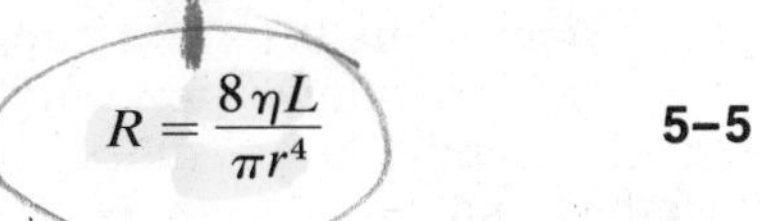

$$R = \frac{8\eta L}{\pi r^4} \qquad \textbf{5–5}$$

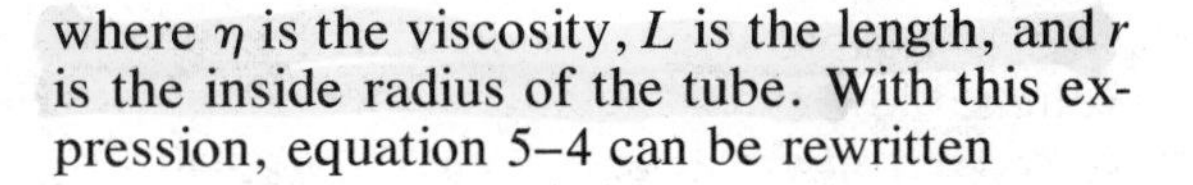

where η is the viscosity, L is the length, and r is the inside radius of the tube. With this expression, equation 5–4 can be rewritten

$$\mathscr{F} = \frac{P_1 - P_2}{\left(\frac{8\eta L}{\pi r^4}\right)} = (P_1 - P_2)\left(\frac{\pi r^4}{8\eta L}\right) \qquad \textbf{5–6}$$

which is referred to as Poiseuille's law. Perhaps most surprising about the above relationship is the fact that the volume flow rate depends upon the fourth power of the tubing radius. This implies that it is much more efficient to increase the size of the tubing than to increase the pressure gradient if an increased volume flow rate is desired. A rather striking example of the application of Poiseuille's law is that of choosing the needle size for a hypodermic syringe. The needle size is much more important than the pressure of the thumb in determining the outflow from a syringe. Doubling the diameter of the needle has the same effect as increasing the thumb pressure by a factor of sixteen. In the administration of I.V. fluids, increasing the size of the needle used is a much more effective means for increasing the volume flow rate than raising the bottle, since the volume flow rate depends so much more strongly upon the aperture radius than upon the fluid pressure.

Before considering the details of units for each of the quantities involved in Poiseuille's law, equation 5–6, it might be helpful to look at a qualitative example of how each of the physical parameters affects the volume flow rate.

Example. Suppose a given tube had an initial volume flow rate of 100 cm³/sec. Calculate the effect on the volume flow rate of doubling each of the other parameters one by one, with the others maintained at their original values.

Solution. The relationship for the initial volume flow rate can be written

$$100\ \text{cm}^3/\text{sec} = \frac{\pi(\text{pressure drop})(\text{radius})^4}{8(\text{viscosity})(\text{length})}.$$

The effect of doubling one parameter can be seen by examining its effect on the above equation. *Doubling the pressure drop* with the other factors remaining constant would double the numerical value of the right side of the equation. To maintain the equality you would

Viscosity unit = Poise (dyne∘sec / cm²)

have to double the left side of the equation (volume flow rate = 200 cm³/sec). *Doubling either viscosity or length* would yield half the volume flow rate (50 cm³/sec). *Doubling the radius* would have a much more dramatic effect since the radius is raised to the fourth power. Doubling the radius would give a factor of $(2)^4$ or sixteen times the volume flow rate (1600 cm³/sec).

Initial volume flow rate = 100 cm³/sec

Double pressure ⇒ 200 cm³/sec

Double viscosity ⇒ 50 cm³/sec

Double length ⇒ 50 cm³/sec

Double radius ⇒ 1600 cm³/sec

For the calculation of volume flow rates, a compatible set of units must be found for equation 5–6. The most common unit for viscosity is the poise, which is equal to a dyne-sec/cm². The viscosity of water at 20°C is about 0.01 poise, compared to a viscosity of about 10 poise for castor oil at the same temperature. The viscosity of the blood at normal body temperature is about .03 poise, or about three times the viscosity of room temperature water. For compatible CGS units the pressure should be measured in dyne/cm², the length and radius in centimeters, and the volume flow rate in cm³/sec.

Example. In Figure 5–12(b), the drop in liquid level between successive manometer tubes is an indication of the pressure drop and could be used to calculate the volume flow rate through the horizontal tube. If the distance between measuring tubes is 20 cm, the fluid height difference is 5 cm for successive tubes, and the inside diameter of the horizontal tube is 0.5 cm, calculate the volume flow rate in cm³/sec. Assume that the liquid is water at 20°C.

Solution. The parameters in equation 5–6 must be expressed in compatible units. From the information given:

$$P_1 - P_2 = 5 \text{ cm of water}$$

$$L = 20 \text{ cm}$$

$$r = 0.25 \text{ cm}$$

$$\eta = 0.01 \text{ dyne-sec/cm}^2.$$

Before the substitution of numerical values is made, the pressure different must be determined in dyne/cm². Using the conversion factor from Table T–2, Appendix A, the pressure difference is

$$P_1 - P_2 = (5 \text{ cm } H_2O)\left(980 \frac{\text{dyne/cm}^2}{\text{cm } H_2O}\right)$$

$$= 4900 \text{ dyne/cm}^2.$$

Now substituting into equation 5–6,

$$\mathcal{F} = \frac{P_1 - P_2}{R} = (P_1 - P_2)\left(\frac{\pi r^4}{8\eta L}\right)$$

$$\mathcal{F} = (4900 \text{ dyne/cm}^2) \times \left(\frac{3.14 \times (0.25)^4 \text{ cm}^4}{8 \times 0.01 \dfrac{\text{dyne-sec}}{\text{cm}^2} \times 20 \text{ cm}}\right)$$

$$\mathcal{F} = 38 \text{ cm}^3/\text{sec}.$$

Since the direct calculation of volume flow rates involves a considerable amount of numerical detail, it should be noted that many applications of liquid flow can be analyzed without recourse to the substitution of all the numerical values. If initial conditions are given, then the effects of subsequent changes can be predicted by a process similar to the previous qualitative example. This approach also makes it possible to retain the common units rather than converting to the CGS units used in the above example.

Example. A gardener uses a 50 ft hose to water his flowers and with the water pressure available at his faucet can get a volume flow rate of 6 gallons per minute through this hose. What volume flow rate would you expect if he added 100 ft of identical hose?

Solution. The pressure drop remains the same since the faucet pressure is the same and the water pressure drops to zero at the open end of the hose. Since the length is three times as great, the volume flow rate would be expected to drop to one third of its original value, or 2 gal/min.

Example. A pressure drop of 1 cm H_2O is maintained across a tube of length 100 cm and inside radius 0.5 cm. A volume flow rate of 24 cm³/sec of water is observed. What would be the flow rate for a pressure drop of 20 cm H_2O? What would you expect to be the volume flow rate through a tube of length 2 cm and inside radius 0.1 cm if supplied with this pressure? What flow rate would be expected if the 2 cm and 100 cm tubes were connected together and the 20 cm H_2O pressure drop placed across the combination?

Solution. Using equation 5–4 we have the relationship

$$24\ \text{cm}^3/\text{sec} = \frac{1\ \text{cm H}_2\text{O}}{R}$$

and can obtain the resistance of the 100 cm tube in units appropriate to this problem:

$$R = \frac{1\ \text{cm H}_2\text{O}}{24\ \text{cm}^3/\text{sec}} = .042\ \frac{\text{cm H}_2\text{O}}{\text{cm}^3/\text{sec}}.$$

The flow rate for a pressure of 20 cm H_2O is

$$\mathcal{F} = \frac{20\ \text{cm H}_2\text{O}}{.042\ \dfrac{\text{cm H}_2\text{O}}{\text{cm}^3/\text{sec}}} = 480\ \text{cm}^3/\text{sec}.$$

We could, of course, have deduced this directly from

$$20 \times \text{pressure} \rightarrow 20 \times \text{volume flow rate}$$

but the above framework will help to handle the case of the smaller tube. To get the flow rate through the 2 cm tube we need to make a comparison between it and the 100 cm tube.

Using equation 5–5 to establish a ratio of resistances

$$\frac{R_1}{R_2} = \frac{\dfrac{\eta_1 L_1}{r_1^4}}{\dfrac{\eta_2 L_2}{r_2^4}} = \left(\frac{\eta_1}{\eta_2}\right)\left(\frac{L_1}{L_2}\right)\left(\frac{r_2}{r_1}\right)^4$$

we obtain for this case

$$\frac{R(2\ \text{cm tube})}{R(100\ \text{cm tube})} = \left(\frac{2\ \text{cm}}{100\ \text{cm}}\right)\left(\frac{.5\ \text{cm}}{.1\ \text{cm}}\right)^4 = 12.5.$$

The resistance of the 2 cm length of tubing is 12.5 times the resistance of the entire 100 cm length of the larger tubing. The flow rate is correspondingly 12.5 times smaller:

$$R = (12.5)\left(.042\ \frac{\text{cm H}_2\text{O}}{\text{cm}^3/\text{sec}}\right) = .52\ \frac{\text{cm H}_2\text{O}}{\text{cm}^3/\text{sec}}$$

$$\mathcal{F} = \frac{20\ \text{cm H}_2\text{O}}{.52\ \dfrac{\text{cm H}_2\text{O}}{\text{cm}^3/\text{sec}}} = 38\ \text{cm}^3/\text{sec}.$$

If the tubes are connected together, neglecting the resistance added by coupling them, the resistances to flow would add since the liquid must flow through one tube and then the other.

$$R_{\text{total}} = (0.042 + 0.52)\ \frac{\text{cm H}_2\text{O}}{\text{cm}^3/\text{sec}} = .56\ \frac{\text{cm H}_2\text{O}}{\text{cm}^3/\text{sec}}$$

The flow rate through the combination caused by a pressure of 20 cm H_2O would then be

$$\mathcal{F} = \frac{20\ \text{cm H}_2\text{O}}{.56\ \dfrac{\text{cm H}_2\text{O}}{\text{cm}^3/\text{sec}}} = 36\ \text{cm}^3/\text{sec}.$$

This is almost the same as that obtained with the small tube alone. The resistance of the larger tube is almost negligible compared to that of the small one even though it is much longer. Because the resistance depends much more strongly on radius than on length, the resistance of a small coupling or a kink or obstruction in a large tube can offer more flow resistance than the entire resistance of the remainder of the tube. Similarly, when a tube is used to supply fluid to a needle in an intravenous infusion apparatus, the resistance of the long tube is often negligible compared to the resistance of the short needle.

Pure liquids such as water obey Poiseuille's law fairly well, but departures from the law may be substantial for suspensions and mixtures of liquids. Blood is a rather complicated fluid since many types of materials are in solution or suspension, and it does not precisely follow the behavior predicted by Poiseuille's relationship. The viscosity of water does not depend upon pressure, and if the pressure gradient in a tube of flowing water is doubled, the volume flow rate will very nearly double. In some experiments with blood, the viscosity is found to decrease with increasing fluid pressure, implying that the flow rate would be more than doubled by doubling the pressure gradient. This departure from Poiseuille's law is thought to be due to the increasing accumulation of red blood cells in the faster axial part of the flow, so that there are fewer red cells near the vessel walls to contribute to wall friction. As discussed in Chapter 7, the viscosity is approximately a constant in the normal range of blood pressures. Anomalous decreases in viscosity with increased pressure have been reported for the synovial fluid which helps in the lubrication of the human joints.

SUMMARY

Although liquids freely take the shape of their containers, they are essentially incompressible and therefore provide efficient means for the transmission of pressure. Pressure is defined as force per unit area. At a depth h in a stationary liquid, any object will experience a pressure owing to the weight of the liquid which is equal to hd_w, where d_w is the weight density of the liquid. Because of this direct dependence of pressure upon the distance from the free surface of a liquid, the height of a liquid column provides a direct measurement of the pressure at its base; therefore, pressures are often stated in millimeters of mercury, centimeters of water, and so on, with reference to the liquid height in a manometer.

Pascal's principle states that any change of pressure in an enclosed fluid is transmitted undiminished to all parts of the fluid. This principle is used to achieve a multiplication of force in the hydraulic press, one of the basic types of machines. Pascal's prlnciple may be illustrated by the behavior of the cerebrospinal fluid, the vitreous humor of the eye, the amniotic fluid, and other enclosed liquids in the body.

Submerged objects experience an upward buoyant force equal to the weight of the liquid displaced. If the density of an object is less than that of the liquid in which it is placed, then it will float. The fraction of the floating object which is submerged gives a relative measurement of the specific gravity of the liquid. This fact is used to advantage in instruments called hydrometers which are used to measure the specific gravities of fluids.

When a liquid flows through a tube there will be a pressure drop as energy is used to overcome the resistance of the tube to the flow process. Poiseuille's law states that the flow rate will be equal to the pressure drop divided by the resistance. The resistance to flow is given by

$$\text{resistance} = \frac{8(\text{viscosity})(\text{length})}{\pi(\text{radius})^4}$$

so that the resistance is most strongly dependent upon the radius of the tube. Poiseuille's law applies to an ideal fluid under conditions of laminar flow; the flow rate will be decreased by any turbulence which is present. Although the blood is not an ideal fluid and departs significantly from Poiseuille's law, the law can nevertheless be useful for a qualitative description of the dynamics of blood flow.

REVIEW QUESTIONS

1. Which of the following factors affect the pressure at the bottom of an open liquid container: volume of liquid, total weight of liquid, weight density of liquid, depth of liquid, shape of container?

2. State Archimedes' principle. What physical factor determines whether or not an object will float? How can Archimedes' principle be used to measure the specific gravity of a liquid sample?

3. Explain how a heavy dentist's chair can be lifted by a small force on a foot pedal connected to a hydraulic jack.

4. When intravenous fluids are administered to a patient, why is the bottle elevated? By what factor would the flow rate change if the height of the bottle above the patient is doubled, with other factors constant? (Assume that the back pressure from the vein is negligible.)

5. Name the factors which affect the flow rate of a fluid through a tube. Which is the most important factor and why?

1 nt/m^2 = Pascal

6. Under what conditions does Poiseuille's law describe fluid flow accurately? Does it adequately describe the flow of blood? Explain.

7. Why does an aneurysm which enlarges an artery sometimes lead to an elevated blood pressure? Explain in terms of the physical factors which affect fluid flow.

PROBLEMS

Worked Example. A swimming pool is 10 ft deep and measures 25 × 40 ft. What is the water pressure in lb/in^2 at the bottom, and what total force is exerted on the pool bottom by the water?

Solution. The water pressure at the bottom is

$$P = hd_w = (10 \text{ ft})(62.4 \text{ lb/ft}^3 = 624 \text{ lb/ft}^2.$$

In lb/in^2,

$$P = (624 \text{ lb/ft}^2)(1/144 \text{ ft}^2/\text{in}^2) = 4.3 \text{ lb/in}^2.$$

The total force on the pool bottom can be found from the pressure definition $P = F/A$:

$$F = P \times A = (624 \text{ lb/ft}^2)(25 \text{ ft})(40 \text{ ft})$$
$$= 624{,}000 \text{ lb}.$$

1. If the pressure exerted by the atmosphere is 14.7 lb/in^2, what total force is exerted upon one side of an 8½ × 11 inch sheet of paper?

2. What is the pressure in lb/in^2 at the bottom of a column of mercury 30 inches high? Mercury has a specific gravity of 13.6.

3. Sea water is about 1.03 times as dense as pure water. The deepest recorded depth in the ocean is about 36,000 ft (6.85 miles) in the Marianas Trench in the Pacific Ocean. What is the pressure in lb/ft^2 at that depth?

4. If the systolic blood pressure is 120 mm Hg, what is the pressure in lb/in^2 and in nt/m^2?

5. A pressure of 0.1 lb/in^2 is to be measured in a clinical setting and it is desirable to make the measurement by measuring the height of a liquid column. Calculate the pressure equivalent in centimeters of mercury (cm Hg) and centimeters of water (cm H_2O) and evaluate which liquid would be most suitable in terms of providing a conveniently measureable column height.

6. The static water pressure at a certain faucet is found to be 50 lb/in^2. If this pressure is produced by the height of the water in a nearby reservoir, what is the height of the water in the reservoir with respect to the faucet?

7. The basement floor of a certain house is 12 ft below the surface of the ground. There is no provision for drainage away from the foundation, so that during extended rainy periods the soil in contact with the house becomes saturated. This is effectively a 12 ft depth of water at the point where the wall joins the basement floor. Calculate the water pressure in lb/in^2 which acts to make the basement leak.

8. The small piston of a hydraulic press has an areas of 2 in^2. The large piston has an area of 16 in^2. What force must be exerted on the small piston to lift a 280 lb load on the large piston?

9. A hydraulic lift is constructed of two connected, liquid-filled cylinders of areas 2 in^2 and 10 in^2 which are fitted with movable pistons. If a force of 20 lbs is exerted on the liquid in the small cylinder by its piston, how much weight can be lifted by the large piston? Assuming 100% efficiency, how far would you have to move the small piston to lift the load one foot?

10. The plunger of a hypodermic syringe which has an area of .8 cm^2 is used to push fluid through a needle which has an inside area of .0008 cm^2. If a force of 4 newtons is exerted on the plunger, what liquid pressure in nt/m^2 would be exerted on the fluid? What minimum force must be exerted on the plunger to force a fluid into a vein which has a venous pressure of 9 mm Hg?

Worked Example. An object having a volume of 5 liters is found to have a mass of 3.5 kg. Will this object float or sink? If it floats, how much of the volume will be submerged?

Solution. The density of the object is

$$d = \frac{3500 \text{ grams}}{5000 \text{ cm}^3} = .7 \text{ gm/cm}^3$$

The object will therefore float. It must displace a weight or mass of water equal to its own in order to float. It must displace 3.5 kg of water, which is 3.5 liters. Therefore, 3.5 liters of the object are submerged, or

$$\left(\frac{3.5 \text{ l}}{5.0 \text{ l}}\right) \times 100\% = 70\%$$

of its volume is submerged.

11. A uniform tube of length 10 cm and specific gravity 1.0 is marked with a scale for determining the specific gravities of liquids. If it floats with 6 cm of length submerged in a certain liquid, what is the specific gravity of the liquid?

12. A certain boat can displace a maximum of 10 ft^3 of water before sinking. If the boat weighs 80 lb, how many 160 lb persons can it carry without sinking?

13. A stone is found to weigh 130 lb in air and 70 lb when it is submerged in water. Find its density, volume, and specific gravity.

14. With a pressure drop of 50 cm H_2O, a tube of length 100 cm and inside radius 1 mm sustains a volume flow rate of 2 cm^3/sec of water.

    a. What pressure would be required to get a volume flow rate of 10 cm^3/sec?
    b. What would be the effect on the volume flow rate of increasing the inside radius to 3 mm, all other factors remaining fixed?
    c. If the length were cut to 20 cm (other factors constant), what flow rate would result?

15. A water pressure of 20 cm H_2O is applied at one end of a tube of length 50 cm. When the other end of the tube is opened, the flow rate is 200 cm^3/min. Assuming laminar flow, calculate the flow rate if:

    a. the pressure were increased to 40 cm H_2O.
    b. the length of the tube were increased to 100 cm.

c. the radius of the tube were doubled.
d. the water were replaced by blood, which has a viscosity approximately four times that of water.

16. A straight pipe sustains a volume flow rate of 20 gal/min when a pressure difference of 20 lb/in^2 is applied to the pipe.

    a. If the needed volume flow rate is 80 gal/min, what pressure drop would be required to produce it?
    b. If the pipe was replaced by one which was the same length but had an internal radius 50% larger, what volume flow rate would be achieved with the original pressure drop, 20 lb/in^2?

17. At a normal blood pressure a certain healthy coronary artery has a 100 cm^3/min volume flow rate of blood through it. If the inside radius of the artery was reduced to 80% of its former radius and all other factors remained the same, what would be the volume flow rate?

18. A 2 meter tube which had an internal radius of 1 cm developed a leak, and was cut and spliced with a 5 cm length of a smaller tube with internal radius 0.25 cm. Find the ratio of the resistance of the small segment to the resistance of the 2 meter length of the larger tube.

Worked Example: For medical applications of Poiseuille's law, it is convenient to express it in the empirical form

$$P = 6.94 \times 10^{-2} \frac{\eta L q}{d^4}$$

where

$P =$ pressure in cm of water

$\eta =$ relative viscosity ($\eta = 1$ for water)

$L =$ length of tubing in cm

$d =$ diameter of tube in mm

$q =$ flow rate in cm^3/min.

Calculate the pressure required to send water at a flow rate of $q = 10$ cm^3/min through:

(a) a small transfusion needle, $L = 3$ cm, $d = 1$ mm.
(b) a hypodermic needle, $L = 2$ cm, $d = 0.3$ mm.

Solution:

$$\text{(a)} \quad P = (6.94 \times 10^{-2}) \left(\frac{3}{1^4}\right)(10) = 2.1 \text{ cm } H_2O$$

$$\text{(b)} \quad P = (6.94 \times 10^{-2}) \left(\frac{2}{0.3^4}\right)(10) = 171 \text{ cm } H_2O$$

19. A bottle of saline solution with viscosity assumed equal to that of water is elevated 60 cm above a needle used for intravenous infusion. If the needle used has a length of 2 cm and a diameter of 0.3 mm, at what rate will the solution flow through the needle? (Neglect back pressure from the vein.)

20. The viscosity of blood is normally about four times the viscosity of water. If a unit of blood is elevated 60 cm above a transfusion needle of length 3 cm and diameter 1 mm, calculate the flow rate through the needle. (Assume that the tube leading to the needle is so large that its flow resistance can be neglected, compared to the resistance of the needle, and assume blood density equal to that of water.)

CHAPTER SIX

The Properties of Gases

INSTRUCTIONAL OBJECTIVES

After studying this chapter, the student should be able to:

1. State the ideal gas law and work problems for each of the following cases: constant pressure, constant volume, constant temperature.

2. Calculate the pressure exerted by a liquid column, given the height of the column and the density of the liquid.

3. Describe the use of a liquid column manometer for the measurement of pressure, such as atmospheric pressure or the pressure of the cerebrospinal fluid.

4. Interchange units of pressure such as lb/in^2, mm Hg, and cm H_2O, given appropriate conversion factors.

5. Describe and sketch the forces which move liquids into syringes and siphons.

6. Use the ideal gas law to explain the origin of the forces which move air into and out of the lungs during respiration.

7. Explain the function of each component of a water-sealed drainage system, given a diagram of the apparatus.

THE NATURE OF AN IDEAL GAS

The molecules of a gas do not experience the strong mutual attraction forces characteristic of the molecules of a solid or liquid. There are small attractive forces in very dense gases and in gases which are near the temperature at which they condense into liquids, but otherwise these forces are often so small that they can be neglected. Many gaseous phenomena can be explained by assuming that the gases are made up of hard spherelike molecules which move randomly at very high speeds and interact with each other only by collisions. The gas which obeys these assumptions is referred to as an *ideal gas*. The discussion in this chapter will be limited to those gases which are close approximations to the ideal gas.

Many of the properties which give different liquids and solids their characteristic properties are the result of the intermolecular forces. Since gas molecules do not experience intermolecular forces except during very brief collisions, they have lost the main cause of differing physical properties. For this reason, all gases which are good approximations to ideal gases have almost identical macroscopic physical properties. For example, the oxygen in the air responds to changes in pressure or temperature in a manner almost identical to that of nitrogen and the other constituents of the air. For this reason, air itself can be treated as an ideal gas without breaking it down to observe the characteristics of its individual gases. Once the properties of the ideal gas are established, they can be applied directly to explain many gaseous phenomena. Of course, distinctions between the gases in the air must be made when microscopic or molecular phenomena are considered, such as the diffusion processes occurring in the lungs. Such processes will be considered in Chapter 9.

THE IDEAL GAS LAW

The state of an ideal gas can be specified by measuring the absolute pressure, the volume, and the absolute temperature of the gas. Temperature scales are discussed in Chapter 10. It will suffice here to say that the zero of the absolute temperature scale is absolute zero and that the degree size is the same as in the Celsius scale. Zero degrees Celsius (the freezing point of water) corresponds to approximately 273 degrees in the absolute scale.

These three state variables, represented by P, V, and T, are found to obey the relationship:

Ideal Gas Law:

$$\frac{PV}{T} = nR = \text{a constant.} \qquad \textbf{6–1}$$

The constant R is referred to as the universal gas constant and has the numerical value R = 8.32 joules/mole-degree. The constant n is the number of moles of the gas, a mole being the mass in grams which is numerically equal to the molecular mass expressed in atomic mass units (e.g., a mole of carbon would be exactly 12 grams, since a carbon atom has a mass of 12.0 a.m.u.).

The quantity PV/T does not change so long as the same mass of gas is kept, even though any one or all of the individual quantities may change. The initial and final states of an ideal gas can then be related as follows:

$$\frac{P_iV_i}{T_i} = \frac{P_fV_f}{T_f} \qquad \textbf{6–2}$$

where the subscripts i and f refer to the initial and final states of the gas.

Many common processes can be described by holding one of the three state variables constant. The three cases are (1) constant temperature (isothermal), (2) constant pressure (isobaric), and (3) constant volume (isovolumetric). For historical reasons, cases (1) and (2) are called Boyle's Law and Charles' Law respectively. These labels may be useful for reference, but it should be kept in mind that they deal with special cases of the ideal gas law.

CONSTANT TEMPERATURE PROCESSES: BOYLE'S LAW

For isothermal processes, $T_i = T_f$ in equation 6–2 and it can be rewritten

$$P_iV_i = P_fV_f \quad \text{or} \quad \frac{P_i}{P_f} = \frac{V_f}{V_i} \qquad \textbf{6–3}$$

Since the product of pressure times volume must remain constant in constant temperature processes, it follows that if the pressure increases, the volume must decrease by the same factor.

Example. If a given mass of gas occupies a volume of one liter when the absolute pressure is 15 lb/in², what will be the volume if the absolute pressure is decreased to 5 lb/in² while keeping the temperature and the number of moles constant?

Solution. When the known quantities are substituted into the equation

$$P_iV_i = P_fV_f$$

it becomes

$$(15\ \text{lb/in}^2)(1\ \text{liter}) = (5\ \text{lb/in}^2)(V_f).$$

The final volume is then V_f = 3 liters. The volume expands by a factor of three when the absolute pressure is reduced by a factor of three. It should be emphasized that the *absolute* pressure and not the gauge pressure must be used when applying the gas law. Unlike equation 6–1, equation 6–3 allows the absolute pressure and volume to be expressed in any convenient units.

This case of the ideal gas law can be used to explain why a glass bottle of fluid used on an IV apparatus must be vented if the fluid is to continue to flow. When the rigid bottle is inverted as shown in Figure 6–1(a), there is a volume of air, V, above the liquid in the bottle at atmospheric pressure, approximately 15 lb/in². When some of the liquid drains out the bottom, the volume of the air increases and the pressure decreases proportionately. If the volume were doubled to $2V$, the pressure inside the bottle would have dropped to 7.5 lb/in² (Figure 6–1(b)). The decrease in pressure in the bottle caused by the increase of the air volume would actually stop the flow before it reached this point. A vent into the bottle would allow air to flow in to keep the air pressure at 15 lb/in² and thus the liquid flow could continue (Figure 6–1(c)). Because of this characteristic behavior of an enclosed gas, a clogged vent tube will stop the flow of the liquid.

CONSTANT PRESSURE PROCESSES: CHARLES' LAW

For isobaric processes, $P_i = P_f$ in equation 6–2 and it becomes

$$\frac{V_i}{T_i} = \frac{V_f}{T_f} \quad \text{or} \quad \frac{T_f}{T_i} = \frac{V_f}{V_i} \qquad \textbf{6–4}$$

If the pressure of a given mass of gas remains constant during a process, then the ratio of the final volume to the initial volume will be the same as the ratio of the final absolute temperature to the initial temperature. This explains the readily observed fact that when the temperature of a gas is raised, it will expand. In most common phenomena, however, both the pressure and volume will increase when the gas temperature rises.

Local processes in the atmosphere are close approximations to constant pressure processes. If the atmospheric pressure remains the same, then this case of the ideal gas law applies. If the temperature of the air is raised, it will expand and in the process will become less dense. It will then experience a net buoyant force upward and will rise. Likewise, cooled air will contract and tend to settle downward relative to the warmer air around

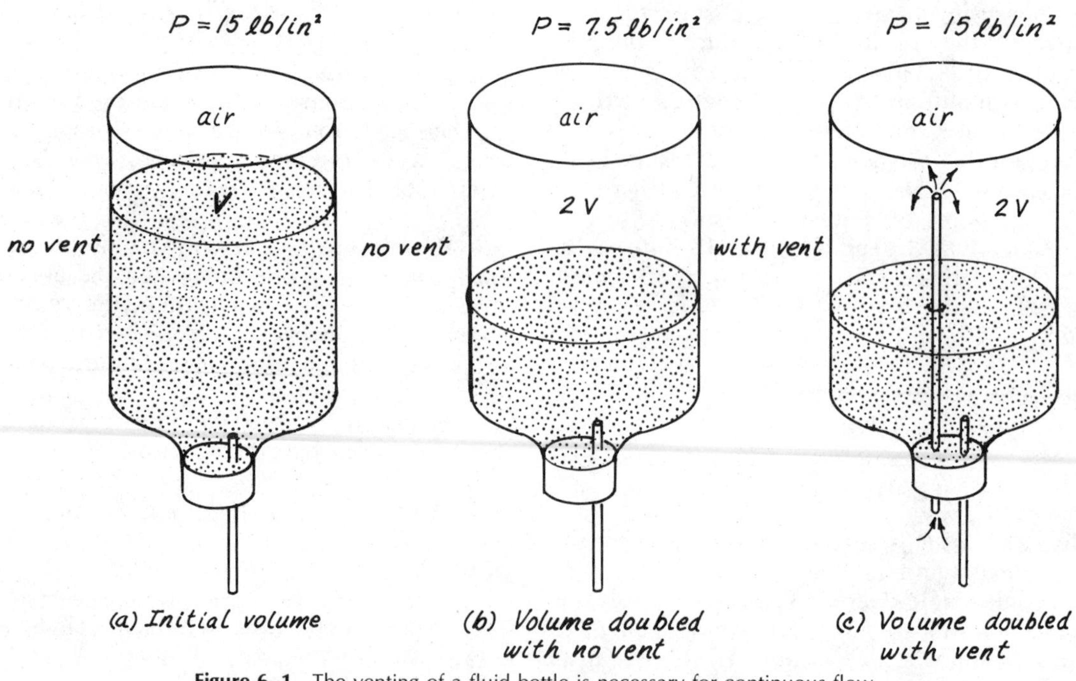

Figure 6–1 The venting of a fluid bottle is necessary for continuous flow.

it. Although weather processes involve simultaneous pressure and volume changes, the examination of constant pressure and constant volume processes can give considerable insight into the movement of air masses in the atmosphere.

According to equation 6–4, the volume will double if the absolute temperature doubles. It must be emphasized that this is the *absolute* temperature; the volume does not double if the temperature is increased from 10° C to 20° C. In the absolute temperature scale this corresponds to an increase from 283°K to 293°K and is thus only a 3.5% increase in absolute temperature. A volume of 1000 cm^3 at 10°C would increase to a volume of about 1035 cm^3 if heated to 20° C at a constant pressure. The temperature scales are discussed in Chapter 10.

CONSTANT VOLUME PROCESSES

If the volume is held constant, then $V_i = V_f$ in equation 6–2 and it can be written

$$\frac{P_i}{T_i} = \frac{P_f}{T_f} \quad \text{or} \quad \frac{T_f}{T_i} = \frac{P_f}{P_i} \qquad \textbf{6–5}$$

Thus, the absolute pressure of a closed volume of gas is directly proportional to the absolute temperature. If the absolute temperature doubles, the absolute pressure will double. The pressure of a constant volume of gas can actually be used as a thermometer.

The effects of this case of the ideal gas law can be observed in an automobile tire. If the pressure in a tire is measured, and the automobile is then driven at high speed for a while, it will be found that the friction will have caused an increase in both the temperature and the pressure of the tire.

THE MEASUREMENT OF ATMOSPHERIC PRESSURE

It has been pointed out that the atmosphere exerts a pressure on every exposed object on the earth. At sea level this pressure is about 14.7 lb/in^2 (15 lb/in^2 is often used as an approximation). The air is a fluid, and the pressure it exerts upon the earth's surface is simply due to its weight. The pressure of 14.7 lb/in^2 implies that a column of air of 1 in^2 cross-sectional area which extended up to the upper edge of the atmosphere would weigh 14.7 pounds. The equation 5–2 which applies to liquids,

$$\text{pressure} = (\text{depth})\ (\text{weight density})$$

could be used to determine the pressure, except that the weight density varies with the height above the earth. Since a gas is compressible, a given mass of gas near the earth's surface is squeezed into a smaller volume by the pressure of the gas above it. Thus the air has its maximum density near the surface and gradually thins out as one goes upward into the atmosphere.

If an open tube is placed into a vessel of mercury, the level of mercury will rise to the same level inside the tube as outside (Figure 6–2(a)). However, if the tube is closed off at the top end and the air is evacuated from it, the mercury will rise in the tube to a height which is proportional to the atmospheric pressure acting downward on the open surface of the mercury. This is an example of the manometer principle discussed in Chapter 5. The pressure acting downward on the liquid surface is transmitted to all parts of the liquid in accordance with Pascal's law. Thus, a pressure of 14.7 lb/in^2 acts upward inside the tube under standard pressure conditions. Since this is not balanced by the air pressure in the tube in Figure 6–2(b), the mercury will rise in the tube until equilibrium is reestablished. This equilibrium occurs when the downward pressure due to the weight of the mercury in the tube exactly balances the upward pressure in the tube due to the atmosphere outside. If the atmospheric pressure increases, the mercury will move higher in the tube until the increase is balanced by the column weight. If the atmospheric pressure is 14.7 lb/in^2 and the vertical mercury column has a cross-sectional area of 1 in^2, the mercury will rise until the column above the open liquid surface weighs precisely 14.7 pounds. The height of this column will be 76 cm = 760 mm high. In the British system this is equivalent to 29.92 inches. Thus, the standard atmospheric pressure is normally quoted as being 760 mm Hg or 29.92 inches of Hg. It is quoted in inches of Hg in weather forecasts in the United States, but most clinical pressure measurements are stated in mm of Hg. One mm of Hg is often referred to as 1 torr, so the standard atmospheric pressure could be said to be 760 torr.

An instrument which measures atmos-

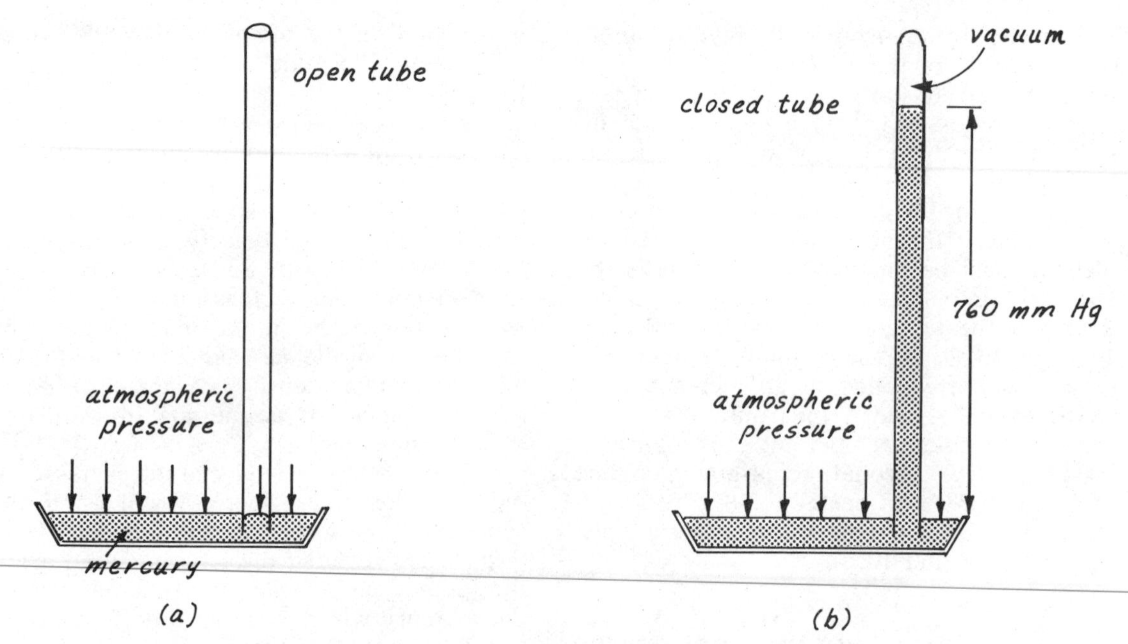

Figure 6–2 Manometer method for measuring atmospheric pressure.

pheric pressure is referred to as a barometer. The mercury manometer system described is one of the most common types of barometer Another common type is the aneroid barometer, which uses the pressure exerted upon a thin-walled vacuum box to move an indicator needle. Liquids other than mercury could be used to construct a manometer type barometer, but mercury is the most convenient. Since it is the liquid with the greatest density, the necessary manometer length is a minimum. If water were used in the manometer, it would be pushed 13.6 times as high as the mercury, since mercury has a specific gravity of 13.6. Thus, if water were used, the column height would be 13.6 × 29.9 inches = 34 feet tall. The pressure of the atmosphere is expressed in a variety of units. For reference, the standard atmospheric pressure at sea level in the most commonly used units is:

Standard Atmospheric Pressure
$= 1.013 \times 10^5$ newtons/m^2
$= 14.7$ lb/in^2
$= 760$ mm Hg
$= 29.92$ inches of Hg
$= 34$ ft of water
$= 408$ inches of water

The pressure of the atmosphere in which we live and breathe is really an enormous pressure. The forces and energy available from manipulation of this pressure are of considerable importance in the operation of various types of apparatus used in health care. Before discussing some of those applications of atmospheric pressure, some examples will be given to illustrate further the nature of this pressure.

Example. Assume that the surface area of an average human being can be approximated by the outside area of a circular cylinder which is 5 feet, 10 inches tall and has a circumference of 30 inches. Use the approximate atmospheric pressure 15 lb/in^2. What is the total force exerted by the atmosphere on the outside of this cylinder?

Solution. The area of the cylinder walls is obtained by multiplying the height by the circumference.

$$P = \frac{F}{A} \quad \text{or} \quad F = P \times A$$

$$\begin{aligned} \text{area} &= 70 \text{ in} \times 30 \text{ in} = 2100 \text{ in}^2 \\ \text{force} &= \text{pressure} \times \text{area} \\ &= (15 \text{ lb/in}^2)(2100 \text{ in}^2) \\ &= 31{,}500 \text{ pounds.} \end{aligned}$$

It is difficult to contemplate the fact that we go about our normal daily activities with a force on the order of 30,000 pounds acting upon us. Of course, the pressure acts symmetrically on all sides, and exactly the same pressure exists inside the body, so that there is no net force. The human body is adapted for activity in this environment of pressure. If suddenly exposed to the vacuum of space by a ripped space suit, the human body would literally explode.

The enormous energy associated with atmospheric pressure changes is demonstrated by the sudden atmospheric pressure change during hurricanes and tornados. A sudden atmospheric pressure drop accompanies a tornado and if a house is well sealed there will be an excess of pressure inside the house. If the atmospheric pressure dropped by 1 lb/in^2, this would cause a net outward force of one pound on every square inch of the interior wall of a house if the pressure could not equalize. There are numerous cases of well-built houses exploding during this quick pressure drop when the owners failed to leave windows or other openings to allow equalization of pressure.

When the pressure of automobile tires or other inflated objects is measured, the pressure quoted is the pressure excess over atmospheric pressure. This is referred to as the *gauge pressure*. The pressure in a tire before it is inflated is 15 lb/in^2. When a tire pressure gauge reads 30 lb/in^2, the actual total pressure is $30 + 15 = 45$ lb/in^2. Since the entire environment has the common pressure of 15 lb/in^2, the most useful way to state pressure is usually in terms of this gauge pressure or excess pressure.

Example. If a spent aerosol can is thrown into a fire, the absolute temperature of the gas inside it can triple. If the original absolute pressure inside the can is assumed to be 1 atmosphere, what will the gauge pressure be after heating? If the interior area of the can is 60 in^2, what is the resulting outward force?

Solution. Applying the constant volume case of the ideal gas law, equation 6–5, it is seen that tripling the absolute temperature will triple the pressure. This would be an absolute pressure

$$P = 3 \text{ atmospheres}$$

$$= 3 \times 14.7 \text{ lb/in}^2 = 44.1 \text{ lb/in}^2,$$

and a gauge pressure

$$P_{\text{gauge}} = 44.1 - 14.7 = 29.4 \text{ lb/in}^2.$$

The resultant outward force would be

$$\text{force} = P_{\text{gauge}} \times \text{area} = (29.4 \text{ lb/in}^2)\,(60 \text{ in}^2)$$

$$= 1764 \text{ pounds}$$

Depending upon a thin-walled, mass-produced aerosol can to withstand such a force is rather risky. The risk is compounded if there is some liquid remaining in the can to vaporize and add to the pressure.

MANOMETERS FOR PRESSURE MEASUREMENT

The mercury manometer used for atmospheric pressure measurement is one of many applications of the manometer principle. The term "manometer" may refer to any vertical tube of liquid used for pressure measurement, but for the measurement of gas pressure they are often in the form of "U-tubes," as shown in Figure 6–3, so that both ends of the liquid column can be seen and the difference in liquid levels can be easily measured.

As mentioned in Chapter 5, the use of the manometer for pressure measurement is based upon the fact that the liquid pressure at a given depth in a static liquid is determined entirely by the liquid density and depth (equation 5–2). Once the density is fixed by the choice of the liquid, the liquid pressure is dependent only upon the depth, which can be easily measured. The measurement of pressure in liquid column height units such as cm H_2O and mm Hg adds to the convenience.

More important than convenience for clinical purposes is the fact that manometers are more nearly foolproof than other types of pressure measurement devices. While all sorts of gauges with mechanical or electronic displays are available, they are all subject to malfunction or miscalibration. If a supplied pressure pushes water 30 cm up an open tube which has been carefully held vertical, the pressure is 30 cm H_2O above the atmospheric pressure; no technician or maintenance expert is required to assure you of its proper operation. Since misdiagnoses and improper treatment might result from pressure measurement errors, this is an important advantage. The reliability of manometer pressure gauges also makes them desirable in research applications.

If an open U-tube manometer is used as in Figure 6–3(a), the difference in height h between the two ends of the liquid is a measurement of the *gauge pressure,* or the excess of the pressure over atmospheric pressure. If the liquid is water and h is 20 cm, the pressure might be referred to as a "positive" pressure

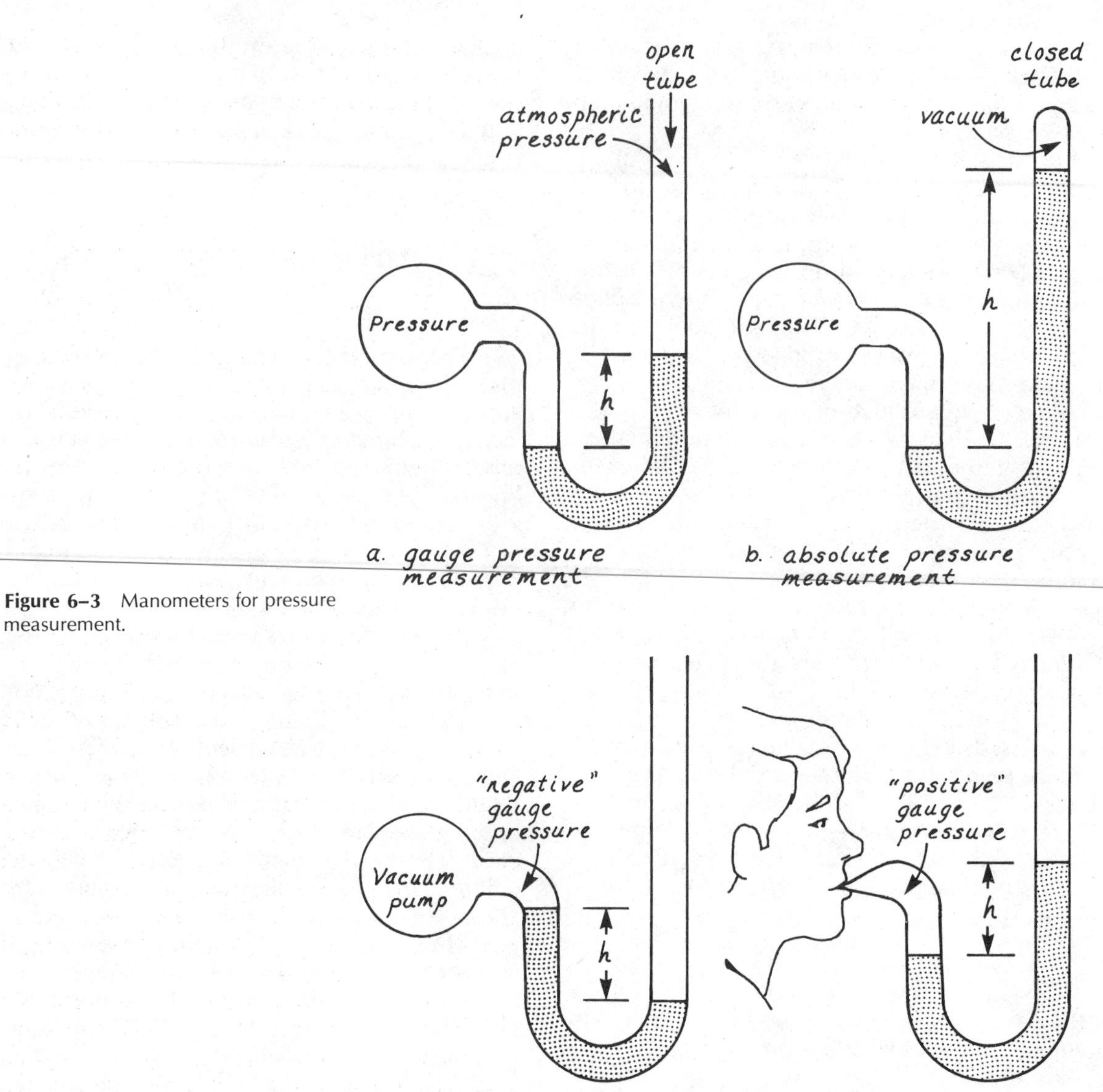

Figure 6–3 Manometers for pressure measurement.

of 20 cm H_2O, with positive indicating an excess over atmospheric pressure. If the manometer is closed and a vacuum produced in the closed end as in Figure 6–3(b), the liquid level difference is a measurement of the *absolute pressure* of the pressure source. (The vacuum must be maintained, of course, to obtain accurate measurements; a leak would destroy the accuracy of the measuring instrument.) Figure 6–3(c) shows the use of a manometer for vacuum measurements. If h is 10 cm, the pressure is said to be -10 cm H_2O or a 10 cm H_2O "negative" pressure since it is below atmospheric pressure. Water manometers are convenient for the measurement of lung pressures. Figure 6–3(d) shows the measurement of the positive gauge pressure during exhalation; the manometer could also be used for measuring the negative inspiratory pressure.

APPLICATION OF ATMOSPHERIC PRESSURE

The combined effects of the atmospheric pressure and the application of the ideal gas law provide the basis for the operation of many common devices. For example, consider the process of filling a medicine dropper or a hypodermic syringe. The pressure which causes the medicine dropper to fill up is that of the atmosphere acting downward on the liquid surface. It can force the liquid into the dropper if the bulb has been squeezed to expel some of the air, thus reducing the pressure inside the bulb. This process is often described as if the lowered pressure "sucked" the liquid in. Though this may be a popular way of talking, it is somewhat misleading. It implies that a vacuum, an absence of matter, can exert a force. The fact is that the pressure of the surrounding atmosphere will force a fluid into any area of lowered pressure.

In the case of the syringe, when the plunger is pulled back the volume of the air below the plunger increases, and thus according to the ideal gas law the pressure will decrease. The pressure of the atmosphere acting on the liquid surface can then force liquid up through the needle to equalize the pressure (Figure 6–4). The process of drinking a liquid through a straw is similar, and is likewise made possible by the atmospheric pressure acting on the liquid surface.

The Siphon

To remove liquid from a large container as shown in Figure 6–5, it is often convenient to use a siphon. If the outlet end of the siphon tube is below the surface of the liquid, a continuous flow can be maintained. When the air is evacuated from the tube, the atmospheric pressure acting on the liquid surface pushes liquid up into the tube. Once the tube is filled with liquid that is allowed to flow, the flow out the bottom under the influence of gravity will maintain a reduced pressure in the tube. If the liquid surface is open to the atmosphere, liquid will continue to be pushed into the tube and to flow out the bottom. If there were a perfect vacuum in a vertical tube and if the liquid were water, the atmospheric pressure would be capable of pushing the water up to 34 ft above the open liquid surface. If air were prevented from entering a filled siphon tube and water flowed out the bottom so that a liquid-free space were created near the top of the siphon, that space would represent a fairly good vacuum and liquid would be pushed by the atmospheric pressure from the open container to fill it. In principle, one could siphon water over a high fence. A more detailed analysis of the pressure variations in a siphon requires an application of the conservation of energy principle to the liquid flow process and will be undertaken in Chapter 8.

In addition to its use for transferring liquids, the siphon can be used to provide suction for removing excess body fluids. For example, suction bottles can be attached as shown in Figure 6–6. When water is siphoned from bottle 2 to bottle 3, the volume of air in bottle 2 increases. According to the ideal gas law, there will be a corresponding decrease in pressure in both bottle 2 and bottle 1, since they are connected. Thus the siphon action produces a suction on the patient while the siphon is running. Though the suction is not easily regulated and not really comparable to the commercial suction machines available, it could be used in emergency situations.

The action of a siphon is also important in the adaptation of the Munro Tidal Drainage apparatus illustrated in Figure 6–7. It is used for irrigation of the bladder. It periodically fills and empties the bladder, using a fluid such as normal saline, a boric acid solution, or other

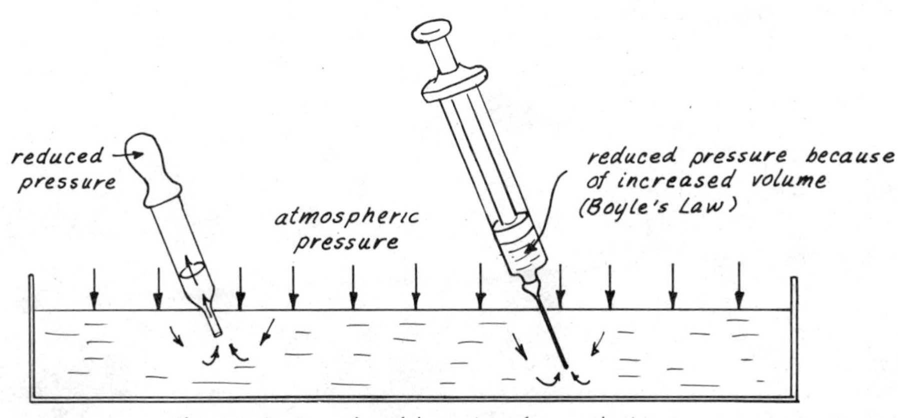

Figure 6–4 Examples of the action of atmospheric pressure.

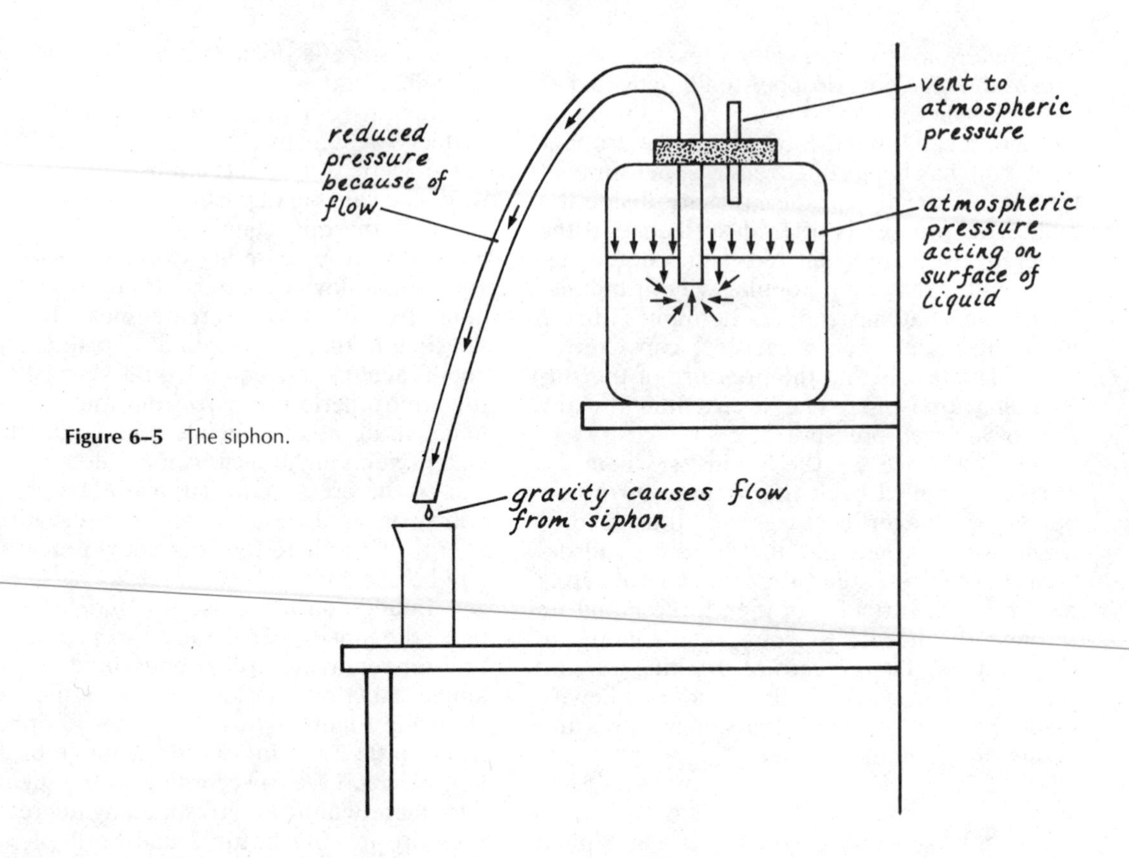

Figure 6–5 The siphon.

commonly used irrigating fluids. Siphon action is initiated by the filling of the bladder and drains the fluid out. When the drainage is complete the siphon is broken and the bladder fills again. The time for filling and emptying the bladder can be controlled and is typically 2 to 3 hours.

As the irrigating fluid moves downward

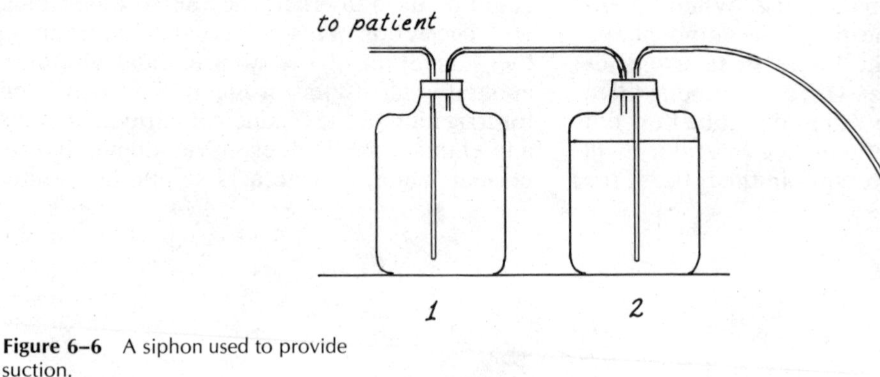

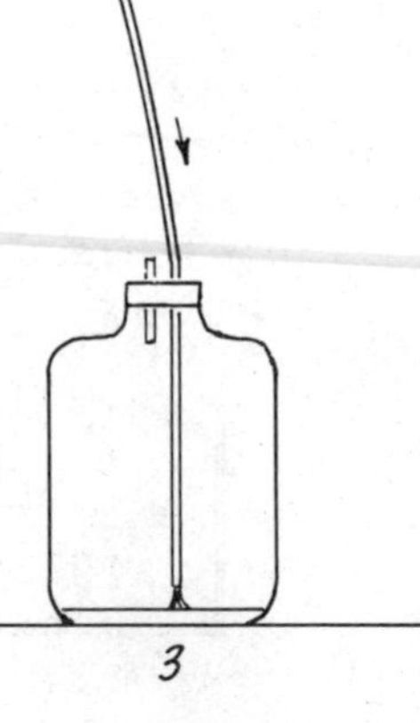

Figure 6–6 A siphon used to provide suction.

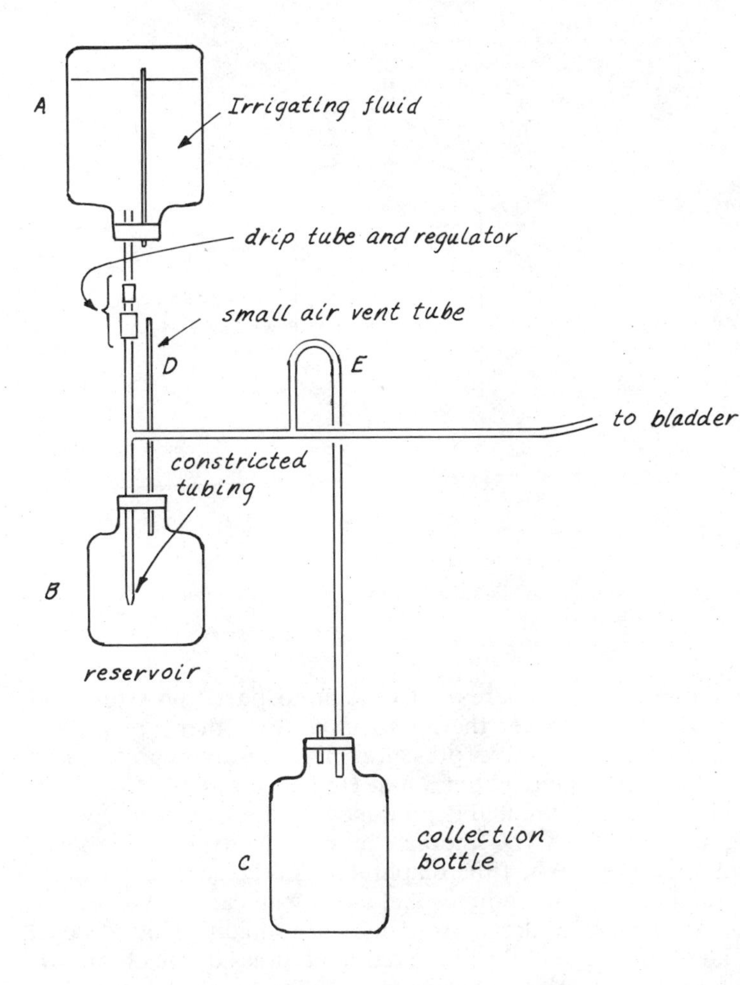

Figure 6–7 An adaptation of the Munro Tidal Drainage apparatus.

from bottle A it has two paths for flow: (1) to the reservoir B and (2) to the bladder. The air vent tube D is necessary to allow flow to reservoir B, because otherwise the air pressure would increase and cause a back pressure to stop the flow. The rate of flow into the reservoir B is limited by a constriction of the end of the tubing to make sure that enough fluid flows to the bladder of the person being treated; the size of this constriction is important because it controls the division of the liquid between the reservoir and the bladder. As the bladder fills and causes a back pressure on the liquid, the liquid level rises to the top of the siphon loop (E) and initiates a siphon action, which will then start to empty both the bladder and the reservoir bottle B. When the bladder and reservoir have been emptied to the point where the input tube in the reservoir (B) is exposed, air from the vent tube (D) will enter the system and stop the siphon action. The filling process then begins again and the cycle is repeated. The rate of filling can be controlled by a clamp on the tubing below the supply bottle.

The Mechanism of Breathing

The process of inhaling and exhaling air is an application of the ideal gas law. Although the temperature of the air will change slightly, the process is closely approximated by the constant temperature case of the law (equation 6–3, Boyle's law). The process can be illustrated by means of a simplified lung model as shown in Figure 6–8. The model consists of a bottle with its bottom covered by a rubber membrane. The bottle has a cork with an inlet tube, and the lungs are represented by two small rubber balloons which are connected to the atmosphere by an open tube.

When the membrane is stretched downward, the volume of the bottle is increased and the pressure inside the bottle decreases. The atmospheric pressure will then push air in to partially inflate the balloons. This approximates the process of inhalation, in which the diaphragm moves downward, increasing the volume of the thoracic cavity.

If the membrane is pushed upward, it de-

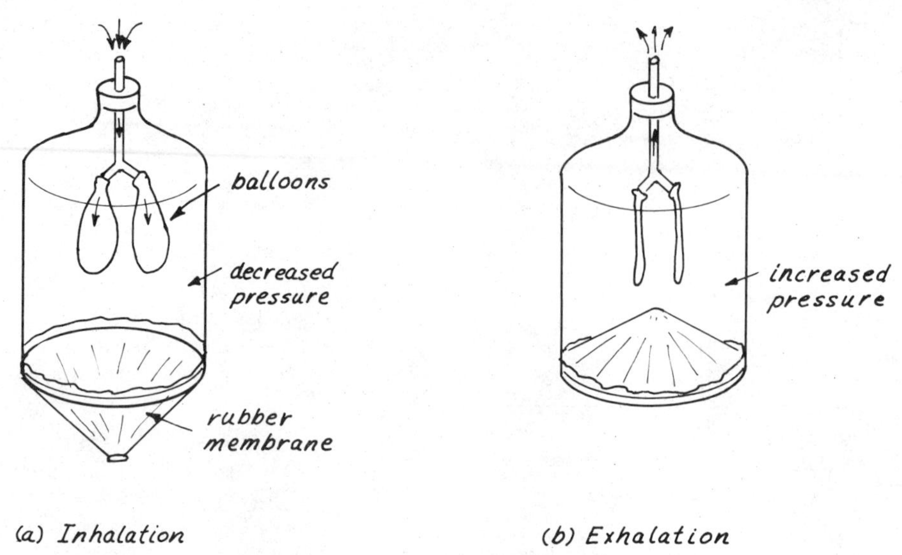

Figure 6–8 Simplified model of the lung system.

creases the volume of the bottle and the pressure must increase. The added pressure flattens the balloons and expels the air to the atmosphere. This is analogous to the exhalation process, in which the diaphragm is relaxed and moves upward, decreasing the volume of the thoracic cavity. The analogy is limited, of course; the lungs do not collapse completely during the expiration process. The process is accomplished mainly by the elastic rebound of the lungs and chest wall and does not require a muscular contraction to squeeze the lungs.

It is clear that the pressure of the thoracic cavity must be below atmospheric pressure during inspiration. This pressure is known as the intrathoracic pressure or intrapleural pressure. This intrapleural pressure drops to about 756 mm Hg, or about 4 mm Hg (5.5 cm water) below normal atmospheric pressure, during inspiration. It is a bit surprising to note that the intrapleural pressure stays 2 to 3 mm Hg below atmospheric pressure even during the exhalation process. Since the normal pressure is always below atmospheric pressure, opening the thoracic cavity to the atmosphere can result in a collapsed lung. This is the reason for using closed drainage systems when fluid must be removed from the thoracic cavity (see section on water-sealed drainage).

The pressure in the lungs (intrapulmonary pressure) must be below atmospheric pressure during inhalation and above atmospheric pressure during exhalation. That is, to get air to move into the lungs there must be a negative pressure relative to atmospheric pressure, and to get the air to move out, there must be a positive pressure. The negative pressure is only about 3 mm Hg below atmospheric pressure and is produced by the increased volume of the thoracic cavity as explained above. When the diaphragm is relaxed, the elasticity of the lungs and chest wall causes the volume of the lungs to decrease rapidly. This causes a momentary excess of pressure over atmospheric pressure (usually about +3 mm Hg) which expels the air.

Respirators and Ventilators

When a patient has difficulty breathing because of paralysis, neurological insufficiency, or other causes, some mechanical means must be employed to produce the pressure variations in the lungs necessary for moving air in and out. It is usually sufficient to aid only the inspiration process, since the elastic contractions of the chest wall and lungs are sufficient to accomplish the expiration of the air.

A variety of mechanisms have been used, from the large "iron lung" type respirators to emergency resuscitators. The iron lung type respirators produce a negative pressure over the patient's entire body with the exception of the head, and thus aid chest expansion and inspiration. The pressure then returns to normal atmospheric pressure to allow expiration.

Recent practice has shown that positive pressure ventilators have many advantages over the older, negative pressure machines. Air or oxygen can be caused to enter the lungs either by expanding the chest via a negative pressure or by applying the gas at a pressure greater than atmospheric pressure (positive pressure) to force it into the lungs. The difficulty with positive pressure ventilation is that it must be cycled according to the patient's needs to allow expiration of the air. This cycling has been accomplished by the development of high precision pressure-sensitive or flow-sensitive valves. The general form of therapy when such devices are used is called intermittent positive-pressure breathing (IPPB). The mechanical devices used to accomplish this ventilation are often called IPPV's (intermittent positive-pressure ventilators).

The development of the field of respiratory therapy has involved the application of an impressive amount of modern technology to the task of aiding the breathing process. Incorporated in the ventilating equipment may be flow-sensitive valves such as the Bennett valve (1) or pressure-sensitive valves such as the Bird valve (2), which are basically cycled by the pressure buildup in the patient's lungs.

Other alternatives are volume-cycled machines, which supply a preset volume of air to the patient's lungs, subject to a maximum pressure. Some ventilators make use of the gas flow effects known as "fluidics," which are described briefly in Chapter 8. The continuing development of microcomputers and the digital electronics industry opens many possibilities for sophisticated time-cycled units, which can be programmed to respond to a variety of clinical conditions. These ventilators make use of the properties of gases described in this chapter; for more detailed descriptions the interested reader is referred to respiratory therapy texts (3,4).

Water-Sealed Drainage

Liquids and gases can be removed from the body by producing a small negative gauge pressure in a drainage tube (suction). Since the interior of the body is at or near atmospheric pressure, fluids will move through the tube in the direction of the lowered pressure. However, drainage of the thoracic cavity is complicated by the fact that this cavity is always below atmospheric pressure, even during exhalation. This negative pressure in the thoracic cavity exerts a suction action which keeps the lung tissue expanded close to the rib cage. Air at atmospheric pressure must be prevented from entering the cavity by way of the drainage tube because that would cause the lung to collapse.

The necessary seal of the drainage tube can be accomplished by submerging the tip of the drainage tube as shown in Figure 6–9. The water in the drainage bottle will prevent air from entering the drainage tube from the outside. If atmospheric pressure enters the drainage bottle, it will push water up into the drainage tube a short distance because of the negative pressure of the thoracic cavity. This column of water will balance the external pressure and prevent air from entering. When a negative pressure is produced in the suction bottle by the suction pump, the water level in the tube will normally drop somewhat below the liquid level in the bottle.

When the chest cavity is opened during thoracic surgery, air at atmospheric pressure collapses the lung. Water-sealed drainage is used post-operatively to remove the excess air as well as collected liquids to restore the normal negative pressure. When the water-sealed drainage is operating properly, this excess air will bubble out through the water in the drainage bottle during expiration. As the patient breathes, the water level in the tube will move up and down with the respiration cycle. When the patient is relieved of unwanted air and fluid, the negative pressure is gradually restored and the collapsed lung expands again.

It is important to regulate the amount of

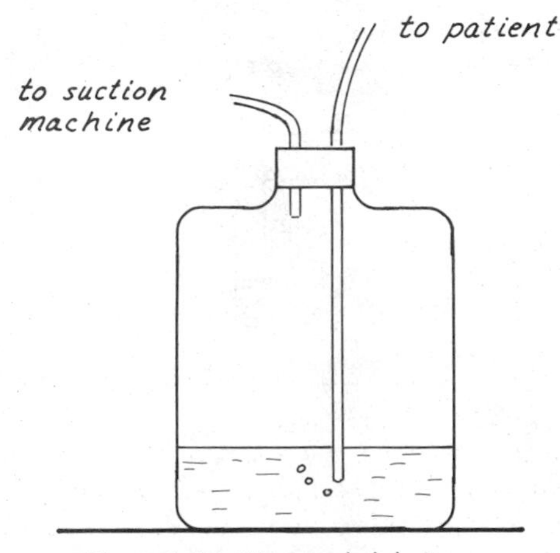

Figure 6–9 Water-sealed drainage.

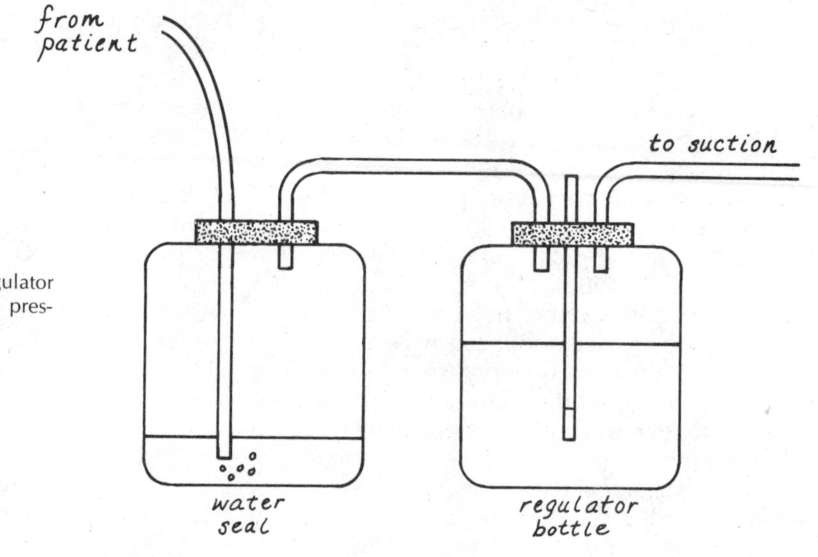

Figure 6–10 Use of a regulator bottle to limit the negative pressure.

negative pressure applied in the suction of the thoracic cavity, since too much negative pressure can be traumatic to the system. Although there are mechanical pressure regulators available, this regulation can be achieved in a straightforward way by the use of a regulator bottle as illustrated in Figures 6–10 and 6–11.

The regulator bottle has openings to allow free flow of air through the bottle to the suction pump, but it is not open to the atmosphere. The control tube is inserted in the bottle with its lower end below the water level and the upper end open to the atmosphere. When a negative pressure is produced in the regulator bottle, the water level in the control tube drops below the level in the bottle, as illustrated in Figure 6–11(c). The negative pressure of suction devices is conveniently expressed in cm of water. If the negative pressure in the bottle is -10 cm H_2O, then the water level in the control tube drops 10 cm below the level in the bottle. If it is desired to limit the negative pressure applied to the patient to -12 cm H_2O, then the control tube is submerged to a depth of 12 cm in the water. If the pump produces a pressure lower than -12

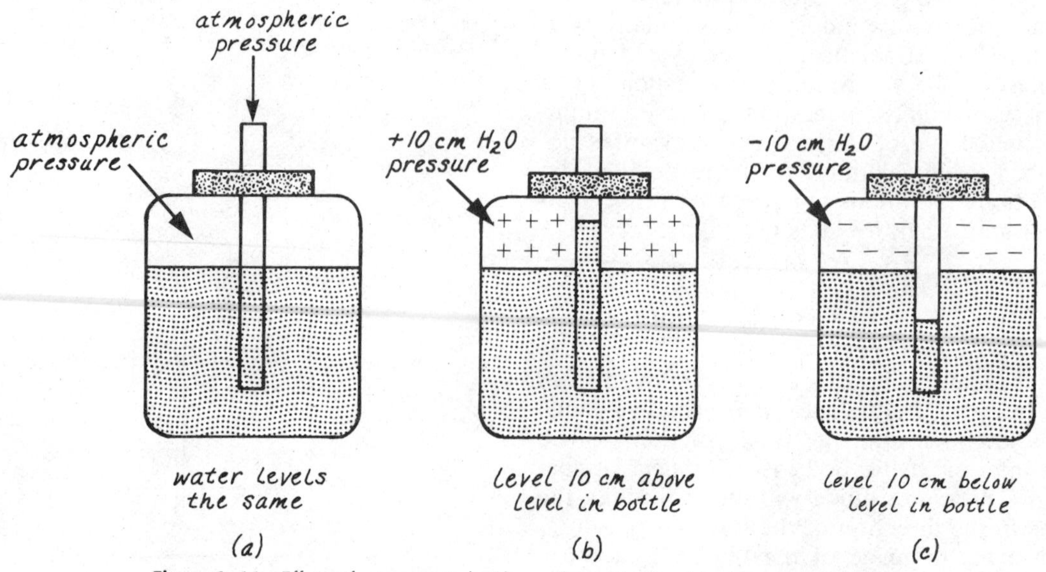

Figure 6–11 Effect of pressure in bottle on the water level in the regulator tube.

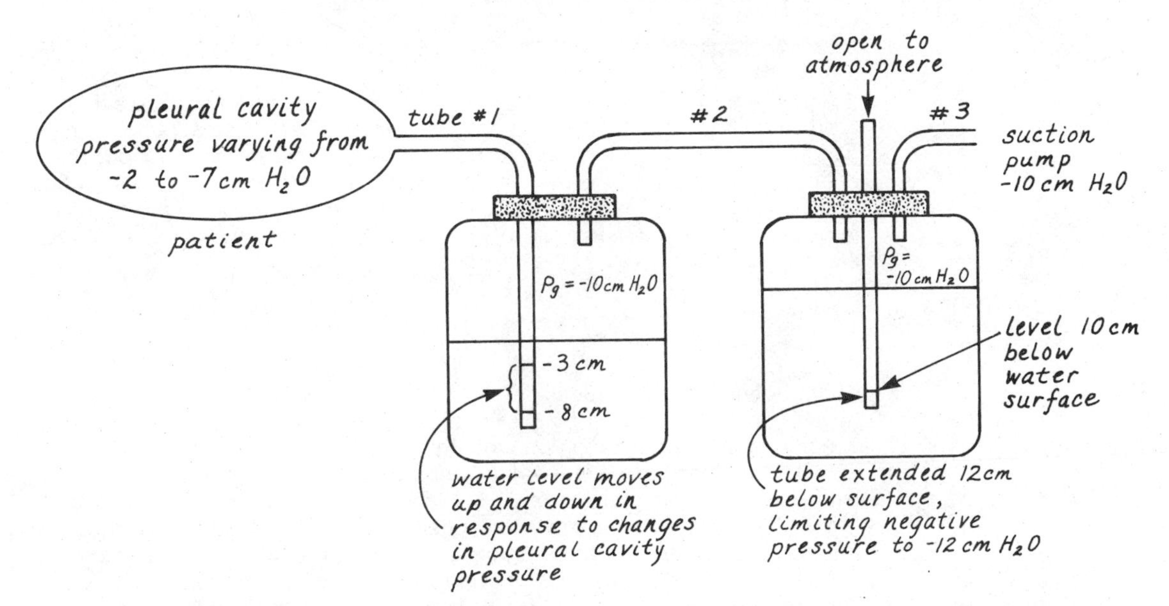

Figure 6–12 Operation of a water-sealed drainage apparatus with a regulator bottle to limit negative pressure.

cm H_2O, the water level in the control tube will drop to the bottom of the control tube and air will bubble in from the atmosphere, raising the pressure back up to the limiting value (-12 cm H_2O). Thus the pump can produce any negative pressure up to -12 cm H_2O, but no more, because air will bubble in from outside to keep it at that level.

The operation of a water-sealed drainage apparatus with a collection bottle and a pressure regulator bottle is summarized in Figure 6–12 with some representative numerical values. It is assumed that a suction pressure of -10 cm H_2O is produced by the pump and that the regulator bottle would prevent any suction pressure lower than -12 cm H_2O. If the pleural cavity pressure varied from -2 cm H_2O to -7 cm H_2O, corresponding exhalation and inhalation, the water level in the water-seal tube would vary from 8 cm to 3 cm below the water level in the bottle. The level in this water-seal tube measures the pleural cavity pressure relative to the suction pressure; if the pleural cavity pressure is -2 cm H_2O the level will be -10 cm $-$ (-2 cm) $=$ -8 cm, and for the -7 cm H_2O inhalation pressure the level will be -10 cm $-$(-7 cm) $=$ -3 cm. This tube is very useful for monitoring the patient's breathing process.

Note that the seal of tube #1 in Figure 6–12 is critical; a break of this seal would allow atmospheric pressure to enter the tube and collapse the lung. The removal of tube 2 or 3 would allow atmospheric pressure air to enter the water-seal bottle but it would not reach the thoracic cavity. The water level in the water-seal tube would now oscillate between 2 cm and 7 cm above the water level in the water-seal bottle, but this would present no threat to the patient.

The regulator bottle has been shown here in conjunction with the closed drainage of the thorax, but it can also be used with open drainage when a limit on the negative pressure is desired. The two bottle system is shown in Figure 6–10 for the explanation of the principles, but many other configurations exist which use the same principle. Some commercially available drainage setups use a molded plastic container with several compartments. These compartments accomplish the regulation of water-seal function and usually have separate compartments for the collection and measurement of exudate.

SUMMARY

The intermolecular attractive forces between gas molecules are extremely small compared to those in liquids and solids. When these attractive forces can be neglected, all gases exhibit the same behavior in response to changes in pressure, volume, and temperature, since the only interactions between the molecules are collisions. When the attractive forces are assumed to be zero, the gas is said to approximate an "ideal gas" and is found to obey the relationship

$$\frac{PV}{T} = nR$$

which is known as the ideal gas law. Most ordinary gases and mixtures of gases can be described by the ideal gas law when they are not near the conditions for condensation into the liquid state.

The weight of the atmosphere exerts a pressure of about 14.7 lb/in^2 on every exposed object at sea level. If a lower pressure is produced by removing air from any volume, this atmospheric pressure will tend to push air or liquid into the volume to produce a pressure equilibrium. This pressure accounts for the effectiveness of syringes, medicine droppers, siphons, and various types of suction devices which make use of the prevailing atmospheric pressure to move liquids. The height to which a liquid will be pushed in an evacuated vertical tube can serve as a measurement of the atmospheric pressure. The standard atmospheric pressure is usually stated as 760 mm Hg, since it will push mercury to a height of 760 mm in a mercury manometer.

The mechanism of breathing can be explained with the aid of the ideal gas law and the concept of pressure equilibrium. The expansion of the pleural cavity lowers its pressure, leading to a lowered pressure in the lungs and subsequent inspiration of air. In the case of breathing difficulty, a respirator may supply air at a pressure greater than atmospheric pressure (positive pressure) to introduce air into the lungs. The elastic contraction of the lungs and chest wall produces a positive pressure in the lungs and accomplishes the expiration of the air. Since the pleural cavity pressure is normally below atmospheric pressure (negative pressure), closed drainage systems must be used to remove liquid from that cavity to prevent the entrance of atmospheric pressure air which would collapse the lungs.

REVIEW QUESTIONS

1. Why does warm air rise?
2. Why does a bubble in water get larger as it rises?
3. What holds the liquid in a medicine dropper?
4. Why is it dangerous to throw an aerosol can in a fire?
5. Why does water not escape freely when a narrow mouthed bottled is suddenly inverted?
6. When you drink through a straw, what is the origin of the force which pushes the liquid up the straw?
7. The atmosphere exerts a force in the neighborhood of 12 tons on your body. Why are you not aware of it?
8. What is the origin of the force which causes water to move through a siphon?

9. What is meant by negative pressure?

10. The lungs are not attached to the pleural walls. Why do the lungs not collapse?

11. If the pleural cavity is always maintained at a negative pressure, how do the lungs achieve the necessary positive pressure for exhalation?

12. Why is water rather than mercury used in a manometer for the measurement of a spinal tap pressure or a central venous pressure?

13. Why is mercury rather than water used in a manometer for the measurement of atmospheric pressure (barometer)?

14. What are the pressure units for the following quantities: (a) a systolic pressure of 120, (b) an auto tire pressure of 30, and (c) a barometric pressure of 30?

PROBLEMS

Worked Example: An automobile tire has a gauge pressure of 30 lb/in^2 at a temperature of 25° C. If the tire is heated to 50° C by high speed driving while the volume remains constant, how much change in tire pressure will occur?

Solution: All such problems involving a constant mass of gas can be solved by application of the ideal gas law in the form:

$$\frac{P_i V_i}{T_i} = \frac{P_f V_f}{T_f}$$

In this case the volume is constant and therefore can be canceled from the equation. The pressure is equal to the gauge pressure plus atmospheric pressure:

$$P_i = 30 \text{ lb/in}^2 + 14.7 \text{ lb/in}^2 = 44.7 \text{ lb/in}^2$$

$$P_f = ?$$

The temperatures must be converted to the absolute temperature scale:

$$T_i = 25° \text{ C} + 273 = 298° \text{ K}$$

$$T_f = 50° \text{ C} + 273 = 323° \text{ K}.$$

Substituting,

$$\frac{44.7 \text{ lb/in}^2}{298° \text{ K}} = \frac{P_f}{323° \text{ K}}$$

$$P_f = (44.7 \text{ lb/in}^2)\left(\frac{323}{298}\right) = 48.5 \text{ lb/in}^2.$$

This is equal to a gauge pressure of 33.8 lb/in^2 and represents a pressure rise of 3.8 lb/in^2 as a result of heating the air.

1. A mass of 32 grams of oxygen (one mole) has a volume of 22.4 liters at atmospheric pressure and 0° C. If it is compressed until it occupies 10 liters, what is the pressure exerted upon it? (Temperature is assumed constant.) What would be the volume occupied by one gram of oxygen at the standard temperature and pressure?

2. The air in a sealed constant volume has a pressure of 1 atmosphere at 0° C. What will be its pressure at 100° C? At what temperature will the pressure be 2 atmospheres?

3. An oxygen cylinder holds one cubic foot of oxygen at a pressure of 2200 lb/in^2. How many cubic feet would the gas occupy at atmospheric pressure (temperature constant)?

4. A fluid bottle is inverted with a tube suspended from the bottle so that the open end of the tube is one meter below the liquid surface in the bottle. The bottle contains 100 cm^3 of air at atmospheric pressure but no vent. Assume that the open end of the tube remains 100 cm below the liquid surface.

   a. Find the equivalent of atmospheric pressure in cm of water.
   b. Find the total pressure acting to push the fluid out of the tube at the bottom.
   c. Assuming that the lower end of the tube is open at atmospheric pressure, calculate the number of cm^3 of liquid which will flow out of the tube before the flow is halted by the partial vacuum in the inverted bottle.

5. Suppose a person can cause a negative pressure in the lungs of −70 mm Hg by a strong inspiratory effort. How far could he suck water up a straw?

6. The gauge pressure of an automobile tire is 30 lb/in^2. If the volume of the tire is one cubic foot, how many cubic feet of air at atmospheric pressure had to be compressed to fill the tire?

7. If two metal hemispheres of diameter 5 inches were placed together so that all the air could be pumped out of the inside of the sphere, how much force would be required to pull them apart? (Hint: only the air pressure components perpendicular to the seal will resist the separation; therefore, the effective area for force exerted by air is just the area of a circle.)

8. Find the force exerted by the atmosphere on a floor 12×15 feet. Why does it not collapse?

9. If a gas occupies a volume of 10 liters at a pressure of 760 mm Hg, what will be its volume at 380 mm Hg if the temperature is held constant?

10. A gram molecular weight (mole) of air has a mass of approximately 29 grams and occupies a volume of 22.4 liters at 0° C and 760 mm Hg pressure.

    a. What would be its volume if it were compressed into a tank at a pressure of 2200 lb/in^2 gauge pressure at 0° C?
    b. What would be the mass of the air in a 100 liter tank under the conditions in (a)?

11. An oxygen cylinder has a gauge pressure of 2200 lb/in^2 at 27° C. What would be its pressure if the cylinder were heated to 77° C (constant volume)?

12. Given 1000 cm^3 volume of gas at 22° C (room temperature), what volume would it occupy at body temperature (37° C) if the pressure is unchanged?

13. Given a 1000 cm^3 volume of gas at 27° C, what temperature would be required for that gas to occupy a volume of 500 cm^3 if the pressure were constant?

14. An inverted container of water is connected to a long tube so that water can drain out but no air can enter. Initially there is a trapped volume of 200 cm^3 of air at the top of the water which is at a pressure of 1000 cm H_2O. Conditions are such that the flow will stop when the pressure of the trapped air drops to 800 cm H_2O. How much water will flow out? (Assume constant temperature.)

15. A volume of 8 liters of gas has a pressure of 15 lb/in^2 and a temperature of 27° C.

    a. If the pressure were increased to 90 lb/in^2 without changing the temperature, what would be the resulting volume?
    b. If the temperature were raised to 627° C while maintaining the original pressure, what would be the final volume?
    c. What would be the resulting volume if the temperature were raised to 627° C and the pressure were increased to 90 lb/in^2 simultaneously?

16. Given 200 cm^3 of gas at 37° C and 760 mm Hg pressure, what would be its volume at 100° C and 850 mm Hg pressure?

17. The standard atmospheric pressure in the units reported by U.S. weather forecasters is 29.9 inches of mercury, which corresponds to 14.7 lb/in^2. If a well-sealed house maintains this air pressure as a tornado comes by, dropping the barometric pressure briefly to 27 inches of mercury, what outward force would be exerted on an 8 ft $\times$ 20 ft wall as a result of the pressure difference?

18. If a 100 liter tank of oxygen has a gauge pressure of 2200 lb/in^2 and a temperature of 27° C, what is the mass of the oxygen contained? (32 grams of oxygen occupies 22.4 liters at 0° C, 760 mm Hg.)

19. If a rising bubble of air has a volume of 1 cm^3 at a depth of 10 meters, what volume will it have just before it reaches the surface if its temperature remains constant? Assume that no gas is added to the bubble as it rises.

20. A pressure gauge at the bottom of an unpressurized water storage tank reads 20 lb/in^2. How deep is the water in the tank?

21. If a 10 ft^3 volume of air at 27° C and 760 mm Hg is taken to a high altitude where the pressure is 400 mm Hg and the temperature is -23° C, what volume will it occupy?

22. A tank of gas at 300° K is under a gauge pressure of 5×10^5 nt/m^2. If the temperature is raised to 600° K, what will the gauge pressure be?

23. An automobile tire is filled to a gauge pressure of 28 lb/in^2 on a cold day when the air and the tire are at 0° C. The automobile is then driven at high speeds and the tire temperature rises to 40° C. Assuming that the tire volume doesn't change, what will be the gauge pressure of the tire?

24. A 250 cm^3 volume aerosol can is empty and has an internal pressure of one atmosphere at 27° C.

    a. If the can is thrown into a fire and heated to 400° C, what will be the pressure in atmospheres?
    b. If the area of the top of the can is 6 in^2, what outward force will act on the top of the can at that temperature because of the difference in gas pressures?

25. A certain water pump can produce a vacuum (or "negative pressure" with respect to atmospheric pressure) of -300 mm Hg. Can this pump raise water out of a 10 ft deep well?

REFERENCES

1. Bennett Respiration Products Inc. *Instruction Manual for Model PR-2 Respiration Unit*. 1639 Eleventh St., Santa Monica, Calif.
2. Bird Corporation. *Training Manual 999–1610*. Palm Springs, Calif. 1970.
3. Egan, D. F. *Fundamentals of Inhalation Therapy*, 3rd ed. St. Louis: C. V. Mosby Co., 1977.
4. McPherson, S. P. *Respiratory Therapy Equipment*. St. Louis: C.V. Mosby Co., 1977.
5. Greenwood, M. E. *An Illustrated Approach to Medical Physics*. Philadelphia: F. A. Davis Company, 1966.
6. Johansen, J. L. *Taking the Mystery Out of Water-sealed Chest Drainage*. RN Magazine, Jan. 1960.

CHAPTER SEVEN

Pressure and the Circulatory System

INSTRUCTIONAL OBJECTIVES

After studying this chapter, the student should be able to:

1. Describe the functions of the various components of the heart during the blood-pumping process.

2. Describe the pressure variations of the blood as it moves through the heart and circulatory system.

3. Name the physical variables associated with the circulatory system which could be changed to increase the blood flow rate, indicate how they should be changed to accomplish the increase, and rank them in order of effectiveness in flow rate control.

4. Describe the conditions under which the blood flow departs from Poiseuille's law.

5. Calculate the velocity of flow through a tube, given the volume flow rate and the area of the tube.

6. Explain how the thin-walled capillaries can withstand pressures comparable to those in the thick-walled aorta.

7. Explain the physical origin of the systolic and diastolic pressures and the implications of a large difference between them.

The circulatory system basically consists of the heart plus the network of arteries, veins, and capillaries. It would be easy to broaden the scope of things included under "circulatory system" since the lungs are a fundamental link in the circulatory process, and so on, but this chapter will deal primarily with the heart and the vessels which carry the blood. It is a drastic oversimplification to view the system as a pump circulating fluid through a closed network of pipes, but such a model will serve as a starting point for discussing pumps and pressure variations in flowing liquid systems. This model will then be expanded to more closely approximate the actual circulatory system.

TYPES OF PUMPS

The function of a pump is to increase the pressure in a fluid to enable the fluid to move in the desired manner. Simple pumps might be classified as either lift pumps or force pumps. Sketches of simple examples are shown in Figure 7–1.

As the piston of the *lift pump* rises, the volume of air below it increases, causing a decrease in pressure. Excess pressure outside this cylinder opens the valve at the bottom and water flows in. The excess pressure is provided by atmospheric pressure acting down on the water surface, so the column below the pump is similar to a water manometer. Since atmospheric pressure provides the force which moves the water into the pump cylinder, the pump cannot operate at heights greater than 34 feet above the water surface. (Atmospheric pressure will push water 34 feet high into a completely evacuated column.) After water is in the cylinder in Figure 7–1(a), the piston is lowered. The pressure of the water opens the valve on the top of the piston and moves water above it. On the next lift of the piston this water flows out the exit pipe.

The *force pump* is more versatile since it need not rely on atmospheric pressure directly. When the piston in Figure 7–1(b) is raised, the fluid pressure tends to decrease and any positive pressure in the fluid in the column below the pump will open the valve and cause flow into the pump cylinder. Such a pump can be used to circulate liquids in a sealed, closed liquid circuit since the remaining liquid pressure of the returning liquid can open the valve for liquid to enter the pump. The force pump does not rely on atmospheric pressure to push the liquid into the pump as does the lift pump. When the cylinder is pushed downward, the increased liquid pressure in the cylinder closes the inlet valve and opens the outlet valve. The liquid is thus pumped out through the exit pipe by the pressure of the piston and there is no height limitation like that on the lift pump. The liquid is prevented from flowing back into the pump cylinder when the piston is lifted because the liquid pressure in the exit pipe closes the exit valve. Variations of the force pump are used in deep wells which pump oil or irrigation water.

The compressed air pump used to inflate such things as air mattresses is a type of force pump even though it uses the atmospheric pressure to get air into it (Figure 7–2). The operation of the compressed air pump can be easily understood by recalling the ideal gas law. The increased volume on the upstroke causes a decreased pressure, and the atmosphere pushes air in. The increased pressure due to the volume-decreasing downstroke pushes the air out.

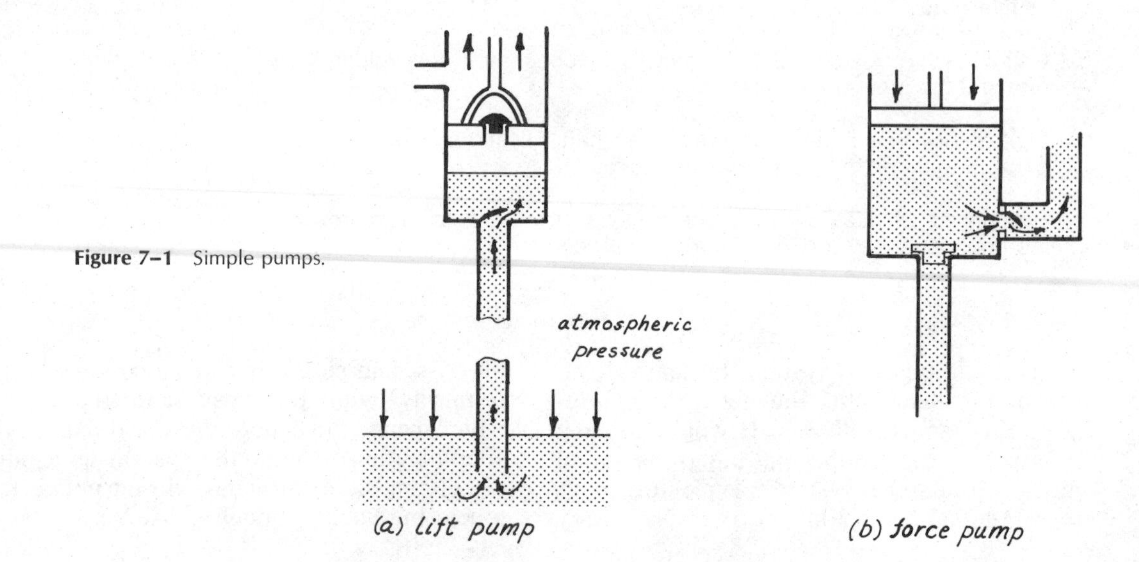

Figure 7–1 Simple pumps.

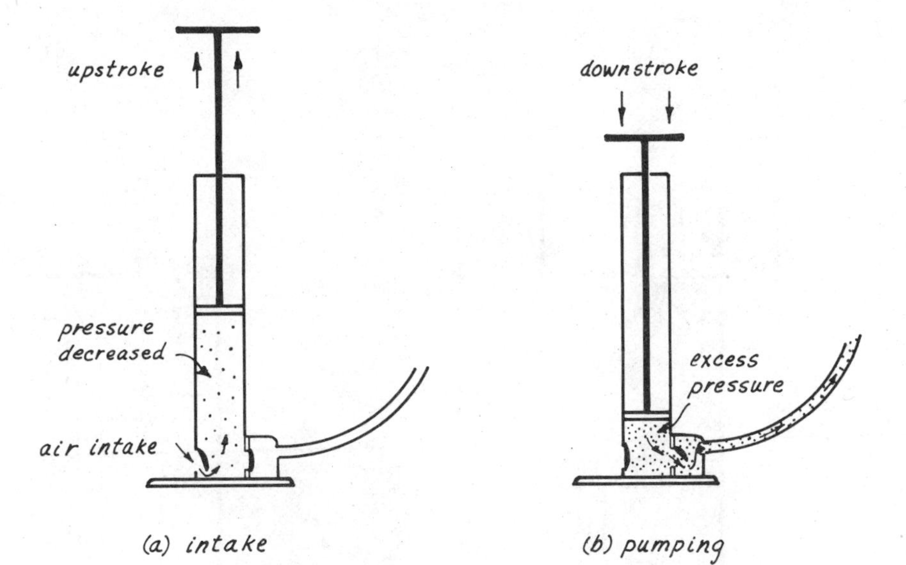

Figure 7–2 Operation of compressed air pump.

THE HEART AS A FORCE PUMP

The heart receives the blood from the venous system and raises its pressure to push it out into the arterial system. The action of the ventricles is that of a force pump which achieves the excess pressure by contraction. A rough analogy can be drawn to the simple force pump shown in Figure 7–3(a) and 7–3(b). In that pump the decreased pressure caused by pulling the plunger down enables the outside fluid pressure to open the snorkel-type ball valve, and fluid flows into the chamber. When the plunger is pushed in, the excess pressure closes the input valve and opens the output valve to force fluid out. During the next input stroke the output valve closes because of the reduced chamber pressure, so that no backflow can occur.

When the right ventricle expands, the valve from the right atrium (the atrioventricular valve) opens to let the venous blood flow into the ventricle, while the output valve (the semilunar valve) remains closed to prevent backflow of previously pumped blood. Upon contraction of the ventricle, the valves reverse and blood is pumped through the pulmonary artery to the lungs. The blood returning from the lungs undergoes a similar pumping process when the left ventricle expands and contracts, pumping the oxygenated blood out through the aorta. So instead of being one pump, the heart is two synchronous force pumps.

From this simplified picture, the effects of faulty heart valves can be seen. Many common faults involve valves which either have holes in them or do not close completely. If the right atrioventricular valve does not close completely, then the contraction of the right ventricle will pump blood back out into the venous system as well as into the pulmonary artery. A central venous pressure measurement would indicate this by a periodic high pressure in the vein. If a semilunar valve did not close properly, then the blood would backflow from the arterial system during the expansion of the ventricle. Either type of faulty valve obviously represents a great impairment of the heart's pumping action.

THE CIRCULATORY SYSTEM

When a pump circulates a fluid in a closed system, it raises the pressure of the fluid to enable it to overcome the system resistance. As the fluid flows through the system and back to the pump, the pressure drops back to the original pressure if the flow has reached a stable uniform rate. For example, if the pump takes water at zero gauge pressure and raises it to 40 lb/in^2 as it goes through the pump, the pressure would

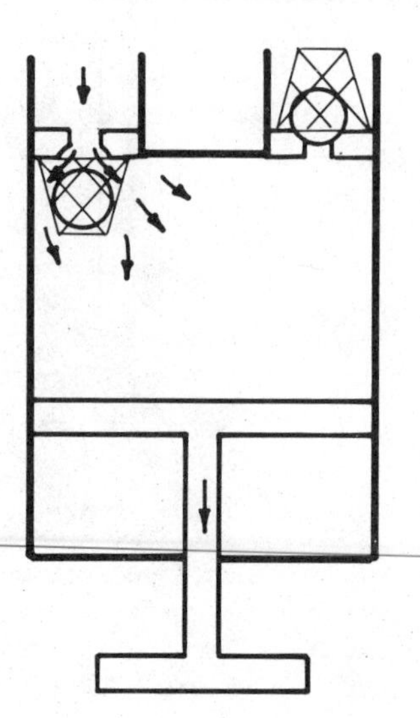

(a) Intake

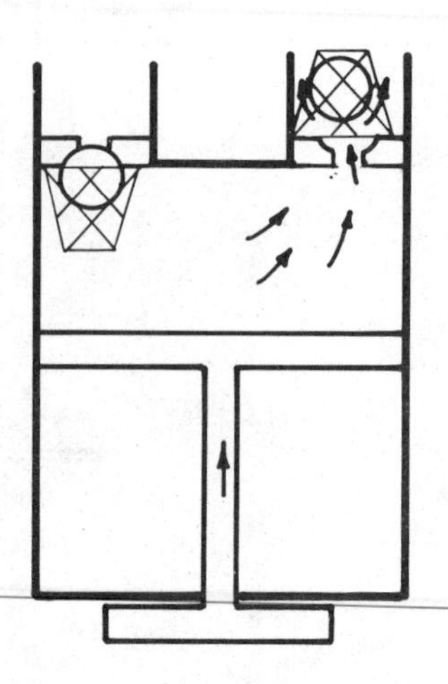

(b) Output

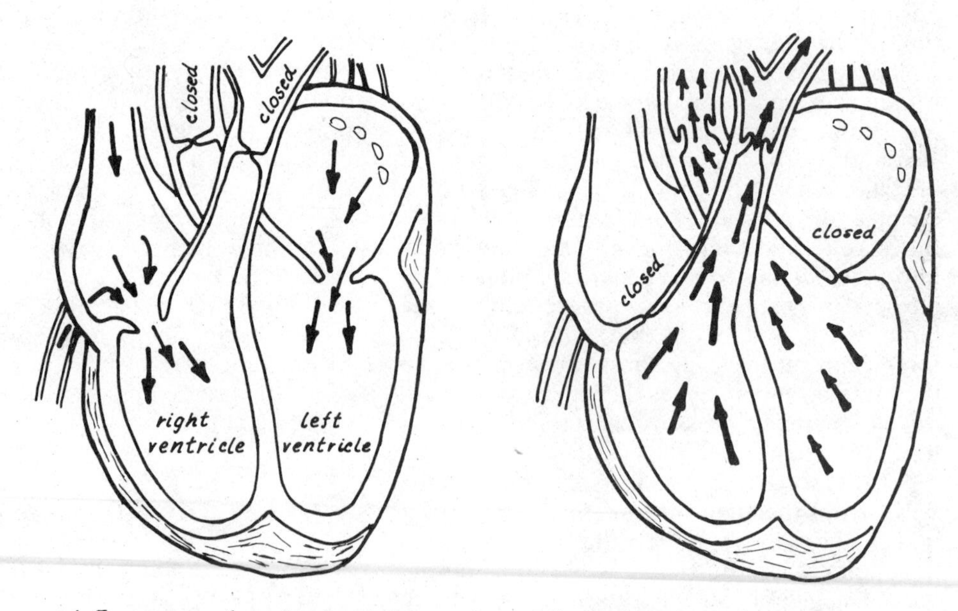

(c) Expansion of ventricles - intake

(d) Contraction of ventricles - output

Figure 7–3 Analogy between the heart and a force pump.

be expected to drop back to zero by the time the water circulates back to the pump. Otherwise, the pressure would continue to build up in the system. When the system reaches an equilibrium flow situation, there will be some base pressure at which the fluid enters the pump and a peak pressure at which the fluid leaves the pump.

In the circulatory system, the pressure is highest as it leaves the left ventricle and lowest as it enters the right atrium. Then the pressure is raised again by the right ventricle and drops to enter the left atrium at a low residual pressure. Figure 7–4 is a sketch showing some representative pressures at various locations in the system. The left ventricle is the main pump and supplies the pressure for the systemic circulation through the body. The amount of pressure drop which occurs in any segment of the circulatory system depends upon the flow rate and the resistance of that segment, according to Poiseuille's law. For a mean arterial pressure of 100 mm Hg, a rough indication of expected pressure drops can be calculated from experimental results using dogs (7): aorta, 4 mm Hg; large arteries, 5 mm Hg; branch arteries to termination ponts at arterioles, 15 mm Hg; arterioles, 39 mm Hg; capillaries, 26 mm Hg; venous system including venules, 7 mm Hg. This leaves a residual pressure of about 4 mm Hg at the entrance to the right atrium. The right ventricle raises the pressure again, e.g., from 4 to 25 mm Hg (Figure 7–4). This pressure is necessary to overcome the resistance in the pulmonary circulation. After passing this resistance, the blood enters the left atrium at a low residual pressure on the order of 8 mm Hg.

The fact that the pressure drop across the arterioles is greater than that across the capillaries is somewhat surprising, since the radius of a typical arteriole may be three times that of a capillary. Recalling Poiseuille's law from Chapter 5, pressure drop equals flow rate times resistance. The resistance is proportional to the tube length divided by the radius to the fourth power. If an individual capillary had a diameter of 8 microns (1 micron = 1 $\mu = 10^{-6}$ meters) and a length of 1 mm compared to a diameter of 20 microns and a length of 2 mm for an arteriole, the capillary would have a resistance to flow almost 20 times as great as that of the arteriole. However, the pressure drop depends upon the volume flow rate as well as the resistance, and since there are many times more capillaries than arterioles, the flow rate through each capillary is much smaller.

The Control of Volume Flow Rate

The heart pumps blood to supply oxygen and nutrients for the cells. If the demands of the cells are not being met, signals are generated which provide "feedback" to the heart and circulatory system. The supply

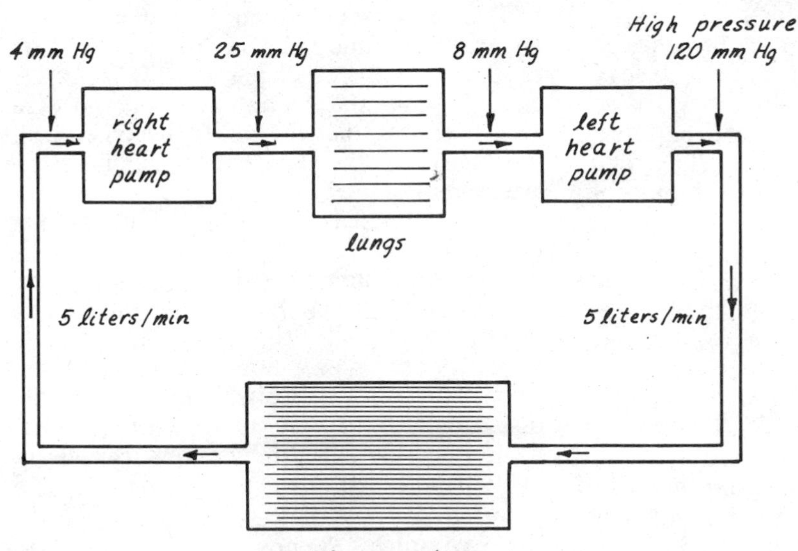

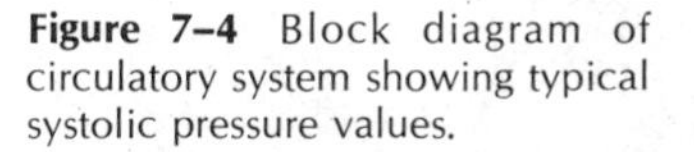
Figure 7–4 Block diagram of circulatory system showing typical systolic pressure values.

is increased by increasing the volume flow rate of the blood. The volume flow rate is then the primary variable to be controlled, and the circulatory system uses all means at its disposal to generate the proper volume flow rate. From Poiseuille's law (equation 5–6) we know that the physical factors which influence the volume flow rate through a tube are the pressure drop, radius, length, and the viscosity of the liquid. Of these variables, the circulatory system must use the blood pressure and the internal radii of the vessels for the short-term control of volume flow rate.

The arterioles are often called the "resistance vessels" of the circulatory system (8) and play an important part in controlling blood flow. They are surrounded by muscular cells which can produce large changes in the vessel diameters. Since the resistance to flow is so strongly dependent upon the diameter, such changes can control local blood flow to the tissues. Given a constant pressure, doubling the diameter of the arteriole increases the flow by a factor of 16, and only a 19% increase in the diameter will double the flow rate. This ability to reduce the resistance of the circulatory system enables the body to respond to the greater demands for blood during exercise without overburdening the heart pump. During moderate exercise the blood flow rate may increase by a factor of three while the blood pressure increases only a small percentage, indicating that the total resistance of the circulatory system has dropped to less than half its previous value. During vigorous exercise, a larger fraction of the blood flows to the muscles because of vasodilation in the muscle tissue. Vessels to the kidneys and digestive tract may actually constrict during this time of high demand for blood for the muscles, so that the body is not only controlling total volume flow rate but also the distribution of the blood (9).

Example: A person in a resting state has a systolic blood pressure of 120 mm Hg. He has a sudden demand for vigorous exercise (e.g., chased by a big dog), and the volume flow rate of his blood must increase to five times the resting value. (Ackerman (3) cites a tenfold increase in volume flow rate, but he doesn't say what is chasing his subject.) If there were no vasodilation, how much would his blood pressure have to increase to provide the blood flow? If no blood pressure increase occurred but all the blood vessels dilated by the same percentage, what percentage dilation would be required to handle the demand?

Solution: To achieve a fivefold increase in volume flow rate by an increase in the blood pressure alone would require a fivefold increase in blood pressure. Starting with 120 mm Hg, this would require a blood pressure of 600 mm Hg, which is physiologically unreasonable. Achieving a factor of five by vasodilation is feasible because the flow rate is proportional to the fourth power of the radius. The Poiseuille's law relationship (equation 5–6) can be written

$$\mathscr{F} = K(P_1 - P_2)r^4$$

for this volume flow rate control process, where K is a constant which includes the viscosity and length, which are presumed to remain the same. To get $5 \times \mathscr{F}$ we must find a radius r' such that

$$(r')^4 = 5r^4.$$

If $r' = cr$ then $(cr)^4 = 5r^4$ and therefore $c^4 = 5$. The multiple c is then

$$c = 5^{1/4} = 1.5.$$

Therefore a factor of 1.5 or a 50% increase in the internal radius of a vessel would give a fivefold increase in volume flow rate. Of course, not every vessel would have to increase 50% to achieve the increase in the actual case of whole body circulation. The larger arteries and veins contribute very little to the overall resistance to flow. As discussed above, the dilation would be necessary only in the smaller vessels, which contribute most of the resistance, such as the arterioles.

The point has been made that vasodilation can produce dramatic increases in the volume flow rate of the blood. It is also true that a reduction in the internal radius of a blood vessel will produce a dramatic *decrease* in the volume flow rate. Occlusion of arteries by fatty deposits, kinks or obstructions, or external pressures which squeeze on the vessels can produce very serious reductions in blood flow.

Example: Suppose a normal coronary artery has a volume flow rate of 100 cm³/min when the person's average blood pressure is 100 mm Hg. Calculate the flow rates if the internal radius of that artery is reduced to 80%, 50%, and then 20% of its normal value. If all blood vessels were similarly affected, what blood pressure would be required to restore the normal volume flow rate of 100 cm^3/min in each case?

Solution: Referring to the expression for volume flow rate in the previous example, the value $0.8r$ can be substituted to evaluate the effect of decreasing the radius to 80% of normal.

Effect on radius: $(0.8r)^4 = 0.41\ r^4$

Effect on volume flow rate:

$$\mathscr{F} = 0.41 \times 100\ \text{cm}^3/\text{min} = 41\ \text{cm}^3/\text{min}$$

To restore a normal volume flow rate you would have to divide the pressure by the factor 0.41 to overcome the effect of the radius reduction,

$$P = \frac{100\ \text{mm Hg}}{0.41} = 244\ \text{mm Hg}.$$

The same calculation for the other two cases yields the following results.

$r' = 0.5r$:
Effect on radius: $(0.5r)^4 = 0.0625\ r^4$
Effect on volume flow rate: $\mathscr{F} = 0.0625 \times 100\ \text{cm}^3/\text{min} = 6.3\ \text{cm}^3/\text{min}$
Pressure required to restore normal $\mathscr{F}$:

$$P = \frac{100\ \text{mm Hg}}{0.0625} = 1600\ \text{mm Hg}$$

$r' = 0.2r$:
Effect on radius: $(0.2r)^4 = 0.0016\ r^4$
Effect on volume flow rate:

$$\mathscr{F} = 0.0016 \times 100\ \text{cm}^3/\text{min} = 0.16\ \text{cm}^3/\text{min}$$

Pressure required to restore normal $\mathscr{F}$:

$$P = \frac{100\ \text{mm Hg}}{0.0016} = 62{,}500\ \text{mm Hg}.$$

Even a reduction to 80% of the normal internal radius requires an increase of blood pressure to physiologically dangerous levels to restore normal flow. With significantly greater reductions in radius, restoration to normal volume flow rates by increased blood pressure is not physiologically possible. Note that the above example assumes that *all* of the vessels of the body are occluded to the same degree when calculating the pressure. If the resistance of only one large artery is increased by obstruction or external pressure, the blood pressure may rise to try to maintain a normal blood volume flow rate through that artery. Then other parts of the arterial system may have to constrict to keep the volume flow rates through them from exceeding the needed levels.

With an understanding of the strong dependence of blood volume flow rates on the vessel radii, one can anticipate the effects of certain drugs that affect the circulatory system. The body will try to keep the blood volume flow rate constant if the demand is constant. If a vasodilator drug is given, a patient should be watched closely for sudden drops in blood pressure, since the body may lower the blood pressure in response to vasodilation to keep the volume flow rate constant. Conversely, vasoconstricting drugs may lead to a rise in blood pressure.

The Applicability of Poiseuille's Law

Since Poiseuille's law has been used as the basis for the above discussion, it seems appropriate to make some comment about its applicability. It was stated in Chapter 5 that blood flow did not obey Poiseuille's law. Studies (7) indicate that the large changes in blood viscosity with pressure which have been reported do not apply to physiological conditions. Though this may yet be subject to debate, it appears that within the normal range of blood pressures and speeds, the viscosity is approximately constant.

Departures from Poiseuille's law might be expected in the microcirculation of the capillaries. The blood plasma alone appears to act very nearly like an ideal fluid (Newtonian fluid), and it is the suspension of particles in it which is responsible for observed departures from Poiseuille's law. The red blood cells are apparently the main cause for such departures in behavior from Poiseuille's law in some of the small vessels. These red blood cells have normal sizes on the order of 8μ ($1\mu = 10^{-6}$ meters) and must move through vessels which may be 4μ or smaller in diameter (10). The red blood cells must then be distorted to get through, and would be expected to behave somewhat differently than when they are suspended in plasma such that the plasma is in contact with the vessel walls.

Other circumstances may arise in circulation to limit the applicability of Poiseuille's law. The law applies only for laminar flow, and it has been shown experimentally that for a given size tube there is a critical speed above which turbulence occurs. This turbulence increases the resistance dramatically, and a large increase in pressure is required to further increase the flow rate. The critical velocity for a long straight tube is given by

$$V_c = \frac{\mathcal{R}\eta}{pr}$$

where η is the viscosity in poise, p is the density of the fluid in gm/cm³, r is the radius of the tube in cm, and $\mathcal{R}$ is an experimental constant called the Reynolds number. For the blood the Reynolds number is about 1000. For the aorta, with a radius of about 0.9 cm, the critical velocity calculated from the equation is about 44 cm/sec. As shown below, the normal average blood speed in the aorta is about 33 cm/sec, but since it increases considerably during exercise, it could be expected that turbulence occurs in the aorta near the heart during exercise. However, even with some turbulence, the resistance of the aorta is quite small compared with the remainder of the circulatory system. Apparently the amount of turbulence at other points in the system is minimal under normal conditions; but, as discussed in Chapter 5, the presence of obstructions or partial occlusions of the vessels can produce turbulence.

Changes in Blood Speed During Circulation

A wide range of blood speeds is observed as the blood moves through the systemic circulation. The speed is a maximum in the aorta, drops to a minimum in the capillaries, and accelerates to a fairly high speed in the major veins leading to the heart. Since the normal volume flow rate of the blood is known to be about 5 liters/min, the average speed of the blood in the aorta can be calculated. The volume flow rate in cm³/sec is equal to the area in cm² times the average speed of flow in cm/sec:

$$\text{volume flow rate } \mathcal{F} = A\bar{v}. \qquad \textbf{7–1}$$

With a typical value of 0.9 cm for the radius of the aorta, the average flow speed is

$$\bar{v} = \frac{\left(5000\ \frac{\text{cm}^3}{\text{min}}\right)\left(\frac{1\ \text{min}}{60\ \text{sec}}\right)}{\pi(0.9\ \text{cm})^2} = \frac{83.3\ \text{cm}^3/\text{sec}}{2.54\ \text{cm}^2} = 32.8\ \text{cm/sec}.$$

As the blood branches out into other pathways, both the total area and the speed of flow change accordingly, but the total flow rate of 5 liters/min remains essentially constant.

The speed of flow of a fluid in a closed system will be inversely proportional to the area of the "pipes" through which it flows. This fact is based upon the observation that the total fluid flowing through the system remains constant. Since the system is closed, if 5 liters/min must flow past point A in Figure 7–5, then 5 liters/min must flow past point B. Since the area at point B is one-half the area at A, then the fluid must flow twice as fast. At point C, the total area is five times as large as at A, so the speed is slower by a factor of five. This phenomenon can be summarized by stating that the product of the area times the speed is a constant in the flow of an incompressible fluid. For the case illustrated in Figure 7–5:

$$A_a v_a = A_b v_b = A_c v_c = 5\ \text{liters/min} = \text{constant}.$$

If the area of flow is decreased, then the speed must increase. This might be taken to imply that the speed of blood flow in the capillaries must be very large compared to the speed in the aorta, since they have diameters as small as 10 microns (micrometers). It must be emphasized that in applying equation 7–1, the *total* area of the system must be used. The total area of the capillary system is on the order of 1000 times as large as the aorta, so the flow in the capillaries can be assumed to be slower by a factor of about 1000. The volume of the circulating fluid is not strictly a constant since nutrients are

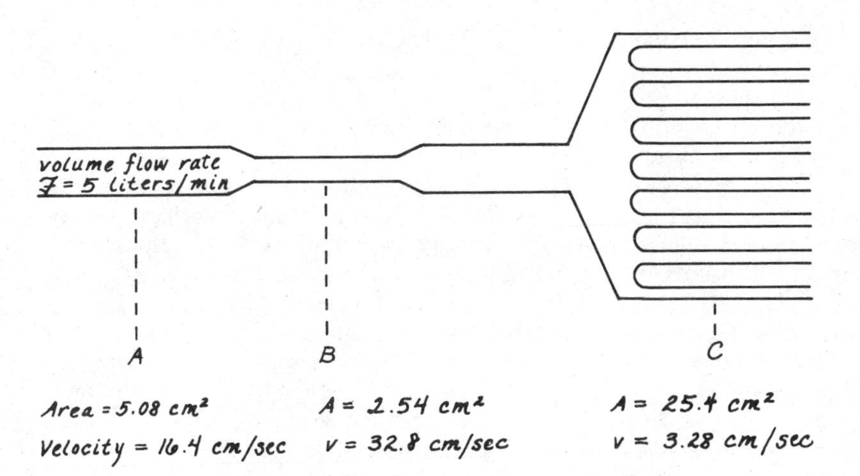

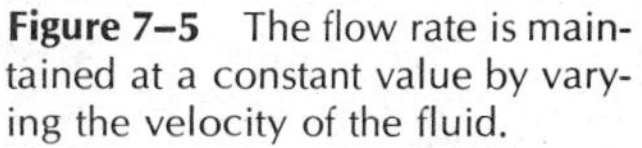
Figure 7–5 The flow rate is maintained at a constant value by varying the velocity of the fluid.

taken from it and wastes are added to and removed from it, but the total volume is approximately constant.

Wall Tension and Laplace's Law

Though the speed of the blood as it enters the capillaries is quite small compared to the aortic speed, the pressure is still a considerable fraction of that in the aorta. The natural question is, "How can the tiny thin-walled capillary withstand a fluid pressure which is approximately 30% of that in the thick-walled aorta?" A French mathematician, Laplace, pondered this question in about 1820 and was able to show that the wall tension required to withstand a given fluid pressure was proportional to the vessel radius. The results for cylindrical and spherical membranes are

$$T = Pr \quad \text{Cylindrical Membrane} \qquad \textbf{7–2}$$

$$T = \frac{Pr}{2} \quad \text{Spherical Membrane} \qquad \textbf{7–3}$$

where r is the radius of curvature of the membrane, T is the tension in the membrane and P is the fluid pressure which acts outward against the membrane. The relationship for a cylindrical membrane like the wall of an artery is illustrated in Figure 7–6.

It is instructive to compare the wall tensions in the aorta and in the capillaries as calculated from Laplace's law. For a mean pressure of 100 mm Hg and a radius of 0.9 cm, the wall tension in the aorta is:

$$T = (100 \text{ mm Hg}) \left(\frac{1333 \text{ dynes/cm}^2}{\text{mm Hg}}\right) (0.9 \text{ cm})$$

$$= 1.2 \times 10^5 \text{ dynes/cm}$$

$$= 1.2 \text{ nt/cm}.$$

A capillary with a radius of 4 microns may be subjected to a pressure of 30 mm Hg, giving a wall tension

$$T = (30 \text{ mm Hg}) \left(\frac{1333 \text{ dynes/cm}^2}{\text{mm Hg}}\right) \times (4 \times 10^{-4} \text{ cm})$$

$$= 16 \text{ dynes/cm},$$

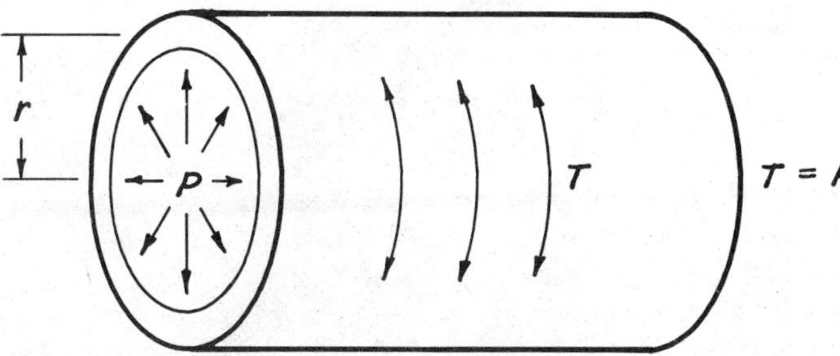

Figure 7–6 Laplace's law: the dependence of wall tension in a cylindrical membrane upon pressure and vessel radius.

a factor of 7500 smaller than the wall tension in the aorta. A smaller vessel can withstand more pressure with a given wall strength.

An ordinary balloon provides an excellent example of Laplace's law. The differences in wall tension at different locations on a cylindrical balloon are illustrated in Figure 7–7. First note that for the enclosed fluid (air), the pressure is the same at all points of the balloon (Pascal's principle) so that changes in wall tension do not arise from differences in pressure. Your experience will verify that the greatest wall tension is at A where the membrane is cylindrical and has the maximum radius (equation 7–2). The tension is somewhat less at B even though the radius of curvature is about the same. At this point it approximates a section of a spherical membrane (equation 7–3) and as such should have half as much tension. (Soap bubbles, etc., tend to take spherical shapes to minimize membrane tension.) At C the membrane tension is even less because the cylindrical membrane has a much smaller radius of curvature. At D the membrane is almost slack because it is in an almost spherical shape with a small radius of curvature.

The example of the balloon may help you to visualize why the walls of an enlarged heart or enlarged blood vessel must provide an even greater tension to withstand the fluid pressure; there is no relief to be had from expansion. If an artery wall expands because it is too weak to provide the required tension, the expansion places an even *greater* tension on the membrane — a classic "vicious cycle." The expansion of arterial walls is normally limited by collagen fibers which circle the artery and limit the expansion. If such limiting mechanisms fail, the membrane will continue to expand until it ruptures.

THE ENERGY SUPPLIED BY THE HEART

The heart does work on the blood flowing through it and in the process gives energy to it. In Chapter 4 it was pointed out that mechanical energy takes the form of potential energy or kinetic energy. In a fluid, potential energy can take the form of gravitational potential energy (weight × height) and the potential energy due to pressure. It is most convenient to express the fluid energy in terms of the energy per unit volume, rather than the total energy. The potential energy per unit volume of the blood as it flows from the heart can be written

$$\frac{\text{P.E.}}{\text{V}} = dgh + P \qquad \textbf{7–4}$$

where d = density of blood, g = acceleration of gravity, h = height of the blood, and P = the liquid pressure. It may be helpful to note at this point that the units of pressure are consistent with those of energy per unit volume:

$$P = \frac{F}{A} \rightarrow \frac{\text{nt}}{\text{m}^2} = \frac{\text{nt m}}{\text{m}^2\ \text{m}} = \frac{\text{joule}}{\text{m}^3}.$$

Pressure can be thought of as stored energy per unit volume. The kinetic energy per unit volume is given by

$$\frac{\text{K.E.}}{\text{V}} = \frac{(1/2 mv^2)}{\text{volume}} = 1/2 dv^2.$$

The total energy per unit volume can then be written

$$\frac{\text{E}}{\text{V}} = 1/2 dv^2 + dgh + P. \qquad \textbf{7–5}$$

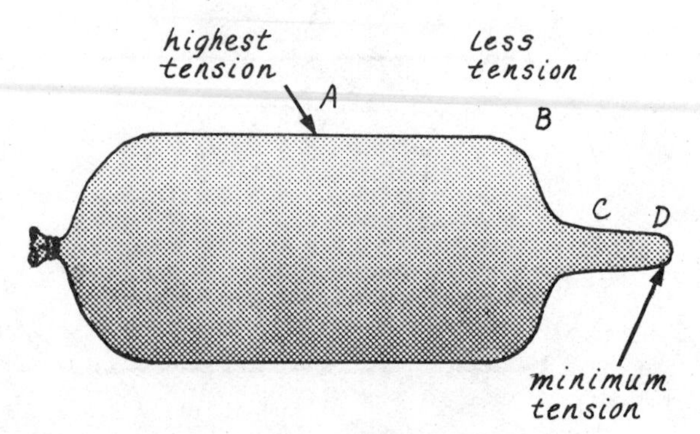

Figure 7–7 Changes in membrane tension in a balloon as an example of Laplace's law.

If it were not for the work done against fluid frictional resistance, this energy per unit volume would remain constant (conservation of energy principle). The heart, in pumping, increases the energy of the blood by increasing the pressure and the kinetic energy. When the blood flows down to the feet, the gravitational potential energy in equation 7–5 is reduced and some of it is converted into pressure and kinetic energy, thus aiding the heart. When the blood flows upward, some of the pressure and kinetic energy is used to overcome gravity, increasing the gravitational potential energy. Therefore, a person's blood pressure at the head when he or she is standing is lower than when in a horizontal position.

The rate of energy supply to the blood is the "output power" of the heart and can be expressed in watts. Part of this power is used to provide the blood pressure and can be called the "pressure" power. The remainder is used to provide the kinetic energy and can be called the "kinetic" power. When the body is active, the blood pressure is normally maintained close to the rest value but the flow rate may increase tenfold, so that the velocity of the blood may be ten times as great. Since the kinetic energy is proportional to the square of the velocity, this represents a hundred-fold increase in the required kinetic power per unit volume. Using typical values for blood pressure and flow rate, Ackerman (3) has calculated some representative values for the output power of the heart. These values are listed below.

	At Rest	Active
"pressure" power	1.0 watt	10 watts
"kinetic" power	0.13 watt	130 watts
total heart power	1.1 watts	140 watts

No "potential power" is listed here because all of the heart's pumping action is done at essentially the same height.

The kinetic energy is sometimes the major type of energy supplied during vigorous exercise. It is remarkable to note the range of power within the capability of the heart. It can range from 1 watt to 140 watts within a short time interval. Compared to the use of electrical power in home appliances, the heart performs its circulatory function with a relatively small amount of power.

Another perspective on the output power of the heart may be obtained with the use of equation 4–6:

$$\text{Power} = F\bar{v}$$

where F is force and v is the speed in the direction of the force. The force can be expressed in terms of the blood pressure acting to push blood through an artery:

$$P = \frac{F}{A} \quad \text{so} \quad F = P \times A$$

$$\text{Power} = PA\bar{v}.$$

But $A\bar{v}$ is equal to the volume flow rate of the blood $\mathscr{F}$ (equation 7–1), therefore

$$\text{Power} = P\mathscr{F}. \qquad \textbf{7–6}$$

Applied to the aorta upon exit from the heart, heart pressure power = (average blood pressure) × (volume flow rate). One direct implication of this is that if a person's blood pressure is elevated, the heart is having to work harder to supply a normal blood volume flow rate. Note that this is just the "pressure" power and does not include the "kinetic" power described above. That kinetic power includes the total heart action, including the pulmonary circulation, and requires quite extraordinary circumstances to reach Ackerman's value of 130 watts. The actual kinetic energy of the blood flow in the aorta is usually small compared to the pressure energy.

THE VARIATIONS OF THE BLOOD PRESSURE

The heart supplies pressure to the blood in the systemic circulation only during the contraction of the left ventricle. If such a pulsating pressure were applied to a fluid in a rigid mechanical pipe system, the pressure would drop to zero between the pulses, as illustrated in Figure 7–8(b). (By Pascal's principle, if the applied pressure were zero, it would be zero throughout the system.) The arteries, however, comprise an extremely elastic system of tubing. When blood leaves the heart, it bulges out the walls of the aorta, storing part of the energy in the form of elastic potential energy. The re-

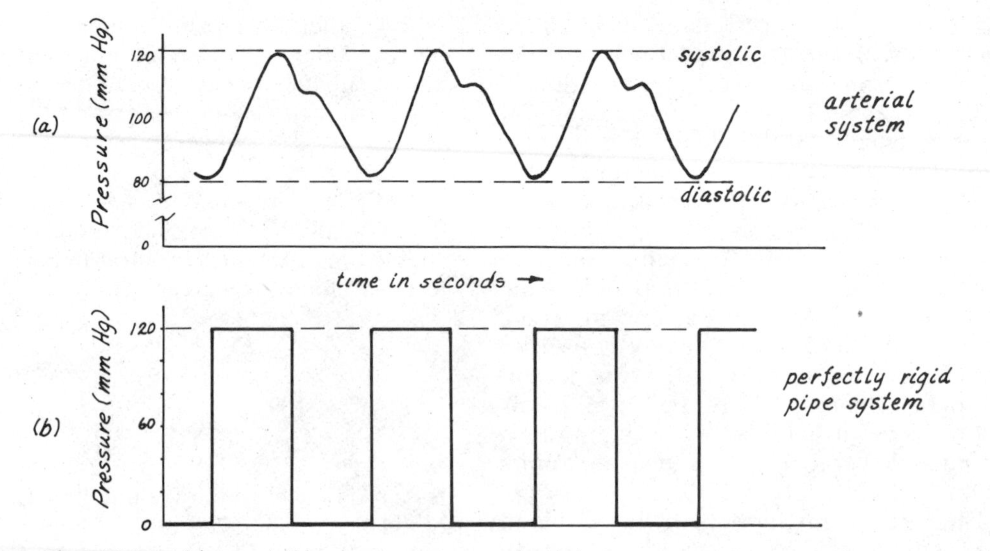

Figure 7–8 The variation of the arterial blood pressure compared with the pressure variation in a rigid system. (See reference 2.)

bound of that elastic wall transmits the energy to the arterial wall further along, which stores part of the energy. This elastic expansion of the wall travels along the arteries like a wave, causing the pulse. The elastic contraction of the arterial walls maintains some pressure in the arterial system even during the expansion cycle of the left ventricle. A typical arterial pressure variation is shown in Figure 7–8(a). The peak pressure produced by the ventricular contraction is called the "systolic pressure," and the minimum pressure maintained by the elastic system is called the "diastolic pressure." If the elasticity of the blood vessels decreases, as in hardening of the arteries, the diastolic pressure drops lower since the elastic rebound is less effective in maintaining the pressure. In this case the difference between the systolic and diastolic pressures becomes greater.

The pressure variation shown in Figure 7–8(a) is typical of the arterial system only. The pressure variations diminish as the blood flows through the arterioles, and there is essentially no periodic variation of the blood pressure in the venous system.

THE MEASUREMENT OF BLOOD PRESSURE

The most common measurement of blood pressure is the indirect measurement of arterial pressure by means of the sphygmomanometer. This method actually measures the effect of an externally applied pressure upon the circulation. Most commonly, an inflatable bag and cuff are wrapped around the patient's left arm at the level of the heart. As the bag is inflated, the pressure acts to stop the blood flow through the brachial artery. While the bag is inflated, the gauge pressure is read on a mercury manometer or a pressure gauge calibrated in mm of Hg. (See Figure 7–9.) The bag is initially inflated to a pressure about 30 mm Hg above the point at which the radial pulse disappears. As the pressure in the bag is slowly released at a rate of about 2 or 3 mm Hg per second, the person making the measurement listens with a stethoscope for sounds in the brachial artery below the inflated bag.

The origin of the sounds heard while the pressure is being released can be explained by reference to Figure 7–8(a). The arterial blood pressure varies between the two pressure limits which are designated systolic and diastolic. When the external pressure is greater than the systolic pressure, there is no blood flow and no sound. As the bag pressure drops below the systolic pressure, blood will start to flow in spurts during that part of the cycle where the arterial pressure exceeds the external resistance pressure of the inflated bag. For example, if the systolic pressure is 120 mm Hg and the bag pressure is 118 mm Hg, blood will flow only during the very brief interval near the peak pressure where the arterial pressure exceeds 118 mm Hg. The turbulence of this pulsating flow pattern produces sounds called the Korotkoff sounds. The pressure noted on the gauge at the onset of these sounds is a measure of the systolic pressure. The sounds change as the pressure is lowered because of changes in the nature of the flow.

When the bag pressure is completely released, no appreciable amount of sound will be heard in the normal artery because the flow will be the streamlined "laminar" flow mentioned in Chapter 5. At first analysis, one might expect the Korotkoff sounds to disappear when the bag pressure drops below the diastolic pressure, since flow will be occurring at all parts of the cycle. While it is true that the pulsating flow pattern caused by cutting off the flow during part of the cycle no longer exists, there is still some constriction on the artery which can cause turbulent flow and some sound. Experimental studies indicate that the best index of the diastolic pressure is the point at which the Korotkoff sounds become muffled, rather than the point at which they become inaudible. This muffling point is usually quite distinct and seems to correspond to the disappearance of the staccato sound associated with the complete closure of the artery. Though there is some doubt about the exact physical significance of the muffling point, it appears to be a more consistent measurement than the inaudibility point. If there is some cause for turbulence in the arterial flow such as a partial occlusion of the artery, or if the patient has undergone vigorous exercise, the audible sounds may persist far below the diastolic pressure. The blood pressure is normally recorded as systolic/diastolic (for example, 120/80), where the units are mm of Hg. If the inaudibility point is significantly below the measured diastolic pressure, it is sometimes recommended that all three pressures be recorded (e.g., 120/80/65). Another advantage of recording the muffling point rather than the inaudibility point as the diastolic pressure is that this point is a change in the *quality* of the sound rather than a change in its *intensity*, and is thus less dependent upon the sensitivity of the hearing of the person making the measurement.

To obtain a direct measurement of the

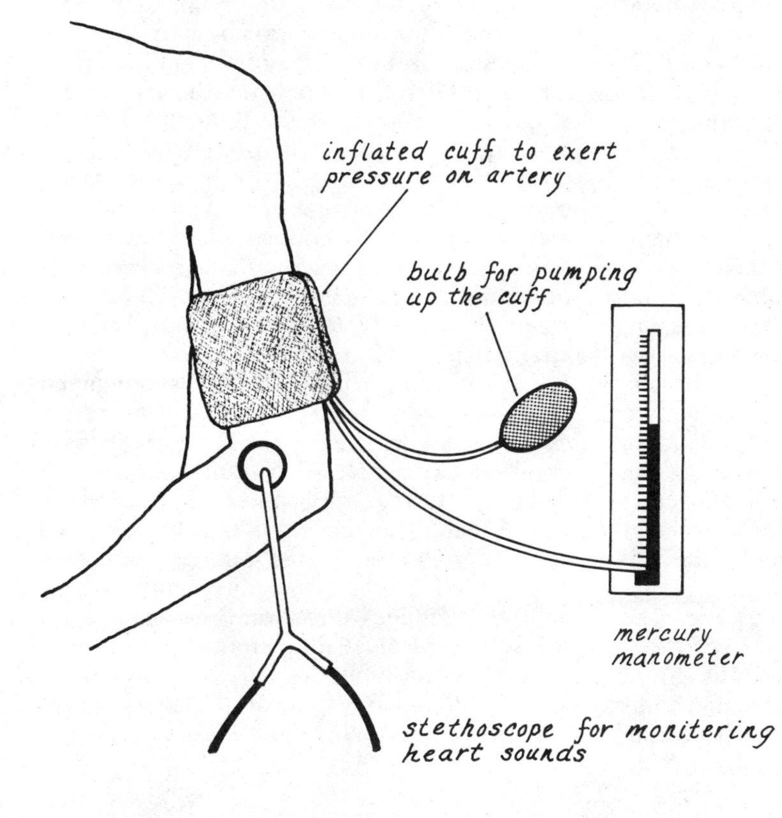

Figure 7–9 Measurement of blood pressure with a pressure cuff and mercury manometer (sphygmomanometer).

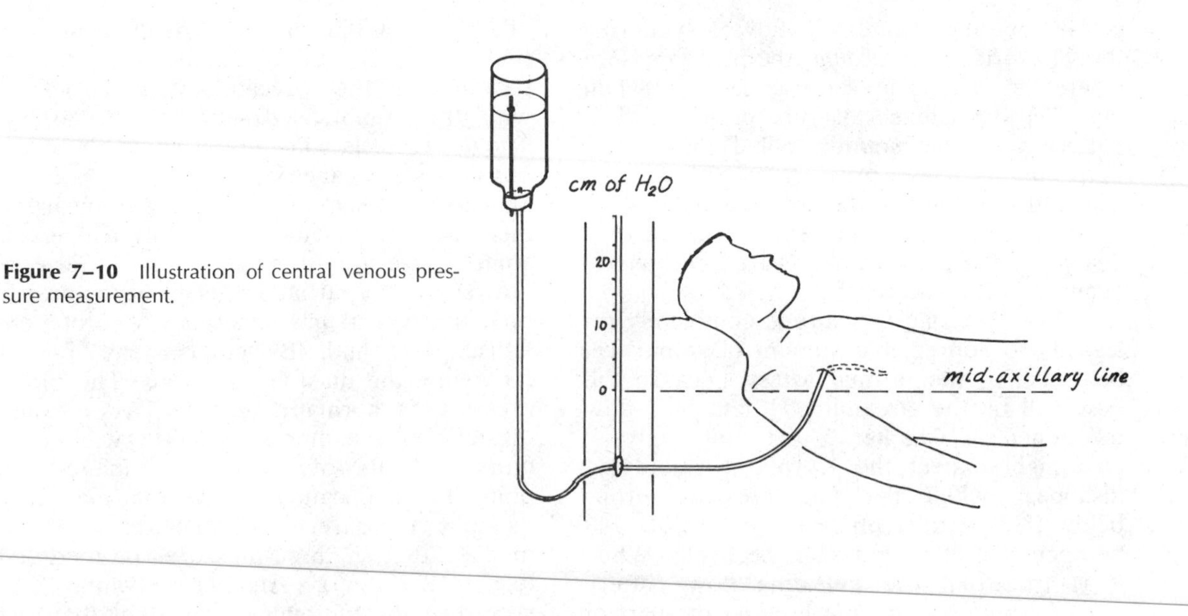

Figure 7–10 Illustration of central venous pressure measurement.

blood pressure, some pressure measuring device must be connected directly into the circulatory system. The measurement of the central venous pressure (CVP) requires the placement of a catheter in the central venous system near the heart. If a water manometer is used to measure the pressure, this catheter is filled with an isotonic saline solution or the intravenous fluid which is being given to the patient. This fluid contacts the blood in the vein and transmits the pressure out to the manometer. This process is based on Pascal's law (Chapter 5); the pressure is transmitted undiminished through the catheter to the pressure-measuring device.

A common method for measuring the CVP is illustrated in Figure 7–10. The intravenous fluid bottle is attached to the manometer and to the patient by means of a threeway stopcock. The fluid bottle is first opened to the manometer, and the pressure rises to a height considerably above that corresponding to the venous pressure. This assures that when the stopcock is opened to the catheter, the column pressure will force fluid into the vein, rather than allowing the blood to flow out of the vein into the catheter. The level of the fluid in the manometer column will drop until the weight of the column of liquid produces just enough pressure to balance the central venous pressure. The height of the column of fluid is a measure of the CVP, and it is usually recorded in cm of water. Since the specific gravity of IV fluids is essentially equal to that of water, the height of the liquid column is measured directly in centimeters.

The normal central venous pressure is generally considered to be 5 to 12 cm of water above the level of the right atrium. The most difficult part of the CVP measurement is the adjustment of the zero level of the manometer so that it is precisely at the level of the right atrium. If the CVP is 5 cm H_2O and the zero level of the manometer is 3 cm too low, a 60% measurement error results. Measurements may be significantly altered if either the patient or the manometer is moved. Because of the difficulty associated with the zero height of the manometer, the trend of the CVP is probably more important than the actual numerical value. The monitoring of CVP values and urine output when a seriously ill patient is receiving intravenous fluids will give indications of how well the heart is tolerating the increase in circulating blood volume.

As an alternative to the water manometer pressure gauge, a pressure transducer can be used which converts the pressure to an electrical signal. Such transducers will be discussed further in Chapter 15. The advantage of such transducers is that the electrical signal can be sent to a remote monitoring station and displayed on an oscilloscope or another voltage measuring device. These transducers are usually external and are used directly as an alternative to the manometer, but a number of very small transducers have been developed which can be introduced directly into the venous system. Such transducers can be moved through the venous system to trace changes in the venous pressure. Tiny catheters and transducers on the order of 1 millimeter in diameter have been developed to measure the blood pressure inside the heart. These are sometimes called "drift" catheters because they are allowed to drift with the blood flow into and even through the heart.

SUMMARY

The ventricles of the heart may be viewed as twin force pumps which provide the pressure necessary to move the blood through the circulatory system. The right ventricle provides the pressure for the pulmonary circulation, and the left ventricle provides a pressure which is four to five times greater for the systemic circulation. For a closed fluid system, the flow rate is the same through any cross section of the system. For the blood system the flow rate through the aorta is about 5 liters/min and is the same through the total cross section of the capillary system if there is no change in the volume of the circulating fluid. The flow rate through a given part of the system is equal to the area of flow times the velocity of flow. Since the total area of the capillary system is up to 1000 times the area of the aorta, the flow velocity through a capillary may only be about one-thousandth of the flow velocity in the aorta. The heart provides energy to the blood in the form of pressure energy and kinetic energy. When the body is at rest, most of the energy supplied is pressure energy, but when vigorous activity demands more energy, it is supplied predominantly in the form of kinetic energy of rapid flow. Upon demand, vasodilation and other mechanisms reduce the flow resistance of the system so that a large increase in flow rate can be achieved by a moderate increase in blood pressure.

The maximum blood pressure in the arterial system occurs when the left ventricle contracts, and is called the systolic pressure. The minimum pressure, which occurs during the rest cycle of the heart, is called the diastolic pressure. The diastolic pressure is a result of the elastic recoil of the blood vessels; in a rigid pipe system there would be no such recoil and the diastolic pressure would be zero. Arterial pressures are measured indirectly by inflating a cuff to a pressure greater than the systolic pressure so that the blood flow in the brachial artery is cut off. As the cuff pressure is lowered, the turbulence of intermittent flow through the artery produces sounds which can be monitored for the measurement of systolic and diastolic pressure. For venous pressure measurements, a catheter is connected to a water manometer or pressure transducer.

REVIEW QUESTIONS

1. Why doesn't the blood velocity increase as the blood enters the small capillaries? What is an approximate ratio of blood velocity in the capillaries to blood velocity in the aorta?

2. Why must the volume flow rate be the same at all stages in a closed liquid circulation system?

3. How does gravity affect the circulation of the blood?

4. If the blood flow rate doubles during exercise, why doesn't the blood pressure double?

5. Why isn't the systolic-to-diastolic pressure variation seen in the venous system as well as in the arterial system?

6. Why doesn't the blood pressure drop to zero during the part of the cycle when the ventricles are not pumping?

7. Explain the terms systolic and diastolic pressure in terms of what is happening in the heart and circulatory system.

8. Why is the point where the Korotkoff sounds become muffled the preferable measurement of the diastolic pressure, rather than the point at which they become inaudible?

9. How is Pascal's principle used in the measurement of central venous pressure?

10. Why is the vertical position of the manometer so critical in a central venous pressure measurement?

PROBLEMS

1. Given a blood volume flow rate of 5 liters/min with a blood pressure of 120 mm Hg, if the required flow were 20 liters/min during exercise and the increased volume flow rate had to be achieved by raising the blood pressure alone, what pressure would be required?

2. If the original volume flow rate through an artery is 800 cm^3/min, what would be the flow rate if the artery is 50% occluded (radius reduced to half the original inside radius)? Assume that the pressure and other factors remain the same.

3. A coronary artery has a normal blood volume flow rate of 100 cm^3/min at an average blood pressure of 100 mm Hg. What would be the volume flow rate if the internal radius were reduced by 10% with the pressure unchanged? If every blood vessel were similarly affected, what average blood pressure would be required to restore the normal volume flow rate. Repeat the calculation for a 30% reduction in radius.

4. The resting blood volume flow rate to the skeletal muscles is typically about 0.75 liter/min (9). If the internal radii of the flow-determining resistance vessels can dilate to 2.2 times their resting radii, what maximum blood flow to the skeletal muscles can be achieved without a blood pressure increase?

5. The salivary glands normally do not require a large blood supply, but must quickly respond when needed. The normal volume flow rate of blood to these glands is only about 20 cm^3/min (9). If the supplying blood vessels can dilate their radii by a factor of 1.9, what blood volume flow rate can be achieved without an increase in blood pressure?

6. The blood volume flow rate to the gastrointestinal tract may vary from 0.7 to 5.5 liters/min (9). If this increase must be achieved entirely by vasodilation, what percentage dilation of the resistance-determining blood vessels is required?

7. A fluid is flowing at the rate of 100 cm^3/sec through a tube of area 2 cm^2. What is the flow velocity? What would be the flow velocity further along the tube where the area constricts to 0.5 cm^2? What will be the flow rate at the constriction?

8. If the inside radius of the aorta is 0.9 cm and the average speed of the blood through the aorta for a resting adult is 33 cm/sec, what is the volume flow rate through the aorta in cm^3/sec and in liters/min? If the total area of the major arteries is 20 cm^2, what will be the average flow velocity of the blood through these arteries?

9. If a capillary has a radius of 2×10^{-4} cm and the average velocity of flow through it is 0.03 cm/sec, what is the volume flow rate through the

capillary? If the flow rate through the aorta is 80 cm^3/sec, how many such capillaries would be required to carry the total blood flow?

10. A continuous tube has a reduction in cross-sectional area from 5 cm^2 to 1 cm^2. At the larger cross section the volume flow rate is 10 cm^3/sec.

    a. What is the average fluid speed at the point where the area is 5 cm^2?
    b. What is the volume flow rate at the point of smaller area?
    c. What is the average fluid speed at the point where the area is 1 cm^2?

11. Assume that the area of the aorta is 2 cm^2 and the blood speed through it is 30 cm/sec. The entire cross-sectional area of the capillaries is assumed to be 2000 cm^2 and the blood is to be viewed as a closed liquid system, i.e., no loss of liquid during circulation.

    a. What is the volume flow rate in the aorta in cm^3/sec?
    b. What is the total volume flow rate through the capillary system?
    c. What is the average fluid speed in the capillaries?

12. The volume flow rate of blood through the aorta is about 5 liters/min for a person at rest. The average speed at a certain point is measured to be 25 cm/sec. What is the cross-sectional area of the aorta at that point?

13. Suppose the aorta has an internal radius of 0.8 cm when at the diastolic pressure of 80 mm Hg and expands to a radius of 1.2 cm momentarily when subjected to the peak pressure of 120 mm Hg from the heart (systolic pressure). Calculate the wall tension of the aorta in each case.

14. A capillary with internal radius of 4 microns can withstand a maximum wall tension of 30 dynes/cm.

    a. What would be the wall tension in the cylindrical wall if the blood pressure at the capillary were 20 mm Hg?
    b. What maximum pressure can it withstand if its radius is unchanged?
    c. If it expanded, what would be the maximum radius it could reach and still withstand the original pressure of 20 mm Hg?

15. A spherical ball with radius of 6 inches can withstand a wall tension of 360 lb/in. What would be the maximum gauge pressure of air which you could put in the ball?

16. The air in a balloon like that shown in Figure 7–7 has a gauge pressure of 1000 mm Hg. The radius of curvature of the large part of the balloon (cylindrical and spherical) is 3 cm and the radius of curvature of the smaller end section is 0.5 cm. Calculate the wall tension in each part of the balloon (four different values).

17. If a central venous pressure was measured to be 12 cm of water, what would the equivalent be in mm Hg? In inches of water?

18. If an intravenous fluid bottle is located 1.5 meters above the patient, what is the pressure in mm Hg of the fluid at the point of entry into the patient? At what minimum height above the patient must the bottle be placed to force fluid into a vein which has a blood pressure of 12 mm Hg?

19. If the heart must pump blood to a height of 35 cm above the heart to the head, how much pressure drop in mm Hg would be experienced in

this process from just the increase in gravitational potential energy, disregarding changes in kinetic energy? Assume a density of 1.05 gm/cm^3 for the blood.

20. How much blood pressure increase would result from the blood flow 130 cm downward from the heart to the feet if there were no changes in the pressure from flow resistance or changes in kinetic energy? (d = 1.05 gm/cm^3 for the blood)

21. Suppose a giraffe which lifts its head 3 meters above its heart must maintain an average blood pressure of 60 mm Hg at its head to keep from fainting while feeding. What minimum average blood pressure must the giraffe's heart produce? (Assume $d = 1.05$ gm/cm^3.)

22. Given an internal radius of 0.9 cm for the aorta and an average blood pressure of 100 mm Hg, what is the average force exerted by the heart on the blood in the aorta? Express the force in dynes and convert to pounds.

23. If the normal blood volume flow rate is 5 liters/min at an average blood pressure of 100 mm Hg, what is the "pressure power" output of the heart in watts?

24. Braunwald (9) quotes a maximum blood volume flow rate of 38 liters/min. If this occurred with an average blood pressure of 120 mm Hg, what would be the "pressure power" output of the heart in watts?

REFERENCES

1. *Van Nostrand's Scientific Encyclopedia,* 5th ed. Princeton, N.J.: D. Van Nostrand Co., 1976.
2. Strong, Peter. *Biophysical Measurements.* Beaverton, Ore.: Tektronix, Inc., 1970.
3. Ackerman, Eugene. *Biophysical Science.* Englewood Cliffs, N.J.: Prentice-Hall, Inc., 1979.
4. American Heart Association. *Recommendations for Human Blood Pressure Determination by Sphygmomanometers.* 44 East 23rd St., New York, N.Y. 10010, 1967.
5. Maier, W.P., and Goldman, L.I. *The Nurses' Role in Central Venous Pressure Monitoring.* AORN Journal, Dec. 1968, pp. 35–37.
6. Clynes, Manfred, and Milsum, John H. *Biomedical Engineering Systems.* Inter-University Electronics Series, Vol. 10. New York: McGraw-Hill, 1970.
7. Burton, Alan C. *Physiology and Biophysics of the Circulation.* Chicago: Year Book Medical Publishers, 1965.
8. Henry, James P., and Meehan, John P. *The Circulation, An Integrative Physiologic Study.* Chicago: Year Book Medical Publishers, 1971.
9. Braunwald, E. *Regulation of the Circulation.* New England Journal of Medicine, Vol. 290, p. 1420, 1974.
10. Skalak, R. Mechanisms of the Microcirculation. In Y.C. Fung et al., eds.: *Biomechanics – Its Foundations and Objectives.* Englewood Cliffs, N.J.: Prentice-Hall, 1972.
11. Cromer, Alan H. *Physics for the Life Sciences,* 2nd ed. New York: McGraw-Hill, 1977.

CHAPTER EIGHT

Further Medical Applications of Pressures in Fluids

INSTRUCTIONAL OBJECTIVES

After studying this chapter, the student should be able to:

1. State how the pressure, velocity, and volume flow rate of a liquid in a tube will change as the fluid moves from a large area to a more constricted area of the tube (Bernoulli's principle).
2. Define entrainment and explain how the Bernoulli principle is employed to accomplish entrainment.
3. Give two or more examples of entrainment.
4. Name the physical variables which influence the rate of flow of a gas through an orifice.
5. Explain why flowmeters for different gases are not interchangeable.
6. Indicate the function of the various parts of a pressure reduction valve, given a diagram of the valve.
7. Describe the Coanda effect and explain how it can be used in a respirator valve.

THE BERNOULLI EFFECT

As a fluid flows in a uniform tube or pipe, there will be a gradual drop in pressure, as discussed in Chapter 5. This gradual drop in pressure is demonstrated by the first three manometers in Figure 8–1. As the pressure drops, the height of the liquid in the manometer will decrease. Note that there is a significant departure from the gradual pressure drop at the point where the tube is constricted. Bernoulli investigated this effect and noted that the pressure in a flowing fluid is lowest where its speed is greatest (the Bernoulli effect). The fluid speed in Figure 8–1 is greatest at the constriction because

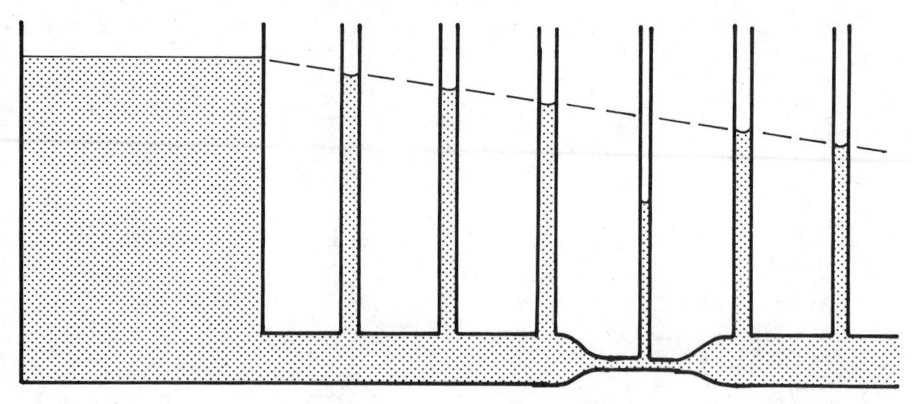

Figure 8–1 Illustration of the Bernoulli effect.

the liquid must flow faster through the restricted area to transport the same volume of fluid in a given time. This dependence of the speed on the tubing area was discussed in Chapter 7 in terms of the circulatory system (equation 7–1). The strict proportionality between the fluid speed and area is true only for the flow of incompressible liquids, but the qualitative features are similar for gas flow.

The energy per unit volume for a flowing fluid may be calculated from equation 7–5:

$$\frac{E}{V} = \frac{1}{2}dv^2 + dgh + P.$$

This is sometimes referred to as Bernoulli's equation. The reduction in liquid pressure in the constricted part of the tube in Figure 8–1 can be shown directly from this equation. Since the tube is level, there is no change in the gravitational potential energy, dgh, during the flow. Since the speed must increase, as described above, the fluid pressure P must correspondingly decrease because the total energy per unit volume cannot increase.

The drop in fluid pressure is directly related to the increase in fluid speed. The fact that the liquid accelerates as it enters the constriction implies that there is a net force acting to the right when it enters the constriction. Since the force is proportional to the pressure, this implies that the pressure is greater to the left of the constriction. The fluid decelerates as it leaves the constriction, which implies a net force to the left. Thus, the pressure rises as the fluid leaves the constriction.

If the narrowing of the tube is sufficient, the fluid pressure resulting may be below atmospheric pressure. If there is an open tube into the constricted area, then the atmospheric pressure will push air into the liquid as shown in Figure 8–2. This is one method for introducing a gas into a liquid.

The energy equation for fluid flow can be used to explain some of the details of siphon action (see Chapter 6). The basic conceptual question in the operation of a siphon is how the liquid can flow upward to get "over the

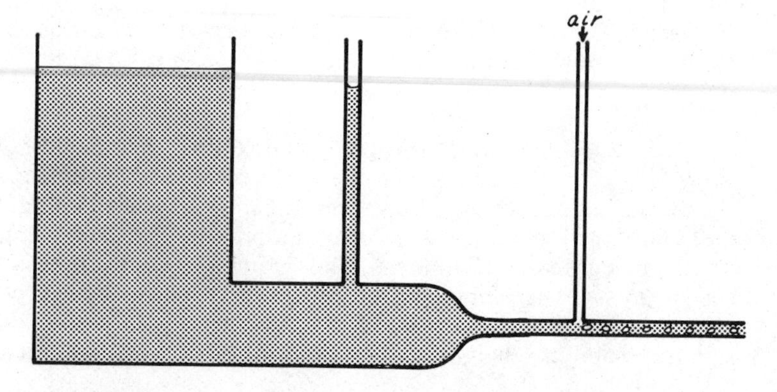

Figure 8–2 The entrainment of gas into a liquid by means of the Bernoulli effect.

hill'' in the siphon. Referring to Figure 8–3, the liquid pressure at B must clearly be lower than that at A to get the liquid to flow uphill. The lowering of pressure can be seen from the application of the conservation of energy principle to the flow process. If the height of the tube at point C is chosen as zero, the energy per unit volume at A can be written

$$\tfrac{1}{2}dv^2 + P_A + dgh_A$$

The point A is at the liquid level and therefore P_A = atmospheric pressure. During the flow process from A to C the fluid must do work against fluid friction, so an amount of energy equal to that work is subtracted from the liquid energy. For a horizontal tube this decrease in energy per unit volume is reflected only in the pressure change (Poiseuille's law)

$$P_1 - P_2 = \mathscr{F}R.$$

In the siphon the energy loss in overcoming the flow resistance must be included in applying the conservation of energy principle. The total energy per unit volume at points A, B, and C when the siphon action has been initiated and the liquid is flowing smoothly can be written

A $\tfrac{1}{2}dv^2 + P_A + dgh_A$

B $\tfrac{1}{2}dv^2 + P_B + dgh_A + dgh + \mathscr{F}R_{AB}$

C $\tfrac{1}{2}dv^2 + P_C + \mathscr{F}R$

Heights h_A and h are identified in Figure 8–3. The total flow resistance of the tube is designated R and R_{AB} refers to the resistance of the segment between A and B. For smooth flow in a tube of uniform cross section, the kinetic energy term is the same at all points in the tube. Equating the total energy per unit volume at A and B yields

$$P_B + dgh_A + dgh + \mathscr{F}R_{AB} = P_A + dgh_A,$$

and pressure P_B is then

$$P_B = P_A - dgh - \mathscr{F}R_{AB}.$$

Since the pressure P_A is equal to atmospheric pressure, this expression shows not only that P_B is less than atmospheric pressure, but that it is less by just the right amount. From this equation we see that atmospheric pressure P_A will overcome the flow resistance between A and B ($\mathscr{F}R_{AB}$) and still supply the necessary gravitational potential energy per unit volume (dgh) to get the liquid to the top of the hill.

Equating the energies at A and C yields

$$\tfrac{1}{2}dv^2 + P_A + dgh_A = \tfrac{1}{2}dv^2 + P_C + \mathscr{F}R.$$

But the kinetic energies are equal and the pressures are equal (both P_A and P_C are equal to atmospheric pressure). This leaves

$$dgh_A = \mathscr{F}R,$$

showing that the gravitational potential energy change per unit volume, dgh_A, during the process of flowing "downhill" a distance h_A is used to overcome the fluid resistance. This change in gravitational energy per unit volume is functionally equivalent to the pressure drop in earlier Poiseuille's law problems.

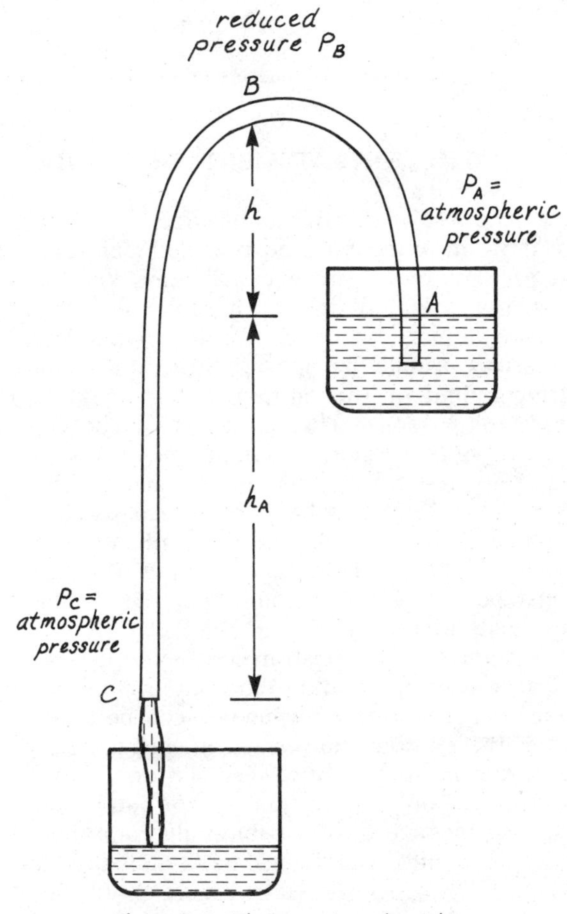

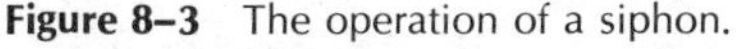
Figure 8–3 The operation of a siphon.

THE VENTURI TUBE

In Figure 8–1, it was indicated that the pressure returned to the normal uniform-pipe

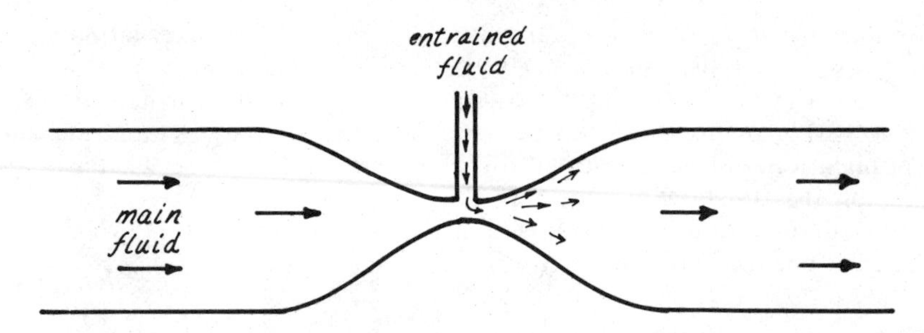

(a) Entrainment of a fluid by means of a smoothly contoured constriction

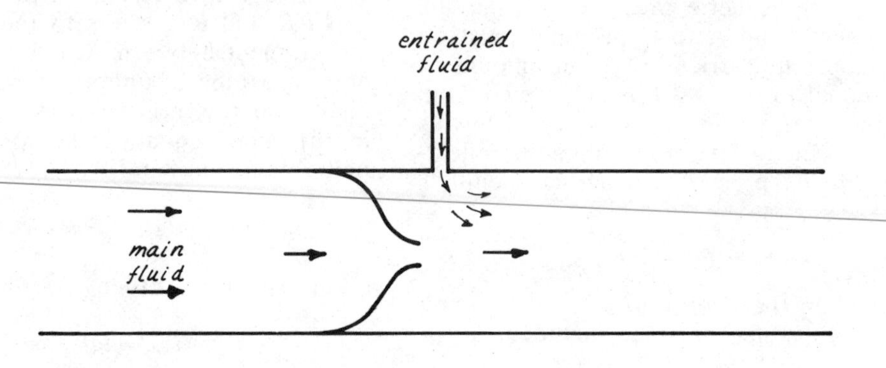

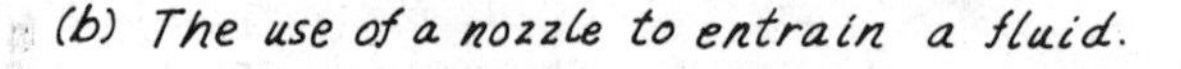

(b) The use of a nozzle to entrain a fluid.

Figure 8–4 Examples of entrainment devices.

Illustration continued on the opposite page.

pressure contour after the fluid passed the configuration. This is true only if viscosity effects and turbulence are negligible. In Chapter 7 it was pointed out that the energy per unit volume of a flowing fluid could be divided into pressure energy and kinetic energy. In the constriction in Figure 8–1, some of the pressure energy is converted to kinetic energy. After it leaves the constriction, some of the kinetic energy will be reconverted into pressure energy, raising the pressure. Some energy will be lost because of friction (viscosity effects). Some of the kinetic energy may go into turbulent flow patterns and thus not be reconverted into pressure.

Venturi showed that in order for a streaming fluid to regain a significant portion of the pressure it had before it entered the constricted area, it is necessary that the tube open up very gradually past the constriction. Basically, this smooth tapering prevents turbulence. Tubes and nozzle arrangements using these design considerations are often called venturi tubes. The reduced pressures and increased speeds obtainable with Venturi tubes have many applications.

THE ENTRAINMENT OF FLUIDS

The use of the Bernoulli effect in a fluid to draw in a second fluid is called "entrainment." With the presence of some kind of nozzle in the fluid stream, there are also propulsion effects which contribute to the low pressure around the nozzle. Either the main driving fluid or the entrained fluid may be liquid or gaseous. That is, a gas can be entrained into a liquid, a liquid into a gas, a gas into gas, or a liquid into a liquid. The Bernoulli effect has been discussed as if it applied to liquids, but the same phenomena occur in gases. The compressibility of gases must be taken into account, but the results are qualitatively similar.

Figure 8–4(a) illustrates the entrainment of a fluid by a smoothly contoured constriction. If the resulting turbulence can be tolerated, the contour can be terminated to form a nozzle as in Figure 8–4(b). To obtain the advantage of a precision nozzle and still keep turbulence down, a smoothly contoured exit pipe called a diffuser can be added as in Figure 8–4(c). As shown in Figure 8–4(c) the

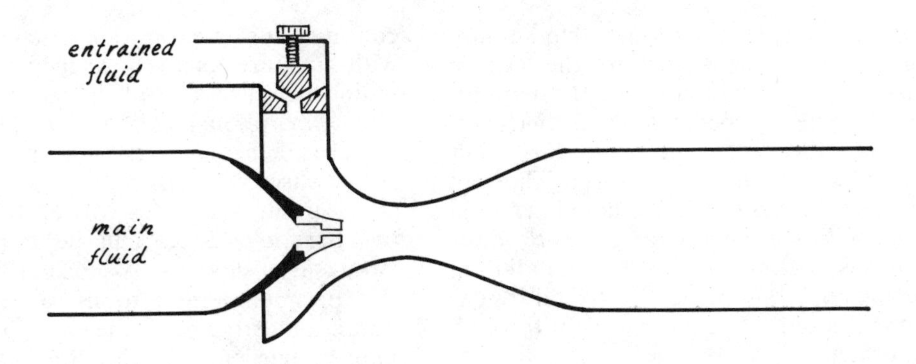

(c) *A more sophisticated entrainment device with a calibrated nozzle and controlled entrainment.*

Figure 8–4 *Continued* Examples of entrainment devices.

amount of entrained fluid can be controlled by varying the size of the orifice leading into the system. This combination of elements is sometimes called an "injector" since varying amounts of a second fluid can be injected into the main flow. Various types of injectors are widely used for mixing gases in the administration of anesthetics.

The details of the contours of the diffuser are quite important. They serve to keep down turbulence and make it possible to transform much of the kinetic energy of the nozzle flow back into pressure energy to maintain the flow. This more efficient flow makes it possible to entrain a larger volume of gas and tends to keep the percentage of entrained gas constant when the total flow rate varies.

Examples of Entrainment

One widely used application of entrainment is the water aspirator. A rapidly moving stream of water entrains air and produces a negative pressure or suction at the entrainment port, as illustrated in Figure 8–5. This type of aspirator is commonly used by dentists to remove saliva from the patient's mouth. A device like this can be used to produce suction in case of a power failure. It could be convenient as an aspirator for use in the home for patient care.

Several types of nebulizers make use of the Bernoulli effect of inject water or a medicated solution into the air stream going to a patient. Oxygen and other therapeutic gases are dry when they are taken from

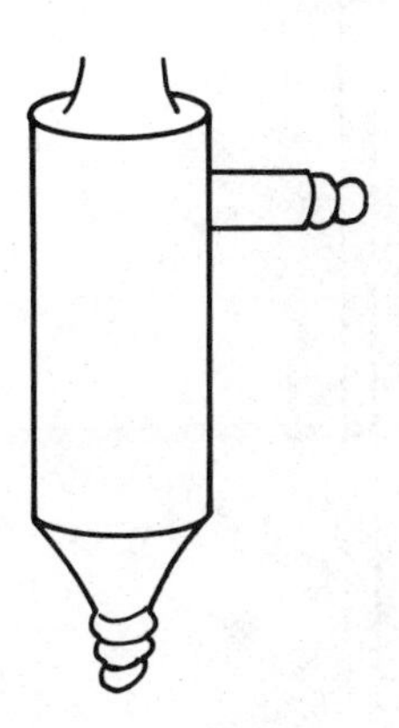

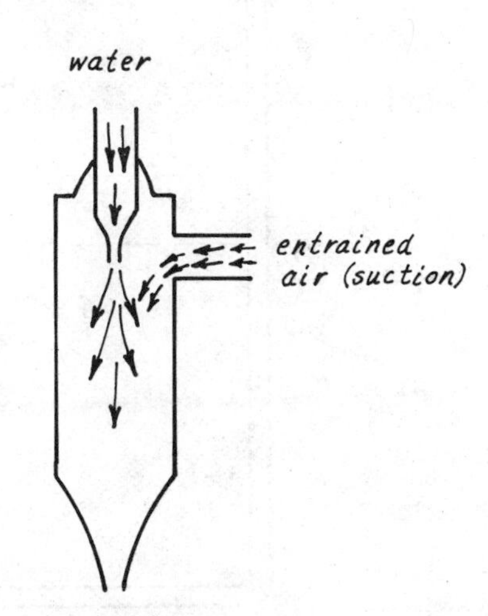

Figure 8–5 The water aspirator.

tanks or supply lines. A considerable amount of moisture must be added to the gas in order to prevent dehydration of the patient. Also, some types of medications are most effective when they enter the respiratory tract in an "aerosol" or nebulized form. The fact that most of the water from a nebulizer is in the form of tiny particles rather than vapor distinguishes nebulizers from humidifiers. The nebulization can be accomplished by entraining the liquid into the air stream through a small orifice.

In Figure 8–6, one type of nebulizer is illustrated. The oxygen or air enters the top through a small orifice at a high speed. By the Bernoulli effect, this high speed results in a lowered pressure near the nozzle; liquid is drawn up from the reservoir and mixed with the supply gas. In this nebulizer the supply gas to the patient is pushed through the jet to accomplish the nebulization. In a number of commercially available nebulizers, the jet is supplied by a separate nebulizer power gas, and the resulting aerosol is then injected into the main flow. This is advantageous for use with IPPB therapy (see Chapter 6) since the aerosol formation can continue during expiration to charge the tube with moisture for the next inspiration. Small, in-line nebulizers operating by the Bernoulli effect are often used for the administration of aerosol medications along with the therapeutic gases.

Further examples of entrainment and the Bernoulli effect can be found in commonly used devices. Atomizers make use of the Bernoulli effect to lift a liquid from a bottle and produce a fine mist. The common bunsen burner uses the Bernoulli effect to entrain air to mix with the gas.

FLOW THROUGH AN ORIFICE: FLOWMETERS

When therapeutic or anesthetic gases are being administered, it is important to know not only the pressure but also the volume flow rate of the gas. When a given volume of gas is administered, both the pressure and the flow rate are necessary for an evaluation of the resistances encountered.

The operation of several types of flowmeters depends upon the nature of fluid flow

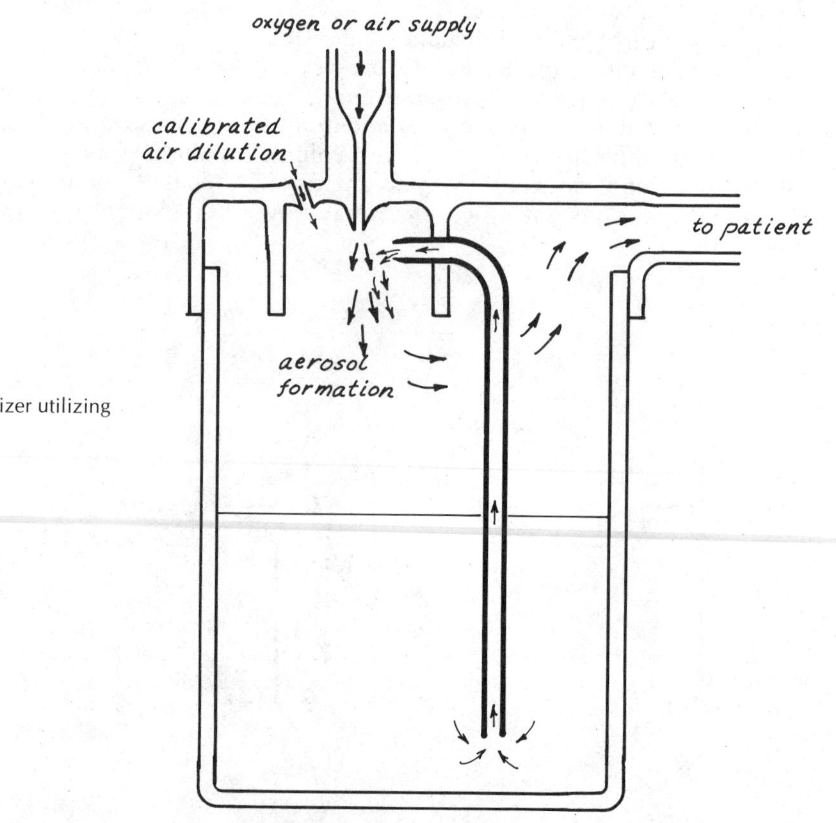

Figure 8–6 Schematic of a nebulizer utilizing the Bernoulli effect.

through an orifice. An orifice is defined as an opening for which the diameter is large compared to the length of the wall material, as with a hole in a thin-walled container. By contrast, a tube has a length which is large compared to its diameter. The rate of flow through a tube is determined by the pressure gradient, the diameter, and the viscosity of the fluid, as long as the flow is free of turbulence. As discussed in Chapter 5, the laminar flow rate through a tube of given length and diameter is then influenced mainly by the viscosity of the fluid. Flow through an orifice, like that through a tube, is dependent upon the pressure drop, or pressure difference between the two sides of the orifice. It is also proportional to the area of the orofice. But flow through an orifice is always turbulent flow, and the kinetic energy of this turbulence becomes more important than viscosity as a determining effect upon the flow rates of gases. Since the kinetic energy is proportional to the density of the gas, gases of different densities will flow at different rates through an orifice, even if their viscosities are the same. The viscosities of oxygen and helium are very similar but the density of oxygen is eight times as large at a given pressure and temperature. With a given orifice under identical conditions, the volume flow rate of the helium will be approximately $\sqrt{8} = 2.8$ times as great. That is, the volume flow rate is approximately proportional to the inverse of the square root of the density. A simplified theory of flow through orifices can be summarized by the proportionality:

$$\text{flow rate} \propto \frac{\sqrt{\text{pressure drop}} \times (\text{area of orifice})}{\sqrt{\text{gas density}}} \qquad \textbf{8–1}$$

This is an approximation, but it agrees fairly well with empirical experiments under clinical conditions.

Equation 8–1 provides the basis for understanding the operation of many types of flowmeters. The "rotameter" type of flowmeter consists of a gradually tapered glass tube containing a small float called a "bobbin." The tube is mounted vertically with the gas flow entering the bottom, as shown in Figure 8–7. The taper of the tube is greatly exaggerated to emphasize that the area for gas flow increases as the bobbin rises. In order to support the bobbin, the difference

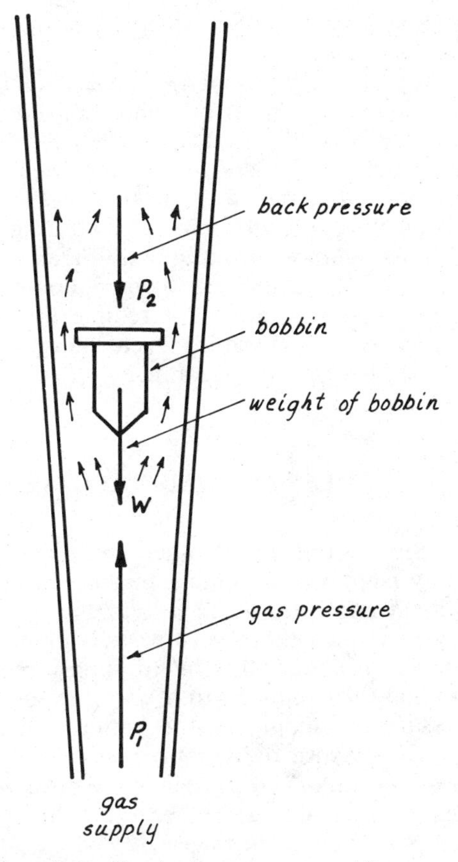

Figure 8–7 Sketch of rotameter type of flowmeter.

between pressures P_1 and P_2 must produce an upward force equal to the weight of the bobbin. For a gas of a given density the pressure difference $P_1 - P_2$ will depend upon the square of the volume flow rate divided by the square of the area of the orifice (equation 8–1). Thus, if the volume flow rate doubles, the area of the orifice must approximately double to keep the bobbin balanced. The balance condition is

$$\frac{(\text{volume flow rate})}{(\text{area of orifice})} = \text{constant}.$$

The area of the orifice is the area between the bobbin and the walls of the tube. Now if the volume flow rate is increased, an unbalanced upward force will be created and the bobbin will move upward until the orifice area increases by the same ratio. The height of the bobbin in the tube can then be calibrated in terms of a convenient volume flow rate scale, usually liters/minute. Other common flowmeters which operate on the same principle use stainless steel balls or other types of "floats."

It is very important to keep in mind that such flowmeters must be calibrated for each individual gas. Since gases differ in density, a flowmeter calibrated for oxygen will give erroneous readings if used to measure the volume flow rate of air or any other gas or mixture. Nurses or other attendants will often have both air and oxygen flowmeters at their disposal and must realize that they are not interchangeable, even though they may be identical in appearance.

PRESSURE REDUCING VALVES

Since oxygen and other gases are commonly supplied in high pressure cylinders with pressures up to 2200 lb/in^2, it is necessary to reduce the pressure. It is commonly reduced to about 50 lb/in^2 to supply respirators and then reduced to a very low positive pressure for supply to the patient. In addition to pressure reduction, it is desirable to have a pressure regulator so that a given pressure is maintained when the volume flow rate is changed. Otherwise, constant adjustment of the pressure reducing valve would be necessary when the resistance to flow varies. The regulating mechanism should also keep the output pressure constant when the pressure in the supply tank drops. The absence of such regulation would make continuous pressure adjustment necessary as the supply tank empties. Although these may seem like stringent requirements, they have been satisfactorily fulfilled by several varieties of valves. The underlying physical principles of such valves will be discussed here. See reference 1 for a more complete discussion of the details of various valves.

Figure 8–8 contains sketches of the basic mechanism of the pressure reducing and regulating valve. It is sometimes referred to simply as a pressure regulator, but it accomplishes both functions. The desired pressure is selected by adjusting the force exerted by the spring (A) on the flexible diaphragm (B). Once this pressure is selected, the valve will intermittently open to admit enough gas from the high pressure source to keep the pressure in the chamber up to the regulated value. The moving parts of the valve are operated by the combined action of three forces: (1) the force F_H exerted downward by the high pressure gas on the J-shaped valve seat, (2) the force F_L exerted upward on the diaphragm by the low pressure gas in the chamber, and (3) the force F_S exerted downward on the diaphragm by the pressure adjustment spring.

Figure 8–8(b) illustrates the equilibrium of forces which exists when there is no output flow from the pressure regulator. The chamber pressure P_L builds up to the preset value and the valve closes. The upward force F_L exerted on the diaphragm is much larger than the downward force F_H exerted by the high pressure. Although the supply pressure P_H is much greater than P_L, the low pressure acts on a much larger area A_L, the area of the diaphragm. Thus the force $F_L = P_L \times A_L$ is much greater than $F_H = P_H \times A_H$, where A_H is the tiny area of the orifice leading from the high pressure supply. The force of the adjustment spring, F_S, acts downward on the diaphragm. When these three forces are at equilibrium, $F_S + F_H = F_L$, then the valve is in the closed position.

Figure 8–8(c) illustrates the condition when the output stopcock is opened and gas flows from the regulated chamber. The pressure P_L is reduced and therefore the force F_L is reduced, upsetting the equilibrium condition described above. A net downward force results which pushes the valve seat down and admits gas from the high pressure source. As flow is maintained, a dynamic equilibrium is established with just enough gas entering the chamber from the high pressure source to maintain the chamber pressure very close to the regulated value. Changes in the output volume flow rate will quickly be compensated by changes in the valve opening. Thus the output pressure, P_L, with remain essentially constant over a wide range of flow conditions. This accomplishes the first requirement for a regulator.

If the regulated pressure were directly proportional to the tank pressure, then when the tank was half-emptied the output pressure would have dropped by 50%. The second requirement for the regulator is that it be independent of the tank pressure so that the regulated output pressure does not change as the tank empties. The fact that this requirement is approximately met by this type of valve can be seen by examining

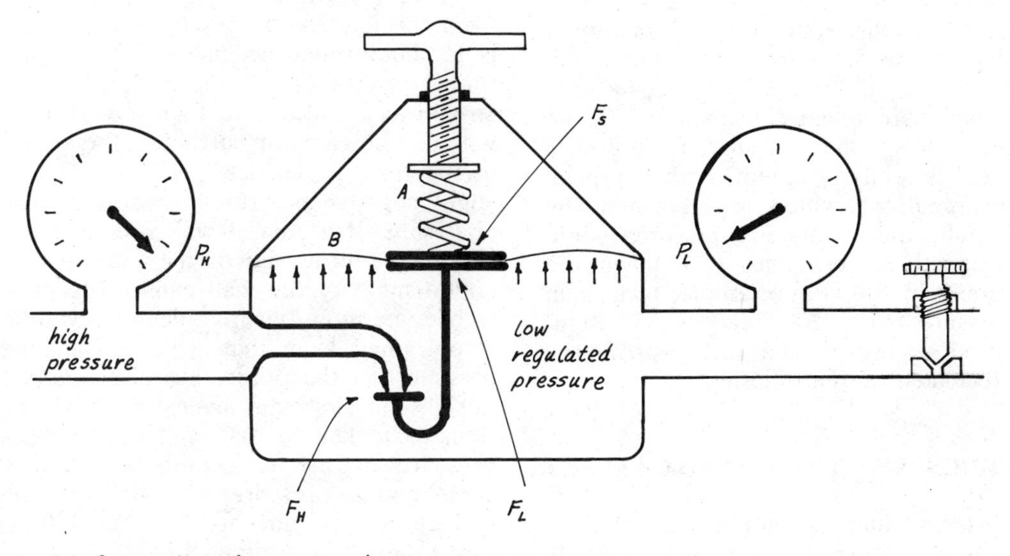

(a) Schematic of valve mechanism

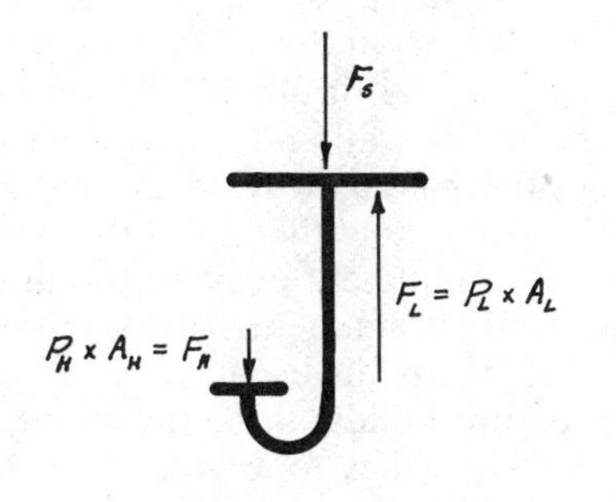

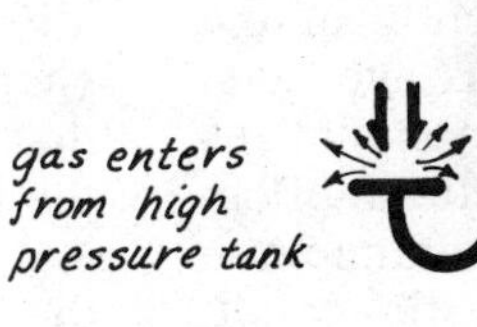

(b) Balance of forces when pressure is at regulated value. The valve is closed.

(c) The flow from the output of the regulator tends to drop the pressure. The valve opens to bring it back to the regulated value.

Figure 8–8 The pressure regulator.

the equilibrium of forces illustrated in Figure 8–7(b):

$$F_L = F_S + F_H.$$

The force F_H exerted by the high pressure gas is quite small compared to the forces F_L and F_S. Since the equilibrium is mainly a balancing of the large forces F_L and F_S, a change in the small force F_H produces a relatively small percentage change in the force F_L and thus the regulated pressure $P_L = F_L/A_L$ is well regulated. With a typical pressure regulator which is set when the tank is full, the regulated pressure would have dropped by less than 10% when the tank is only ¼ full. For example, a drop in tank pressure from 2200 lb/in² to 550 lb/in² might produce less than a 10% drop in the preset regulated output pressure.

FLUIDICS AND THE COANDA EFFECT

The term ''fluidics'' applies to the class of devices which uses the dynamics of fluids in enclosed passageways to accomplish a number of tasks such as switching, pressure and flow sensing, and amplification. One major characteristic of this rapidly expanding area of technology is the fact that all these functions are carried out with no moving parts (3). The devices utilize only the driving force of the fluid and the interaction of the fluid with precisely contoured passageways. The absence of moving parts minimizes maintenance and makes possible the fabrication of the devices from glass or ceramic materials which can withstand caustic conditions.

One of the basic phenomena involved in the operation of fluidic devices is the ''Coanda effect'' or the ''wall attachment effect.'' This effect occurs when air or another fluid is forced through a small orifice or jet which has properly contoured surfaces downstream from the jet. As discussed previously, if air is forced through a small orifice, it will entrain surrounding air into the driving air stream as illustrated in Figure 8–9(a). There will also be some turbulence in the air stream in the form of vortices (eddies). There is a slight negative pressure on each side of the airstream. If a curved wall is added on one side of the jet as shown in Figure 8–9(b), the confinement of the wall causes the pressure to become more negative near the wall as air is entrained from that region. The ambient pressure on the other side of the airstream pushes the airstream against the wall, and it remains locked to that wall by the negative pressure at the wall until interrupted by some counterpressure. Coanda was able to deflect an airstream through a full 180-degree turn by extending the wall contour.

The application of the Coanda effect to a respirator valve (4) is illustrated in Figure 8–10. When air or oxygen is directed through the nozzle, it quickly becomes attached to the lower wall and follows the contour through the passageway to the patient's airway. It is remarkable that even though the top channel is completely open to the atmosphere, the wall attachment is so effective that very little of the air escapes. Eventually the back pressure builds up to the point that it causes the airstream to break away from the contour and become attached to the more remote wall in the top channel, initiat-

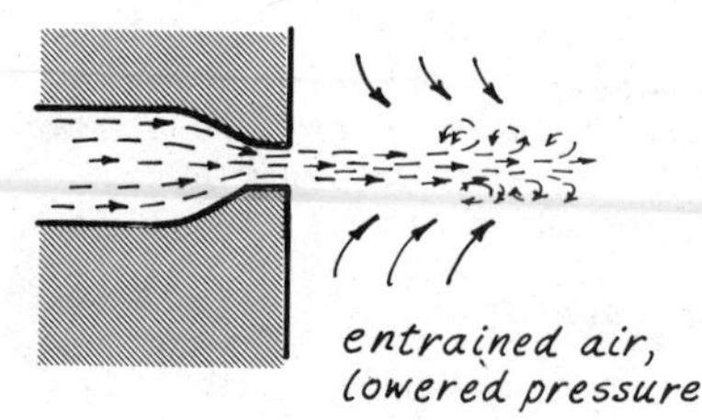

a. Entrainment into the driving airstream

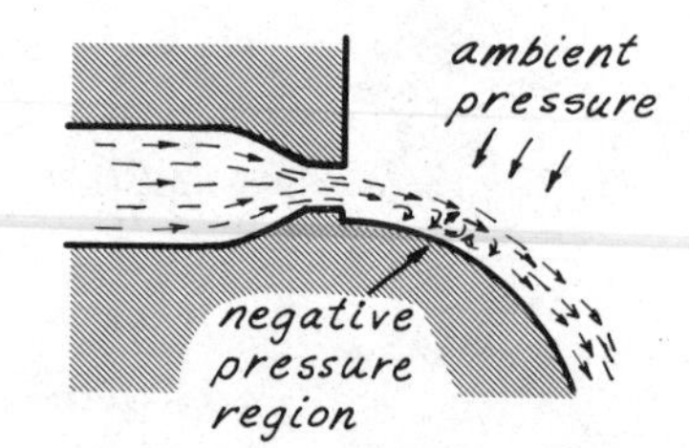

b. Wall attachment initiated by negative pressure near wall.

Figure 8–9 The Coanda wall effect.

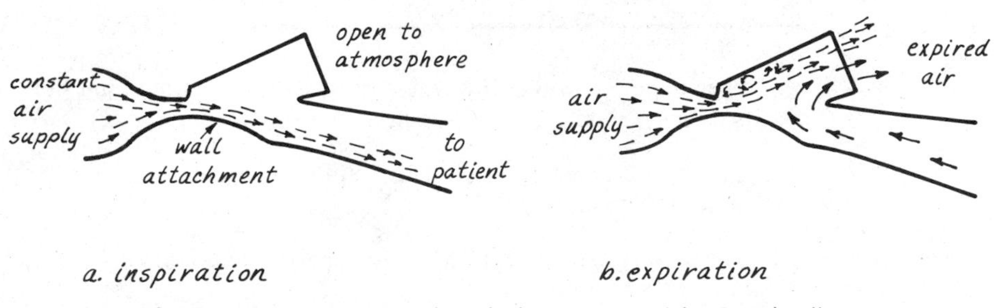

Figure 8–10 A respirator valve which makes use of the Coanda effect.

ing the expiration phase. When the backflow of exhalation ceases, the airstream will again become attached to the preferred contour in the lower channel. Thus, the periodic respiration process can take place with a steady, uninterrupted flow of the supply gas and with only the interaction of the airstreams and the channel contours to accomplish the switching process. Therefore, no moving parts are used.

The fluidic respirator valve is incorporated with timers, flow rate controls, and other devices, all fluidically operated, to construct complete respirators which have no moving parts. If a jet is interfaced with two identical contours, a fluidic switch is created, since a small puff of air from a control orifice can switch the flow from one channel to the other. Fluidic devices can be constructed in complete analogy to many electronic devices. With multiple contours and interacting air channels, fluid amplifiers, fluid memory devices, and even fluidic computers are feasible.

SUMMARY

A float device with a variable orifice can be calibrated to measure the volume flow rate, but since the calibration depends upon the gas density, flowmeters creased velocity, with the accompanying lowered pressure in the constriction, is referred to as the Bernoulli effect. If the tube is gradually contoured to a very small constriction or nozzle, the pressure may be lowered below atmospheric pressure. Devices using this principle, often called Venturi tubes, are used to provide suction for the entrainment of a second fluid into the mainstream flow. Such devices find many medical applications in nebulizers, anesthesia gas mixers, and so forth.

The nature of fluid flow through an orifice is used to measure volume flow rates of gases. The volume flow rate through an orifice is approximately described by the proportionality

$$\text{volume flow rate } \alpha \sqrt{\frac{\text{pressure drop}}{\text{gas density}}} \times \text{area of orifice.}$$

A float device with a variable orifice can be calibrated to measure the volume flow rate, but since the calibration depends upon the gas density, flowmeters for different gases are not interchangeable.

The forces acting on the diaphragm in a pressure reducing valve open and close an orifice to keep the output gas pressure nearly constant over a wide range of input pressures. This makes practical the use of high pressure cylinders to supply the low gas pressures needed for anesthesia and respiratory therapy.

REVIEW QUESTIONS

1. What happens to the flow velocity, volume flow rate, and fluid pressure as a fluid enters a constricted area in a tube?

2. How can a constricted area in a tube be used to entrain another fluid into a flowing fluid?

3. Name some examples of the use of the Bernoulli effect for entrainment.

4. Why can't an oxygen flowmeter be used to regulate the flow of air to a patient?

PROBLEMS

1. Liquid pressure represents potential energy per unit volume. Calculate the energy per unit volume represented by a pressure of 20 cm H_2O in the CGS unit dynes/cm_2 (which is equivalent to dynes · cm/cm^3 or erg/cm^3). Calculate the gravitational potential energy per cm^3 of water at a height of 20 cm.

2. If a 60 cm tube with flow resistance 950 dynes sec/cm^5 were used to make a siphon for removing water from a large container, and the lower end of the tube were 20 cm below the surface of the water, what would you expect the volume flow rate through the siphon to be once it was established and flowing smoothly?

   Note: Problems 3 to 7 are designed to lead the interested student through a calculation of the pressure and energy relationships in a tube with a constriction in it (Bernoulli effect). Problems 3, 6, and 7 can be worked independently of the others.

3. Use the resistance expression from Chapter 5 (Poiseuille's law) to calculate the resistance to water flow for a 60 cm tube of internal radius 0.2 cm. Calculate the resistance of a 5 cm long tube of internal radius 0.1 cm. The appropriate CGS unit is dynes · sec/cm^5.

4. Suppose the 60 cm tube above were cut and the 5 cm tube inserted so that the flow goes through 40 cm of the large tube, then through the small tube, and then through the final 20 cm of the large tube.

   a. What total resistance to flow of water would this system represent if the resistance of the connections could be ignored and no turbulence was produced?
   b. What volume flow rate would result if a pressure of 5 cm H_2O was supplied to this system?
   c. What would be the average liquid speed in the large and small tubes?

5. If the tubing system described in problem 4 above is assumed to have a volume flow rate of 2.2 cm^3/sec when supplied with a pressure of 5 cm H_2O, what would be the pressure at the end of the 40 cm length of large tubing before entering the constriction?

6. The calculation of kinetic energy per unit volume $\frac{1}{2}dv^2$ for a liquid flow process is complicated by the fact that the speed v is not constant across the tube. To calculate the average kinetic energy per unit volume one must evaluate $\frac{1}{2}d(\overline{v^2})$ and for varying speed $(\overline{v^2})$ is not equal to $(\overline{v})^2$, i.e., the average of the square of the speed is not equal to the square of

the average speed. For a certain model of the speed variation (parabolic speed distribution), $(\overline{v^2}) = 1.2(\overline{v})^2$ and the energy per unit volume is

$$\tfrac{1}{2} d(\overline{v^2}) = \frac{1.2}{2} d\,(\overline{v})^2 = \frac{1.2}{2}\frac{d\mathscr{F}^2}{A^2}$$

where A is the inside area of the tube. Using this model, calculate the energy per unit volume in the 0.2 cm radius tube and the 0.1 cm radius tube if $\mathscr{F} = 2.2$ cm³/sec.

7. If the kinetic energy per unit volume in the tubing system described in problem 4 is 185 erg/cm³ in the large tube and 1470 erg/cm³ in the small tube, how much pressure drop will be experienced as the water flows from the larger tube into the smaller tube? Express the pressure drop in cm H_2O. (Ideal conditions with no turbulence are required for such a calculation to approach experimental results.)

8. A pressure regulator like that described in Figure 8–8 has a low pressure membrane area A_L of 6 in² and the area of the high pressure outlet orifice is $A_H = 0.01$ in².

   a. If the tank pressure $P_H = 2200$ lb/in² and an outlet pressure 50 lb/in² is desired, what force F_S must be exerted by the regulator spring?
   b. If the tank pressure drops to 550 lb/in² (¼ tank) and no changes are made in the regulator setting, what will be the outlet pressure?
   c. How low will the tank pressure be when the outlet pressure drops to 47 lb/in²?

REFERENCES

1. Macintosh, R., Mushin, W. W., and Epstein, H. G. *Physics for the Anaesthetist*, 3rd ed. Oxford: Blackwell Scientific Publications, 1964.
2. Egan, Donald F. *Fundamentals of Inhalation Therapy*. 3rd ed. St. Louis: C. V. Mosby Co., 1977.
3. Angrist, Stanley W. *Fluid Control Devices*. Scientific American, December 1964. p. 80.
4. Burns, Henry L. *A Pure Fluid Cycling Valve for Use in Breathing Equipment*. Inhalation Therapy. Vol. 14, p. 11, 1969.

CHAPTER NINE

Molecular Phenomena Related to Biological Processes

INSTRUCTIONAL OBJECTIVES

After studying this chapter, the student should be able to:

1. Define diffusion, osmosis, and dialysis and explain the origin of the energy which enables such processes to proceed.
2. Define osmotic pressure and explain the meaning of the terms hypotonic, hypertonic, and isotonic.
3. Explain the importance of using isotonic or near-isotonic solutions for injection into the blood.
4. Describe with a sketch the balance of osmotic and hydrostatic pressures which accomplishes the supply of nutrients on one end of the capillaries and the collection of waste materials on the other.
5. Describe the role of diffusion processes in moving molecules across living cell membranes and give examples of processes which act to transfer molecules in the opposite direction.
6. Explain how intermolecular attraction acts to produce capillary action, surface tension, and viscosity.

There are numerous macroscopic processes in living organisms which depend rather directly upon the properties of molecules. Many of these processes are based upon two physical facts:

(1) Molecules at ordinary temperatures have a large amount of kinetic energy.

(2) In a liquid or a solid there are large attractive forces between molecules.

The large kinetic energy of molecules is related directly to such phenomena as diffusion, osmosis, and dialysis. The attractive forces between molecules account for such phenomena as cohesion and adhesion, adsorption, viscosity, surface tension, capillarity, and others.

THE KINETIC ENERGY OF MOLECULES

A glass of water or a piece of metal may appear to be at rest, but on a molecular scale this is far from being true. At ordinary room temperatures, the molecules of solids, liquids, and gases move continuously and at very high speeds. It is necessary to gain some mental picture of this ceaseless activity to understand the processes of diffusion, osmosis, and dialysis.

Though the phenomena discussed in this chapter are primarily liquid and solid state phenomena, it is instructive to take a look at the nature of the motion of gas molecules. Free of the attractive forces which hold liquids and solids together, gas molecules travel randomly and at very high speeds. As an example, consider nitrogen, the major constituent of the air. A nitrogen molecule is extremely small, having a diameter of about 1.8×10^{-10} meters and a mass of about 4.7×10^{-26} kilograms. But with its diminutive size it has an excessive amount of activity. At the freezing temperature of water (0°C) it has an average speed of about 490 m/sec. This is equivalent to about 1610 ft/sec or 1100 miles/hour, which is about the speed of a bullet from a high-powered rifle. Since there are about 3×10^{19} other molecules with similar speeds and dimensions in every cubic centimeter of air, it will collide with about 5 billion other molecules each second.

If the gas is confined in a container, the gas molecules will make billions of collisions with the walls each second. Each collision will exert a force on the wall. The average effect of these collisions causes the pressure of a gas on the walls of its container.

In the solid and liquid states, the motion of molecules is limited by their mutual attractive forces and by the fact that there is less empty space in which to move. But this does not imply that they are at rest — that is far from being true. The molecules in a solid are limited to a fairly small volume of space, because they are generally held in periodic array called a lattice. The limitation of motion by the attractive forces of the surrounding molecules in the lattice gives solids their characteristic rigidity. But within the confines of their limited space they move or vibrate back and forth very rapidly. This oscillating motion is so similar to the motion of a mass on a spring that a system of springs is often used as a model for the forces in a solid lattice. The molecules of a liquid are less limited since there is no periodic structure or "lattice" in a liquid. They move about each other more or less like sticky spheres with a mutual attraction but no preferred order.

It might seem that the nature of these constraints upon the molecules of solids and liquids would keep the speeds of individual molecules very low. But when the average kinetic energies of the molecules are analyzed, they correspond to speeds of several hundred miles per hour in solids and liquids at room temperature. Thus, despite the attractive forces, the molecules are extremely agitated and have a large amount of kinetic energy.

DIFFUSION

Diffusion refers to the process by which molecules intermingle as a result of their random motion. It is convenient to discuss the diffusion across an interface or boundary between two different types of substances. As an example, consider two containers of gas, *A* and *B*, which are separated by a partition as shown in Figure 9–1(a). The molecules of both gases are in constant motion and make numerous collisions with the partition separating the gases.

If the partition is removed as in Figure 9–1(b), the gases will mix because of the random velocities of their molecules. In time a uniform mixture will be formed with the same number of *A* type molecules in each unit of volume, and likewise a uniform distribution of *B* type molecules. This tendency toward a uniform distribution is due to the nature of the random statistical process. The tendency toward a uniform distribution might be better understood if the diffusion of a single gas were considered. If only the *A* type molecules were placed in the container in Figure 9–1(a) and the other side were a perfect vacuum, the *A* type molecules would diffuse toward the right side when the partition was removed. For the sake of discussion, assume that there are 20 of the *A* molecules. Now at the time when there are 15 molecules on the left side and 5 on the right side, the diffusion will still be proceeding to-

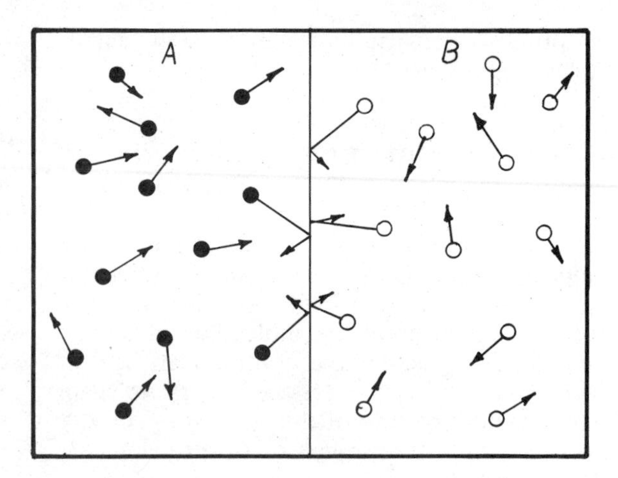

(a) Gases separated by a partition

Figure 9–1 Diffusion.

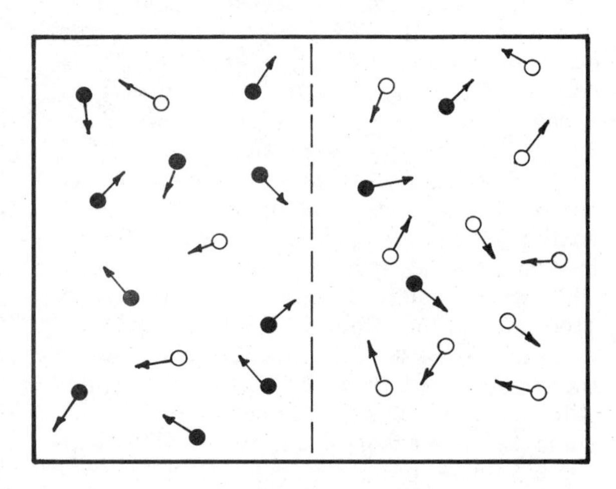

(b) Diffusion of gases across an interface

ward the right side. Although one of the five molecules might cross the boundary to the left, it is three times as likely that one of the left-hand molecules will cross over to the right. Once there are 10 molecules on each side, the diffusion will stop because there will be no further net transfer of *A* type molecules across the interface. This does not imply that none are crossing; it just means that as many are crossing to the left as to the right. Therefore an equilibrium condition exists.

This example illustrates the very basic and important physical fact that molecules will tend to move from an area of larger concentration to an area or lesser concentration of that type molecule. The energy which accomplishes this net transfer of molecules is the random kinetic energy of the molecules. Although the process occurs most rapidly in gases, it also occurs in liquids and even solids. It is safe to say that when any two dissimilar substances are in contact with each other, diffusion is taking place.

Since diffusion is the result of molecular motion, anything which increases the motion will speed the diffusion. Heating and mechanical stirring will hasten the diffusion process. For example, if a lump of sugar is dropped into a cup of coffee, the sugar molecules will begin to diffuse into the coffee and vice versa. Soon, there will be a thin layer of coffee around the sugar; that layer is saturated with sugar. In time, the sugar molecules would diffuse throughout the cup to provide a uniform solution. But we are usually impatient and stir the coffee rather than wait for the slow diffusion process. This example not only demonstrates the tendency of all substances to diffuse into an area of lesser concentration, but also exposes some of the exceptions to the simplified picture of diffusion presented here. It is easy to demonstrate that if a sufficient amount of sugar is put into the cup, there will be some which cannot be forced into solution. The solution is said to be saturated, and there will be no further diffusion of sugar into the liquid. It has reached the point where the cohesive forces tending to hold the sugar together as a solid are stronger than the remaining dynamic forces which tend to hold it in solution.

OSMOSIS

While there is some disagreement about a precise definition of the word osmosis, it is usually used to describe the diffusion of a solvent across a membrane while the dissolved matter, or solute, is left behind. In biological processes, this solvent is usually water. The process is perhaps best described by example. In Figure 9–2(a), two solutions of differing concentration are separated by a membrane. Both the solvent molecules and the solute molecules possess a large amount of kinetic energy. If the membrane were removed, the solute molecules would diffuse to the left. A homogeneous solution would result.

Both solvent and solute molecules are bombarding the membrane and trying to penetrate. If the membrane were permeable to both types of molecules, diffusion would occur in both directions and the two solutions would approach equal concentrations. If the membrane is permeable to the solvent molecules but blocks the passage of the solute molecules, it is said to be *semipermeable*. Mainly because of their kinetic energy, the solvent molecules will diffuse across the membrane in *both* directions. Since there are more solvent molecules per unit volume on the left, more of them will successfully penetrate the membrane, resulting in a decrease in volume on the left side and an increase on the right (Figure 9–2(b)). The transfer of a given type of molecule is always from the location where it is more concentrated to an area of lesser concentration. It can be said to transfer down the "concentration gradient."

The common terminology used with solutions sometimes causes some confusion at this point. For example, a solution of salt in water is said to be a "concentrated" salt solution if there is a large amount of salt and thus a smaller amount of water per unit volume. In Figure 9–2 the solution on the right would be said to be more concentrated, since there is a larger number of solute molecules per unit volume. In discussion of solutions, the word "concentration" usually refers to the solute molecules, whereas the concentration referred to in discussion of diffusion and osmosis is the concentration of the molecules which are diffusing. In osmosis with an aqueous salt solution, the water molecules diffuse from where they are most concentrated to the point where they are less concentrated. In solution terminology, this is from the less concentrated *salt* solution to the more concentrated. Keeping in mind the basic physical process which is taking place will help to avoid this confusion. If agitated molecules are bombaridng a membrane from both sides, it is obvious that the net transfer will be from the side where there are many to the side where there are few molecules.

As mentioned above in reference to Figure 9–2, the volume on the right side will increase as a result of the net transfer of solvent molecules. Since the liquid level on the right is higher, the fluid pressure acting on the right side of the membrane is greater than that on the left. This fluid pressure tends to cause the solvent molecules to move back across the membrane and thus opposes the net diffusion from left to right. The kinetic energy of the solvent molecules on the left provides the energy required to raise the fluid level on the right. When the difference in height of the two liquids becomes great enough, the hydrostatic pressure will prevent further net transfer across the membrane and an equilibrium condition will have been established. The maximum difference in height is a measure of the difference in "osmotic pressure" between the two solutions.

The osmotic pressure of solutions is a very important concept in physiology since it is related to the transfer of fluids in the

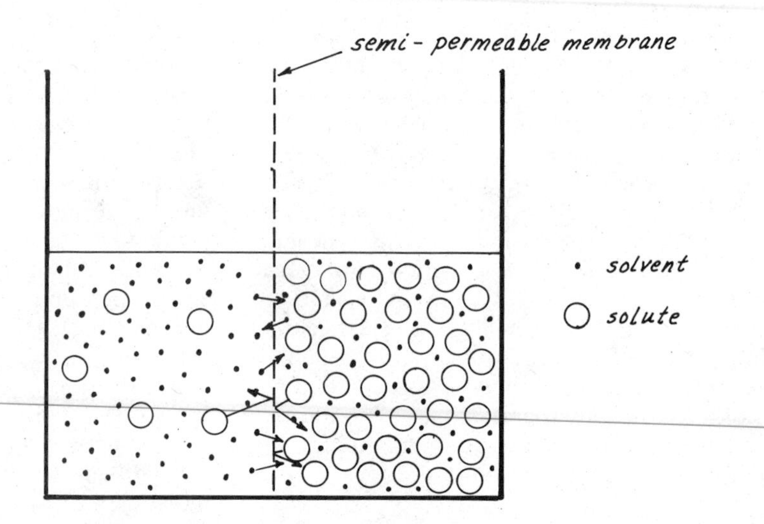

(a) The solvent molecules diffuse from an area of high concentration to an area of lower concentration but the solute is blocked by the membrane.

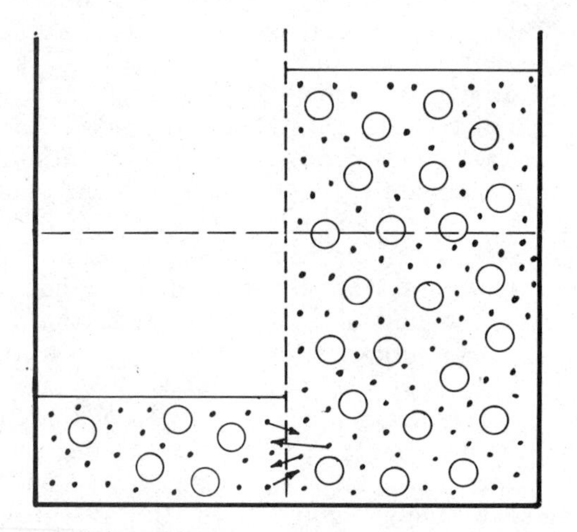

(b) The net migration of the solvent molecules to the right yields a difference in the height of the liquids which is proportional to the osmotic pressure difference.

Figure 9–2 Illustration of osmosis.

body. Yet it is often inconsistently used in medical literature. The osmotic pressure of a solution is defined with respect to the pure solvent. If the pure solvent and a solution were separated by a semi-permeable membrane, there would be a net diffusion from the pure solvent to the solution. However, if a sufficiently large hydrostatic pressure were exerted on the solution side, the net transfer would be stopped, or even reversed. The *osmotic pressure* of the solution can be defined as the pressure required to prevent the diffusion from a pure solvent into that solution. It is usually measured in mm of Hg, cm of water or some other unit related to the height of a liquid column. It could be measured in atmospheres, lb/in^2 or any other convenient pressure scale.

If the solution is made more concentrated by adding more solute molecules and decreasing the number of solvent molecules, the concentration gradient across the membrane is increased. There will be a stronger tendency toward net transfer of solvent molecules from the pure solvent, and a larger pressure will be required to overcome that tendency. Thus, a solution with more dissolved matter has a higher osmotic pressure. It must be recognized that the osmotic pressure is a measure of the tendency for solvent transfer *into* the solution across a membrane. The awkwardness of this concept is probably sufficient explanation of its confused use in the literature.

If two solutions, *A* and *B*, which are separated by a semi-permeable membrane have the same solvent concentration, there will be no net transfer across the membrane. The solutions are said to be *isoosmotic* or *isotonic*, and it could be said that their osmotic pressures are balanced. If solution *A* is more concentrated than *B*, then it has a higher osmotic pressure and is said to be *hyperosmotic* or *hypertonic* with respect to *B*. Solution *B* is *hypoosmotic* or *hypotonic* in reference to *A*. Since *B* is less concentrated and thus has a larger number of solvent molecules, diffusion will move solvent molecules from *B* to *A*. The solvent will move from the hypotonic to the hypertonic solution. Figure 9–3 may help with the visualization of the direction of solvent motion during osmosis. Since all of these terms are used widely in medical and biological literature, it is important for the life-sciences student to be thoroughly familiar with them.

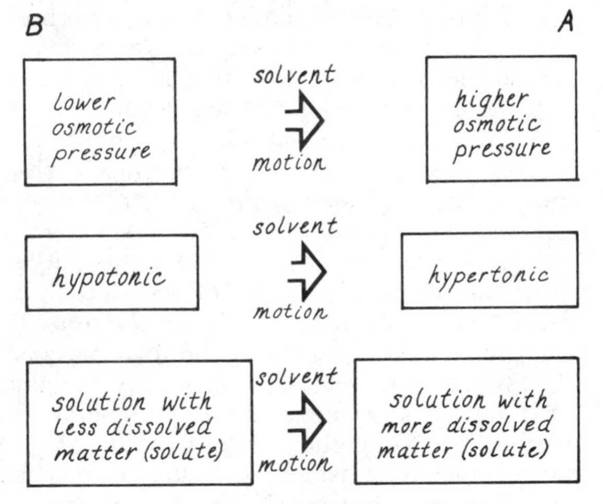

Figure 9–3 The direction of solvent motion in osmosis.

The terms isotonic, hypotonic, and hypertonic are relative terms and must be used with respect to some reference solution or solvent. When they are used for fluids in the body, the plasma is usually the reference fluid. An isotonic saline solution is one which would cause no water transfer across a membrane if normal plasma were on the other side of the membrane. The fluid inside red blood cells is isotonic with respect to the plasma. If an osmotic imbalance were to occur, water would cross the membrane until the solvent concentrations were equalized. The diffusion across a membrane always proceeds toward the isotonic condition. (Other processes can maintain an unbalanced condition, as discussed below.) A solution of about 0.9% sodium chloride is isotonic with the plasma and thus with the red blood cells. If a red blood cell were placed in such a solution, there would be no net transfer of water across the membrane. If a red blood cell were placed in pure water, the water would rapidly cross the membrane into the cell and it would quickly burst. On the other hand, if the red blood cell were placed in a 1.5% sodium chloride solution (hypertonic), water would pass out of the cell and it would shrivel up.

The osmotic pressures of solutions can be quite large. For example, the osmotic pressure of blood is about seven atmospheres. When the effects of one atmosphere of pressure are considered (Chapter 6), this is seen to be an enormous pressure. One atmosphere of pressure can push water to a height of 34 feet in an evacuated column. If the dilution effects could be counterbalanced, pure water diffusing across a semipermeable membrane into blood could push it to a height of over 225 feet (blood has a specific gravity of about 1.05). In the body, of course, the blood is never interfaced with pure water; the osmotic pressure differences across body membranes are always fairly small. Considering the above example, it is clear that osmotic pressure offers a possible mechanism for getting water to the top of trees. The largest known tree in total mass is the General Sherman sequoia, having a height of 276 feet. If osmotic pressure alone were responsible for getting water to the top, it would require an osmotic pressure of about 8 atmospheres. This is well within the range of osmotic pressures of solutions. A molar solution of a nonionizable solute will have an osmotic pressure around 22.4 atmospheres.

DIALYSIS

The term osmosis is usually limited to those processes in which only one material, the solvent, is transported across the membrane. In many cases in living organisms there are membranes which are permeable to several types of molecules. For example, certain membranes in the body are permeable to water, salts, glucose, urea, and other small organic molecules. However, they normally block the passage of larger molecules such as hemoglobin, globulin, albumin, and other large protein molecules. Such membranes might be said to be "selectively permeable." This process, involving the diffusion of several types of molecules through a selectively permeable membrane, is referred to as "dialysis."

The most common use of the term "dialysis" is with reference to the kidney function. The removal of waste materials from the blood takes place in small bodies called nephrons in the human kidney. There are about a million nephrons in each kidney. (References 2 and 3 are suggested for thorough descriptions of the kidney.) For our purposes, a nephron can be considered to have two main parts, the glomerulus and the tubule (Figure 9–4). The glomerulus is a coiled ball of thin-walled capillaries which is enclosed by a section of the tubule called Bowman's capsule. Most of the constituents of the blood diffuse into the tubule from the glomerulus, crossing two membranes. Only the very large molecules such as proteins are left behind in the capillaries.

This process might be thought to be dialysis in the sense that selectively permeable membranes are involved, but it does not proceed by diffusion. It is more properly termed "glomerular filtration" since it is caused by the hydrostatic pressure difference across the membrane and actually works against the preferred diffusion direction. Once in the tubules, however, about 99% of this solution is reabsorbed through the tubule walls into the bloodstream. Most of the water and almost 100% of the glucose are reabsorbed, along with a large percentage of the salts. The percentage of reabsorption for each important constituent has been calculated, based on the relative concentrations of the substances in the urine and in the plasma. A glomerular filtration rate of 125 cm^3/minute was assumed in order to calculate the percentages shown in Figure 9–4. This remarkable process of reabsorption of the vital body fluids, with the resultant concentration of the urine, proceeds largely by diffusion and could be referred to as dialysis. However, diffusion alone does not seem to be sufficient to explain fully the reabsorption percentages, particularly for sodium and potassium. Apparently some active transport mechanisms are also operating, as discussed below in the section on living membranes.

The term dialysis is perhaps most appropriately used with respect to the function of the artificial kidney, which is commonly referred to as a "dialysis machine." Basically, the artificial kidney operates by removing blood from the body and passing it along one side of a selectively permeable membrane. Large volumes of a fluid called the "dialysate" flow on the other side of the membrane. The dialysate is a carefully controlled solution of electrolytes (sodium, potassium, etc.) and other chemicals which approximate the normal blood in concentration. Urea, creatinine, and other waste molecules diffuse across the membrane and are carried away, while the red blood cells and large particles and molecules in the blood are retained. If the ion concentration in the blood is abnormally high, there will be a net diffusion of these ions out of the blood so that the electrolyte concentration can be adjusted during dialysis. The dialysate is often subjected to a negative pressure to speed the removal of

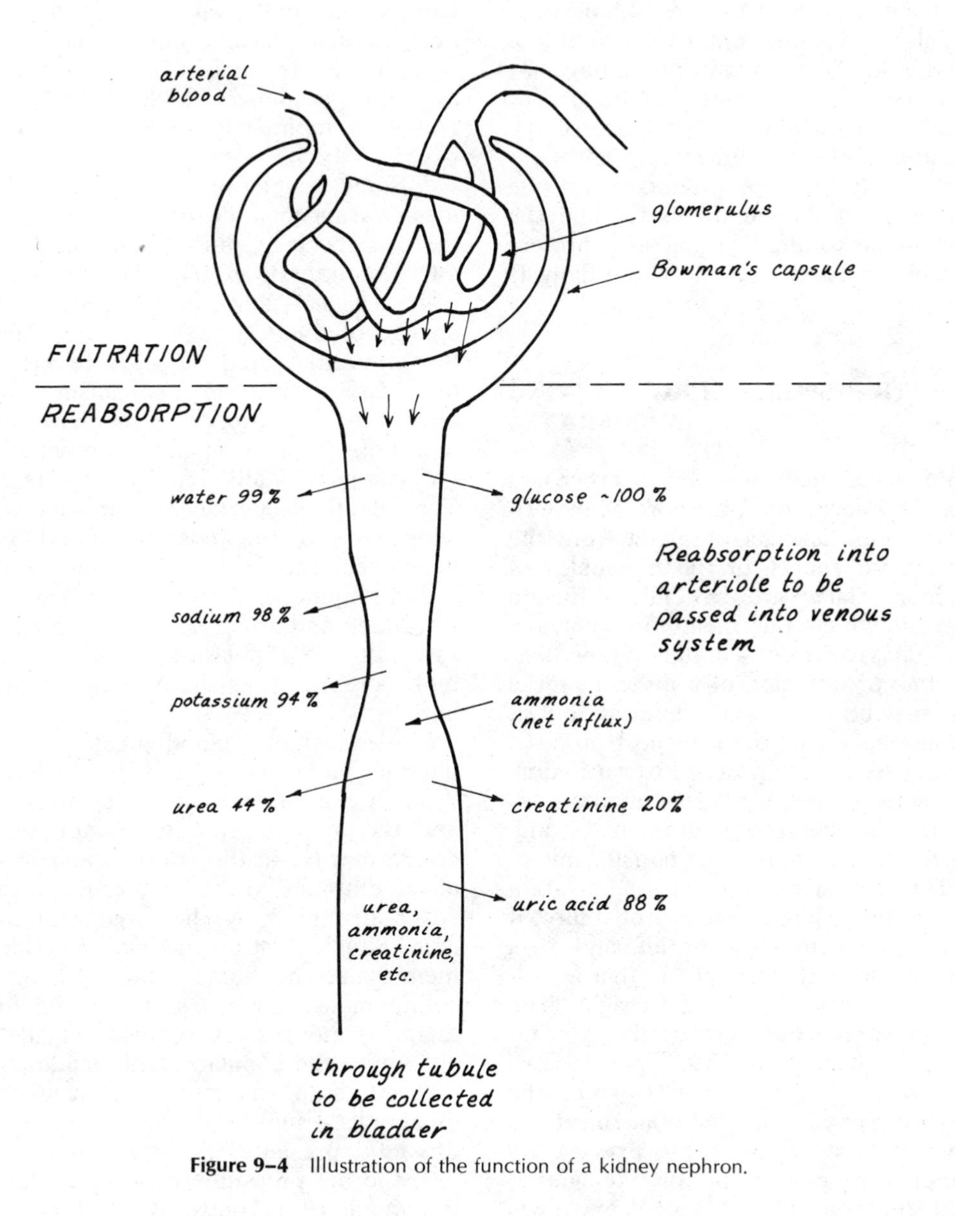

Figure 9–4 Illustration of the function of a kidney nephron.

excess water from the blood. The dialysate is carefully regulated so that the osmotic pressure difference across the membrane does not become too large. For example, if pure water were substituted for the dialysate, the water would diffuse into the blood extremely rapidly. The blood would become hypotonic with respect to the red blood cells and enough water would enter them to cause them to burst. The dialysis procedure has proven to be quite successful when proper care is taken to maintain the balance of the concentrations. The careful mixing and control of the dialysate, the production of the negative pressure on the dialysate, and the operation of elaborate warning and protection devices are the functions of the dialysis machine.

TRANSPORT ACROSS LIVING MEMBRANES

Osmosis and dialysis refer to processes in which the energy for transport of materials across membranes is obtained from the inherent kinetic energy of the molecules of the materials. They are basically diffusion processes in which the membrane plays a completely passive role. The membrane controls the rate of diffusion of a given material by its permeability to that material, but it supplies no energy for the transport process. Such processes always proceed toward equal concentrations of a given material, i.e., toward a zero concentration gradient of any material which can move through the membrane. This tendency toward equalization operates on the basis of simple probability; if there are 10 molecules on the left and 5 on the right, then it is twice as likely that a molecule will penetrate from left to right. The net transfer approaches zero as the concentrations equalize. If these diffusion processes were the only transport processes, then the only way to maintain unequal concentrations across a membrane would be to exert a hydrostatic pressure on the fluid on one side.

When the concentrations of some types of molecules in living organisms are examined, it becomes evident that the "passive transport" mechanisms of osmosis and dialysis are insufficient to explain them. Unequal concentrations of certain substances are maintained across cell walls and other membranes without appreciable hydrostatic pressure differences. Often molecules are transported across membranes in the direction opposite to that dictated by diffusion probability. Such transport requires that energy be supplied to the molecules by the membrane, since they are being moved "uphill" against the normal osmosis or dialysis process. This is referred to as "active transport" and can be accomplished only by living membranes which have a source of energy. Although the mechanisms for active transport are not completely understood, there are many examples of such transport in the body.

Some insight into the various transport mechanisms may be obtained by examining the transport of fluids from the capillaries into the interstitial fluid and then into the cells. The first stage of this process, the exchange of fluids between the capillaries and the interstitial fluid, appears to proceed by the passive diffusion mechanism. The plasma, with the exception of the very large protein molecules, is capable of moving through the capillary walls into the tissue spaces. The selectively permeable capillary walls are subjected to the positive blood pressure which tends to force the plasma out. Opposing this hydrostatic pressure is the tendency for water and small solute molecules to diffuse into the capillaries because of the "osmotic pressure" of the proteins in the capillaries.

Although the blood pressure in the capillaries varies widely, there is a consistent drop in pressure between the arteriolar end and the venular end. Characteristic values are 35 mm Hg at the arteriole and 15 mm Hg at the entrance to the venule. Because of the space occupied by the large protein molecules which exist on the capillary side of the membrane, the number of water and small solute molecules is less there and diffusion tends to move more of these smaller molecules into the capillary. This tendency is expressed as an "osmotic pressure difference" of about 22 mm Hg which opposes the hydrostatic pressure. As illustrated in Figure 9–5, the net pressure at the arteriolar end of the capillary is outward. A representative pressure is 35 mm Hg − 22 mm Hg = 13 mm Hg. On the venular end, the osmotic pressure is predominant and an inward pressure of 22 mm Hg − 15 mm Hg = 7 mm Hg causes fluid transfer back into the capillary.

The blood albumin constitutes a considerable fraction of the "large molecules" which stay in the capillaries and which contribute to the osmotic pressure difference illustrated in Figure 9–5. If a person loses albumin from the blood, then the balanced capillary processes described above are upset. The fluid balance is shifted toward the tissue, since more fluid leaves the capillaries on the arteriolar end and less returns to the capillaries. The resulting collection of fluid in the tissues is referred to as edema.

The balance can also be upset by a loss of salt from the blood. The blood and the interstitial fluid become hypotonic with respect to the cells. Water molecules tend to diffuse into the cells at a greater than normal rate, depleting the interstitial fluid.

The diffusion mechanism is always present, but is not sufficient to explain the transfer of fluids from the interstitial fluid into the cells. The various cells of the body have different needs and maintain different concentration gradients with respect to the plasma. The cell membranes meticulously regulate the interchange of substances between the interstitial fluid and the cytoplasm of the cell. Particularly in the case of electrolytes, a large concentration gradient may be maintained across the cell wall. For example, the red blood cell contains about 20 times as much potassium as sodium, whereas the plasma surrounding it has about 20 times as much sodium as potassium. Part of this capability can be explained by the electrical properties of the electrolytes and the cell walls. The electrolytes separate into

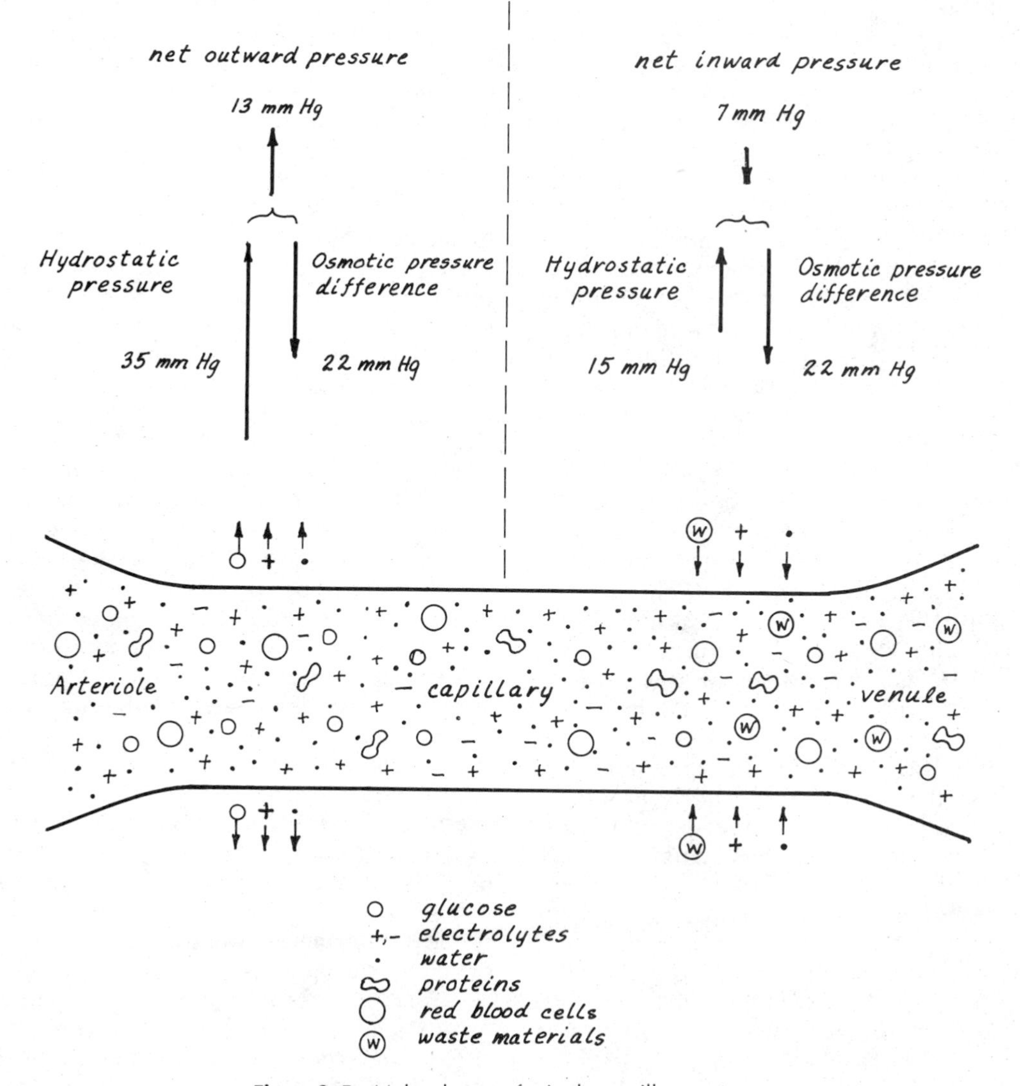

Figure 9–5 Molecular transfer in the capillary system.

charged ions (e.g., Na^+ and Cl^-), and cells apparently have the capacity to trap charged particles on their walls. As will be discussed in Chapter 12, unlike electrical charges attract each other strongly and like charges repel. If a layer of positively charged molecular ions were trapped on the inside of a cell membrane, this would inhibit the entrance of positive ions and enable the cell to maintain a concentration gradient. The unequal distribution of ions across living membranes gives rise to electrical potentials (voltages) which have important physiological consequences (see Chapter 16). The combined effects of electrical forces and the tendency for diffusion are often discussed in terms of an "electrochemical gradient."

Many cells demonstrate the ability to "pump" ions and molecules against an electrochemical gradient. That is, the thin cell membrane can overcome both the electrical forces and the diffusion forces to take in or eject materials. The "active transport" mechanisms which accomplish this feat are not well understood, but it is clear that they require a source of energy to operate the "pump." Active transport appears to be necessary to explain the sodium transfer from the kidney tubules as mentioned above. It has been demonstrated to occur in the membranes of the intestines as well. In other areas the rate of transfer of certain ions and molecules seems to be too rapid to be explained by the passive processes and therefore active transport is assumed.

The purpose here has been to present briefly some of the physical factors underlying the incredibly intricate processes for maintaining fluid balance in the body. The human body may contain something like 100 trillion (10^{14}) cells, and each of these must be supplied with nutrients and allowed to excrete waste material. Much of this process proceeds by diffusion or is aided by diffusion. For example, when a cell uses its supply of glucose, a concentration gradient with respect to the surrounding fluid is produced and more glucose enters. Other processes require more complicated mechanisms to accomplish the particular interior balance of water, salts, and organic matter on which life depends. Much remains to be understood about these mechanisms. The ability of the hundreds of different types of cells to obtain the chemical balance necessary for their function is a topic for continued active research.

DALTON'S AND HENRY'S LAWS

The physiological applications up to this point have been limited to the diffusion of solvents and solutes which are normally liquid or solid. Of course, the diffusion of gases in the respiration process is crucially important as the basis for getting oxygen into the body and removing carbon dioxide. Before discussing the actual diffusion processes, it is necessary to introduce some further concepts about the behavior of gases.

The large scale behavior of gases as a function of pressure, volume, and temperature was discussed in Chapter 6. In large scale physical phenomena where the ideal gas law can be applied, the air is treated as a single gas which under standard conditions exerts a pressure of 760 mm Hg on the surface of the earth. Since diffusion is a molecular phenomenon, the behavior of the individual constituents of the air must be considered when diffusion occurs. The first step is to analyze the pressure exerted by each component of the air.

Microscopically, the pressure exerted by a gas is the result of the collisions of the gas molecules with the walls of the container. If there were only a few collisions, the gas pressure would be experienced as a series of sharp impulses. But as explained earlier, there is an astronomical number of rapidly moving gas molecules in a cubic centimeter of normal air, so the number of collisions is very large and is experienced as a steady "pressure." Since the energy is distributed uniformly among the gas molecules, the pressure exerted by a particular type of gas will be proportional to the number of molecules of that type present to collide with the walls. Thus, it is found that the *partial pressure* exerted by a given component of the air is directly proportional to its percentage of the number of molecules of the air. Since a given volume of any ideal gas will have the same number of molecules, this percentage is also equal to the volume percentage. For example, since oxygen makes up 20.95% of the volume of dry air, the pressure exerted by the oxygen molecules on the walls of a container will be 20.95% of the total pressure. If the atmospheric pressure is 760 mm Hg, then the partial pressure of the oxygen will be 0.2095×760 mm Hg = 159 mm Hg. This relationship between percentage composition and partial pressure is referred to as *Dalton's law*. It follows that the total pressure is equal to the sum of the partial pressures. For example, assume that a volume of oxygen is put into a container and found to produce a pressure of 200 mm Hg. If an equal volume of nitrogen were put in the container, it would likewise exert a pressure of 200 mm Hg. If both gases are put in the container simultaneously, the pressure will be the sum of the partial pressures, or 400 mm Hg. The approximate percentages and partial pressures for dry air at 760 mm Hg pressure are given in Table 9–1. Normally, water vapor will be present and will contribute a partial pressure, reducing the partial pressures for the constituents listed in Table 9–1.

When a gas or a mixture of gases is in contact with the surface of a liquid, a certain number of gas molecules will go into solution. *Henry's law* states that at a given temperature the amount of a gas which will go into solution is proportional to the partial pressure of that gas. This is just a reflection of the fact that if the partial pressure is twice as high, twice as many molecules of that gas will hit the liquid surface, and on the average twice as many will be captured and go into solution.

TABLE 9–1 The Constituents of Dry Air.

Component	Percentage	Partial Pressure (mm Hg)
Nitrogen (N_2)	78.08	593.4
Oxygen (O_2)	20.95	159.2
Argon (A)	0.93	7.1
Carbon dioxide (CO_2)	0.03	0.2
	99.99%	759.9 mm Hg

THE TRANSPORT OF RESPIRATORY GASES

The exchange of gases in the lungs is a process which proceeds by diffusion and which therefore is based upon the kinetic energy of the gas molecules and the concentration gradient at the interface where exchange takes place. The interior of the lungs contains very small, saclike structures called alveoli which serve as the interface for the exchange. Oxygen is taken into the alveoli because the partial pressure of oxygen in the air is higher than in the alveoli. Similarly, carbon dioxide is given off because the alveolar partial pressure of CO_2 is higher than that in the inspired air.

A sketch of the gas exchange process is shown in Figure 9–6 with representative figures for the partial pressures of oxygen and carbon dioxide. If completely dry air is taken into the lungs, the partial pressures are $P(O_2) = 159$ mm Hg and $P(CO_2) = 0.2$ mm Hg, according to Table 9–1. The partial pressure of oxygen in the alveoli is less, largely because the air there is completely saturated with water vapor which exerts a partial pressure of about 47 mm Hg. The partial pressure of CO_2 is about 38 mm Hg, much higher than that in the atmosphere; thus, it diffuses out rapidly.

The blood in the pulmonary capillaries is separated from the air in the alveoli by very thin membranes. The oxygen diffuses into the blood at a rate proportional to its partial pressure, according to Henry's law. The partial pressure of O_2 in the blood quickly approaches 100 mm Hg and is nearly saturated. Saturation would occur if the partial pressure reached the alveolar level of 103 mm Hg, since equal diffusion would occur in both directions and the blood would gain no more oxygen. The oxygen is carried by hemoglobin to the capillaries, where the concentration gradients are favorable for the transfer of O_2 into the tissue and the outward transfer of CO_2 to the blood. References 2 and 3 are recommended for further details about the respiratory process. The partial pressures quoted here are representative numbers for the discussion of the physical process and are subject to considerable variation.

COHESION AND ADHESION

Many aspects of the behavior of liquids can be attributed to the strong attractive forces between individual molecules. Both the liquid state and the solid state are characterized by strong internal attractive forces which tend to hold the molecules together. When the forces are between like molecules, they are referred to as *cohesive* forces. For example, the molecules of a water droplet are held together by cohesion. When the attractive forces are between unlike molecules, they are said to be *adhesive* forces.

One of the common examples of adhesion and cohesion is the formation of the curved surface or meniscus observed when a liquid is placed in a tube. As illustrated in

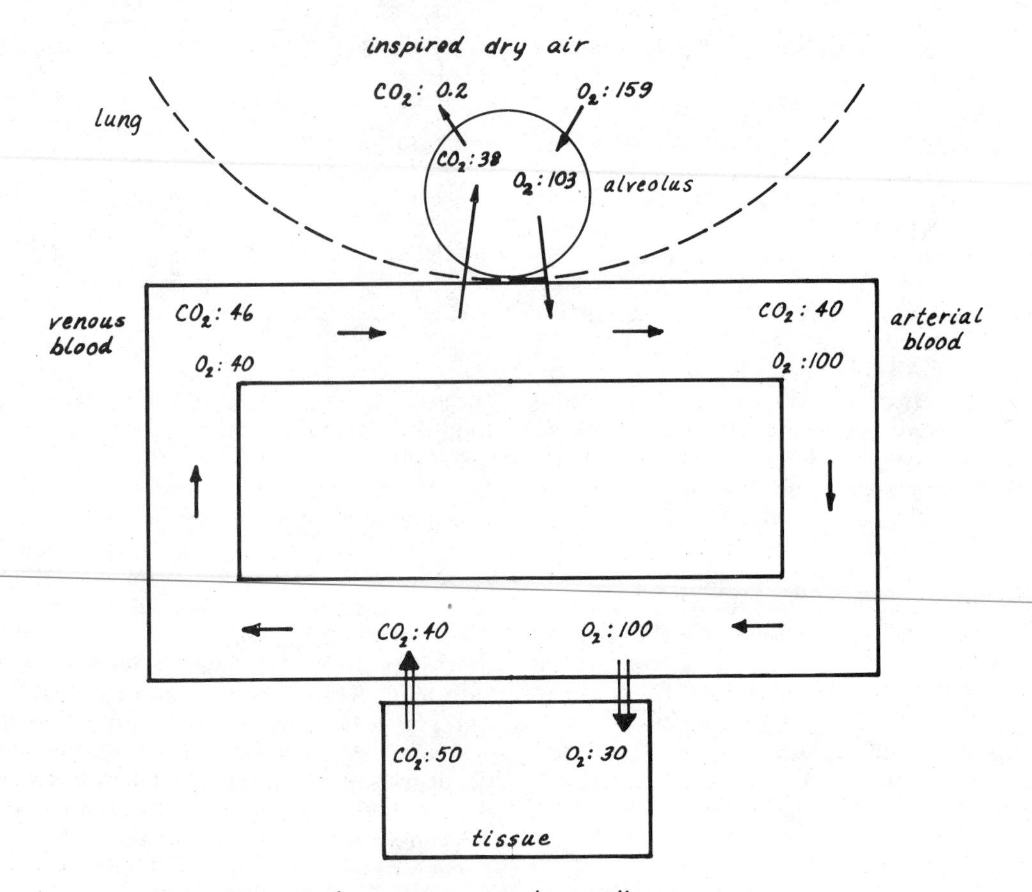

Figure 9–6 The exchange of gases in the respiratory process.

Figure 9–7(a), the meniscus of water turns up because the water molecules at the edge adhere to the glass wall more strongly than they cohere to each other. By contrast, in Figure 9–7(b), the meniscus of mercury turns down because the cohesion of mercury atoms is much stronger than their adhesion to the container wall.

The meaning of adhesion is easy to remember with reference to adhesive tape, which will stick to almost anything. Glues and cements operate on the basis of strong adhesion. The strength of the adhesive forces which can be exerted by water can be demonstrated by wetting two flat plates of glass and sticking them together. A very large force is required to separate them. This can be a nuisance in the laboratory when wet microscope slides stick together.

SURFACE TENSION

The cohesive forces between liquid molecules are responsible for the phenomenon known as surface tension. The molecules at the surface of a liquid do not have other like molecules on all sides of them and consequently they cohere more strongly to those directly associated with them on the surface. This forms a surface "film" which makes it more difficult to move an object through the surface than to move it when it is completely immersed. If carefully placed on the surface, a small needle can be made to float on the surface of water even though it is several times as dense as water. If the surface is agitated to break up the surface tension, the needle will quickly sink.

Figure 9–7 The shape of the meniscus depends upon the relative strengths of adhesion and cohesion. (a) Water–adhesion stronger than cohesion. (b) Mercury–cohesion stronger than adhesion.

The surface tension is responsible for the shape of liquid droplets. Although easily deformed, droplets of water tend to be pulled into a spherical shape by the cohesion forces of the surface layer. The relatively high surface tension of water accounts for the ease with which it can be nebulized, or placed in an aerosol form. Low surface tension liquids tend to evaporate quickly and are difficult to keep in an aerosol form. All liquids display surface tension to some degree. The surface tension of liquid lead is utilized in the manufacture of various sizes of lead shot. Molten lead is poured through a screen of the desired mesh size at the top of a tower. The surface tension pulls the lead into spherical balls, and it solidifies in that form before it reaches the bottom of the tower.

The cleaning properties of a liquid are strongly dependent upon its surface tension. Normally, a liquid with a low surface tension will be a good cleansing agent because it will spread out and "wet" the surface readily. The function of soaps and detergents is largely that of lowering the surface tension of water so that it will penetrate pores and fissures rather than tending to bridge such openings by means of surface tension. Soaps fit into a broader class of substances known as "wetting agents" which reduce the surface tension of the liquid in which they are dissolved. Since an increase in the temperature of a liquid will agitate the molecules at the surface and reduce their cohesive forces, an increase in temperature will lower the surface tension. This explains why hot water is a better cleaning agent than cold water. Figure 9–8 illustrates the reduction of the surface tension of water with temperature. The surface tension of boiling water is about 80% of the surface tension at normal room temperatures. The units of surface tension are dynes/cm. The water surface tension of 72 dynes/cm at 25°C means that a force of 72 dynes would be required to break a surface film which was 1 cm long.

There is a strong correlation between the surface tension of a liquid and its effectiveness as an antiseptic or disinfectant. Other things being equal, the antiseptic with the lowest surface tension will generally be the most effective. The antiseptic hexylresorcinol has a surface tension of 37 dynes/cm, half that of water, and is marketed under the name ST-37 to indicate that property. The enhanced effectiveness of an antiseptic with low surface tension is partly due to the fact that it will wet the surface more thoroughly and thus have more contact with the area to be disinfected. It is also partly due to the fact that the molecules of antiseptics which lower the surface tension of a solution will be more concentrated at the surface and at any interface with another substance. The film formed at the interface with some foreign substance is a result of interfacial tension, which is in all physical respects the same as surface tension. Bacteria in a liquid will be surrounded by such an interfacial film, and antiseptic molecules will tend to concentrate at this film. This may partially account for the effectiveness of very dilute antiseptic solutions.

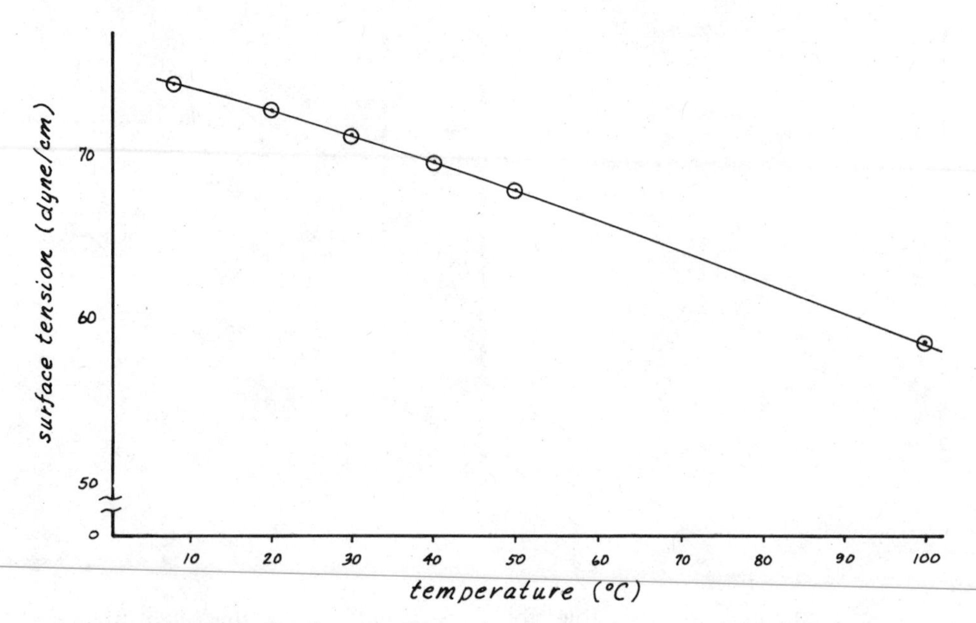

Figure 9–8 The decrease of the surface tension of water when it is heated.

CAPILLARY ACTION

The movement of fluids in small tubes against pressure gradients and against the force of gravity is the result of attractive forces on the molecular level. More specifically, it is the result of adhesion and surface tension. As illustrated in Figure 9–9(a), the adhesion of water to the walls of a vessel will cause an upward force on the liquid at the edges and result in a meniscus which turns upward. The surface tension acts to hold the surface intact, so instead of just the edges moving upward, the whole liquid surface is dragged upward. The net upward force is proportional to the length of the edge in contact with the tube wall. An equilibrium will be reached when the weight of the water lifted is equal to the net upward force. This would seem to imply that a larger radius capillary tube would lift the water higher because the outside film in contact with the tube is longer and thus there is a larger upward force. The opposite is the case, however, because the weight of the water lifted is proportional to the square of the tube radius and thus increases faster than the upward force when the radius is increased. Because of this fact, the smaller the capillary, the higher the liquid will rise, as illustrated in Figure 9–9(b). The height can be calculated by the first equilibrium condition discussed in Chapter 3, setting the upward force equal to the downward force (weight):

(surface tension)(circumference of tube)

= (g)(density)(volume of column)

$$(T)(2\pi r) = gd(h \times \pi r^2).$$

Solving for the height of the column, h:

$$h = \frac{2T}{drg} \qquad \textbf{9–1}$$

where r = radius of tube, d = density, g = acceleration of gravity, and T = surface tension.

Although the mathematical details are not greatly significant for the purposes here, it is important to note that the ability to lift fluids depends on having small "tubes" or columns in which the fluids can rise. Absorbent cotton or gauze and the "wick" type surgical drains operate by capillary action. This capillary action is, of course, one of the forces which aid the heart in moving blood through the capillary system. An extremely large pressure would be required from the heart to move the blood through the capil-

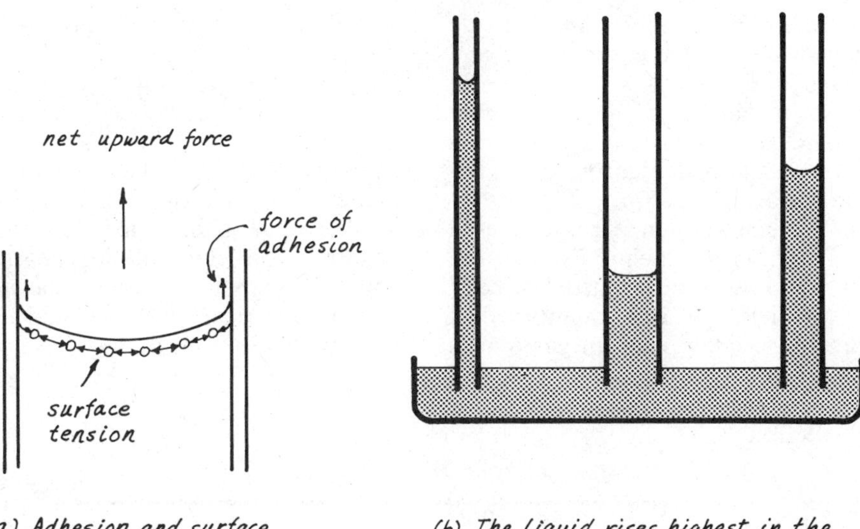

Figure 9–9 The basis of capillary action.

laries if hydrostatic pressure alone were responsible.

There are many examples of capillary action in nature. Capillarity is responsible for the motion of water upward through the ground from depths of several feet to the surface where it can be utilized by plants. If there are many small pores in the surface, then water will come to the surface and evaporate rapidly. Cultivation of the soil to break up these capillaries tends to conserve water in the soil. Capillary action probably aids in raising water to the leaves of trees, although it can be conclusively demonstrated that capillarity alone cannot accomplish this task.

VISCOSITY

As discussed in Chapter 5, one of the main factors which determine the flow rate of a fluid through a tube is the viscosity or resistance to flow. Viscosity may be thought of as arising from fluid friction, but basically it is due to the mutual cohesive forces of the fluid molecules. The stronger the cohesive forces, the greater the resistance to flow. An increase in temperature weakens these cohesive forces and thus decreases the viscosity.

Normal human blood is from two to five times as viscous as water. An increase in the viscosity of the blood will place a greater load on the heart, since more pressure will be required to maintain the normal flow rate. Some diseases can increase the blood viscosity by a factor of five or more. Muscular activity increases the required blood flow, but the accompanying heat also reduces the viscosity of the blood so that the heart does not have to increase its work quite as much as might be expected.

When fluids are being administered to a patient, it is important to maintain the normal viscosity of the blood as well as the osmotic pressure balance. While an isotonic saline solution would maintain the normal osmotic pressure of the blood, it would significantly reduce the viscosity. Often a solution of a substance such as dextran is administered in an isotonic saline solution when it is necessary to increase the blood volume, since such a solution helps to maintain the normal viscosity.

ADSORPTION AND ABSORPTION

Molecular attractive forces can cause outside substances to become attached to a liquid or solid. When the outside substance adheres to the surface, the process is referred to as *adsorption*. When the outside substance penetrates into the substance and distributes itself, the process is referred to as *absorption*. Absorption is the more common term which can be applied to the formation of solutions of solids in liquids, gases in liquids, gases in solids, and so on.

Adsorption occurs when a gas or liquid condenses or otherwise attaches itself to the surface of a solid. Charcoal particles have the capacity to adsorb large quantities of gases and are thus widely used in gas masks and as odor removers. Fine particles provide the maximum surface area and thus improve the efficiency of adsorption processes.

Since colloidal suspensions provide a very large effective surface area of the supended substance, they are often good adsorbers. Medications which are normally insoluble can sometimes be caused to be adsorbed on the colloidal particles. Some colloidal suspensions have the ability to adsorb toxins from the intestinal tract and are thus medically useful.

SUMMARY

The molecules in a liquid or gas at ordinary temperatures have a large amount of kinetic energy. The random high velocities of the molecules cause a net transfer from areas of high concentration to areas of lower concentration (diffusion). Osmosis is the selective diffusion process which occurs when two solutions of differing concentration are separated by a membrane which is permeable to the solvent but impermeable to the solute molecules. The solvent molecules will diffuse down their concentration gradient, causing a net transfer of volume into the solution with the largest concentration of solute molecules. The osmotic pressure of a solution could be defined as the pressure required to prevent diffusion into the solution from the pure solvent. The osmotic pressure increases with the concentration of solute molecules. Given any two solutions separated by a semi-permeable membrane, osmosis will transfer solvent molecules toward the solution with the highest osmotic pressure. If the osmotic pressures are equal, no transfer will occur and the solutions are said to be isotonic. For body fluids, the osmotic pressures are stated with reference to normal plasma. Fluids are said to be hypertonic or hypotonic if their osmotic pressures are higher or lower, respectively, than that of the plasma. Often several types of solute molecules will diffuse with the solvent through a selectively permeable membrane while leaving larger molecules behind. This type of process may be referred to as dialysis. Besides these processes which depend directly upon concentration gradients and molecular kinetic energy, other processes occur in the membranes of living cells which may either aid diffusion or actively transport molecules or ions against the concentration gradients.

Many processes which are vital to life depend to some extent on diffusion. In the kidney, selective diffusion along with active transport mechanisms rids the body of wastes. In the capillaries, osmotic pressure aided by blood pressure accomplishes the transfer of nutrients and the collection of wastes. In the lungs, diffusion processes transfer oxygen to the blood and remove carbon dioxide. The relative diffusion rates of the gases depend upon their partial pressures in the air, which (according to Dalton's law) are proportional to their percentages of abundance in the air. Henry's law states that these gases will enter into solution at rates proportional to their partial pressures. These laws provide the basis for understanding the transport of respiratory gases.

The strong attractive forces between molecules in the liquid state account for such phenomena as cohesion, adhesion, surface tension, capillary action, and the viscosity of liquids. They are also related to the processes or absorption and adsorption.

REVIEW QUESTIONS

1. How can you account for the fact that liquids which are apparently at rest can move freely across certain membranes?

2. What can be done to speed the dissolving of a solid material in water?

3. What is the difference between osmosis and dialysis?

4. Why are osmosis and dialysis said to be diffusion processes?

5. Define osmotic pressure. Does a solvent migrate toward a higher osmotic pressure or toward a lower osmotic pressure?

6. What is meant by the terms isotonic, hypotonic, and hypertonic?

7. What would happen to the red blood cells if a hypotonic solution were injected into the bloodstream? If a hypertonic solution were injected?

8. How does surface tension affect the properties of an antiseptic?.

9. Why do falling liquid drops tend to take on spherical shapes?

10. Why doesn't the mercury column in a clinical thermometer fall?

11. If you produced drops of water, alcohol, and soap solution using the same medicine dropper, would the drops of the three substances be the same size? Explain.

12. Which of the following contribute to capillary action: cohesion, adhesion, surface tension, adsorption, viscosity? Explain.

13. What factors affect the height to which a liquid will rise in a capillary tube?

PROBLEMS

1. If the partial pressure of oxygen in the alveolar air is 103 mm Hg, what percentage of the air does it occupy? (The alveolar pressure may be 3 mm Hg below atmospheric pressure, but the atmospheric pressure, 760 mm Hg, may be used as an approximation here.)

2. If the partial pressure of CO_2 in the alveoli of the lungs is 38 mm Hg, what is its percentage of the composition of alveolar air (assume total pressure = 760 mm Hg)?

3. If the partial pressure of water vapor is 47 mm Hg in the lungs at 760 mm Hg pressure, what percentage of the air does the water vapor constitute?

4. If alveolar air had partial pressures 47 mm Hg for water vapor and 38 mm Hg for CO_2 but otherwise had normal relative pressures of the other gases (i.e., the same as in the atmosphere), what percentages of the air would be made up of nitrogen and oxygen? What would be the partial pressures of nitrogen and oxygen under these conditions?

REFERENCES

1. Krauskopf, K. B., and Beiser, A. *The Physical Universe,* 3rd ed. New York: McGraw-Hill, 1973.
2. Langley, L. L., *Physiology of Man,* 4th ed. New York: Van Nostrand-Reinhold, 1971.
3. Bell, G. H., Davidson, J. N., and Scarborough, H. *Textbook of Physiology and Biochemistry,* 7th ed. Baltimore, Md.: Williams and Wilkins Co., 1976.
4. Egan, D. F.: *Fundamentals of Respiratory Therapy,* 3rd ed. St. Louis: C. V. Mosby Co., 1977.
5. Holter, H. *How Things Get into Cells.* Scientific American, Sept. 1961, p. 167.
6. Solomon, A. K. *Pumps in the Living Cell.* Scientific American, Aug. 1962, p. 100.
7. Hokin, L. E., and Hokin, M. R. *The Chemistry of Cell Membranes.* Scientific American, Oct. 1965, p. 78.
8. Ackerman, E. *Biophysical Science.* Englewood Cliffs, N.J.: Prentice-Hall, 1979.
9. Flitter, H. H. *An Introduction to Physics in Nursing,* 7th ed. St. Louis: C. V. Mosby Co., 1976.

CHAPTER TEN

Internal Energy, Heat, and Temperature

INSTRUCTIONAL OBJECTIVES

After studying this chapter, the student should be able to:

1. Give a definition of heat.
2. Convert Fahrenheit temperatures to the Celsius and Kelvin scales and vice versa.
3. Describe the molecular origins of thermal expansion and calculate the change in the dimensions of an object, given the change in temperature and the appropriate thermal expansion coefficient.
4. Define specific heat.
5. Define the physical calorie and the dietary calorie and state the difference between them.
6. Calculate the energy required to produce a given temperature change in an object when supplied with the mass and specific heat of the object.
7. Calculate the equilibrium temperature which will occur when a hot object is placed in a liquid, given the specific heats, masses, and original temperatures.
8. Describe the process of evaluating the energy content of foods.

INTERNAL ENERGY

At ordinary temperatures the molecules of solids, liquids, and gases are in ceaseless, rapid motion. The energy associated with this random, disordered molecular motion is referred to as *internal energy*. The internal energy is popularly associated with "heat energy," but, as we shall see, "heat energy" is an imprecise term which leads to some conceptual difficulties (1). The condition under which the molecular motion ceases is referred to as the absolute zero temperature. If you look at a solid object or a glass of water at room temperature, they may appear to be completely at rest, but a microscopic view would reveal the chaotic molecular activity which contributes to the internal energy. If the temperature of an object is raised, the level of this activity is increased. If a person touches the hot object, he experiences collisions with the energetic molecules of the object, which may transfer enough internal energy to his finger to cause pain and physiological damage.

In Figure 5–1, a simplified model of the solid, liquid, and gaseous states was given. In Figure 10–1 that simple model is again presented to illustrate the type of molecular motion associated with internal energy. In the solid, the motion is restricted to back-and-forth motion about an equilibrium position, analogous to the motion of a mass on a spring. In the liquid state, the molecules have unrestricted positions and move around each other under the constraint of strong mutual attractive forces. In both the solid and the liquid there is a considerable amount of potential energy as well as kinetic energy associated with the state. In the gaseous state the internal energy is essentially all in the form of the kinetic energy of motion, since the intermolecular attractive forces can usually be neglected.

It is important to keep the words "random" and "disordered" associated with the concept of internal energy as the energy of molecular motion. Otherwise, internal energy becomes associated with any type of molecular motion. Certainly the molecules of a speeding bullet are in motion, but that motion is not classified as internal energy. The motion of the bullet is an example of *ordered* motion, since all the molecules are moving in the same direction. The ordinary, macroscopic motion of objects can always be classified as ordered motion, since there will be a large number of molecules moving together in a certain direction. Whether an object as a whole is in motion or at rest, however, the microscopic or molecular view of the object will reveal the random agitation of the molecules.

THE DISTINCTION BETWEEN INTERNAL ENERGY AND TEMPERATURE

The terms "hot" and "cold" are useful relative terms, but they are imprecise and sometimes deceptive when judged by our senses. The concept of temperature as an objective physical measurement is necessary for reproducible experiments, diagnoses, and treatments. To demonstrate that our senses are not dependable for temperature measurements, consider a block of metal and a block of wood which have been immersed in boiling water and removed. When touched with a finger, the metal will seem considerably "hotter" than the wood. If the two blocks are then immersed in ice water and allowed to come to equilibrium, the metal block will now seem considerably "colder" than the wood. Yet in both experiments the temperatures of the blocks were the same. The difference as perceived by our sense of touch is a result of the widely differing thermal conductivities of wood and metal, as will be discussed later. The point here is that differing

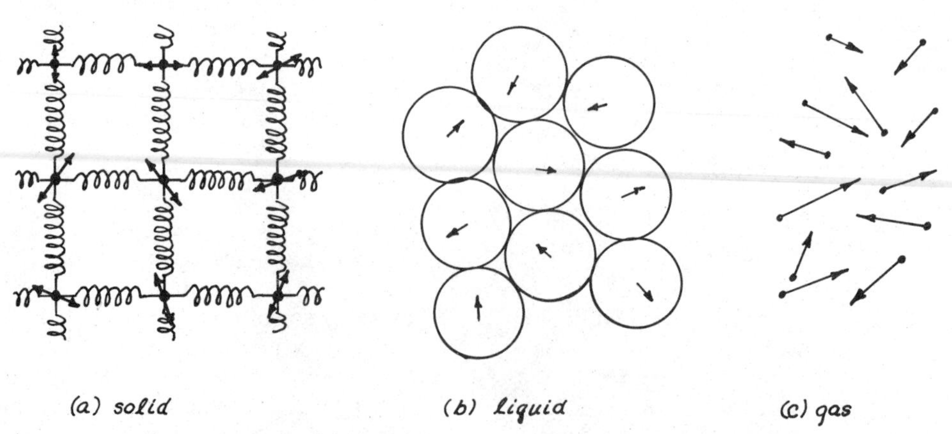

Figure 10–1 Simplified models of the solid, liquid, and gaseous states.

sensations of hot and cold may have causes other than temperature difference.

What then is temperature? The temperature of an ideal gas is a direct indication of the average kinetic energy of the molecules. This statement about gases can be made because all of the molecular energy is in the form of kinetic energy (motion energy). In solids and liquids the relationship is more complicated because of the potential energy associated with intermolecular forces, but the temperature is still related to the internal energy. The close similarity of temperature and internal energy for a given material has sometimes led to confusion between the concepts "internal energy" and "temperature." The reason that distinction must be made is that two objects which have the same temperature do *not* necessarily have the same amount of internal energy per gram. As will be seen, a gram of water and a gram of a metal such as copper which are at the same temperature will have drastically different internal energy. Some materials can "hold" more internal energy than others at a given temperature.

Two objects which are at the same temperature are said to be in thermal equilibrium. Regardless of how much internal energy they possess, there will be no transfer of internal energy between them. Temperature is usually measured by allowing a substance to come to thermal equilibrium with a thermometer which is calibrated in degrees. Before discussing the principles behind these thermometers, the temperature scales will be introduced.

TEMPERATURE SCALES

In order to define a reproducible temperature scale, there must be at least two standard reference points which can be used to calibrate a thermometer. The two most convenient reference points are the freezing point and boiling point of water. The freezing temperature of pure water is essentially independent of its environment within the normal range of atmospheric pressure. The boiling point depends upon the atmospheric pressure, so the pressure must be specified. The universal standard for temperature scales is the boiling point at 760 mm Hg, the standard atmospheric pressure.

The Celsius or centigrade scale is the most logically developed scale for the measurement of ordinary temperatures. As illustrated in Figure 10–2(b), the freezing point of water is chosen to be 0°C and the boiling point at standard pressure is 100°C. The convenient division into 100 parts or "cents" is the reason for the commonly used designation "centigrade scale." The Kelvin or absolute scale maintains the same degree size as the Celsius scale, so that there are 100 degrees Kelvin between freezing and boiling; but the zero of this scale is chosen to be absolute zero. Absolute zero is about −273.15°C, so the zero on the Celsius scale corresponds to 273.15°K, usually rounded to 273°K for practical purposes. To convert any temperature from Kelvin to Celsius degrees, you simply subtract 273 degrees. For example, the temperature of liquid nitrogen is about 77°K. The Celsius equivalent is 77°K − 273°C = − 196°C.

The Fahrenheit scale is a much older temperature scale and much less convenient than the other two scales. About the turn of the 18th century, Gabriel Fahrenheit devised this temperature scale by designating the coldest temperature he could produce in a mixture of salt, ice, and water as zero degrees Fahrenheit. Body temperature was said to be about 100°F, but perhaps he had a chill from his work with the cold ice mixtures, since this doesn't agree with present day measurements of body temperature. His standard points, as accepted today, make the freezing point 32°F, body temperature 98.6°F, and the boiling point 212°F. The Fahrenheit scale is no longer used appreciably in scientific work and, it is hoped, will be discarded in general use as the United States joins the rest of the world in adopting the more logical Celsius scale and metric unit systems.

A direct comparison of the Fahrenheit and Celsius scales is shown in Figure 10–3. To convert from one scale to the other it is necessary to take into account the different degree size as well as the fact that the zeros of the scales are not at the same temperature. Note in Figure 10–2 that the temperature difference between freezing and boiling is 180°F but only 100°C. Thus, the Celsius degree represents a larger temperature change by the ratio 180/100 = 9/5.

$$1\,°C = \frac{180}{100}\,°F = \frac{9}{5}\,°F.$$

A temperature of 10°C represents 10 Celsius degrees above the freezing point of water.

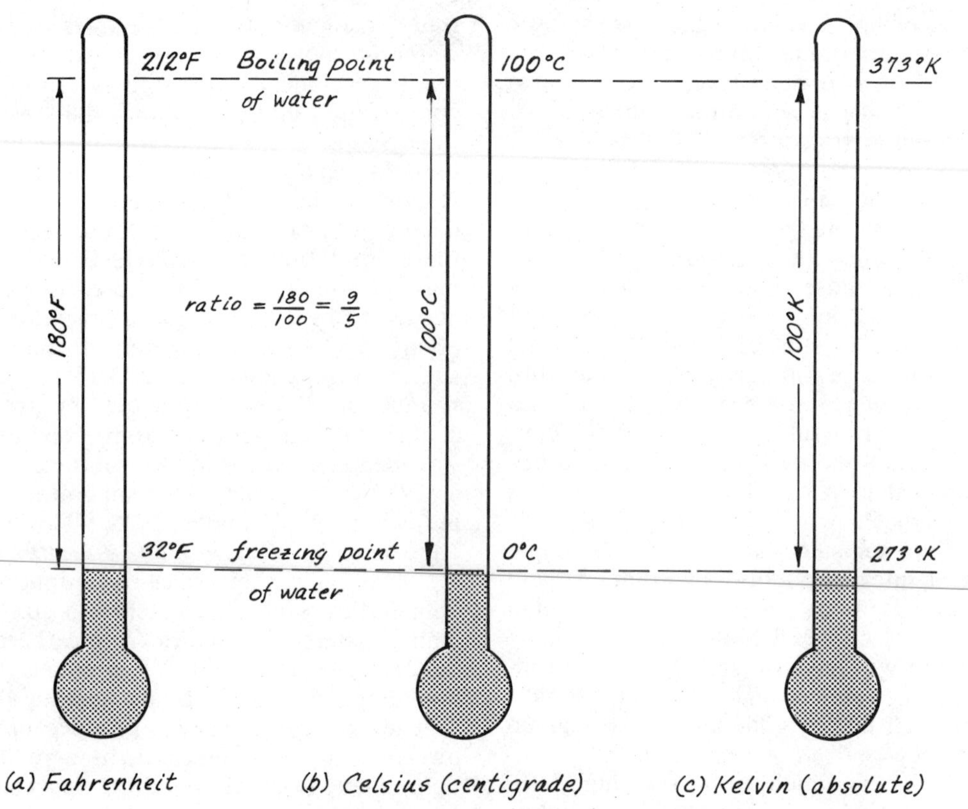

Figure 10–2 The commonly used temperature scales.

This would be the same as (9/5) × 10° = 18°F above freezing, but freezing is 32 degrees on the Fahrenheit scale. Thus, the Fahrenheit equivalent to 10°C is 18°F + 32°F = 50°F. This conversion from Celsius to Fahrenheit can be reduced to the formula

$$T_F = (9/5)\,T_C + 32. \qquad \textbf{10–1}$$

To convert the temperature 77°F to Celsius degrees, it is necessary to get the number of degrees *above freezing* before multiplying by the conversion factor 5/9. The conversion formula is then

$$T_C = (5/9)\,(T_F - 32) \qquad \textbf{10–2}$$

where the subtraction of the 32 must be done before multiplying. If T_f = 77°F, then

$$T_C = (5/9)\,(77°\text{F} - 32°\text{F}) = (5/9)\ (45°\text{F}) = 25°\text{C}.$$

The most difficult thing to remember about the conversion is where to put the 32. It is

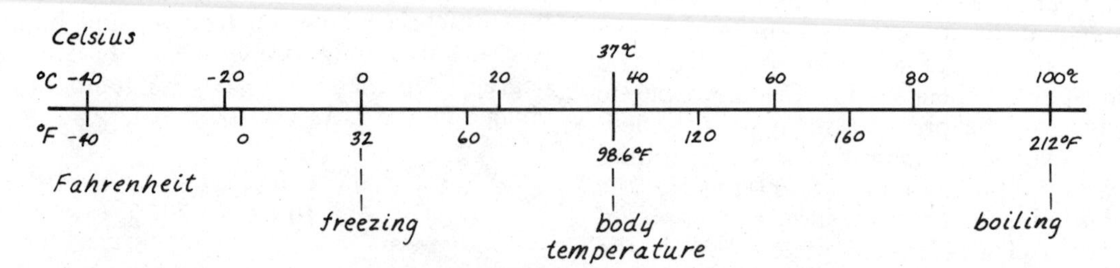

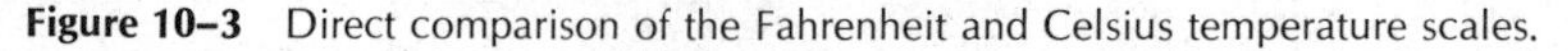
Figure 10–3 Direct comparison of the Fahrenheit and Celsius temperature scales.

perhaps helpful to note that when multiplying by the factor you are converting the number of degrees above freezing. The Celsius temperature gives you this directly, but in the Fahrenheit scale you must subtract 32 from the temperature.

HEAT AND THE FIRST LAW OF THERMODYNAMICS

The conservation of energy principle is a basic law of nature, and must apply to the internal energy that matter has because of random molecular agitation as well as to the large-scale motions discussed in Chapter 4. The internal energy as a characteristic of some object will be represented by the symbol U, and it will remain the same unless there is an energy transfer to or from the object which affects the molecular motion. The internal energy U may be changed by doing work on the object, as in compression of a gas, or by putting it in contact with a higher temperature object so that energy is transferred to it. The change in internal energy can be expressed as

$$\Delta U = Q + W \qquad \textbf{10–3}$$

where W represents the work done *on* the object and Q represents the energy transferred *to* the object from a higher temperature body. This relationship is called the *first law of thermodynamics*. The quantity Q may be referred to as *heat,* and may be further defined as "internal energy in transit" or as the energy transferred by virtue of a temperature difference. This energy transfer Q is a microscopic process, whereas the work done on the system, W, is a macroscopic process.

The reason that the word "heat" must be treated carefully here is that the same change in the internal energy U could be accomplished by heat transfer Q, by doing work W on the system, or by a combination of heat and work, with the final results being indistinguishable. For example, suppose a volume of gas at temperature 20°C is compressed by a piston to a volume of 200 cm^3 and in the process its temperature is observed to rise to 40°C. This sort of thing is commonly observed in the process of pumping up a basketball or a bicycle tire; the temperature rises because of the work done on the gas by the pump. In the case of an ideal gas, a temperature increase is a direct indication of an increase in internal energy.

Now, as an alternative, suppose we start with 200 cm^3 of gas at 20°C and hold its volume constant while placing it over a flame until the temperature rises to 40°C. The temperature change in this case has been accomplished entirely by heat transfer Q. The final state of the gas (200 cm^3 of gas at 40°C) gives no clue about the process used to get it to that state; it could have been accomplished by heat or work. Heat and work are methods of energy transfer, and once that transfer has been accomplished, those terms are no longer meaningful. You can't accurately speak of the "heat energy" of the gas, because it may have received its energy by having work done on it. However, we can use the *internal energy,* U, for characterizing the energy state of the gas.

An increase in the temperature of an object indicates that its internal energy has increased, since temperature is related to molecular kinetic energy. In most cases the temperature will rise when energy is transferred to a body by heat (Q), but this is not always true. As will be seen in Chapter 11, when you are adding heat to a block of ice to melt it, its internal energy is increasing, but its temperature will remain the same (0°C) until it is melted. This further illustrates the need for care in the use of the word heat. In this case the energy transferred by heat is used to alter the bonds which hold the ice in its solid configuration. During the melting process the average molecular kinetic energy does not increase, so we observe that the temperature remains the same.

As another example of the first law of thermodynamics, suppose an amount of heat Q is added to a gas which is allowed to freely expand. The expanding gas may do an amount of work W on its surroundings which is exactly equal to the energy transferred to it by heat. This work would have a negative sign in equation 10–3 since it is work done *by* the system and not *on* it. Therefore

$$\Delta U = Q + W = 0$$

and the internal energy (and temperature) of the gas will remain the same. This is called an *isothermal* expansion of the gas; the energy transferred by heat enables the gas to do work but does not raise its temperature.

THERMAL EXPANSION

One of the consequences of a change in the temperature of a body is that it expands or contracts. If a solid is pictured as a periodic lattice of atoms as in Figure 10–1, the result of an increase in temperature is that the atoms vibrate back and forth over greater distances and with greater average speeds. This larger range of motion causes the atoms to have a larger average distance between them, and the material expands. Since the interatomic distances all expand by the same amount, on the average, then all dimensions of a solid body will increase by the same percentage. The expansion of a solid is conveniently expressed in terms of the fractional expansion $\Delta L/L$, where L is the original length and ΔL is the change in length. For a given change in temperature, the fractional expansion will be the same for a material, regardless of the length L. That is, if a 10 foot aluminum rod expanded 1 inch, then a 20 foot aluminum rod would expand 2 inches under the same conditions. In both cases the fractional expansion is 0.008 (1 inch/120 inches or 2 inches/240 inches). Over small temperature ranges the fractional expansion is found to be proportional to the change in temperature. That is, if a rod expanded 1 inch when heated 10°C, then it would expand 2 inches when heated 20°C. These characteristics of linear expansion (expansion in one dimension) can be summarized by the equation

$$\frac{\Delta L}{L} = \alpha \, \Delta T \qquad \textbf{10–4}$$

where ΔT is the change in temperature and α is the linear expansion coefficient of the particular material. The linear expansion coefficient is the fractional expansion per degree temperature change and thus has the units 1/°C or 1/°F. Table 10–1 gives some characteristic expansion coefficients. Table T–5 in Appendix A contains a more extensive listing of expansion coefficients.

Most ordinary solids will undergo a fractional expansion in the range from 1 to 25 parts per million for each 1°C rise in temperature. The expansion coefficient α is characteristic of the particular material, and the 25-fold range in expansion coefficients is a reflection of the bond strengths, or the strengths of the springs in the schematic model in Figure 10–1. The actual dimension changes for most objects are quite small within the normal range of temperatures.

TABLE 10–1 Typical Thermal Expansion Coefficients.

	Fractional Expansion per Degree at 20°C (68°F)	
Material	**(1/°C)**	**(1/°F)**
Aluminum	24×10^{-6}	13×10^{-6}
Brass	19×10^{-6}	11×10^{-6}
Copper	17×10^{-6}	9.4×10^{-6}
Glass, ordinary	9×10^{-6}	5×10^{-6}
Glass, Pyrex	4×10^{-6}	2.2×10^{-6}
Quartz, fused	0.59×10^{-6}	0.33×10^{-6}
Iron	12×10^{-6}	6.7×10^{-6}
Steel	13×10^{-6}	7.2×10^{-6}
Platinum	9×10^{-6}	5×10^{-6}
Tungsten	4.3×10^{-6}	2.4×10^{-6}

Example. If a steel meter stick is precisely one meter long at 20°C, how much error will be made by using it for measurement at the freezing temperature, 0°C?

Solution. The fractional change in length is given by equation 10–4:

$$\frac{\Delta L}{L} = \alpha \, \Delta T$$

$$= (0.000013/°C)\,(20°C) = 0.00026,$$

where $\alpha = 13 \times 10^{-6}$°C is obtained from Table 10–1. The change in length is then

$$\Delta L = (0.00026)(L) = 0.026 \text{ cm}.$$

The error in measurement will be only about 0.26 mm for this typical temperature change.

When the expansion of two- and three-dimensional objects is considered, a good model to keep in mind is that thermal expansion is similar to a photographic enlargement. Every dimension of the object increases by the same fraction, so that no distortion occurs if an object is made of only one type of material. If a flat plate with a hole in it is heated, the plate will get larger and the hole will get larger, as illustrated in Figure 10–4(b). With two dimensions the area of the plate will increase approximately 2% if the length of a side increases 1%. The area expansion coefficient is approximately twice the linear expansion coefficient:

$$\frac{\Delta A}{A} = 2\alpha\,\Delta T. \qquad \textbf{10–5}$$

Similarly, if a three-dimensional object of any shape is heated, it will expand uniformly. A 1% linear expansion will result in about 3% volume expansion, and the volume expansion coefficient β is about three times the linear expansion coefficient:

$$\frac{\Delta V}{V} = 3\alpha\,\Delta T = \beta\,\Delta T. \qquad \textbf{10–6}$$

The increase in volume owing to thermal expansion can be calculated by equation 10–6 regardless of the shape of the container. For example, the change in volume of an irregularly shaped metal teapot could be calculated by using the change in temperature and the linear expansion coefficient of the metal (Figure 10–4(c)).

The thermal expansion of liquids can be specified in terms of a volume expansion coefficient (see Table T–5). The volume expansion coefficients of liquids are often considerably larger than those of solids. This fact is often observed when a person fills the gasoline tank of his car on a cool morning and leaves the car sitting in the sun. The gasoline has a larger volume expansion coefficient than the metal tank, and it will expand and overflow. On a hot sunny day the loss of

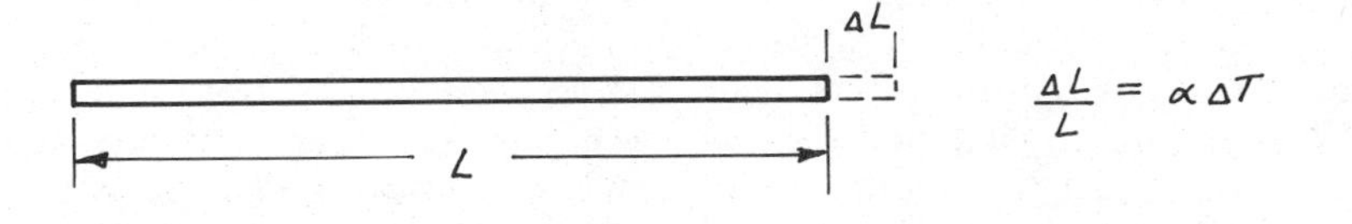

(a) linear expansion

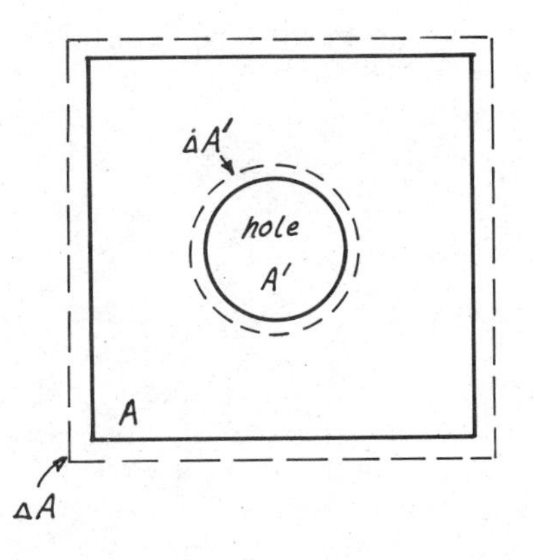

(b) area expansion

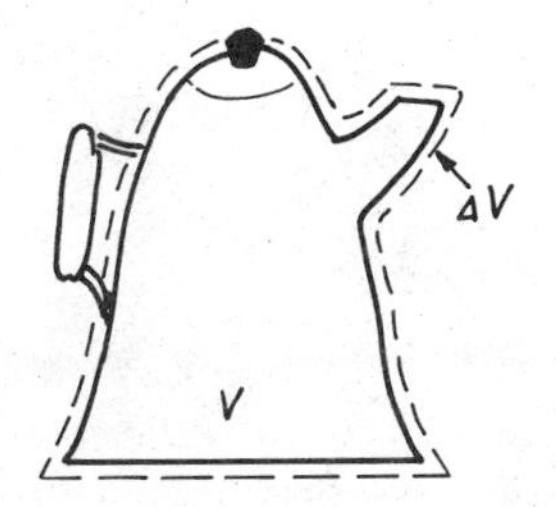

(c) volume expansion

Figure 10–4 Illustrations of thermal expansion.

over a quart of gasoline from a full tank would not be unusual.

The fact that solid objects expand when heated can be used in the assembly of tight-fitting nozzles or other metal parts. If a nozzle must fit into a tight orifice, the insertion can be made easier by cooling the nozzle or heating the orifice, or both.

The thermal expansion of solid objects can create destructive stresses. When the bottom of a glass vessel is heated, it will often crack because the bottom expands more than the top, creating large forces which break the brittle glass. Pyrex is less likely to crack under unequal heating because its linear expansion coefficient is less than half that of ordinary glass. Quartz tubing is often used rather than glass in conditions of extreme temperature changes because its expansion coefficient is about one-fifteenth that of glass (see Table 10–1).

When solid objects are constructed from more than one type of material, the different rates of thermal expansion may create stresses. A considerable amount of research went into developing the mercury-silver amalgams used by dentists to fill cavities. The thermal expansion coefficient of the filling material must be very nearly equal to that of the tooth. Otherwise, painful stresses are created by unequal thermal expansion or contraction when the tooth is exposed to high or low temperatures.

METHODS FOR TEMPERATURE MEASUREMENT

In order to measure temperature conveniently, some easily observable physical property must be found which changes with temperature in a precise and reproducible manner. Most common thermometers either produce an electrical signal or operate on the basis of thermal expansion. The electrical devices will be discussed in Chapter 15.

Liquid Expansion Thermometers. If a full reservoir of liquid is attached to a small capillary tube, then any increase in temperature will cause expansion and force some of the liquid into the small tube. Since the expansion is proportional to the temperature, the height of the column in the capillary can be calibrated directly in degrees Celsius or Fahrenheit. This is the basis for the mercury-filled clinical thermometers and the alcohol-filled ordinary household thermometers.

Bimetallic Strip Thermometers. If two metals A and B which have different thermal expansion coefficients are bonded together in the form of a thin bimetallic strip, the behavior of this strip can be used to measure temperature. If equal lengths of metals A and B are bonded at 0°C, the strip will be straight at that temperature. If metal A has a larger thermal expansion coefficient than B, then at a temperature higher than 0°C it will be longer than B. The strip must then bend toward B to keep them bonded together (Figure 10–5(b)). The amount of bending is proportional to the temperature. Most oven thermometers and other high-temperature thermometers and thermostat controls are bimetallic strips, because they are quite rugged and can tolerate wide ranges of temperatures. The bimetallic strip in Figure 10–5 can also be used to measure low temperatures, since when it is cooled below 0°C, strip A contracts faster and becomes shorter than B. The strip then bends toward A as shown in Figure 10–5(c).

Constant Volume Gas Thermometer. The constant volume gas thermometer is based upon the ideal gas law discussed in Chapter 6. When the volume is constant, the pressure of an enclosed gas is directly proportional to the absolute or Kelvin temperature:

$$P = KT$$

where T is the temperature and K is a numerical constant. If a pressure gauge is attached to the closed volume, then the pressure gauge face can simply be marked off in degrees Kelvin rather than mm of Hg or some other pressure unit. This kind of thermometer is easy to calibrate, but it is large and cumbersome compared to the tiny liquid-in-glass thermometers.

INTERNAL ENERGY AND SPECIFIC HEAT

As mentioned earlier, two substances which are at the same temperature will in general possess different amounts of internal energy per gram. The *specific heat* of a substance is defined as the amount of heat in calories required to raise the temperature of

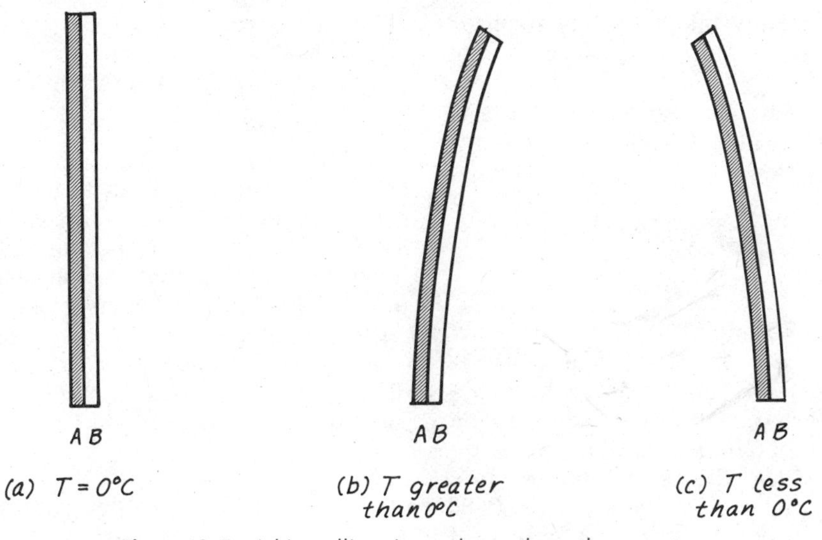

Figure 10–5 A bimetallic strip can be used as a thermometer.

one gram of the substance by one degree Celsius. The specific heat of water is 1.0 calorie/gm-°C. This is, of course, no coincidence because water is used as the standard in setting the size of many physical units. The calorie as a unit of heat or internal energy can be defined as the heat or internal energy required to raise the temperature of one gram of water by one degree Celsius. The dietary Calorie is 1000 of these calories, or one kilocalorie. It is usually spelled with a capital letter (Calorie) to distinguish it from the physical unit (calorie) just defined, but the confusion remains. Another energy unit, the British Thermal Unit (BTU), is defined, as the amount of energy required to raise the temperature of one pound of water by one degree Fahrenheit. Since the BTU and calorie are both defined with water as the standard, the specific heat of water is 1.0 in either unit system; in fact, the numerical specific heats of all substances are the same in the two systems. Lists of specific heats of various substances are given in Tables 10–2 and T–6 (Appendix A).

When heat is added to an object, its change in temperature depends upon its mass and its specific heat. The relationship can be expressed as

$$Q = sm\,\Delta T, \qquad \textbf{10–7}$$

where Q is the heat in calories (not food Calories), s is the specific heat in calories/gm-°C, m is the mass in grams, and ΔT is the temperature change in °C. (These quantities could also be expressed in the BTU system, but that system will not be used here.)

Example. How much heat is required to raise the temperature of 250 grams of water by 10°C? (250 grams is about one cup.)

Solution. The specific heat of water is 1 cal/gm-°C. Using equation 10–7:

$$Q = sm\,\Delta T = (1.0 \text{ cal/gm-°C})(250 \text{ gm})(10\text{°C})$$
$$= 2500 \text{ cal.}$$

TABLE 10–2 Specific Heats of Selected Materials.

MATERIAL	SPECIFIC HEAT (cal/gm-°C or BTU/lb-°F)
Water at 15°C	1.00
Ice at 0°C	0.51
Steam (at constant pressure)	0.48
Aluminum	0.217
Brass	0.09
Copper	0.092
Gold	0.031
Iron	0.11
Lead	0.030
Silver	0.056
Alcohol (ethyl)	0.60

Example. How much heat is required to raise the temperature of 250 grams of brass by 10°C?

Solution. The specific heat of brass is 0.09 cal/gm-°C (Table 10–2). The heat required is

$$Q = (0.09 \text{ cal/gm-°C})(250 \text{ gm})(10\text{°C})$$
$$= 225 \text{ cal.}$$

These two examples illustrate the fact that water has a very large heat capacity compared to metals and most other substances. It takes over 10 times as much heat to raise the temperature of water as that required to produce the same temperature change in an equal mass of brass. If the 2500 calories were given to the 250 grams of brass, it would raise its temperature by 111°C. The large specific heat of water makes it a good temperature regulator, since a great deal of heat is required to change its temperature. The temperature near the ocean or other large bodies of water has a smaller range than inland temperatures because of this regulating effect. The body temperature is a valuable diagnostic aid, since a fever or a severe chill indicates a significant metabolic change. A great deal of internal energy is required to produce a fever of 104°F, since the body is mostly water and has a large specific heat.

Example. If a 165 lb person were assumed to have a specific heat of 0.8 cal/gm-°C, how much extra released internal energy would be required to raise his temperature from 98.6°F to 104°F?

Solution. The change in internal energy ΔU equals the heat Q since no work is done (equation 10–3). So Q must be obtained from equation 10–7, $Q = sm\,\Delta T$, but the mass must be in grams and the temperature change in °C. Since it takes 9/5°F to equal 1°C, the temperature change is

$$\Delta T = (5/9)(104 - 98.6) = 3\text{°C}.$$

From Table T–2, Appendix A, it is found that a weight of 2.2 lb has a mass of 1 kg. The mass is then

$$M = \frac{(165 \text{ lb})}{(2.2 \text{ lb/kg})} = 75 \text{ kg} = 75{,}000 \text{ grams.}$$

The energy required is then

$$Q = (0.8 \text{ cal/gm-°C})(75{,}000 \text{ grams})(3\text{°C})$$
$$= 180{,}000 \text{ cal.}$$

This example serves to illustrate the fact that a fever represents quite a large amount of extra released internal energy. It also illustrates the fact that the physical calorie is too small a unit in which to express the energies involved in body metabolism. The dietary Calorie (kilocalorie) is more practical. The 180,000 calories in the example above are equivalent to 180 food Calories.

HEAT OF COMBUSTION: THE DIETARY CALORIE

The energy value of foods taken into the body is expressed in terms of the Calorie, which is a heat or internal energy unit. This is appropriate because the process of using the foods is basically an oxidation process similar to ordinary burning in terms of the chemical energy released. An approximation of the nutritional energy available from a food may be obtained by measuring the heat of combustion of the food. This is normally done by burning the food in a pure oxygen atmosphere inside a calorimeter. The calorimeter is composed of an oxidation chamber surrounded by a measured volume of water. When the food has burned completely, the heat of combustion is given to the water. Applying the principle of conservation of energy, the energy given off by the combustion must equal the heat energy gained by the calorimeter if it is sufficiently isolated so that no heat escapes. From the measurement of the calorimeter temperature before and after the combustion, equation 10–7 can be used to calculate the energy released by the combustion. This heat of combustion is usually tabulated in kilocalories per gram.

The heats of combustion for some common foods are listed in Table 10–3. To obtain accurate measurements of the energy available to the body, corrections would have to be made for substances such as cellulose which will burn in a calorimeter but which are not metabolized by the body. The three basic types of foods (proteins, carbohydrates, and fats) have been shown to pro-

TABLE 10–3 Energy Values of Selected Foods in Food-type Calories (Kilocalories).

FOOD	KCAL/GM	FOOD	KCAL/GM
Apples, raw	0.64	Meat, lean	0.27
Beans, navy	3.54	Milk, whole	0.72
Bread, white	2.66	Oatmeal, cooked	0.63
Butter	7.95	Orange juice	0.43
Buttermilk	0.37	Potatoes, boiled	0.97
Cheese, cheddar	3.93	Rice, cooked	1.12
Chocolate	5.70	Salmon, broiled	1.70
Cream, 40%	3.81	Spinach, cooked	0.58
Egg, boiled	1.62	Sugar, granulated	3.94
Ice cream, plain	2.10	Tomato	0.23
Lard	9.30	Turnip	0.27
Lettuce, leaf	0.20		

duce about equal amounts of heat when burned outside the body and when utilized within the body. A small correction must be made for proteins, since they are incompletely oxidized by the body, and energy-containing residues of the proteins are excreted in the urine.

Example. A certain calorimeter has a heat capacity equivalent to that of 5000 grams of water. When a 50 gm slice of white bread is burned in the calorimeter, the water temperature rises from 25°C to 51.6°C. Calculate the heat of combustion of white bread in kcal/gram.

Solution. From equation 10–6, the heat gained by a calorimeter is calculated:

$$Q = sm\,\Delta T$$
$$= (1 \text{ cal/gm-°C})(5000 \text{ gm})(51.6\text{°C} - 25\text{°C})$$
$$= 133{,}000 \text{ calories}$$
$$= 133 \text{ kilocalories (food Calories).}$$

The heat of combustion is then

$$\frac{Q}{M} = \frac{133 \text{ kcal}}{50 \text{ grams}} = 2.66 \text{ kcal/gm.}$$

Although the process of converting food into available energy for the body is similar to combustion, not all of this energy is in the form of heat. This is obvious, since the body does mechanical work; but this fact can be demonstrated conclusively by calculating the temperature rise of the body which would result if all the energy went into heat.

Example. Suppose a 165 lb man consumed a normal diet of 2500 kilocalories, which was completely released in the form of heat. If the body were assumed to have a specific heat of 0.8 cal/gm-°C and no heat were allowed to escape from the body, by how much would the temperature of the body rise?

Solution. The 165 lb is equivalent to 75 kg. Applying the energy equation, $Q = sm\,\Delta T$:

$$2{,}500{,}000 \text{ calories}$$
$$= (0.8 \text{ cal/gm-°C})\,(75{,}000 \text{ grams})\,(\Delta T).$$

Solving for the unknown quantity,

$$\Delta T = 41.7\text{°C} = 75\text{°F}.$$

A 75°F rise in the body temperature is preposterous, of course, but the exercise serves to illustrate how large an amount of energy an adult's diet represents. It also emphasizes the fact that most of the energy either is used in forms other than heat or is exhausted from the body as heat during the 24 hour period to which the diet applies.

THE MECHANICAL EQUIVALENT OF HEAT

At the time when the calorie as a unit of heat began to be used, heat was thought to be a fluid called "caloric" which flowed into and out of objects. It was not directly associated with energy of the mechanical type. About 1840 a British physicist, James Joule, demonstrated that when mechanical energy was expended on a paddle system which stirred some water, the temperature of the water was raised, indicating that energy was being transferred to the water in the form of heat. Careful experiments led him to the measurement of the mechnical equivalent of heat. He found that about 4.186 joules were equal to one calorie. (The metric energy unit was subsequently named in his honor.)

This equivalence implies that any heat transfer measurement could be expressed in terms of the standard energy unit, the joule. Further, the rate of production of heat could be expressed in joules per second or watts.

This equivalence has been very important in demonstrating the conservation of energy principle in processes in which heat is produced.

Example. If a person completely utilizes a diet of 2500 kilocalories in 24 hours, how many joules has he used, and what is the equivalent mechanical power in watts?

Solution. Since one calorie is equal to 4.186 joules, 2500 kilocalories is equal to

$$(2{,}500{,}000 \text{ calories})(4.186 \text{ joules/cal}) \times 1.05 = 10^7 \text{ joules},$$

or over 10 million joules. To get the power in watts:

$$P = \frac{W}{T} = \frac{10{,}500{,}000 \text{ joules}}{86{,}400 \text{ seconds}} = 121 \text{ watts}.$$

The human machine then operates on an average power rather close to that of an ordinary light bulb.

Example: In Chapter 7 it was stated that the resting power output of the human heart is about 1.1 watts. What food energy in kilocalories would be required to operate the heart for 24 hours at this resting level, assuming the efficiency of the body for food energy use to be 25%?

Solution: The total energy used in 24 hours is

$$(1.1 \text{ joule/sec})\ (24 \text{ hours})\ 3600 \text{ sec/hour}) = 95{,}040 \text{ joules}.$$

Since 1 kilocalorie = 4186 joules, the total energy output in kilocalories is

$$\frac{95{,}040 \text{ joules}}{4186 \text{ joules/kcal}} = 22.7 \text{ kcal}.$$

But with a 25% utilization efficiency, the body would have to use 4 kcal to get 1 kcal of useful output energy. Therefore food energy required

$$= 4 \times (22.7 \text{ kcal}) = 91 \text{ kcal}.$$

SUMMARY

Internal energy is the energy associated with the random, disordered motion of molecules. When the temperature of an object is raised, this implies that the molecules of the object are more active and that it therefore possesses a larger amount of internal energy. Though the temperature is proportional to the internal energy for a given material, different materials will in general have differing specific heats and possess differing amounts of internal energy per gram at a given temperature. The most common temperature scales are the Fahrenheit and the Centigrade (Celsius) scales. A third scale, the absolute of Kelvin scale, is often used in scientific work. When heated, ordinary materials expand so that the percentage change in any dimension is proportional to the change in temperature. This regular expansion is used to construct thermometers, such as the ordinary mercury-in-glass thermometers and bimetallic strip thermometers.

Heat and internal energy can be measured in joules, the basic metric energy unit, but the more common unit is the calorie, which is the heat required to raise the temperature of one gram of water by one degree Celsius. The kilocalorie, one thousand calories, is the unit commonly referred to as the food Calorie. The specific heat, a relative measure of the heat capacity of a material, is defined as the number of calories required to raise the temperature of one gram of the material by one degree Celsius. The change in heat of an object, Q, may be determined from the relationship $Q = sm\,\Delta T$ if the specific heat, s, and the change in temperature, ΔT, are measured. The energy equivalents of foods are determined from the measurement of their heats of combustion, since the metabolic processes in the body are similar to combustion in energy release.

REVIEW QUESTIONS

1. When internal energy is defined as the energy associated with the random, disordered motion of molecules, why must the words "random" and "disordered" be included?

2. Will substances with the same mass and same temperature always have the same amount of internal energy? Explain.

3. Why is the sense of touch an unreliable test for temperature?

4. When a metal object with a hole in it is heated, will the hole get bigger or smaller?

5. Why does a bimetallic strip bend when heated? How can this effect be utilized to make a thermometer?

6. What is the advantage of Pyrex over ordinary glass for handling hot substances?

7. What are the advantages of the Celsius scale over the Fahrenheit temperature scale? Under what conditions would the absolute or Kelvin scale be preferred?

8. Why is water a good substance to use in a hot water bottle?

9. What significance does the large specific heat of water have in relation to the maintenance of a constant body temperature?

10. What effect will the presence of a large body of water have upon the temperature extremes experienced in its vicinity? Explain.

11. How is the dietary Calorie related to the physical calorie?

12. The dietary Calorie equivalent of foods is determined by burning them in a calorimeter. Explain why this is feasible as a method for determining food value.

PROBLEMS

1. Find the Celsius temperatures corresponding to 68°F, 0°F, 100°F.

2. Find the Fahrenheit temperatures corresponding to 37°C, 20°C, −40°C.

3. If a person with a fever has a body temperature of 104°F, what is the temperature in degrees Celsius?

4. If the temperature of an object is increased by 20°C, what is the temperature change in degrees Fahrenheit?

Worked Example. A car gasoline tank has a volume of 20 gallons and is made of brass. If it is filled to the brim with gasoline, which has a volume expansion coefficient of 0.001/°C, and then is allowed to sit in direct sun so that its temperature rises by 20°C, how much gasoline will overflow from the tank?

Solution. From Table 10–1 it is found that the linear expansion coefficient of brass is 19×10^{-6}/°C, so the volume expansion coefficient is

$$\beta = 3 \times 19 \times 10^{-6}/°C = 57 \times 10^{-6}°C.$$

The volume expansion of the gas tank is then

$$\Delta V = \beta \Delta T V = (57 \times 10^{-6}/°C)(20°C)(20 \text{ gal}) = 0.023 \text{ gal.}$$

The volume expansion of the gasoline is considerably greater:

$$\Delta V = (0.001/°C)(20°C)(20 \text{ gal}) = 0.4 \text{ gal.}$$

The amount of overflow is then

$$0.400 \text{ gal} - 0.023 \text{ gal} = 0.38 \text{ gal}$$

or a loss of about 1.5 quarts of gasoline.

5. A river is spanned by successive 80 ft steel beams. If the maximum temperature to which the bridge will be subjected is 60°C, what size gap must be left between the beams if the bridge is constructed when the temperature is 10°C?

6. A 50 ft steel measuring tape is accurately 50 ft long at a temperature of 25°C. How long will it be at 0°C?

Worked Example: (a) How much work would be done by a 50 kg woman in climbing a mountain which is 1600 meters (approximately 1 mile) high if only the lifting work is considered?

(b) If the woman must use 4 kcal of food energy for every kcal converted to mechanical energy for climbing, how many kcal of food energy are consumed in this process?

(c) If the amount of energy given off as heat is 75% of the food energy used, how much would her body temperature rise if she were unable to give off this energy? (Assume a specific heat of 0.8 cal/gm-°C for the body.)

(d) If the climb is made in one hour, what is the average power expended in watts?

Solution: (a) The minimum amount of work done would be the work required to lift the woman to the top of the mountain.

$$\text{work} = \text{weight} \times \text{height} = (50 \text{ kg})(9.8 \text{ m/sec}^2)(1600 \text{ m}) = 784{,}000 \text{ joules.}$$

(b) The energy equivalent in calories may be found from the conversion factor

$$4.186 \text{ joules} = 1 \text{ calorie}$$

$$784{,}000 \text{ joules} = 187{,}000 \text{ calories} = 187 \text{ kcal.}$$

The food consumption required is:

$$\text{food calories} = 4 \times 187 \text{ kcal} = 749 \text{ kcal.}$$

(c) If the energy converted to heat, 562 kcal, is kept in the body, the temperature with rise according to the relationship $Q = sm\,\Delta T$:

$$\Delta T = \frac{Q}{sm} = \frac{562 \text{ kcal}}{(0.8 \text{ kcal/kg-°C})(50 \text{ kg})} = 14°C = 25°F.$$

(d) If the work, 784,000 joules, is accomplished in one hour, the average power is

$$P = \frac{W}{t} = \frac{784{,}000 \text{ joules}}{3600 \text{ sec}} = 218 \text{ watts.}$$

7. How many calories would be required to raise the temperature of a 1200 gram iron skillet by 80°C?

8. A 500 gram copper pipe is heated to 300°C in the process of soldering it. If it is dropped into 4 kg of water (approximately a gallon) which is at 25°C, what will be the final temperature?

9. If a 300 gram aluminum pot contains 400 grams of water at 20°C, how much heat energy will be required to heat both pot and water to 100°C?

10. If 720 calories of heat energy are required to raise the temperature of a 400 gram block of metal by 20°C, what is the specific heat of the metal? What kind of metal is it? (Probable metal composition can be guessed from this data.)

11. A 100 gram mass of metal with specific heat 0.2 cal/gm-°C has a temperature of 20°C. What would be its final temperature if 100 calories of energy were added to it?

12. How much energy is given up when 200 grams of water cools from 90°C to 30°C?

13. How much energy must be added to raise the temperature of 250 cm^3 of water from 10°C to 30°C?

14. How much water at 10°C must be added to 100 grams of water at 90°C so that the mixture will have a final temperature of 30°C?

15. Suppose 3000 dietary calories of energy is added to a bathtub full of water (100 kg) which is near room temperature (30°C). How much will the temperature rise?

16. Calculate the food energy available from a 50 gram chocolate bar. If this amount of energy were added to a baseball-sized chunk of lead (2000 grams) in the form of internal energy, how much would its temperature rise?

17. A 200 watt immersion heater is used to heat 250 grams of water from 25°C to 100°C to make a cup of tea. How long will the process take if no energy is lost to the surroundings?

18. A kilogram piece of lead is dropped from a height of 100 meters. If all of the gravitational potential energy of the lead were converted into internal energy during the fall and impact, and if all of that energy were imparted to the lead, how much would its temperature rise?

19. A one kilogram piece of copper has an initial velocity of 20 m/sec and is sliding along a rough surface. It quickly comes to rest under the influence of friction. If half of the energy is transferred to the copper by frictional heating and the other half transferred to the surface, how much will the temperature of the copper change?

20. A residential hot water heater holds 160 kg of water. It takes this water in at 20°C and heats it to 60°C.

    a. How many kilocalories of energy are required for this heating process?
    b. If the heating is done electrically and the energy costs 10¢ per kilowatt-hour, how much will it cost to heat one tank of water?

21. If a slice of bread supplies 100 kcal to a person of mass 50 kg and specific heat 0.8 kcal/kg-°C, how much will his body temperature rise if no heat is given off? (All the food energy is assumed to be converted

to internal energy.) Find his resulting body temperature in degrees Fahrenheit if it was originally normal.

22. A 150 lb man running a mile in 8 minutes produces mechanical energy at the rate of about 280 watts. Assuming that the body is 25% efficient in the process of converting food energy to mechanical energy, how many kcal of food energy must be consumed to provide the energy for the mile run? If a slice of white bread has a mass of 50 grams, how many slices of bread would be required to provide this energy?

23. Assume that 0.04 calorie is required to move each pound of body weight a distance of one foot horizontally, as in walking. If a 120 lb nurse walks 4 miles during her time on duty, how much energy is used? Express the answer in kcal (dietary Calories). Since the body efficiency is only about 25%, how many kcal of food must be consumed to supply this energy?

24. Assuming that walking requires a total food energy of 0.16 calorie per foot of horizontal distance for each pound of body weight, how far would a 120 lb person have to walk to burn up ¼ lb of candy (450 kcal)?

25. The heat capacity of a certain calorimeter is such that it is equivalent to 500 grams of water. If 10 gm of a certain food is burned in the calorimeter, the temperature rises 35° C. What is the energy value of the food in kcal/gm? To what food in Table 10–3 is it most nearly equivalent?

26. Guyton (2) reports that one gram of fat will be stored for each 9.3 kcal of food energy intake in excess of energy output.

    a. How many excess kilocalories is represented by one pound of excess body weight?
    b. Assuming that walking requires a total food energy of 0.16 calorie per foot of horizontal distance per pound of body weight, how far would a 150 lb person have to walk to lose one pound of body weight? (This would be only the direct result of energy consumption, assuming direct reversibility of the storage process. It does not include any effects on the body metabolism which might lead to further weight loss after the exercise is finished, and therefore may be grossly misleading as an evaluation of exercise as a means to weight loss.)

REFERENCES

1. Zemansky, M. W. *The Use and Misuse of the Word "Heat" in Physics Teaching.* The Physics Teacher, Sept. 1970, p. 295.
2. Guyton, A. C. *Basic Human Physiology,* 2nd Ed. Philadelphia: W. B. Saunders Co., 1977.

CHAPTER ELEVEN

The Effects of Heat Energy

INSTRUCTIONAL OBJECTIVES

After studying this chapter, the student should be able to:

1. Explain why large amounts of energy are required to accomplish the solid-liquid and liquid-gas phase transitions.
2. State the steps involved in the use of the liquid-gas phase transition in a refrigerator.
3. Explain the high efficiency of perspiration evaporation as a cooling mechanism for the body compared to other forms of cooling.
4. Name and describe the three basic methods for heat transfer.
5. Name the physical variables which influence the rate of heat loss by conduction and calculate the rate of heat loss, given values for these variables.
6. Name the physical variables which influence the rate of heat loss by radiation.
7. Define saturation vapor pressure and relative humidity.
8. Calculate the relative humidity at a certain temperature, given the actual humidity and the saturated vapor density at that temperature.
9. Explain why closed, heated rooms become "dry" and must be humidified.
10. Define "boiling point" and explain why it changes with altitude above sea level.
11. Define "dew point."

CHANGES OF PHASE

Giving internal energy to a solid body increases the kinetic energy of its molecules. This added internal energy takes the form of more agitated vibrations about the equilibrium positions of the molecules, as sketched in Figure 11–1. It also has the effect of stretching and weakening the bonds, resulting in thermal expansion as discussed in Chapter 10. If enough internal energy is added, the molecular activity will become great enough to break the characteristic solid bonding forces and free the molecules from their rather rigid relative positions. This can be pictured as breaking the springs in our simple model of a solid. This breakdown of the rigid structure corresponds to melting. It is one example of a "change of phase" or "phase transition," in this case from the solid to the liquid phase. The change of phase requires a large amount of internal energy compared to the process of raising the temperature of the solid. In the case of water, it requires 80 calories per gram to change ice to water at 0°C, as noted in Figure 11–1.

Continuously adding heat to a liquid after the solid-liquid phase change will gradually raise its temperature up to the boiling point, which is the point where the liquid-to-gas phase change occurs rapidly. When the solid-to-liquid phase change occurs, the intermolecular attraction is reduced, but it remains very strong. The liquid-to-gas phase change requires enough internal energy to completely overcome the attractive forces and free the molecules almost completely from any mutual attraction. This transition takes considerably more internal energy than the solid-liquid transition. It requires 540 cal/gm to convert water to steam at 100°C. This is almost seven times as much energy as that required to change ice to liquid water.

The energy required to melt one gram of a solid once it has reached its melting temperature is commonly referred to as the *latent heat of fusion* or *latent heat of melting* for that substance. We will represent the latent heat of fusion by h_f and the energy required to melt a mass m is given by

$$Q = mh_f. \qquad \textbf{11–1}$$

For example, the heat required to melt one kilogram of ice is

$$\begin{aligned} Q &= (1000 \text{ grams})(80 \text{ cal/gram}) \\ &= 80{,}000 \text{ calories.} \end{aligned}$$

The energy required to vaporize one gram of a liquid once it has reached its boiling temperature is called the *latent heat of vaporization*. This latent heat h_v is 540 calories per gram for water at 100°C as noted above. The energy required to vaporize a mass m of a substance at its boiling point is given by

$$Q = mh_v, \qquad \textbf{11–2}$$

which for one kilogram of water at 100°C would be

$$\begin{aligned} Q &= (1000 \text{ grams})(540 \text{ cal/gm}) \\ &= 540{,}000 \text{ calories.} \end{aligned}$$

The latent heats of fusion and vaporization of some common substances are given in Table 11–1.

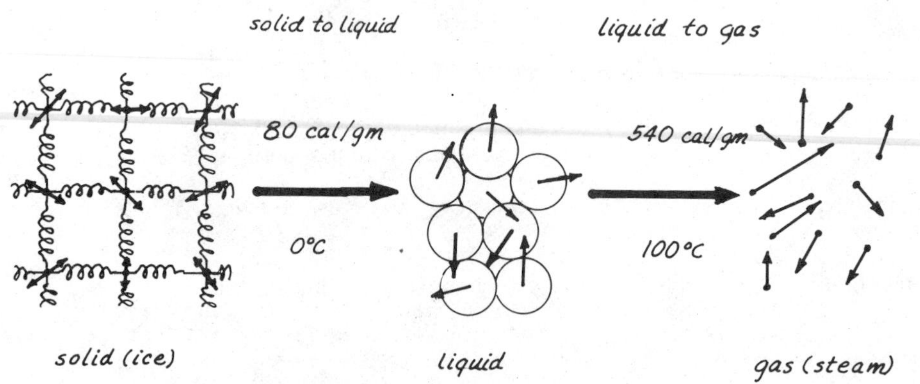

Figure 11–1 The phase changes of water.

TABLE 11–1 Latent Heats of Fusion and Vaporization of Some Common Substances.

	MELTING TEMPERATURE (°C)	LATENT HEAT OF FUSION h_f (CAL/GM)	NORMAL BOILING TEMPERATURE (°C)	LATENT HEAT OF VAPORIZATION h_v (CAL/GM)
Water	0	80	100	540
Ammonia	−75	108	−33	327
Ethyl Alcohol	−114	26	78	204
Ether			35	84
Freon-12			−29.8	39.5
Nitrogen	−210	6.2	−196	48
Oxygen	−219	3.3	−183	51
Lead	327	5.9	1620	208
Mercury	−39	2.7	357	68

When it is considered that only one calorie is required to change the temperature of water 1°C, even with its large heat capacity, the internal energies involved in the phase transitions are seen to be enormous. To illustrate the energies involved, consider the process of taking a very cold 1 gram piece of ice at −50°C and adding heat to it at the rate of one calorie per second. The rate of temperature rise can be calculated using the energy equation, equation 10–7. The specific heat of water is 1 cal/gm-°C, but the specific heat of ice and the specific heat of steam at a constant pressure are both about 0.5 cal/gm-°C. Applying the equation

$$Q = sm\Delta T$$

with s = 0.5 cal/gm-°C for ice, one calorie is seen to raise the temperature of the ice by 2°C:

$$Q = 1 \text{ cal} = (0.5 \text{ cal/gm-°C})(1 \text{ gram})\Delta T$$

$$\Delta T = \frac{1}{0.5} \text{ °C} = 2\text{°C}.$$

The ice will then rise in temperature up to the melting point, 0°C, in 25 seconds with the addition of 25 calories of internal energy as illustrated in Figure 11–2, since its temperature will be rising at the rate of 2°C per second. At 0°C it must absorb 80 calories in order to change into water, *without raising*

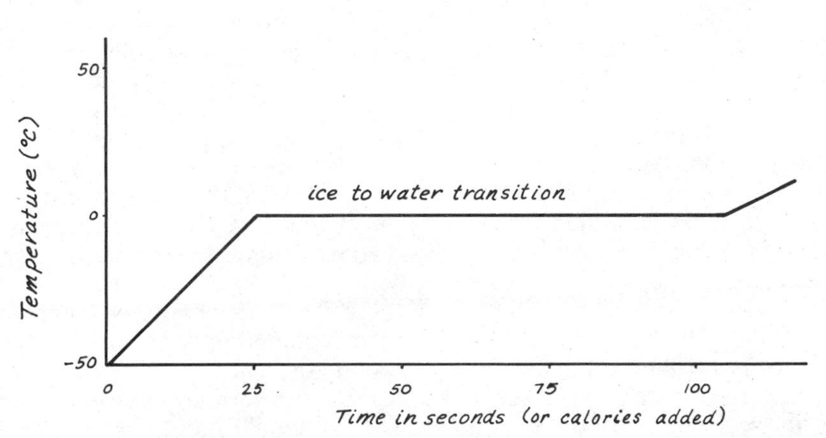

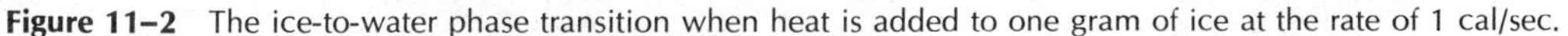

Figure 11–2 The ice-to-water phase transition when heat is added to one gram of ice at the rate of 1 cal/sec.

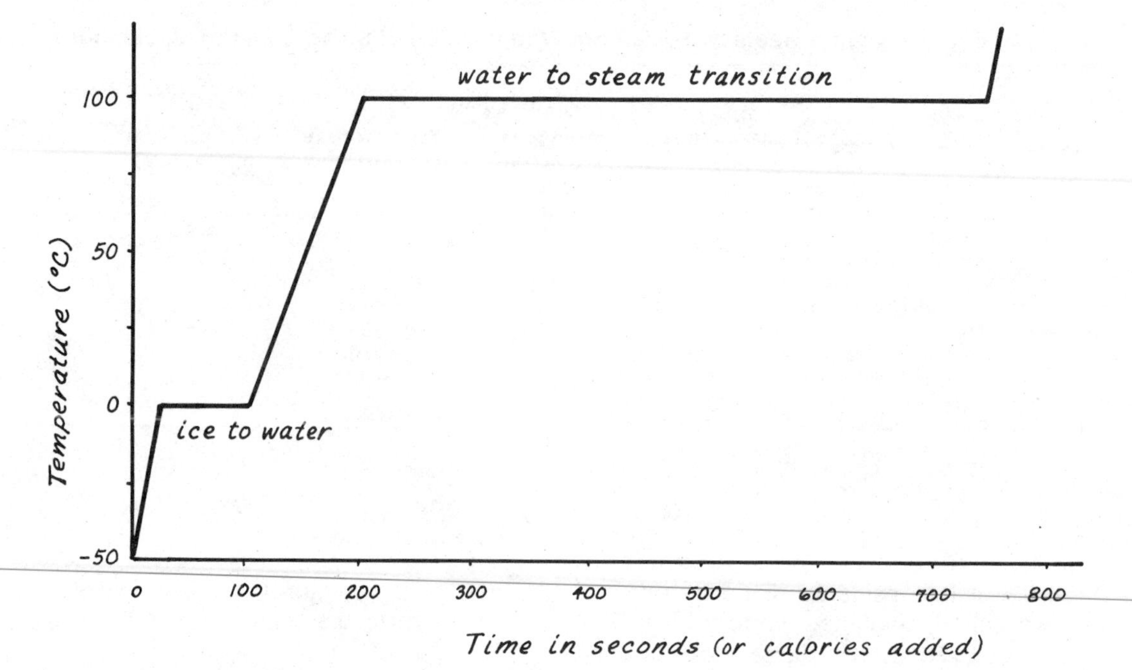

Figure 11–3 The temperature as a function of time for one gram of water to which heat is added at the rate of 1 cal/sec.

its temperature. With heat added at the rate of 1 cal/sec, it will take 80 seconds to convert the ice to water, during which time its temperature will remain constant at 0°C (see Figure 11–2). The fact that you can have ice at 0°C or water at 0°C is a good illustration of the statement in Chapter 10 that two objects which are at the same temperature can have drastically different amounts of internal energy.

Since it takes 1 calorie to raise the temperature of water 1°C, the water temperature will rise at the rate of 1°C/sec after all the ice has melted. It will then take 100 sec for the temperaturé to rise to 100°C, the boiling point. The whole process of changing from ice at −50°C to water at 100°C will have taken 25 + 80 + 100 = 205 seconds as shown in Figure 11–3. But now the gram of 100°C water must absorb 540 calories to break the attractive bonds and free the molecules in the form of steam at 100°C. This will require 540 seconds, during which the temperature remains constant at 100°C. This gives the long constant-temperature plateau in Figure 11–3. Only after all the water is converted to steam will the temperature of the steam start to rise at the rate of 2°C/sec.

The temperature plateaus in Figure 11–3 illustrate why the freezing and boiling points of water are so convenient as calibration points for thermometers. If a vessel contains a well stirred mixture of ice and water at thermal equilibrium, then one can be confident that the temperature of the mixture is 0°C. If a vessel of water which is open to the atmosphere at 760 mm Hg pressure is brought to a vigorous boil and stirred, its temperature will be 100°C. The stirring is important in both cases, since otherwise sizable temperature gradients can exist.

Because of the constant temperature behavior during phase transitions and their relative independence of surrounding conditions other than pressure, most standard temperature points are phase transitions in some substance. Solid carbon dioxide (dry ice) undergoes a direct solid-to-gas phase transition (sublimation) at −78.5°C. A mixture of dry ice and a solvent such as acetone is often used to get a constant temperature bath at −78.5°C. Liquid nitrogen undergoes a liquid-to-gas phase transition at 77°K (−196°C) and can be used as a temperature reference at that point. For temperatures above 0°C the melting points of various organic compounds can be used. The melting point of lead (327°C) is often used for a higher temperature reference point.

It should be noted that the temperatures

where phase changes occur represent points where the equation $Q = sm\Delta T$ (equation 10–7) cannot be used, i.e., they are points of discontinuity. If a substance is heated from a temperature below its melting point to a temperature above it, the calculation of Q requires a combination of equations 10–7 and 11–1. In general, the heat required to raise the temperature of a mass m from T_i to T_f when a phase change occurs at some intermediate temperature T_o would be

$$Q = s_1 m(T_o - T_i) + mh + s_2 m(T_f - T_o) \quad \textbf{11–3}$$

where the latent heat h involved could be the latent heat of fusion h_f or of vaporization h_v. The subscripts are used on the specific heats because the specific heat at temperatures above the phase change point will generally be different from that below the phase change.

Example: How much heat is required to raise the temperature of 200 grams of ice from −20°C to 60°C?

Solution: The specific heat of water is 1.0 cal/gm°C, the specific heat of ice is about 0.5 cal/gm°C, and the latent heat of fusion at 0°C is $h_f = 80$ cal/gm. Using equation 11–3, the heat required is

$$Q = (0.5\ \text{cal/gm–°C})(200\ \text{gm})(0°\text{C} - [-20°\text{C}]) + (200\ \text{gm})(80\ \text{cal/gm}) + (1.0\ \text{cal/gm–°C})(200\ \text{gm})(60°\text{C} - 0°\text{C})$$

$$Q = \underset{\text{to raise temp. of ice}}{2{,}000\ \text{cal}} + \underset{\text{to melt ice}}{16{,}000\ \text{cal}} + \underset{\text{to raise temp. of water}}{12{,}000\ \text{cal}}$$

$Q = 30{,}000$ calories.

When the amount of heat added is known and the final temperature must be calculated, the energy for phase changes must be included.

Example: Five kilocalories of energy are added to a 50 gm cube of ice which has an original temperature of −20°C. What will the final temperature be?

Solution: It is acceptable to try the equation $Q = sm\Delta T$, as long as it is recognized that if the temperature obtained is above the melting point, it will be incorrect because the heat of fusion and the change in specific heats will have been left out of the calculation. In this example that equation would yield

$$\Delta T = \frac{Q}{sm} = \frac{5000\ \text{calories}}{(0.5\ \text{cal/gm°C})(50\ \text{grams})} = 200°\text{C}.$$

This is clearly incorrect because it would lead to a temperature above the boiling point, ignoring two phase transitions! It is best to proceed step-by-step when a phase transition is involved.

1. Heat required to raise the temperature to 0°C:

$$Q_1 = (.5\ \text{cal/gm°C})(50\ \text{gm})(0°\text{C} - [-20°\text{C}]) = 500\ \text{cal}.$$

2. Heat required to melt ice:

$$Q_2 = (50\ \text{gm})(80\ \text{cal/gm}) = 4000\ \text{cal}.$$

3. Remaining energy:

$$5000\ \text{cal} - Q_1 - Q_2 = 500\ \text{calories}.$$

4. Temperature increase of 50 gm of water:

$$\Delta T = \frac{500\ \text{calories}}{(1\ \text{cal/gm°C})(50\ \text{gm})} = 10°\text{C}.$$

The final temperature of the resulting water is therefore 10°C. If the energy remaining after step 1 had been less than 4000 calories, you could conclude that not all of the ice would melt and that the final temperature would be 0°C. By using $Q = mh_f$ you could calculate how many grams of the ice would be melted by the remaining energy.

APPLICATIONS OF PHASE CHANGES

There are numerous ways in which the large internal energies involved in phase changes can be utilized. The increase of internal energy involved in the liquid-gas phase transition is the latent *heat of vaporization, h_v*. If a liquid can be forced to evaporate, then it will extract this heat of vaporization from its surroundings and act as a cooling agent. On the other hand, if a vapor is forced to condense, it will release the latent heat of vaporization and give energy to its environment. Similarly, the solid-liquid phase transition can be utilized. Melting solids extract internal energy, and freezing liquids give up internal energy to their environments. This energy is the latent *heat of fusion*, h_f.

REFRIGERATION. A refrigerator must extract heat from a cold area and exhaust it in an area of higher temperature. This is, of course, opposite to the normal flow of heat and requires work to "pump" the internal energy out of the interior of the refrigerator. The process can be accomplished by making use of the large latent heat of vaporization of some substance, as shown schematically in Figure 11–4. A refrigerant fluid is circulated in a closed system of coils. By means of an expansion valve and reduced pressure in the cooling coils, the refrigerant can be caused to evaporate and extract the latent heat of vaporization from the interior of the refrigerator. This energy is carried out to the compressor, where increased pressure forces the condensation of the refrigerant. In the process of condensation, the refrigerant gives off the latent heat of vaporization to the surrounding air. The fact that the latent heat of vaporization represents a large amount of internal energy makes this an efficient method for transferring heat. Though refrigerators differ widely in the types of refrigerant gases used and the temperature attainable, most of them use this basic principle.

COOLING BY PERSPIRATION. When perspiration evaporates, it extracts the latent heat of vaporization from the skin. This latent heat of vaporization is about 580 cal/gm at normal body temperature. Since the specific heat of the body is somewhat less than 1 cal/gm-°C, the evaporation of one gram of perspiration could cool over 600 grams of the body material by 1°C. The evaporation of 250 cm^3 of perspiration would cool the entire body of a 120 lb person by about 3°C. Circulating air helps to cool the body because it speeds the evaporation process. It is difficult to keep cool on a humid day because the level of moisture in the air inhibits the evaporation of perspiration. Alcohol sponges are used to cool feverish patients because alcohol evaporates much more quickly.

APPLICATION OF THE MELTING TRANSITION. Ice is, of course, much more effective in cooling a solution than just cold water because of the internal energy required to melt the ice.

Example. If 20 grams of ice at 0°C are placed in 30 grams of water at 100°C, what will be the final temperature of the mixture?

Solution. This is an application of the heat equation, $Q = sm\Delta T$, and the principle of conservation of energy. Thermal equilibrium will occur when the heat gained by the ice is equal to the heat lost by the 30 grams of water. It is assumed that all the ice will melt and that the temperature of the result-

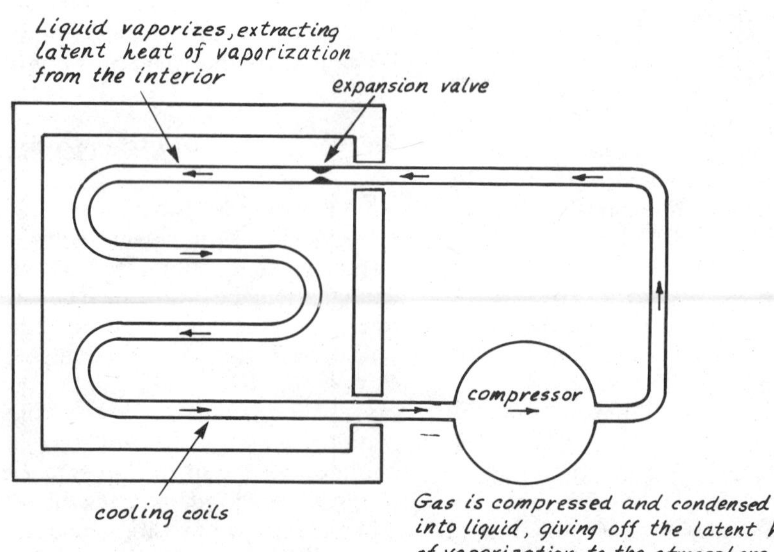

Figure 11–4 Schematic of refrigeration cycle.

ing 20 grams of water will be raised to some final temperature T_f

energy gained by 20 gm of ice
$= $ energy lost by 30 gm of water

$$Q_{ice} = \text{energy gained by 20 gm of ice} = \text{energy lost by 30 gm of water} = Q_{water}$$

$$m_i \times h_f + s m_i \, \Delta T_i = s m_w \, \Delta T_w$$

$$(20\text{ gm})(80\text{ cal/gm}) + (1\text{ cal/gm-°C})(20\text{ gm})(T_f - 0\text{°C}) = (1\text{ cal/gm-°C})(30\text{ gm})(100\text{°C} - T_f)$$

$$1600 + 20\ T_f = 3000 - 30\ T_f$$

$$50\ T_f = 1400$$

$$T_f = 28\text{°C}.$$

The final temperature of the 50 grams of water is 28°C, a drop in temperature of 72° for the hot water.

Example. If 20 grams of water at 0°C are mixed with 30 grams of water at 100°C, what will be the resulting temperature?

Solution. The approach is identical to that of the previous example except for the 1600 calories required to melt the ice.

$$\text{heat gained} = \text{heat lost}$$

$$Q_{\text{gained}} = Q_{\text{lost}}$$

$$(1\text{ cal/gm-°C})(20\text{ grams})(T_f - 0) = (1\text{ cal/gm-°C})(30\text{ grams})(100\text{°C} - T_f)$$

$$20\ T_f = 3000 - 30\ T_f$$

$$T_f = 60\text{°C}.$$

The cooling is, of course, much less effective, cooling the hot liquid by only 40°C compared to 72°C in the previous example.

EVAPORATION AND VAPOR PRESSURE

The phase change of water from the liquid to the gaseous state has been discussed mainly in terms of boiling at 100°C, but of course this phase transition occurs at all temperatures above 0°C in the form of evaporation. Because of higher than average kinetic energy, some of the molecules will escape from the liquid into the air (Figure 11–5(a)). If the temperature of the liquid is increased, more molecules will have the energy necessary to escape and the evaporation rate will therefore be greater. If the container is closed, the evaporation process will reach an equilibrium when the number of molecules bouncing back into the surface is equal to the number leaving. The vapor is then said to be saturated and the pressure exerted on the container walls is the *saturation vapor pressure*. The mass of water vapor per unit volume under this condition is the *saturation vapor density*. The saturation vapor pressure of water at body temperature is about 47 mm Hg. The saturation vapor density and pressure increase with temperature as shown in Table 11–2, but they are independent of the volume of the container. When the temperature reaches 100°C, the vapor pressure is equal to atmospheric pressure, 760 mm Hg. The precise definition of the *boiling point* of any liquid is in fact the temperature at which the vapor pressure

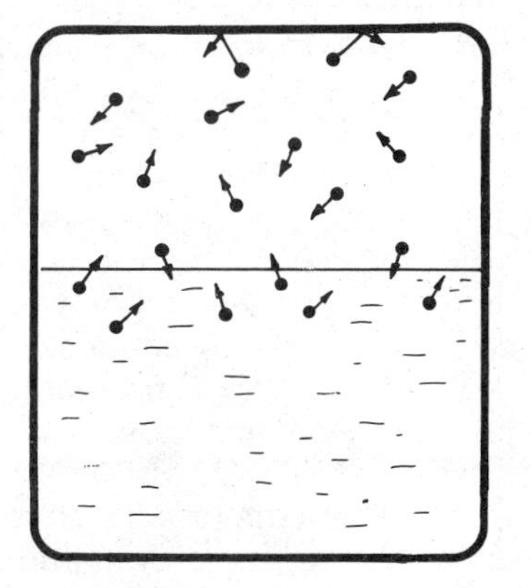

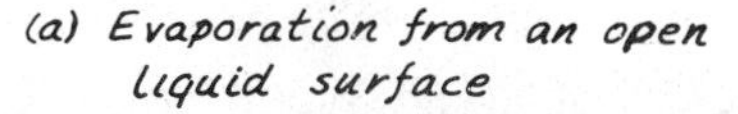
(a) Evaporation from an open liquid surface

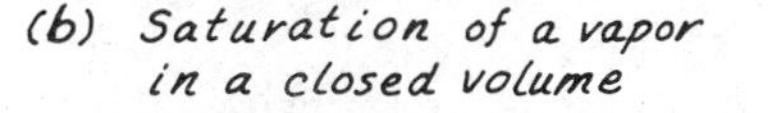
(b) Saturation of a vapor in a closed volume

Figure 11–5 Molecules with higher than average kinetic energy will escape from the liquid surface.

TABLE 11–2 Saturation Vapor Pressure and Density.

Temperature °C	°F	Vapor Density gm/m³	Vapor Pressure mm Hg
−10	14.0	2.36	2.15
0	32.0	4.85	4.58
5	41.0	6.80	6.54
10	50.0	9.40	9.21
11	51.8	10.01	9.84
12	53.6	10.66	10.52
13	55.4	11.35	11.23
14	57.2	12.07	11.99
15	59.0	12.83	12.79
20	68.0	17.30	17.54
25	77.0	23.0	23.76
30	86.0	30.4	31.8
37	98.6	44.0	47.07
40	104.0	51.1	55.3
60	140.0	130.5	149.4
80	176.0	293.8	355.1
95	203.0	505	634
96	204.8	523	658
97	206.6	541	682
98	208.4	560	707
99	210.2	579	733
100	212.0	598	760
101	213.8	618	788
200	392.0	7840	11,659

equals atmospheric pressure. By reference to Table 11–2, it can be seen that if the atmospheric pressure is lowered to 733 mm Hg, the boiling point of water is decreased to 99°C. An increase in pressure raises the boiling point above 100°C and allows the water to be superheated, as in a pressure cooker or autoclave.

When water is boiled in a container, the vapor pressure of the water exceeds atmospheric pressure and can therefore push the air out of the container. The magnitude of the atmospheric pressure can be demonstrated by boiling water in a can and then sealing the can as it cools. The water vapor which replaced the air in the can will condense as the can cools. Since the vapor pressure of the cooled liquid will be much less than atmospheric pressure, the unbalanced pressure will crush the can. This experiment demonstrates why autoclaves must be constructed of sturdy material to withstand the unequal pressure when they are cooled. The partial vacuum caused by the condensing vapor makes it impossible to open the door of the cooled autoclave until air has been admitted to equalize the pressure.

RELATIVE HUMIDITY

The amount of water in the air is usually less than the saturation density listed in Table 11–2. The percentage of saturation humidity at the given temperature is referred to as the *relative humidity*. It can be calculated from the relationship

11–4

$$\text{R.H.} = \frac{\text{actual vapor density}}{\text{saturation vapor density}} \times 100\%.$$

The most common units for the vapor density are grams per cubic meter (gm/m³).

Example. If the actual vapor density is 10 grams/m³ at 68°F, what is the relative humidity?

Solution. From Table 11–2, the saturation vapor density at 68°F is found to be 17.3 gm/m³. The relative humidity is then

$$\text{R.H.} = \frac{10.0\ \text{gm/m}^3}{17.3\ \text{gm/m}^3} \times 100\% = 57.8\%.$$

Since the saturation vapor density increases with the temperature, the same actual vapor density will represent a smaller relative humidity if the temperature of the air is increased. Since the membranes of the body tend to be sensitive to the relative humidity rather than the absolute humidity, the air will seem dryer if it is heated without increasing the number of grams/m³ of water vapor in the air. Central heating systems which heat and circulate a closed volume of air will reduce the relative humidity in the process unless water is added to the air by means of a humidifier.

Example. If the air in the previous example is heated to 77°F without increasing the actual vapor density (10 gm/m³), what will be the relative humidity after heating?

Solution. From Table 11–2, the saturation vapor density at 77°F is 23.0 gm/m³. The relative humidity is

$$\text{R.H.} = \frac{10\ \text{gm/m}^3}{23.0\ \text{gm/m}^3} \times 100\% = 43\%.$$

This is just above the 40% relative humidity recommended as a minimum for breathing.

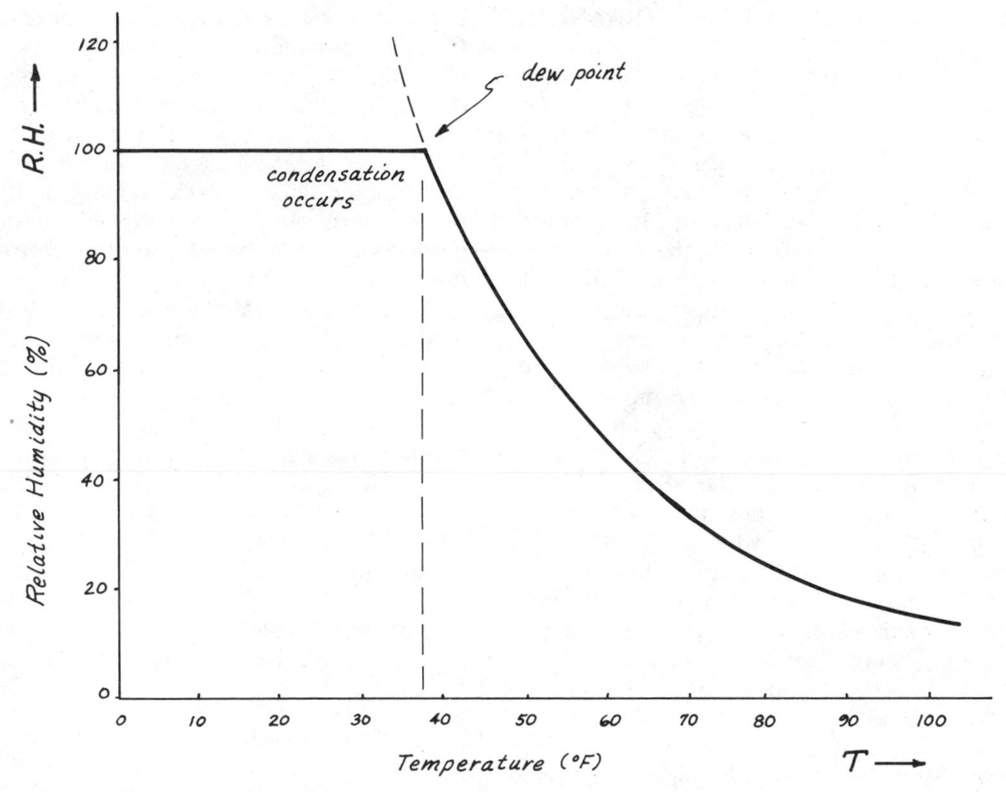

Figure 11–6 Relative humidity as a function of temperature for an actual humidity of 6 grams/m³.

If the temperature of the air is decreased while holding the actual moisture content constant, the relative humidity will increase.

Example. If the relative humidity is 50% and the temperature is 77°F on a given day, what is the relative humidity in a cool basement which has the same actual vapor density but a temperature of 59°F?

Solution. If the relative humidity is 50%, then the actual vapor density is one-half the saturation vapor density at 77°F:

$$\text{actual humidity} = 0.5 \times 23.0 \text{ gm/m}^3$$
$$= 11.5 \text{ gm/m}^3.$$

Since the saturation vapor density at 59°F is 12.8 gm/m³, the relative humidity is

$$\text{R.H.} = \frac{11.5 \text{ gm/m}^3}{12.8 \text{ gm/m}^3} \times 100\% = 90\%.$$

If the air is gradually cooled while maintaining the moisture content constant, the relative humidity will rise until it reaches 100%. This temperature, at which the moisture content present in the air will saturate the air, is called the *dew point*. If the air is cooled further, some of the moisture will condense. For example, if the air contains 6 gm/m³ at 86°F, the relative humidity is 20%. If the air is cooled, the relative humidity will increase as shown graphically in Figure 11–6. At 50°F that 6 gm/m³ will represent about 64% R.H., and at about 37°F it will saturate the air (R.H. = 100%). If it is cooled to 32°F, then 1.1 gm/m³ must condense since the air can hold only about 4.9 gm/m³ at that temperature. A pitcher of ice water will usually collect condensed water on its outer surface, since the pitcher will be below the dew point in the room. The fogging of eyeglasses and automobile windshields when their temperatures are below the dew point of the surrounding air is a common phenomenon. Cold refrigeration coils can be used to dehumidify a room, since they will collect condensate from the air. The formation of clouds in the atmosphere occurs when the air temperature drops below the local dew point.

HEAT TRANSFER

When a thermal gradient exists, heat will tend to move from the high temperature area to the low temperature area. This internal energy transfer may occur by one or more of the following three mechanisms: (1) conduction, (2) convection, and (3) radiation of heat. It can also occur by the transfer of the latent heat of fusion or vaporization.

Conduction is the primary method of heat transfer in solids. If a silver spoon is placed in a hot cup of coffee, the other end of the spoon will quickly become hot because of the conduction of internal energy through the handle of the spoon. Conduction is the transfer of heat by the direct interaction of molecules in a hot area with molecules in a cooler area. Microscopically, this interaction is in the form of collisions between molecules, with the more rapidly moving molecules in the high temperature area giving internal energy to molecules in a colder area. The efficiency of heat conduction depends upon the number of collisions and the amount of energy transferred during each collision. The *thermal conductivity* of a material is a measure of this efficiency. The thermal conductivities of a number of common substances are listed in Table 11–3. Note that metals are generally much better heat conductors than nonmetals. Gases are usually poor heat conductors because of the smaller number of molecular collisions which occur in the gaseous state. The difference in thermal conductivity between metals and non-metals like wood explains the results of the earlier example in which metal and wood blocks at 0°C and 100°C were touched. At 100°C the metal block feels hotter since it is a better conductor and conducts heat to the finger more rapidly. At 0°C the metal feels colder since it conducts heat away from the finger more rapidly.

TABLE 11–3 Thermal Conductivities in (cal/sec)/(cm² × °C/cm).

MATERIAL	THERMAL CONDUCTIVITY (*k*)
Silver	1.01
Copper	0.99
Aluminum	0.50
Iron	0.163
Lead	0.083
Ice	0.005
Glass, ordinary	0.0025
Concrete	0.002
Water at 20°C	0.0014
Asbestos	0.0004
Hydrogen at 0°C	0.0004
Helium at 0°C	0.0003
Snow (dry)	0.00026
Fiberglass	0.00015
Cork board	0.00011
Wool felt	0.0001
Air at 0°C	0.000057

Conduction is the main mechanism for heat loss from a building in winter. The rate of heat loss is proportional to the outside area of a house, the thermal conductivity of its walls, and the difference between the inside and outside temperatures. It is inversely proportional to the wall thickness, since a thicker wall would reduce the heat loss. These factors can be combined to give the conduction equation

$$\text{rate of heat loss} = \frac{\left(\begin{matrix}\text{wall}\\\text{area}\end{matrix}\right)\left(\begin{matrix}\text{thermal}\\\text{conductivity}\end{matrix}\right)\left(\begin{matrix}\text{temperature}\\\text{difference}\end{matrix}\right)}{\text{wall thickness}} \qquad \textbf{11–5}$$

$$\text{or} \quad \frac{Q}{t} = \frac{Ak(T_2 - T_1)}{d},$$

where Q = heat loss in calories, t = time in seconds, A = area in cm², T_2 = inside temperature, T_1 = outside temperature (°C), k = thermal conductivity of the wall expressed in (cal/sec)/(cm² × °C/cm), and d is the wall thickness in cm. The quantity $(T_2 - T_1)/d$ is referred to as the temperature gradient or thermal gradient. The above relationship is the general equation for conduction through a flat barrier and can be used qualitatively to describe more general examples of heat conduction.

For example, from equation 11–5 it can be seen that doubling the thickness, *d,* of a garment will halve the rate of heat loss from the body. However, wearing a garment with the same thickness but with one-half the thermal conductivity, *k,* would accomplish the same result and perhaps be more comfortable. It can be seen that keeping the inside of a house at 40°F when the outside temperature is 20°F will require the same amount of energy as keeping the house at 70°F when the outside temperature is 50°F, since the heat loss depends upon the temperature difference rather than upon the actual temperatures involved.

Since the thermal conductivity of air is quite small, air spaces between walls provide insulation against heat loss. The heat loss can be reduced further by inserting an insulating material such as fiberglass in the walls. This material traps air spaces so that there is very little air circulation and therefore a smaller loss by convection (see below). The same principle applies to clothing and blankets made of wool or some other material which traps air spaces. The warmth of a sweater is dependent not so much upon the type of fibers used, as upon the low thermal conductivity of the small air pockets trapped in the porous material. This principle is used by birds, which in winter fluff up their feathers to entrap air spaces for insulation.

In applying heat to the body, the rubber hot water bottle serves as an insulator which slows the rate of heat transfer from the water to the body to prevent burns and to allow the available internal energy from the hot water to be used over a longer period of time. A flannel cloth can further insulate the water to slow the heat transfer, but its insulating properties are destroyed if it is wet, since water is a conductor of heat. These insulators between the water and the patient will cause a large temperature gradient so that the temperature at the patient's skin is much lower than the water temperature.

Whereas conduction of heat through a liquid or gas refers to the transfer of heat by molecular collisions without movement of the fluid as a whole, *convection* refers to heat transfer by the movement of the fluid. Air is a poor conductor, but it can efficiently transfer heat by convection if the air is heated in one location and then circulated to carry the heat elsewhere. The movements of fluids which carry heat are called convection currents.

The origin of convection currents in gases can be understood from the relationship between the temperature, volume, and pressure of gases (the ideal gas law). As air rises in temperature, it expands and therefore becomes less dense than the surrounding air. The resulting buoyant forces cause it to rise; the more dense cool air will tend to move down to replace it. In a room heated by a so-called "radiator," these influences will set up a continuous cyclic convection current as illustrated in Figure 11–7. Heat-generated convection currents are major factors in atmospheric air movements. Any hot object exposed to air will generate such currents, and a considerable part of the cooling process of such objects can be attributed to convection. Clothing helps to prevent such currents of air from touching the skin and therefore minimizes convection loss.

Radiation is fundamentally different from the other two types of heat transfer. In conduction, internal energy is transferred from molecule to molecule. In convection, the moving molecules carry the heat with them. Both of these require material of some kind to transport the energy, but radiant energy is of the same basic nature as light and can travel through a vacuum. As will be discussed in Chapter 17, light is a traveling wave phenomenon which carries radiant heat with it. The earth receives all of its energy from the sun by the process of radiation. The radiant energy given off by objects near room temperature is in the infrared range, but when an object becomes "red hot," it is radiating some of its energy in the form of visible light.

The heat lost by radiation from a hot object at absolute temperature T_2 when it is surrounded by an environment with uniform temperature T_1 is given by

$$\frac{Q}{t} = e\sigma A\,(T_2^4 - T_1^4) \qquad \textbf{11–6}$$

where A is the area radiating, σ is the Stefan-Boltzmann constant ($\sigma = 5.67 \times 10^{-8}$ watts/m^2 · °K^4).

The symbol e is the "emissivity" of the object, a measure of its effectiveness as a radiator. For a perfect radiator, $e = 1$. This ideal radiator is also an ideal absorber, absorbing 100% of the radiant energy incident upon it. (Since a body that absorbs all of the

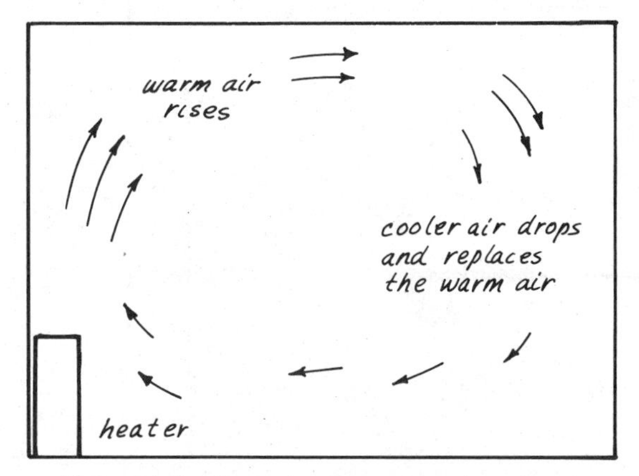

Figure 11–7 Convection currents in a heated room.

light incident on it would appear completely black, the ideal emitter-absorber is called a "black body.") The emittance is strongly correlated to the color of an object. A light-colored object will be both a poor absorber and a poor emitter of radiant heat. A perfect reflector would be represented by $e = 0$.

Good reflectors such as thin films of gold, silver, or aluminum may be well over 90% efficient in reflecting heat energy. Such coatings are used on artificial satellites to prevent overheating by absorption of the sun's radiation. The ordinary vacuum bottle is an excellent example of the control of all three types of heat transfer. As shown in Figure 11–8, the vacuum flask is a double-walled glass bottle which has a good vacuum between the walls. This vacuum makes heat transfer by conduction and convection impossible, since both processes require a medium for the transport of energy. The walls of the bottle are covered with a very thin reflective metallic film to prevent heat transfer by radiation. Often the bottle is further isolated from the environment by suspending it on springs or other small supports in a closed air space. Vacuum flasks are often referred to as dewar flasks after the inventor, James Dewar. Specially constructed dewar flasks are used widely in low temperature research and applications, since they are capable of holding cold liquefied gases such as liquid nitrogen (77°K) and liquid helium (4°K) for considerable lengths of time.

PHYSIOLOGICAL APPLICATIONS OF HEAT TRANSFER

As discussed in Chapter 10, the body obtains energy by the oxidation of foods. All such processes produce some energy which cannot be utilized and which is given off to the environment. Besides the processes of conduction, convection, and radiation from the skin, the process of evaporation from the skin carries off a significant amount of heat, as discussed previously. A large amount of water vapor is added to the expired air, and the latent heat of vaporization of this water is lost from the body.

The discussion here will be limited to the physical processes involved in giving off heat from the body. For a discussion of the wide variations in the amount of heat produced by the body, a physiology text should be consulted (1, 2). For the present purposes, the basal metabolic rate will serve as a convenient reference for examining the heat production of the body. This is the amount of internal energy which must be produced to maintain body functions such as respiration and heart beat when a person is at rest (see reference 1 for a more complete

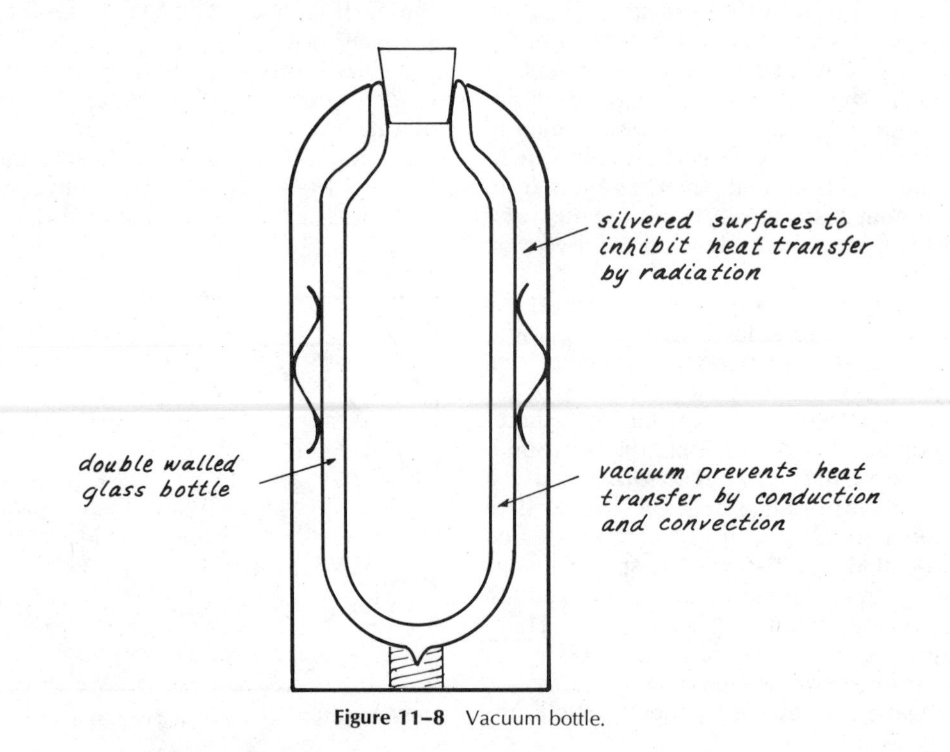

Figure 11–8 Vacuum bottle.

definition). For an adult male the heat loss from the body associated with the basal metabolism is about 40 kilocalories/hr per square meter of square area. This rate of energy loss is equivalent to 46.5 watts, so for a person with a surface area of about 2 m^2, the basic rate of energy loss is about 90 watts. Guyton (4) reports that the energy expenditure during strenuous exercise may be 600 kcal/hr (700 watts) for a short time and that a laborer may average as high as 290 kcal/hr (340 watts) over a 24 hour period.

As shown in one of the examples in Chapter 10, the body temperature would rise very rapidly if there were not efficient means of transferring the excess energy away from the body as heat. The heat transfer mechanisms must be flexible enough to handle the extremes of activity cited above, but as a base for comparison, the various methods of heat transfer from the body will be related to the nominal 90 watt energy loss rate which the body must maintain at all times.

Radiation is a surprisingly effective mechanism for the transfer of energy from the body. This is in contrast with room heaters called "radiators" or "radiant heaters," which often actually transfer most of their energy by conduction and convection. But the skin is apparently an almost ideal absorber and radiator in the infrared region, absorbing approximately 97% of the infrared radiation which strikes it (4). The radiation efficiency is equal to the absorption efficiency and this leads to an emission coefficient e = .97. This is about the same for white and black skin. Black skin is a more efficient absorber of visible light and would absorb energy faster in direct sunlight, but skin color makes very little difference in the infrared region where most of the radiation of energy occurs. A model of radiation heat loss should help with its evaluation.

Example: The skin temperature of a resting person is about 34°C. If the person is nude in a room with uniform surrounding temperature 23°C, what would be the rate of heat loss if the total skin area is 2 m^2?

Solution: From equation 11–6:

$$\frac{Q}{t} = e\sigma A\,(T_2^4 - T_1^4)$$

$$e = .97$$

$$\sigma = 5.67 \times 10^{-8}\ \text{watts/m}^2\cdot{}^\circ\text{K}^4$$

$$A = 2\ \text{m}^2$$

$$T_2 = 307^\circ\text{K}$$

$$T_1 = 296^\circ\text{K}$$

$$\frac{Q}{t} = (.97)(5.67 \times 10^{-8}\ \text{watts/m}^2\cdot{}^\circ\text{K}^4)(2\ \text{m}^2) \times [(307^\circ\text{K})^4 - (296^\circ\text{K})^4]$$

$$\frac{Q}{t} = 133\ \text{watts.}$$

Remarkably, this radiation loss alone is more than enough to accomplish the minimal energy transfer of 90 watts, so a person would begin to feel cool in this situation even if no other heat loss mechanism were in operation. This radiative loss could be reduced by lowering the skin temperature. The skin temperature is controlled by the blood flow to it, and this blood supply can be controlled by vasoconstriction and vasodilation (see Chapter 7). Since changes in the internal radii of the vessels are so effective in volume flow rate control, the arterioles which feed the vessel network near the skin can change this blood supply from nearly zero to almost a third of the body's blood supply (4). If vasoconstriction produced a lowering of the skin temperature to 30°C in the above example, the radiation heat loss rate would drop to 83 watts. Otherwise, activity which increased the rate of internal energy production could make the person more comfortable. If no better options exist, shivering is a muscular activity which can triple the body's production of internal energy compared to the resting state.

It is reported (4) that 60% of the body's energy loss in circumstances such as in the above example may occur by radiation. It is apparently a significant energy loss mechanism even through normal clothing. Clothing which is highly reflective of infrared radiation could help prevent this loss by reflecting the energy back to the body. A popular item with backpackers is a very light blanket which is coated with a highly reflective metallic coating. It is surprisingly warm since it reflects the radiation loss back to the body.

As mechanisms for removing body heat, conduction and convection are closely related. Without air movement, heat loss by conduction would be quite inefficient because of the low thermal conductivity of air, but heat must be conducted to the layer of air immediately surrounding the body before it can be carried away by convection currents. The importance of the thermal conductivity of the air has been demonstrated by undersea experimenters living in a helium and oxygen atmosphere. Helium is a much better heat conductor than the nitrogen which is replaced in the air. This changes the body's rate of heat loss to the extent that the experimenters feel chilled even in an ambient atmospheric temperature of 80°F.

Example: As a model of the conduction heat loss process, we will consider the same circumstances as in the radiation example above (nude person, skin temperature 34°C, surrounding air temperature 23°C). The further assumption is made that there is no significant large scale air circulation and that the air 5 cm away from the skin is at the ambient temperature 23°C.

Solution: Using equation 11–5:

$$A = 2\ \text{m}^2 = 2 \times 10^4\ \text{cm}^2$$

$$k = .000057\ (\text{cal/sec})/(\text{cm}^2 \times {}^\circ\text{C/cm})$$
from Table 11–3

$$T_2 - T_1 = 11^\circ\text{C}$$

$$d = 5\ \text{cm}$$

$$\frac{Q}{t} = (2 \times 10^4\ \text{cm}^2) \times$$

$$[5.7 \times 10^{-5}\ (\text{cal/sec})/(\text{cm}^2 \times {}^\circ\text{C/cm})]\ \left(\frac{11^\circ\text{C}}{5\ \text{cm}}\right)$$

$$\frac{Q}{t} = 2.5\ \text{cal/sec} = 9.0\ \text{kcal/hr} = 10.5\ \text{watts.}$$

This may not be a realistic model, but it yields a result fairly close to the experimental result of 10 to 12% loss by conduction and convection when compared to the basal metabolic loss rate of 90 watts. The conduction-convection loss could be increased greatly by circulating room temperature air close to the skin; if the distance from the skin to room temperature air in the example above were reduced by a factor of ten to 0.5 cm, the conduction-convection loss rate would increase tenfold. Reducing the skin temperature by vasoconstriction as discussed above would reduce heat loss, and vasodilation to raise the skin temperature would increase the heat loss rate.

Conduction is the main mechanism by which heat is transferred from the core of the body to the skin. To facilitate this transfer, the skin temperature is normally maintained at temperatures 4 to 5°C below internal temperatures, depending upon the blood supply to the skin, as discussed above.

Cooling the body by evaporation of perspiration is an efficient process because during evaporation the large latent heat of vaporization, $h_v = 580$ cal/gm at 37°C, is extracted from the skin. Even when no perspiration is evident, evaporation from the skin and exhaled water vapor are on the order of 600 grams per day (4). The associated energy loss rate is

$$\frac{Q}{t} = (600\ \text{gm/day})(580\ \text{cal/gm}) \times$$

$$(4.186\ \text{joule/cal}) \times \left(\frac{1\ \text{day}}{24\ \text{hr}}\right)\left(\frac{1\ \text{hour}}{3600\ \text{sec}}\right)$$

$$\frac{Q}{t} = 17\ \text{watts.}$$

Therefore, the perspiration heat loss is a significant fraction of the basic 90 watt energy loss rate.

A fact of major significance about perspiration heat loss is that it will remove heat from the body even when the ambient temperature is higher than body temperature. It is the body's only major heat loss mechanism under this condition since radiation, conduction, and convection can transfer heat only from high temperature areas to lower temperature areas. They transfer energy into the body rather than out of it at ambient temperatures above body temperature. It is fortunate that the perspiration heat loss mechanism is very flexible. After several weeks of acclimatization to tropical climates, a person can produce perspiration at the prodigious rate of 3.5 liters/hr (4)! If all of this were evaporated and thereby extracted its latent heat of vaporization from the skin, the cooling rate would be

$$\frac{Q}{t} = (3500\ \text{gm/hr})(580\ \text{cal/gm}) \times$$

$$(4.186\ \text{joule/cal}) \times \left(\frac{1\ \text{hour}}{3600\ \text{sec}}\right)$$

$$\frac{Q}{t} = 2360\ \text{watts.}$$

This would appear to be enough cooling even during strenuous activity in the tropics. Unfortunately, it is very difficult to evaporate that much perspiration. The evaporation rate may be increased by circulating air, but if the relative humidity is high, the evaporation is inhibited.

SUMMARY

With the addition of a sufficient amount of internal energy, a solid will undergo a "change of phase" to the liquid state and eventually to the gaseous state. At a given pressure the phase transitions of a material occur at definite temperatures. For example, the phase transitions of water of 0°C (ice to water) and 100°C (water to steam) at atmospheric pressure serve as the calibration points for most thermometers, since they are precisely reproducible. During phase transition, the temperature of a material remains constant while it absorbs or expels a large amount of internal energy. These large energy changes have many practical applications. In the process of absorbing the large "latent heat of vaporization," a liquid will extract large amounts of internal energy from its environment while evaporating. This fact is used in most refrigeration processes and in the cooling of the skin by the evaporation of perspiration. The energy involved in the solid-to-liquid phase transition is also useful in many applications.

The liquid-gas phase transition occurs at temperatures below the boiling point through the process of evaporation. In a closed volume the resulting gas will reach a saturation vapor pressure at which there is an equilibrium between evaporation and condensation. The saturation vapor pressure increases with temperature. The boiling point is defined as the temperature at which the vapor pressure reaches atmospheric pressure (vapor pressure of water = 760 mm Hg at 100°C). Under ordinary conditions the water in the air does not reach saturation, and the percentage of saturation is referred to as the relative humidity. Body tissues tend to be sensitive to the relative humidity rather than the absolute humidity.

Heat can be transferred by the processes of conduction, convection, and radiation. Conduction is the direct transfer of heat energy in solid objects and depends upon the area, thickness, and thermal conductivity of the solid and upon the temperature difference between different parts of the solid. Convection is the transfer of heat by thermally generated currents in the air or in a liquid. Radiation is heat energy transfer in the form of emitted "light-like" waves such as infrared radiation. The body gives off heat by all three of these methods and also loses heat through the evaporation of perspiration and the exhalation of evaporated water.

REVIEW QUESTIONS

1. Why is a quantity of ice at 0°C a more effective coolant than the same quantity of water at 0°C?

2. Will foods cook more quickly in vigorously boiling water than in gently boiling water?

3. What happens to the heat taken from food in a refrigerator?

4. Explain why boiling points and freezing points are particularly suitable as calibration temperatures for thermometers.

5. In the "freeze-drying" process of taking moisture out of foods, the pressure around the food is lowered by means of a vacuum pump so that the water will evaporate. The residue left behind will often be frozen. Why?

6. Why is a pressure cooker needed for cooking at very high altitudes?

7. What factors influence the rate of evaporation from an open liquid?

8. Why does the saturation vapor pressure of water increase with temperature?

9. What will happen to the relative humidity in a closed house at night when the air is heated? Explain.

10. Why does a cool basement tend to be damp?

11. What is meant by the "dew point"? Why does moisture form on the outside of a cold pitcher of water?

12. What is the main advantage of a high pressure steam sterilizer (autoclave) over simple boiling water for sterilizing instruments?

13. When a glass stopper in a bottle is stuck, it may sometimes be loosened by holding the neck of the bottle under a hot water faucet. Explain the physical principles responsible for this observed phenomenon.

14. By what methods is heat given off by the body? Which of these methods are effective when the temperature surrounding the body is higher than body temperature?

15. What physical property makes a metal cup less suitable for hot coffee than a china cup?

16. A house is insulated to keep it warm in winter. Will this insulation make it harder or easier to keep cool during the summer?

17. Why are the freezing units of most refrigerators placed at the top of the cabinet?

18. Will the temperature of a closed room be lowered by the operation of an electric fan?

19. Why is a vacuum flask silvered?

PROBLEMS

Worked Example: A cup of coffee containing 250 gm of coffee is at a temperature of 100°C. If the specific heat of the coffee is assumed to be 1 cal/gm-°C, compare the cooling effects of (a) adding 50 grams of ice and (b) forcing 50 grams of the liquid to evaporate.

Solution: (a) If 50 gm of ice melts, it will extract the necessary energy from the coffee (80 cal/gm). The resulting 50 grams of water will be heated to come to equilibrium with the coffee.

$$\text{energy gained by ice} = \text{energy lost by coffee}$$
$$(M_{ice})\, h_f + SM_{ice}\, \Delta T_{ice} = SM_{coffee}\, \Delta T_{coffee}$$

$$(50 \text{ gm})(80 \text{ cal/gm}) + (1 \text{ cal/gm-°C})(50 \text{ gm})(T_f - 0°\text{C})$$
$$= (1 \text{ cal/gm-°C})(250 \text{ gm})(100°\text{C} - T_f)$$
$$4000 + 50\ T_f = 25{,}000 - 250\ T_f$$
$$300\ T_f = 21{,}000$$
$$T_f = 70°\text{C}.$$

(b) If 50 gm of the coffee is forced to evaporate (for example, by putting it in a partial vacuum), that 50 grams will extract from the remaining 200 grams

$$(50 \text{ grams})(540 \text{ cal/gm}) = 27{,}000 \text{ calories}.$$

If the change in temperature is calculated,

$$\Delta T = \frac{Q}{SM} = \frac{-27{,}000 \text{ cal}}{(1 \text{ cal/gm-°C})(200 \text{ gm})} = -135°\text{C},$$

a final temperature of -35°C is obtained. This cannot be accurate, because the coffee would be ice at -35°C and we haven't accounted for the energy of the water-ice transition. If we assume that it is cooled to 0°C, we can calculate the energy extracted for that process:

$$\Delta T = 100°\text{C}$$

$$Q \text{ extracted} = (1 \text{ cal/gm-°C})(200 \text{ gm})(100°\text{C}) = 20{,}000 \text{ cal}.$$

But the evaporating liquid extracted 27,000 calories, so the other 7000 calories extracted will actually freeze part of the liquid. The mass of ice resulting will be

$$M_{ice} = \frac{7000 \text{ cal}}{80 \text{ cal/gm}} = 87.5 \text{ gm at } 0°\text{C}.$$

So the evaporation process not only cools the coffee to 0°C, but it also turns 87.5 gm of it into ice. This illustrates the large amount of energy extracted during the evaporation process.

1. A 10 gram mass of ice is at the temperature 0°C. If 1000 calories of energy are added, what will be its final temperature?

2. The cooling of a refrigerator occurs by vaporization of a circulating refrigerant substance. Commonly used refrigerants have heats of vaporization around 40 cal/gm. How much refrigerant would have to be evaporated to freeze 300 cm³ of water to form ice if the water was initially at a temperature of 25°C?

3. A pan containing 300 grams of ice at 0°C is placed over a low burner which supplies 50 calories per second to the ice in addition to any heat given to the pan.

    a. How long will it take to completely melt the ice to form water at 0°C?
    b. How long will it take to heat the resulting water to 100°C?
    c. How long will it take to boil all the water out of the pan after reaching 100°C?

4. A 100 gram mass of ice at -10°C is to be changed to 100 grams of steam at 100°C. How much energy in calories will be required?

5. If 30 grams of ice are added to 200 grams of water at 30°C, what is the final temperature?

6. How much heat energy is extracted from the environment when 500 grams of ice melt?

7. How much energy will be required to melt 50 grams of ice at 0°C, heat the resulting water to 100°C, and vaporize it into steam at 100°C?

8. If 20 grams of ice at 0°C are placed in 100 grams of water at 100°C, what final temperature will result if no heat is lost to the environment?

9. If 5 grams of water evaporate from a container which originally held 200 grams of water at 100°C, what final temperature will result under the following assumptions: heat capacity of container negligible, no heat loss to surroundings, heat of vaporization of all 5 grams approximately equal to the value at 100°C.

10. Air conditioners are usually rated either in "tons" or in terms of the number of BTU per hour that they will extract from your house. The BTU, British thermal unit, is a heat energy unit. It takes 144 BTU of energy to melt one pound of ice. A "one ton" air conditioner will extract enough energy to melt one ton of ice in 24 hours. How many BTU are required to melt a ton of ice? What is the equivalent of a one ton air conditioner when expressed in BTU/hr?

11. If the water vapor density in the air is 12 gm/m^3 at a temperature of 68°F, what is the relative humidity? What is the dew point temperature?

12. If the dew point is 12°C and the temperature is 25°C, what is the relative humidity?

13. If outside air at a temperature of 41°F and a relative humidity of 50% were taken into a closed house and heated to 68°F, what would be the relative humidity after heating? What would be the relative humidity when heated to 98.6°F, body temperature?

14. The saturation vapor density is 10 gm/m^3 at 52°F and approximately 50 gm/m^3 at 103°F. If the relative humidity were 60% in a volume of air at 52°F and that air was subsequently heated to 103°F without changing the amount of moisture in the air, what would be the relative humidity after heating?

15. If a person breathes 10 liters per minute of air at 68°F, 50% relative humidity, how much water per minute must the internal membranes supply to saturate this air at 98.6°F? If all of this moisture is subsequently exhaled, how much water per day is given off by the body in this process? If each gram of water extracts 580 calories of energy as it is vaporized, how much daily heat loss in kilocalories does this represent?

16. If the relative humidity is 50% at a temperature of 68°F, what would be the relative humidity of this same air if it were heated to 86°F without changing the amount of water in the air?

17. Calculate the rate of heat loss in cal/sec and watts for a window which consists of 2 square meters of glass area with thickness 3 millimeters if the outside temperature of the glass is at 0°C and the inside surface is at 10°C. (The inside of the glass will not be as warm as the room temperature.)

18. Calculate the rate of heat loss in cal/sec and watts for a window area 2 m^2 which consists of two layers of glass of thickness 3 mm each, which enclose a sealed volume of air with thickness 6 mm, if the inside temperature of the glass is 20°C and the outside temperature is 0°C. In applying the conduction equation, assume that the temperature drop across the 6 mm of glass and the 6 mm of air has the ratio

$$\frac{\Delta T_{\text{glass}}}{\Delta T_{\text{air}}} = \frac{k_{\text{air}}}{k_{\text{glass}}},$$

i.e., the larger the conductivity, the smaller the temperature drop. Then the heat flow through either the glass or the enclosed air may be calculated using the conduction equation, since that which flows through the glass must flow also through the sealed air once an equilibrium flow is established.

19. At body temperature, the heat of vaporization is about 580 cal/gm instead of the 540 cal/gm required at 100°C. An average human body may give off about 2000 kcal of heat per day. If all of this were given off by the evaporation of perspiration, how much water would be evaporated per day?

20. A person with skin area 2 m^2 and 97% radiation efficiency is at rest and has a skin temperature 28°C. He or she is in an environment with uniform temperature 22°C.

    a. Calculate the rate of radiation heat loss in watts.
    b. Calculate the rate of heat loss by radiation if the skin temperature is raised to 36°C by vigorous exercise.

21. A person with skin area 2 m^2 is nude in a room of still air at 22°C and it is assumed that the temperature near the skin drops to 22°C at a distance of 5 cm from the skin so that this 5 cm of air can be considered to be the "wall" that the heat is penetrating.

    a. What will be the rate of heat loss in watts if the skin temperature is 28°C?
    b. What will be the rate of heat loss in watts if the skin temperature is 36°C?

22. Use the same circumstances as in problem 21, except that for experimental purposes the nitrogen in the air has been replaced by helium, so that the thermal conductivity of helium is approximately correct for the mixture.

    a. Calculate the conduction heat loss rate in watts if the skin temperature is 28°C.
    b. If the conduction loss rate must be limited to 20 watts for comfort, what is the minimum temperature which would be acceptable in the room?
    c. Suppose a pressure of 5 atmospheres of helium mixture must be maintained for experiments in a diving bell, and that the thermal conductivity increases to five times the value at one atmosphere. Recalculate the minimum comfortable temperature to limit the conduction heat loss rate to 20 watts for a 28°C skin temperature.

23. A certain marathon runner must give off energy at a rate of 200 watts over and above that which he can give off by conduction, convection,

and radiation. How much perspiration must evaporate from his skin per hour to maintain this cooling rate?

24. A 60 kg runner on a hot, humid day cannot give off enough heat to maintain a normal body temperature. If the rate of production of internal energy is 30 watts greater than the heat transfer rate out of the body, how much would the runner's body temperature rise during a 30 minute run? Assume a specific heat of .8 cal/gm°C and assume the body to be at a uniform temperature.

REFERENCES

1. Langley, L. L. *Physiology of Man, 4th ed.* New York: Van Nostrand Reinhold, 1971.
2. Bell, G. H., Davidson, J. N., and Scarborough, H. *Textbook of Physiology and Biochemistry, 9th ed.* Baltimore, Md.: The Williams and Wilkins Co., 1976.
3. Glasser, O., editor. *Medical Physics,* Vol. I. Chicago: Year Book Medical Publishers, Inc., 1944.
4. Guyton, A. C. *Basic Human Physiology, 2nd ed.* Philadelphia: W. B. Saunders, 1977.

CHAPTER TWELVE

Introduction to Electricity and Magnetism

INSTRUCTIONAL OBJECTIVES

After studying this chapter, the student should be able to:

1. Calculate the force between two electric charges, given appropriate numerical data.
2. Define the terms conductor and insulator.
3. Define the terms volt and ampere, and describe what is measured by these two units.
4. Define capacitance and give an example of an application of a capacitor.
5. Sketch the basic elements contained in the tube of an oscilloscope and describe the function of each.
6. Describe the magnetic properties of iron, nickel, and cobalt, indicating how they differ from other materials.
7. Name the elements required to construct a strong electromagnet.

THE ELECTRICAL NATURE OF MATTER

All atoms are made up of protons, neutrons, and electrons. The classical model of the atom consists of a positively charged nucleus made up of protons and neutrons, and a number of negatively charged electrons in orbit about that nucleus. The simplest form of that model pictures electrons as tiny particles which circle the nucleus in definite orbits similar to the orbits of the planets about the sun. Though this model is useful for visualizing electrical processes, it must be pointed out that it is quite inadequate for describing the details of atomic structure.

These details can be explained with the methods of quantum mechanics, the branch of physics used to describe molecular and atomic phenomena. The quantum theory of the atom will be discussed further in Chapter 20. The usefulness of the orbital model of the atom, the Bohr model, is based upon the fact that the physical parameters of electrons are "quantized" and can take on only certain discrete values. For example, an electron has a definite mass ($m_e = 9.1 \times 10^{-31}$ kg) and a definite charge (1.6×10^{-19} coulombs) and can conveniently be considered a particle.

The electron is considered to have one "quantum" of charge. The proton also has one quantum of charge, but it is of the opposite polarity. The electron charge is designated "negative" and the proton charge "positive." The neutron has no charge and is "neutral." The basic unit of charge is the coulomb. The quantum of charge, positive or negative, has the value 1.6×10^{-19} coulombs. The experimental evidence to date indicates that there is no smaller unit of charge, and that all charged objects have an integer multiple of this quantum of charge.

The electron is the primary charge carrier in most electrical phenomena involving metal wires because it is the lightest and most mobile of the constituents of the atom. A brief look at the relative masses of the atomic constituents will make this clear. The mass of a proton is about 1836 times the mass of an electron, and the mass of a neutron is about 1839 times the electron mass. The nucleus of an aluminum atom, for example, has 13 protons and 14 neutrons bound tightly together so that the total nuclear mass is nearly fifty thousand times the mass of a single electron. Since the electrons in outer orbits are light and not very tightly bound, they can be made to move around in bulk material much more easily than the massive nuclei. The discussion of electron orbits will be resumed in Chapter 20, and nuclear structure will be discussed in Chapter 21.

THE BEHAVIOR OF ELECTRIC CHARGES

Charges of the same polarity repel each other, while unlike charges experience a strong attractive force. The force between two charged particles is found to be proportional to the product of their charges and inversely proportional to the square of the distance between them. This experimental relationship is known as Coulomb's law:

$$F = \frac{Kq_1q_2}{r^2} \qquad \textbf{12–1}$$

where q_1 and q_2 are the two charges in coulombs, and r is the distance between them in meters. The parameter K is Coulomb's constant with the value

$$K = 9 \times 10^9 \text{ nt-m}^2/\text{coul}^2$$

in the MKS system. The electrostatic force, F, is the force on each charge. The forces on the two charges are, of course, equal in magnitude and opposite in direction, as required by Newton's third law, and are directed along the line joining the two particles as shown in Figure 12–1.

Example. What would be the force experienced by two charges of one coulomb each, which are one meter apart?

Solution. Applying Coulomb's law,

$$F = \frac{Kq_1q_2}{r^2} = \frac{9 \times 10^9 \text{ nt-m}^2}{\text{coul}^2} \times \frac{1 \text{ coul} \times 1 \text{ coul}}{1 \text{ m}^2}$$

$$= 9 \times 10^9 \text{ newtons}$$

From Table T–2 in Appendix 1, it is found that 4.45 nt = 1 pound. Therefore, the force between the charges is

$$F = (9 \times 10^9 \text{ nt})\left(\frac{1 \text{ pound}}{4.45 \text{ nt}}\right)$$

$$= +2 \times 10^9 \text{ pounds,}$$

which is about a million *tons* of force. The positive sign indicates a repulsive force. Each of the charges is repelled with a force of 1 million tons, as illustrated in Figure 12–1(b). If one of the charges had been −1 coulomb, then the force would be a 1 million ton attractive force on each charge, as illustrated in Figure 12–1(a). The force calculated from Coulomb's law would then have a negative sign, indicating attraction. This example shows that the coulomb is a very large unit; the free charges normally encountered

(a) Unlike charges attract

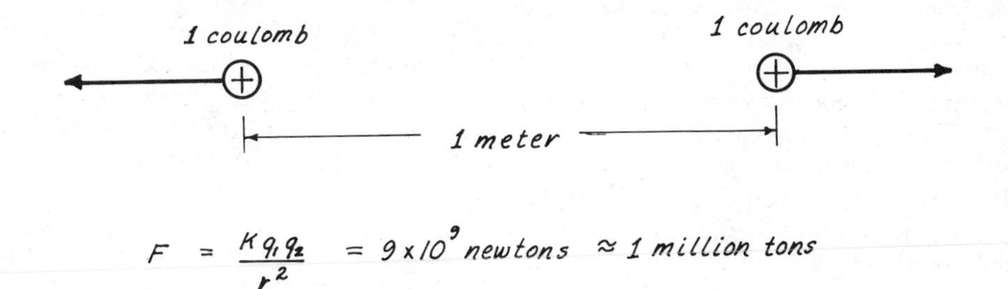

$F = \frac{Kq_1q_2}{r^2} = 9 \times 10^9$ newtons $\approx$ 1 million tons

(b) Like charges repel

Figure 12–1 Coulomb's law.

are more conveniently measured in microcoulombs (10^{-6} coul).

The form of Coulomb's law is like that of the universal law of gravitation, equation 3–2:

$$F_{\text{gravity}} = \frac{Gm_1m_2}{r^2}$$

and

$$F_{\text{electric}} = \frac{Kq_1q_2}{r^2}.$$

Both depend upon products of the appropriate quantities (mass or charge) and both are "inverse square" laws since the force is inversely proportional to the square of the separation distance. It is interesting to compare the magnitudes of these forces. A hydrogen atom is composed of one proton and one electron. There is an electrostatic attraction force between them since they have opposite charges, and also a gravitational attraction force. The electrostatic force is found to be more than 10^{39} times as strong as the gravitational force! Therefore, the gravitational force inside an atom is negligible compared to the electrostatic force, and the forces which hold atoms together are electrical forces. In fact, chemical bonding forces, frictional forces, forces experienced during collisions, and most other common interaction forces are found to be electrical in nature when viewed on an atomic scale.

When the strength of the electrostatic attractive forces is considered, it is not surprising that atoms are normally found in a neutral state with the same number of protons and electrons and thus no net charge. When an atom loses an electron and becomes a positively charged "ion," it will attract any available electron to it to neutralize that positive charge. Therefore, bulk matter is usually electrically neutral.

This basic electrical neutrality can be disturbed if electrons are removed from some of the atoms, or if extra electrons are placed in the material. If a glass rod is rubbed with silk, some of the electrons are removed from the rod and it then has a net positive charge. The silk is then negatively charged, since it has an excess of electrons. When a rubber rod is rubbed with fur it becomes negatively charged because excess electrons are transferred to the rod. This type of charge transfer is quite common. Synthetic materials used in clothing, carpeting, and blankets often become charged when they are rubbed against other substances. The resulting electrical forces cause them to cling together and can cause sparks when they are separated. Though these forces are significant, the comparison with the force between the two charges illustrated in Figure 12–1 indicates that the fraction of the electrons removed in such processes is extremely small. The following example shows that the total amount of electric charge in bulk matter is very large.

Example. How many coulombs of positive charge are contained in one cubic centimeter of copper?

Solution. Copper has an atomic weight of 63.54. Therefore, one mole of copper, 63.54 grams, contains Avogadro's number of atoms (6×10^{23}). Since the density of copper is 8.9 gm/cm^3, one cubic centimeter contains 0.14 mole and therefore 8.4×10^{22} atoms. Each copper nucleus contains 29 protons. The total charge is then

$$q = (8.4 \times 10^{22} \text{ atoms})\left(29 \frac{\text{protons}}{\text{atom}}\right) \times \left(1.6 \times 10^{-19} \frac{\text{coulombs}}{\text{proton}}\right)$$

$$q = 3.9 \times 10^{5} \text{ coulombs.}$$

Very large forces would be exerted upon surrounding matter if it were not for the fact that this large positive charge is neutralized by the 3.9×10^5 coulombs of negative charge of the associated electrons.

THE FLOW OF ELECTRIC CHARGE

The outer electrons of metal atoms are very loosely bound and can be easily detached to move through the material. Metals are said to be good *conductors* of electricity because of the availability of charge carriers. Metals offer very little resistance to the flow of electrons through them. On the other hand, the electrons in rubber, ceramics, and other materials are very tightly bound to their parent atoms and it is very difficult for electric charge to move through these materials. Such materials are called *insulators*. There is a limited class of materials which offer intermediate resistance to charge flow and are known as semiconductors. They are used widely for the manufacture of transistors and other solid-state electronic devices.

The movement of charges can be illustrated with the use of a simple device called an electroscope. The device is composed of a metal rod with two strips of very thin metal foil attached to the bottom, as shown in Figure 12–2(a). The lower part of the rod is usually enclosed in glass to protect the delicate foil strips, and the rod is separated from the case by means of an insulator such as rubber. If the electroscope is uncharged, as in Figure 12–2(a), the two foil strips will hang loosely. If the electroscope has a net negative or positive charge, the foil strips will repel each other and stand out as shown in Figures 12–2(b) and 12–2(c).

If a positively charged rod (deficient in electrons) is brought close to the top of an uncharged electroscope as in Figure 12–3, the net positive charge will attract some electrons from other parts of the rod, causing the top of the electroscope to be negatively charged. The bottom of rod and the foil strips will then be positively charged, since they will have a deficiency of electrons. The repulsion of like charges will cause the foils to separate.

Because of the strong mutual repulsion of like charges, any excess charge will spread out on a conducting surface in such a way that maximum distances between charges can be maintained. Even though the electrons are the mobile particles, an excess positive charge will distribute itself as readily as an excess negative charge. Since a positive charge in bulk matter represents a deficiency of electrons, the available mobile electrons will redistribute themselves on a conducting surface so that the points of electron deficiency are as far apart as possible. If the metal chassis of an electrical appliance becomes charged, the entire chassis will have an excess charge since the charges will be distributed by their mutual repulsion. Now if an additional conducting path is provided in the form of a wire connected to the earth, the excess charges will move along that wire to "ground" since the earth is a conductor and the charges can move much further apart on the surface of the earth. This removal of excess charge is one of the functions of a "ground" wire, as discussed in Chapter 13. In terms of the electroscope, Figure 12–4 shows that the excess electrons on a negatively charged electroscope will move off to the earth through a ground wire and the foil strips will quickly collapse to their uncharged positions.

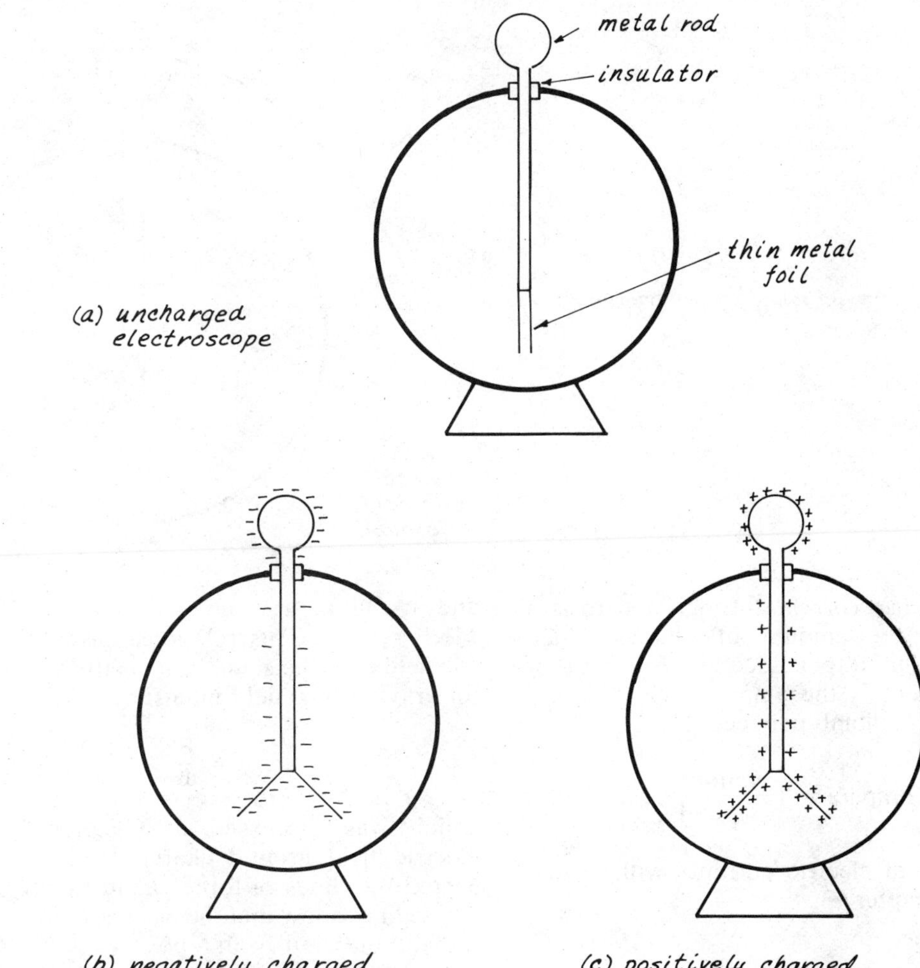

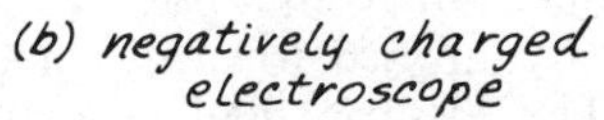

Figure 12–2 The electroscope.

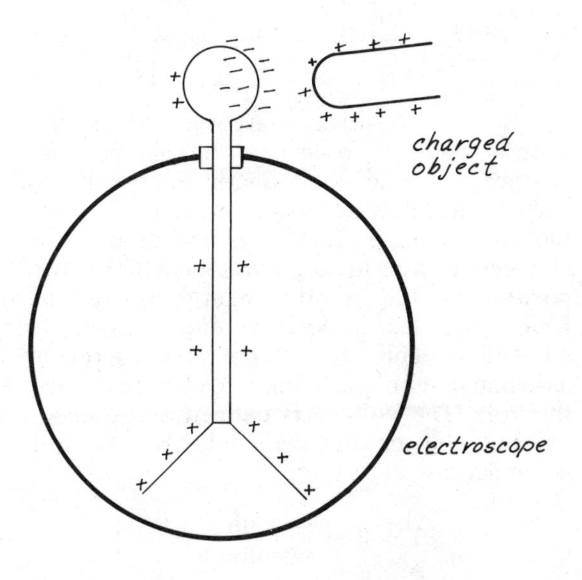

Figure 12–3 The effect of a charged object on an electroscope.

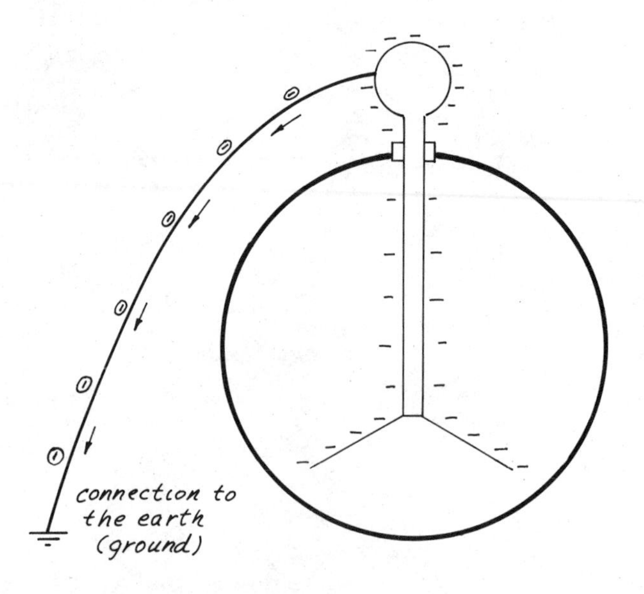

Figure 12–4 The flow of excess charge through a ground connection.

The *electric current* through a wire is a measure of the amount of charge which passes through it per second. The unit of electric current is the ampere, which is defined as one coulomb per second:

$$1 \text{ ampere} = 1 \frac{\text{coulomb}}{\text{second}}.$$

Applications of electric currents will be discussed in Chapter 13.

ELECTRIC FIELDS AND VOLTAGES

If you brought a small positive test charge near another concentration of positive charge, the test charge would experience a repulsive force. You could say that there is an "electric force field" around the concentration of charge which will repel any other positive charge but attract a negative charge. The electric field at any point is defined as the force per unit charge exerted upon any point test charge* at that point. The direction of the electric field is defined as the direction of the force on a positive test charge. The electric field is then defined by the relationship

$$E = \frac{F}{q} \qquad \textbf{12–2}$$

and its units are newtons/coulomb in the MKS system. This relationship for the electric field is of the same form as the expression for gravitational field intensity

$$g = \frac{F}{m}$$

which was discussed in Chapter 3. The electric field around charges can be represented by "lines of force" as in Figure 12–5. These lines show that the electric field extends radially outward from a positive point charge and radially inward toward a negative point charge. The electric field strength drops off rapidly with the distance from the charge. This can be seen by applying equation 12–1 to calculate the electric field due to a point charge Q:

$$E = \frac{KQ}{r^2} \text{ for a point charge.} \qquad \textbf{12–3}$$

Since a positive charge would experience a repulsive force near another positive charge, it would gain kinetic energy if released and allowed to accelerate away from the outer charge. Just as in the case of an elevated object in a gravitational field, the position of the positive charge is said to have some electric potential energy associated with it. This electric potential energy is associated with the common electrical unit, the volt. The voltage is defined as the electric potential energy per coulomb. The volt is defined by:

$$1 \text{ volt} = 1 \frac{\text{joule}}{\text{coulomb}}.$$

*A "point charge" is a concentration of charge at a mathematical point (i.e., having no physical length in any direction). For all practical purposes, an electron is a point charge.

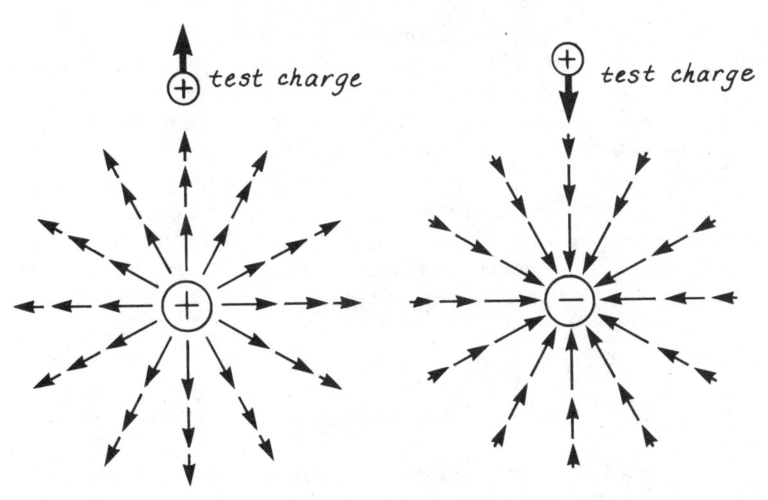

Figure 12–5 The electric fields around positive and negative point charges.

Then 100 volts means that each coulomb of charge under consideration has 100 joules of electric potential energy.

It is instructive to apply the ideas of electric field and voltage to a set of parallel metal plates as shown in Figure 12–6. If one plate has an excess positive charge and the other an excess negative charge, then there is an electric field between the plates directed toward the negative plate, since a positive charge placed between the plates would experience a force toward the negative plate. The symbols V_1 and V_2 represent the voltages associated with the two plates. The positive plate represents a high electric potential energy for a positive charge, since the charge is repelled by the positive plate and attracted by the negative plate. Since voltage is the measurement of electric potential energy per unit charge, the positive plate has a high voltage. The negative plate has a lower voltage. Positive charge will tend to flow from a high voltage region to a low voltage region, and this direction is chosen as the standard direction for electric current flow. That is, the standard direction for all electric currents is the direction which positive charge would flow, downhill from high voltage to low voltage. This is a bit awkward in the light of our present knowledge, because the mobile, negatively charged electrons are known to be the charge carriers in most cases, but the universally accepted convention is maintained. Transferring one coulomb per second of positive charge from plate A to plate B is referred to as a one ampere electric current from A to B. Transferring one coulomb per second of *negative* charge from B to A accomplishes the same electrical result and is also referred to as a one ampere electric current from A to B, *as if* the positive charge were moving. The conventional current direction is opposite to the direction of electron motion.

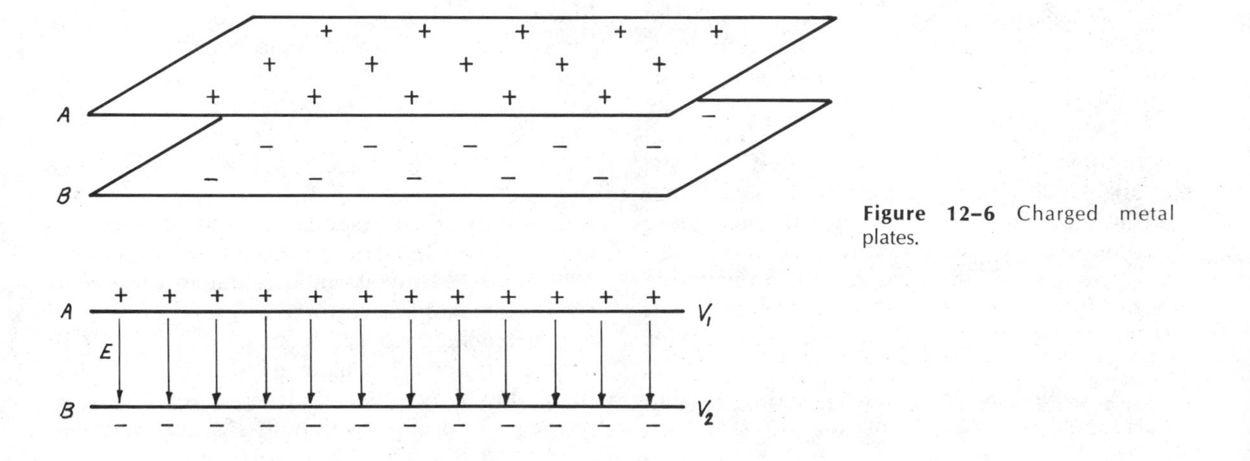

Figure 12–6 Charged metal plates.

The parallel charged plates are introduced here because there are many applications of electricity which make use of this configuration. Parallel plates can be used to store electric charge in a state of elevated electric potential energy (i.e., high voltage). The stored charge can then be released to do work. In this context the set of parallel plates is referred to as a *capacitor* and has wide application in electronics as an electrical energy storage device. The capacitance is defined as the charge which can be stored per volt of electric potential,

$$C = \frac{Q}{V} \qquad \textbf{12–4}$$

where Q is the charge in coulombs stored on *each* plate and V is the voltage difference between the two plates. The unit of capacitance is the coulomb/volt, which is called a farad. For example, if 0.001 coulombs can be stored on a capacitor by generating a voltage of 1000 volts between the plates, the capacitance is

$$C = \frac{0.001 \text{ coul}}{1000 \text{ volts}} = 0.000001 \text{ farad}$$
$$= 1 \text{ microfarad.}$$

The charge will remain on the capacitor indefinitely if there is no conducting path available to "discharge" the capacitor. If a wire is connected between the two charged plates, electrons will move quickly from the negative plate to the positive plate to neutralize the unbalanced charge.

The process of discharging a capacitor releases energy. The defibrillator is essentially a large capacitor. An electrical energy of several joules may be stored in the defibrillator. When the electrodes are connected across the patient's body, the body forms a conducting discharge path and a large current flows for an instant.

There are many cases in which cell membranes maintain unbalanced charge layers which can store electrical energy and aid in the transport of charged electrolyte ions across the membranes. Similar biological "capacitors" in nerve cells can discharge to produce the electrical impulses involved in the transfer of nerve signals (see Chapter 16).

A configuration similar to the parallel plates can be used to create an "electron gun." If an electron is released from the negative plate, it will be accelerated toward the positive plate. Since the mass of the electron is quite small, it can attain very high speeds in such processes. As the electron travels between the plates, its electric potential energy is converted into kinetic energy. Since the voltage difference, V, between the plates is a measure of the electric potential energy *per unit charge,* then

$$\text{P.E.} = qV \qquad \textbf{12–5}$$

where q is the charge in coulombs.

Example. If the voltage difference between two charged metal plates is 100 volts, what speed will an electron attain while traveling from the negative to the positive plate?

Solution. From Table T–3 in the Appendix, the mass and charge of the electron are

$$m_e = 9.1 \times 10^{-31} \text{ kg}$$

$$q_e = 1.6 \times 10^{-19} \text{ coul.}$$

The electric potential energy of the electron, q_eV, is converted into kinetic energy, $\frac{1}{2}m_ev^2$. Using the conservation of energy principle

$$\text{P.E.}_{\text{after}} + \text{K.E.}_{\text{after}} = \text{P.E.}_{\text{before}} + \text{K.E.}_{\text{before}}$$

$$\frac{1}{2} m_ev^2 = q_eV$$

$$v^2 = \frac{2q_eV}{m_e}$$

$$= \frac{2(1.6 \times 10^{-19} \text{ coul})(100 \text{ joule/coul})}{(9.1 \times 10^{-31} \text{ kg})}.$$

To simplify the units, refer to Table T–1. Since the volt is defined as a joule/coulomb, the charge units cancel. This becomes

$$v^2 = 3.5 \times 10^{13} \text{ m}^2/\text{sec}^2,$$

$$v = 5.9 \times 10^6 \text{ m/sec} = 1.3 \times 10^7 \text{ mi/hr.}$$

Therefore, an electron speed of over 13 million miles/hr can be obtained with just 100 volts. If a slit or hole is cut in the positive plate, some electrons will pass through, forming a high speed beam of electrons. Of course, the positive plate would tend to slow them down by an attractive force, but this problem is easily overcome by keeping the

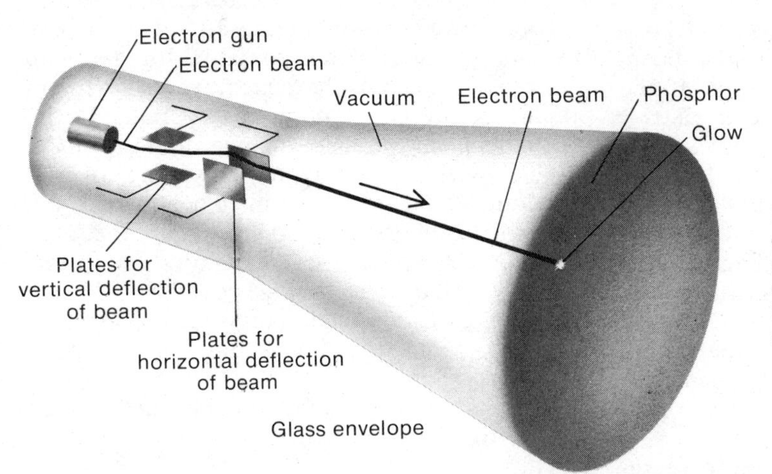

Figure 12–7 An oscilloscope tube. An electron beam passes between two sets of plates. The electric field between them changes the direction of the beam so that it can be made to hit anywhere on the phosphorescent face of the tube.

area surrounding the electron beam at the same voltage as the positive plate, so that no deceleration occurs. Electron guns operating on this principle are used in oscilloscopes and television receivers.

CATHODE RAY TUBES; THE OSCILLOSCOPE

A cathode ray tube is a device which forms a visual display of the electron beam from an electron gun such as that described above. The electron gun is placed in an evacuated glass tube and aimed so that the high speed electron beam strikes a phosphor coating on one end of the tube (see Figure 12–7). When struck by the beam, the phosphor coating gives off light, forming a visual image of the electron beam. Because of the historical practice of calling the negative electrode the "cathode," the electron beam is often called a "cathode ray."

The great usefulness of the cathode ray tube stems from the fact that the position of the electron beam on the phosphor screen can be moved easily and very rapidly to form a moving display of electrical signals, such as an ECG signal. The movement of the beam is accomplished by two sets of deflection plates similar to the parallel charged plates discussed above. The deflection plates are shown schematically in Figure 12–7. If the deflection plates are charged, then the path of the moving electrons is bent toward the positive plate. The amount of deflection of the beam is proportional to the voltage between the deflection plates. Therefore, the deflection of the beam can be calibrated to give a visual measurement of the voltage of the deflection plates.

The oscilloscope is basically a cathode ray tube with associated amplifiers and other electronics. The horizontal deflection of the electron beam on an oscilloscope is usually used as a time measurement by causing the beam to sweep across the phosphor screen at a uniform, predetermined speed. A time and voltage measurement from an oscilloscope display is illustrated in Figure 12–8. The time base is created by putting a gradu-

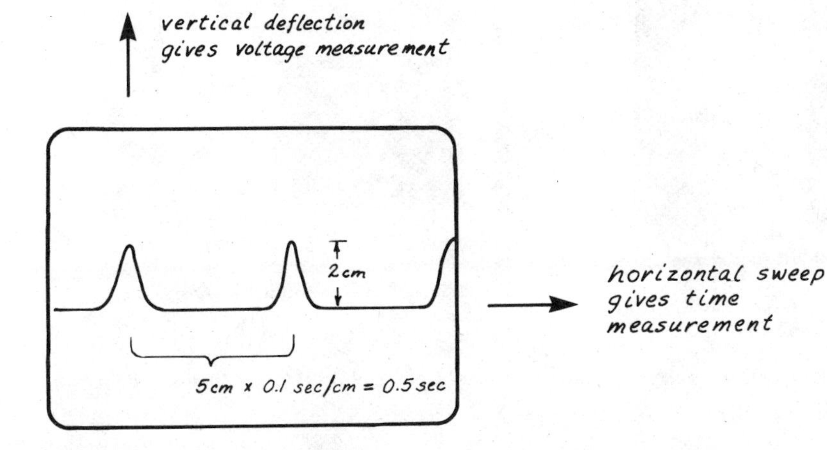

Figure 12–8 Sample oscilloscope display illustrating time and voltage measurements.

ally increasing voltage "ramp" on the horizontal deflection plates to move the beam across the screen at a uniform rate. If the entire sweep across a 10 cm screen takes 1 second, then the time control indicates a sweep rate of 0.1 sec/cm. In Figure 12–8 the peaks are separated by a horizontal distance of 5 cm which corresponds to a time interval of 0.5 second between events. The vertical deflection is calibrated in volts/cm. If the sensitivity associated with Figure 12–8 is 2 volts/cm, then the 2 cm vertical deflection corresponds to a 4 volt electrical signal. A typical oscilloscope monitor is shown in Figure 12–9.

The oscilloscope is an extremely versatile display device. Some of the typical controls are shown in Figure 12–10. Both the sweep time and the vertical gain on a general purpose oscilloscope can be varied over a wide range. The input signal is used to deflect the electron beam in the vertical direction, and the vertical gain or vertical sensitivity control determines how much this input signal is amplified.

Typical settings for the observation of an ECG signal are shown in Figure 12–10. If the sweep time is set to 0.2 sec/cm and the heart rate is 60 beats/min (1 second between beats), the peaks of the ECG pattern for successive beats will be 5 cm apart. If the heart rate increased, the peaks would be closer together. If the sweep speed of the oscilloscope beam were increased, the display would spread out horizontally; for a sweep rate of 0.1 sec/cm the peaks would be 10 cm apart for a 60 beat/min heart rate. If the peak-to-peak voltage change for the ECG signal were 1 millivolt, a setting of 0.2 mv/cm would give a display height of 5 cm on the oscilloscope screen. Often the controls on hospital ECG monitors are much more limited. The voltage level may be set by the ECG amplifier before going into the monitor, and the horizontal sweep speed may have only one or two settings so that the heart rate can be quickly determined from the horizontal distance between ECG peaks.

Multiple sweep oscilloscope monitors, such as that shown in Figure 12–9, make possible the monitoring of the ECGs of several patients at the same time from a central monitoring location. Other information such as blood pressure, respiration rate, etc. can be supplied to different traces on the monitor screen for continuous display during surgical procedures.

MAGNETS AND MAGNETIC FIELDS

Simple magnets have two "poles" which are designated as north and south poles. If the north poles of two magnets are

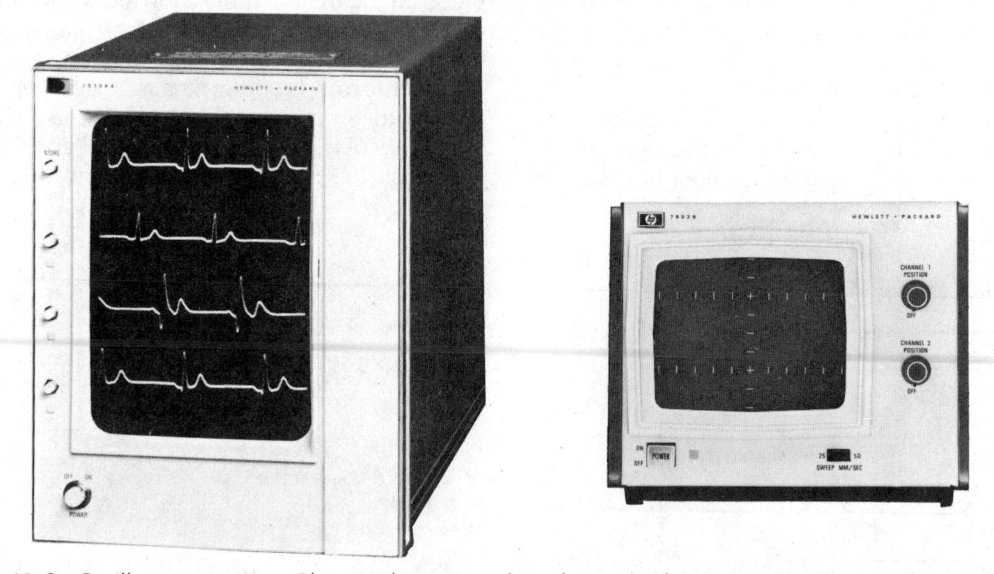

Figure 12–9 Oscilloscope monitors. (Photograph courtesy of Hewlett-Packard Company, 1501 Page Mill Road, Palo Alto, California 94304.)

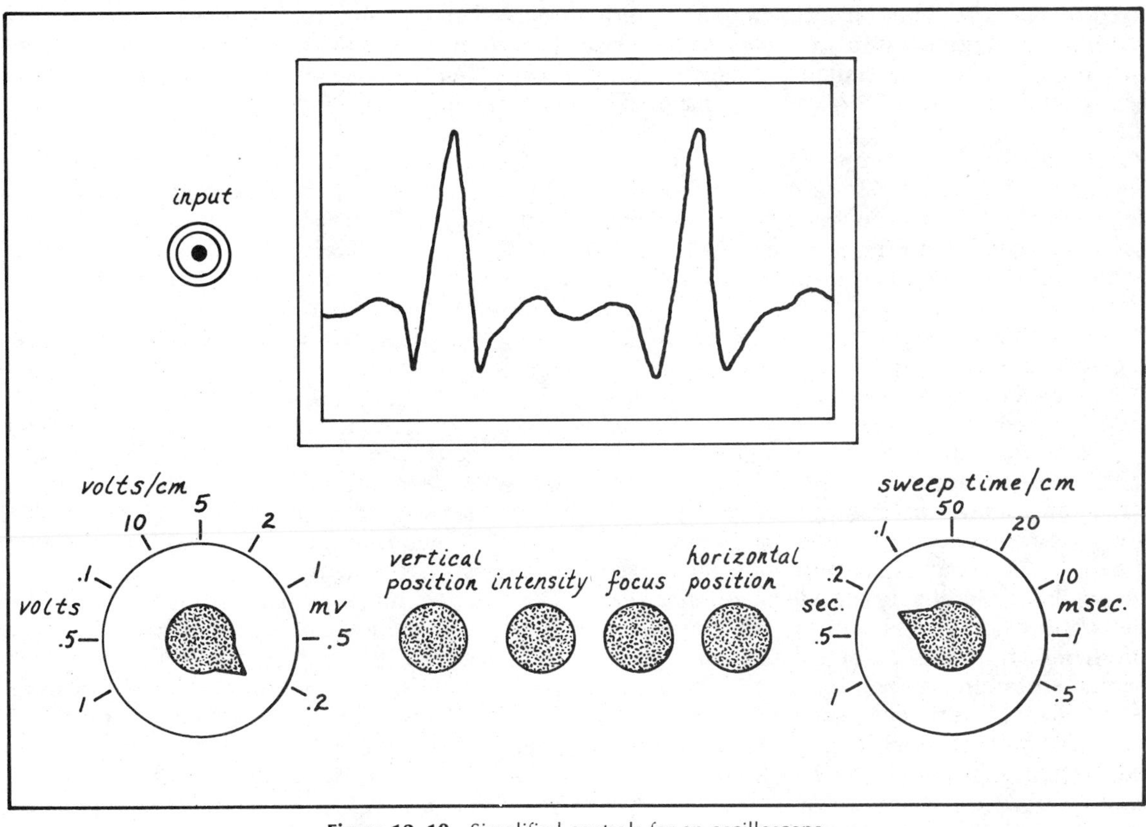

Figure 12–10 Simplified controls for an oscilloscope.

brought close together, they will repel each other, but the north pole of one magnet will attract the south pole of another. These poles always occur in pairs; there is apparently no magnetic analog to the point charge as a source of electric fields. The behavior of magnets can be described in terms of a magnetic field, but this field is not directed radially outward from the source like the electric field due to a point charge (Figure 12–5). The magnetic field of a simple bar magnet is sketched in Figure 12–11. Arrows directed along the field lines indicate the direction of the magnetic field, and the density of the lines indicates the relative strength of the magnetic field. Note that the field is strongest at the poles and that the field is directed outward from the north pole of the magnet. A small test magnet would tend to line up with this magnetic field with its north pole directed along the direction of the magnetic field. This is the principle of operation of the ordinary magnetic compass. The compass needle is a small, freely suspended magnet which can rotate to align its north pole with the direction of the earth's magnetic field. A tiny compass needle can be

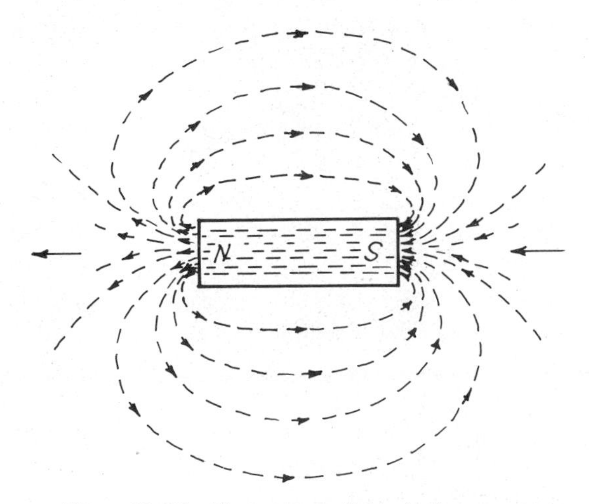

Figure 12–11 The magnetic field of a bar magnet.

used to map the direction of a magnetic field as shown in Figure 12–12.

It is general knowledge that only a few types of materials are attracted by magnets. These materials exhibit the property of "ferromagnetism" or "iron-like" magnetism. The only common materials which exhibit ferromagnetism are iron, nickel, cobalt, and some alloys. Some other rare elements such as gadolinium and dysprosium also exhibit the property.

Ferromagnetism occurs as a result of long-range ordering of the atoms of a solid such as iron into small ordered regions called "domains." The orientation of the atoms in these domains is such that the domain as a whole acts like a tiny magnet. A piece of unmagnetized iron will have many such domains, as illustrated in Figure 12–13(a). These magnetized domains will normally have random orientations so that the magnetic fields cancel and the bulk material produces no magnetic field. If the material is subjected to an external magnetic field, however, the tiny domains tend to line up with the field. The material then yields a net magnetic field and is said to be "magnetized" (Figure 12–13(b)). This alignment of the domains with an external magnetic field accounts for the fact that a piece of iron will be attracted by either pole of a bar magnet. If the north pole of the magnet is brought close to the iron, the domains are aligned with their fields pointing away from the magnet so that the surface of the iron represents a south pole and is attracted. When the south pole of a magnet is brought close, the domains align so that the surface of the material constitutes a north pole and attraction again occurs.

When the external magnetic field is removed, the domains do not immediately resume their random orientations, so the material retains part of its magnetization for a time. Certain iron alloys retain their magnetization for long periods and are useful for the manufacture of the so-called "permanent magnets." Other alloys, such as the alnico (aluminum-nickel-cobalt) alloys, are superior to iron in many applications. The thermal agitation of the molecules of the material tends to randomize the orientations of the domains and thus demagnetize the material. Heating a magnet above a certain temperature (the Curie temperature for the material) will destroy the magnetic ordering and the material no longer exhibits the ferromagnetic property. A magnet will not pick up a red-hot piece of iron, since iron does not exhibit ferromagnetism at that temperature.

Most common materials can be classified magnetically as either ferromagnetic, paramagnetic, or diamagnetic. Diamagnetism is the result of the natural tendency for bulk matter to oppose or exclude a magnetic field. It would cause a slight repulsion of the material by a magnet, but the effect is so small that it is negligible in practical applications. All bulk materials exhibit diamagnetism, but the effect is so weak that a material is called diamagnetic only if the stronger paramagnetic and ferromagnetic effects are absent. Although it is stronger than diamagnetism, the paramagnetic effect is still on the order of a million times weaker than the ferromagnetism of iron. It results from the intrinsic magnetic properties of the electrons orbiting around the atoms of the material. Associated with the quantum mechanical property called "electron spin" (1, 2), each

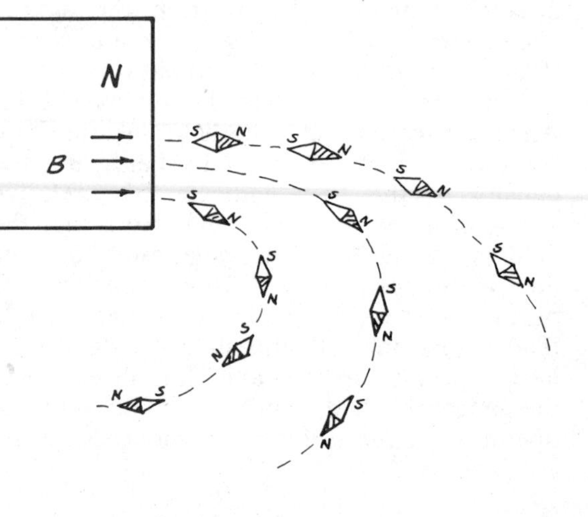

Figure 12–12 Use of a compass to map the direction of a magnetic field.

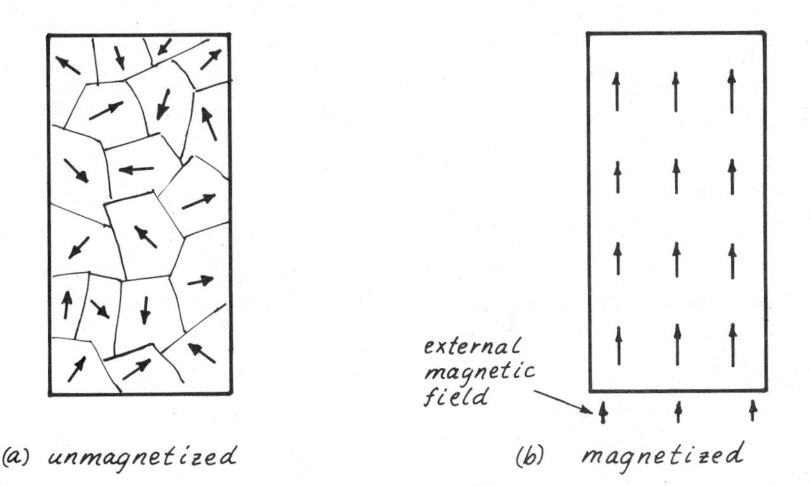

Figure 12–13 Ferromagnetic domains and the magnetizing effect of an external magnetic field.

electron has some of the properties of a magnet and can be aligned with a magnetic field. In most atoms these "magnets" are paired with opposite orientations so that their magnetic fields cancel. In the atoms of paramagnetic materials there is an "unpaired electron spin," so that a small magnet is available to line up with external magnetic fields. Paramagnetic materials can therefore experience a small attractive force in the field of a strong magnet. For medical applications the most important paramagnetic substance is oxygen. The paramagnetic property of oxygen is used in some oxygen analyzers. Since the other constituents of the air are diamagnetic, the small magnetic force on a sensitive dumbbell-type pendulum apparatus is used to measure the oxygen content of a gas in the Beckman oxygen analyzer (3).

ELECTROMAGNETS

If an electric current, I, flows through a straight wire, a magnetic field will be generated by that current. The magnetic field is represented by a letter B to agree with generally used notation. The field will circle the current-carrying wire as indicated in Figure 12–14(a). The direction of the magnetic field associated with any current may be determined by the "right-hand rule." Point the thumb of the right hand in the direction of the current, and curl the index finger around the wire. The direction of the index finger gives the direction of the magnetic field. It should be noted here that the direction of the electric current I in Figure 12–14 is the conventional electric current direction and is opposite to the actual direction of electron movement in the wires. It is the standard practice in most practical applications of electricity to talk about the electric current as flowing from high voltage to low voltage, as if positive charges were flowing.

If the current-carrying wire is bent into a loop, the magnetic fields add together inside the loop to give a stronger magnetic field. A further strengthening of the magnetic field can be obtained by winding a long coil of wire as shown in Figure 12–14(c). This coil arrangement is often referred to as a "solenoid." Note that the magnetic field configuration is like that of the bar magnet in Figure 12–11. The solenoid can properly be called an "electromagnet" and could be used for the same functions as a bar magnet. If the field directions of Figure 12–11 are compared to those in Figure 12–14(c), the top of the solenoid is seen to be a "north pole" and the bottom a "south pole."

The principal disadvantage of the solenoid of Figure 12–14(c) is that very large currents are required to produce useful magnetic fields. Practical electromagnets are made by adding an iron core to the solenoid (Figure 12–14(d). The magnetization of the iron core greatly enhances the magnetic field attainable with a given electric current. The small magnetic field produced by the current causes the domains of the ferromagnetic core to line up and produce an additional

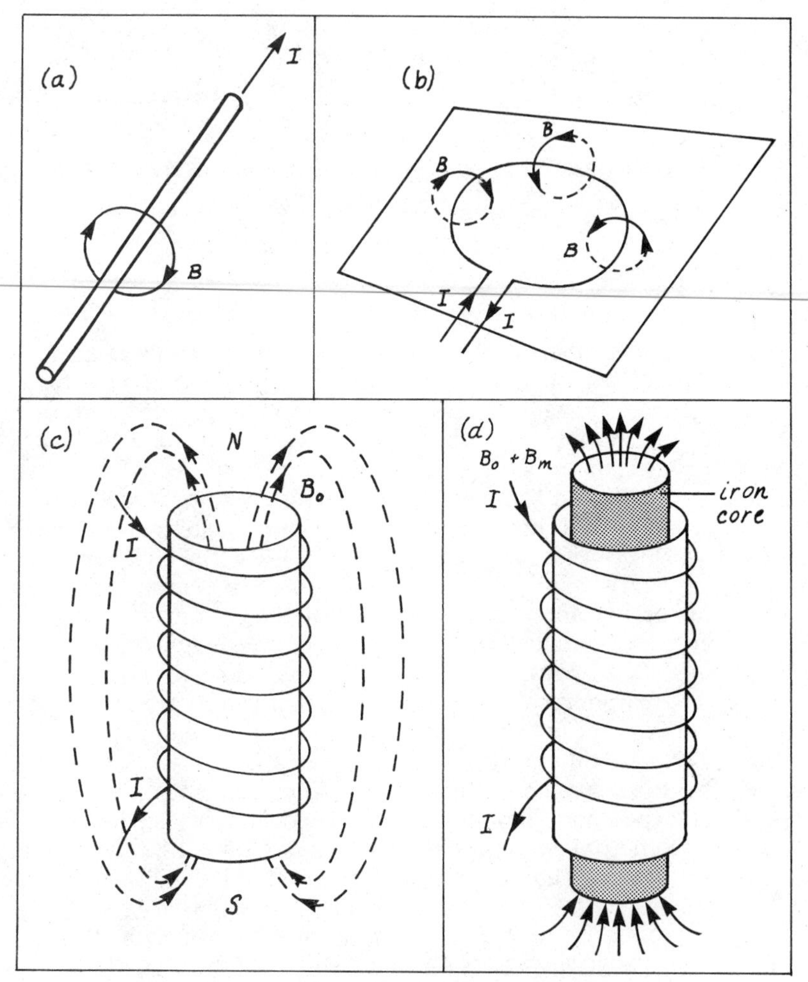

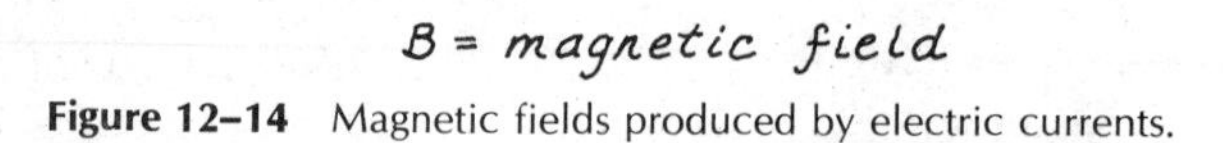

Figure 12–14 Magnetic fields produced by electric currents.

magnetic field, B_m. Instead of just the field B_o in Figure 12–14(c), a field $B_o + B_m$ is produced with the iron core. With common iron alloys, B_m may be several hundred times as large as B_o. Therefore, in a good electromagnet, the electric current acts primarily to magnetize the core, which produces fields many times larger than the magnetic field of the current alone. Field amplification factors of over 20,000 have been obtained with special ferromagnetic alloys (2).

Solenoids with movable iron cores are used widely to operate valves for the control of liquid or gas flow (solenoid valves). In the unenergized position, the iron core drops or is pushed part of the way out of the coil. When current flow is initiated in the solenoid, it magnetizes the core and pulls it into the center of the solenoid, opening or closing a fluid pathway. Such valves are convenient for remote control of fluids and automatic operation of safety valves.

SUMMARY

Electrical phenomena have their origins in the interactions between positive and negative charges. Like charges repel and unlike charges attract with a force which is proportional to the product of the charges and inversely proportional to the square of the distance between them (Coulomb's law). The force between the positive nuclear charges (protons) and the negative electrons around the nucleus is the basic force which holds matter together on the atomic scale. The light and mobile electrons are responsible for most electric currents in metallic wires. Matter may be roughly divided into insulators and electrical conductors, depending upon whether the electrons in a material are free to move. The nature of an electric charge is such that any other charge in its vicinity will experience a force from it, so that the charge can be said to be surrounded by an electric force field. The electric field at any point in space may be defined as the force per unit charge exerted on a charge at that point. The presence of an electric field implies that any charge in the field possesses some electric potential energy. This electrical potential energy is commonly measured in volts, which is by definition joules/coulomb. If a voltage is maintained between two oppositely charged plates, a capacitor is formed which can store electrical energy. Living cell membranes often act like capacitors in storing electrical energy.

A free electron will be repelled by a negatively charged electrode and attracted to a positively charged electrode. This fact is used to advantage in the oscilloscope when high voltage electrodes accelerate electrons to very high speeds so that they will produce light when striking a phosphor coated screen. Other charged plates along the electron path can deflect the electron beam in a controlled manner so that the beam can display a time-varying electrical signal on the oscilloscope face.

The sources of magnetic fields are called magnetic poles; such poles always occur in pairs called north and south poles. Many useful magnetic phenomena involve iron, nickel, or cobalt, the ferromagnetic materials. Such materials have internal ordering effects which can produce strong magnetic fields. Any electric current produces a magnetic field, but such fields are normally too weak to produce a good electromagnet unless the field is enhanced by an iron-like material.

REVIEW QUESTIONS

1. Why are atoms normally found to have an electron number equal to the number of protons in the nucleus?

2. What are the distinctive differences between ions, atoms, and molecules?

3. If the distance between two electric charges is doubled, by what factor does the electric force change?

4. Why does excess charge from a charged object flow through a ground wire to the earth to neutralize the object?

5. What is an ampere? What is a volt? Which is proportional to electric energy?

6. When an electric current flows in a metal, what is actually moving?

7. A piece of metal foil will be attracted to a charged object, but after touching it will be repelled. Explain.

8. What causes the bright line or dot on an oscilloscope or television screen?

9. What components are used to make an electromagnet? Could iron be replaced by other materials in such magnets?

PROBLEMS

Worked Example: What is the force between two 10 μcoul charges which are 5 cm apart?

Solution: The charges must be expressed in coulombs and the distance in meters. The force is then

$$F = (9 \times 10^9 \text{ nt-m}^2/\text{coul}^2)\,\frac{(10 \times 10^{-6}\text{ coul})(10 \times 10^{-6}\text{ coul})}{(0.05\text{ m})^2}$$

$$= 360 \text{ newtons}$$

Worked Example: If electric charges were used to suspend a 100 kg mass in the air at a height of 1 meter, what charge would be required, assuming the charge concentrations to be equal on the mass and on a point on the ground?

Solution: To support the 100 kg mass, a force equal to its weight is required:

$$F = (100\text{ kg})(9.8\text{ m/sec}^2) = 980\text{ nt}$$

The charge q is given by the equation

$$980\text{ nt} = (9 \times 10^9\text{ nt-m}^2/\text{coul}^2)(q^2/1\text{ m}^2)$$

$$q^2 = 10.9 \times 10^{-8}\text{ coul}^2$$

$$q = 3.3 \times 10^{-4}\text{ coul}$$

1. If a charge of 10 μcoul (10×10^{-6} coulombs) and a charge of -30 μcoul were placed 10 cm apart, what would be the magnitude of the force exerted on each? Would it be an attractive or repulsive force?

2. A cubic centimeter of copper contains about 3.9×10^5 coulombs of positive charge and the same amount of negative charge. If the electrons were all stripped off and taken a distance of one meter away, what

would be the magnitude of the attractive force tending to bring them back together?

3. Common static charges can produce forces large enough to pick up bits of paper and other light objects. If a 1 gram piece of paper can be lifted by a charged metal disc made of 1 cm^3 of copper at a distance of 1 cm, what is the net charge on each object (assuming their charges are equal)? Express this charge as a fraction of the total charge of the copper nuclei (3.9×10^5 coulombs).

4. Calculate the velocity of electrons when they have been accelerated by a voltage of 20,000 volts in a television picture tube or oscilloscope.

5. If the sweep rate on an oscilloscope monitor was set at 0.2 sec/cm and the horizontal distance between the peaks of an ECG display was 3 cm, what heart rate in beats per minute would be indicated? If a timer is set to measure the time between peaks and to sound an alarm when the heart rate exceeds 120 beats/min, what time interval would be set?

REFERENCES

1. Krauskopf, K. B., and Beiser, A. *The Physical Universe.* New York: McGraw-Hill, 1967.
2. Blackwood, O. H., Kelly, W. C., and Bell, R. M. *General Physics,* 4th ed. New York: Wiley, 1973.
3. Beckman Instruments, Inc. *Instructions, D2 Oxygen Analyzer.* Fullerton, Calif.

CHAPTER THIRTEEN

Practical Electric Circuits

INSTRUCTIONAL OBJECTIVES

After studying this chapter, the student should be able to:

1. State Ohm's law and use it to solve problems.
2. Calculate any one of the quantities (voltage, current, and resistance) associated with an element in an electric circuit, given numerical values for the other two quantities.
3. Explain the meaning of the expressions "short circuit" and "open circuit" and state the hazards associated with each.
4. Calculate the electric power in watts associated with a given piece of electrical apparatus, given appropriate data.
5. Calculate the voltage, current, or power for a single element in an electric circuit, given a diagram and appropriate numerical data.
6. Explain the explicit functions of the fuse and the ground wire and state the hazards created by their omission.
7. Explain the difference between AC and DC and state some advantages of AC electrical supplies.

The purpose of this chapter is to provide a basic operational understanding of electric circuits. Emphasis is placed upon those principles which will aid in the safe and efficient operation of the electrical appliances and machinery involved in modern health care. This chapter provides the background for Chapter 14, which will deal directly with electrical safety.

ANALOGY WITH A HYDRAULIC CIRCUIT

To aid in the understanding of the basic electrical quantities voltage, current, and resistance, the analogy with a water circuit is sometimes useful. It must be said at the outset that this analogy is limited and arguments by analogy are precarious, but the similar-

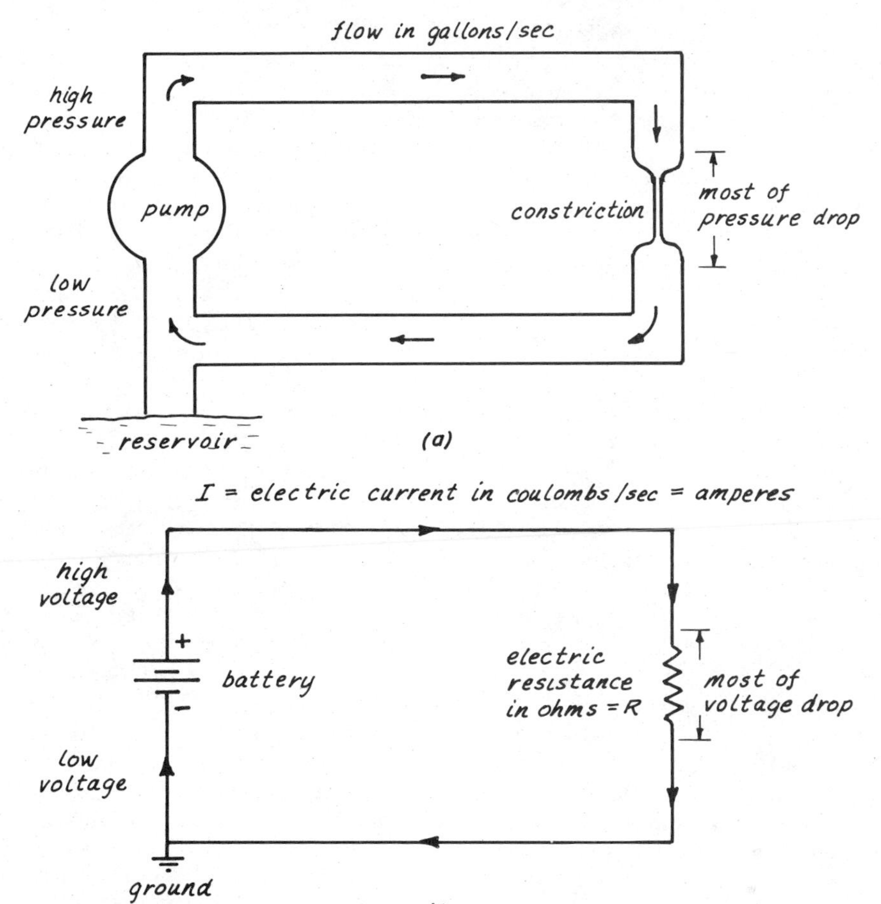

Figure 13–1 Analogy between an electric and a hydraulic circuit. (a) Hydraulic circuit. (b) Direct current electric circuit.

ities between these two types of circuits can be helpful in giving an intuitive feeling for some of the electrical quantities.

Figure 13–1(a) is a sketch of a closed water circuit and Figure 13–1(b) is the analogous electric circuit. In order to get water to flow uphill in the water circuit, there must be a pump: a device that does work on the water in order to overcome gravity. In the absence of a pump the water might flow momentarily, depending on the original position of the water, but there would be no continuous circulation. After the water has been given a positive pressure on the high side of the pump, it can flow downward through the constriction.

Even in a horizontal water circuit, in which gravitational potential energy is not a factor, the pump is required to do the work necessary to overcome the resistance of the circuit. In that case the amount of flow depends upon the pressure supplied by the pump and the nature of the constriction. As discussed in Chapter 5, the flow rate in gal/sec will be proportional to the pressure and inversely proportional to the resistance of the pipe system (Poiseuille's law).

ELECTRIC CURRENT, VOLTAGE, AND RESISTANCE

The electric current is the amount of charge per second which moves past a given point in the circuit. It is measured in amperes (1 amp = 1 coul/sec). In order to maintain a continuous flow of charge through the resistance, there must be a battery to provide energy to the charge. The battery "pumps" the charge "uphill," giving each unit of charge a certain amount of energy. This energy may be measured in joules per coulomb of charge. By definition, a 1 volt battery is a battery which supplies 1 joule of energy to each coulomb of charge which flows through it from the low voltage to the high voltage side (1 volt = 1 joule/coul). The electrical resistance is measured in ohms. The ohm unit is usually represented by the Greek letter Ω. The relationship between electric current, voltage, and resistance is given by Ohm's law:

$$I = \frac{V}{R}, \qquad \textbf{13–1}$$

where I = current, V = voltage, and R = resistance. Note that this is identical in form to the liquid volume flow rate relationship, equation 5–4. The electric current in amps (amperes) can be increased by either increasing the battery voltage or decreasing the resistance. This is analogous to increasing the water flow by increasing the pressure or decreasing the hydraulic resistance by using a larger pipe.

Note that in the water circuit (Figure 13–1(a)), most of the path is made up of large pipe which offers very little resistance to flow. Most of the pressure drop in the circuit occurs at the constriction, and for all practical purposes the resistance to flow offered by the large pipe is negligible. By analogy, the power cord to an electrical appliance is like a large conduit; it offers essentially negligible electrical resistance compared to the appliance itself in normal operation. Most of the voltage drop occurs across the appliance's resistance.

In addition to Ohm's law, electric circuits are governed by two other laws which may be referred to as the *voltage law* and the *current law*.

Voltage law: The sum of the voltage drops around any closed path must add to zero.

Current law: The sum of the currents into any junction must equal the sum of the currents out of that junction.

The voltage law is an application of the conservation of energy principle to the closed path. Voltage is electric potential energy per unit charge, and the voltage law says in essence that there is no net energy gained by the charge as it flows around the closed path. Whatever energy is gained by the charge from a battery or other source is used as it completes its path around the circuit. An electric circuit is a convenient means for transforming energy. Chemical energy from a battery may be converted into electric energy, producing a rise in voltage for the charge, and that energy may be transformed into heat, light, or mechanical work at a distant location as the charge flows through the circuit. The current law is just an expression of conservation of charge; whatever amount of charge flows into a given part of a circuit must also flow out of it because no charge can be lost or gained.

The voltage law applies to any closed path and is a powerful tool for analyzing circuits in which there are alternate paths for current flow. For example, if there is a 12 volt battery in a circuit and a closed path is taken in the direction of current flow, the battery represents a 12 volt rise in voltage (or a drop of −12 volts). Regardless of the path taken through the remainder of the circuit, there must be a 12 volt drop along that path. In Figure 13–2, a 12 volt battery is used to supply electric energy to a light with a resistance of 6 ohms, and a heater with a resistance of 2 ohms as shown. If a path around the circuit goes through the battery and the 2 ohm resistance, then there must be a 12 volt drop in potential across the 2 ohm resistor in order for the voltage drops to sum to zero. The current is given by Ohm's law:

$$I = \frac{V}{R} = \frac{12 \text{ volts}}{2 \text{ ohms}} = 6 \text{ amperes.}$$

(The connecting wires are assumed to have zero resistance in this calculation.) Similarly, there is a 12 volt drop across the 6 ohm resistance of the light, resulting in a current.

$$I = \frac{12 \text{ volts}}{6 \text{ ohms}} = 2 \text{ amperes.}$$

The light and heater are said to be wired in *parallel*. The voltage drop will be the same across all circuit elements which are wired parallel as a result of the voltage law. Since the voltage law is essentially a form of the conservation of energy principle, the energy used by a charge in moving around a circuit must be equal to the energy given to it by the battery or other source of electric energy regardless of which parallel path is followed.

As an application of the current law, note the currents in the circuit in Figure

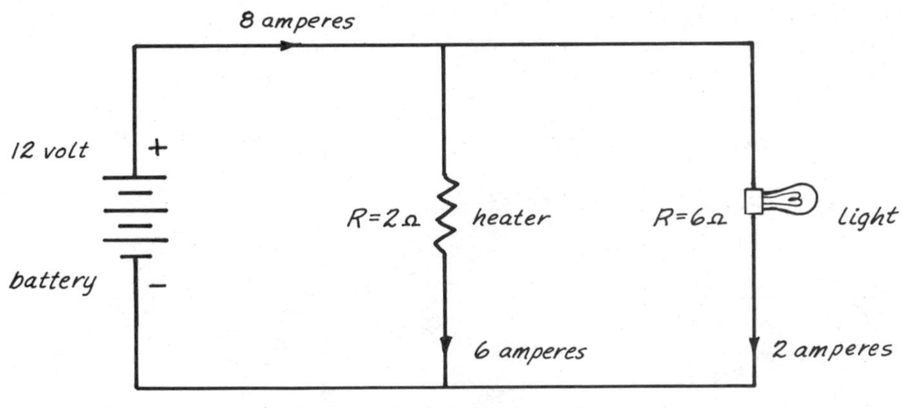

Figure 13–2 Resistances in parallel.

13–2. The fact that there is a current of 6 amps through the heater and 2 amps through the light implies that there is a total of 8 amps flowing from the battery. No charge is lost in the junction; therefore the current into the junction must equal the current out of the junction.

The discussion here has dealt with direct current (DC) circuits. Ohm's law and the voltage and current laws form the foundation for the understanding of DC circuits. The voltage and current laws are general and apply also to alternating current (AC) circuits. Ohm's law in a modified form also applies to AC circuits when the circuit elements are resistances, but changes must be made for non-resistive circuit elements as discussed in a later section. The major emphasis here is on DC circuits because the physical principles can be presented in a more straightforward manner using DC applications. These principles can then be applied to many common AC circuits.

Many of the malfunctions of electrical equipment could be classified as *short circuits* or *open circuits*. An open circuit occurs if a wire breaks and no electrical contact is being made. The circuit is interrupted and the resistance to flow is essentially infinite. From Ohm's law it is clear that no current will flow. However, the apparatus may represent a shock hazard as discussed below. A short circuit occurs when the normal resistance is bypassed by a low resistance path. For example, the two wires of a power cord may touch because of faulty insulation. Like two large pipes being joined to bypass a small one, this situation bypasses the resistance represented by the appliance because the path of least resistance will be followed. Since the electrical resistance, R, is very small, the electric current will become very large. In the absence of a fuse or circuit breaker to limit the current, it will become large enough to heat the wires and perhaps cause a fire.

If electric current must pass through two or more resistances in series, the equivalent resistance is the sum of the individual resistances (Figure 13–3(a)).

$$\textbf{Series: } R_{equiv} = R_1 + R_2 + R_3 + \ldots \qquad \textbf{13–2}$$

(All current passes through each resistor.)

However, if two or more resistances are placed in parallel, they provide alternate paths for current flow and the effective resistance is decreased (Figure 13–3(b)). The equivalent resistance is given by the following relationship.

$$\textbf{Parallel: } \frac{1}{R_{equiv}} = \frac{1}{R_1} + \frac{1}{R_2} + \frac{1}{R_3} + \cdots \qquad \textbf{13–3}$$

(The voltage is the same across each resistor.)

As shown in Figure 13–3, resistances of 2 Ω and 6 Ω in series are equivalent to an 8 Ω resistance, but in parallel they are equivalent to a resistance of 1.5 ohms. The equivalent parallel resistance may be verified by applying Ohm's law to Figure 13–2. The 12 volt battery is supplying 8 amperes to the parallel 2 Ω and 6 Ω resistances. By Ohm's law this is equivalent to a resistance

$$R_{equiv} = \frac{V}{I} = \frac{12 \text{ volts}}{8 \text{ amp}} = 1.5 \text{ ohms.}$$

Two 10 Ω resistors would be equivalent to a 20 Ω resistor if placed in series, but equiva-

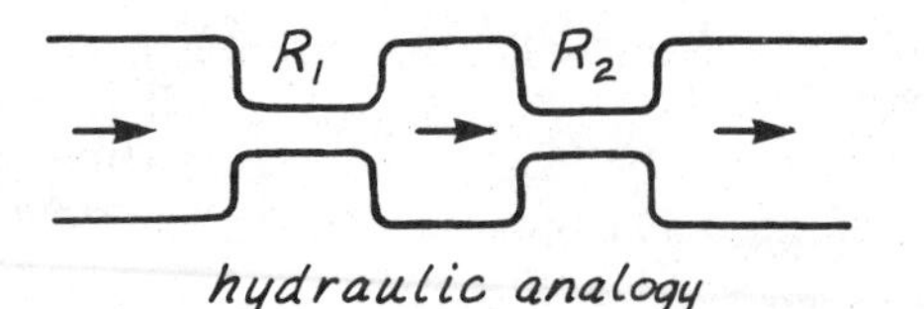

$$R_{equiv} = 2\,\Omega + 6\,\Omega = 8\,\Omega$$

(a) Resistances in series add.

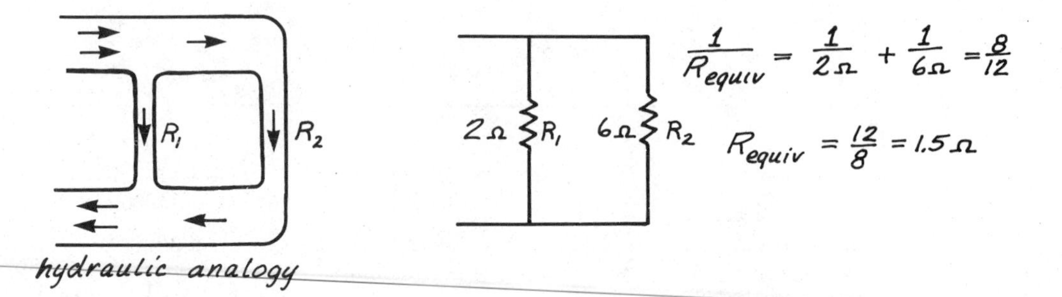

(b) Resistances in parallel offer alternate paths.

Figure 13–3 Combining resistances in series and parallel.

lent to a 5 Ω resistor if placed in parallel. Any appliance or piece of electrical apparatus represents some electrical resistance. These resistances are almost universally connected in parallel. Household appliances connected to 120 volt wall outlets are in parallel so that each appliance has the same voltage supplied to it. Each electrical device therefore represents an alternate path for electric current. If too many devices are connected to the same circuit, the sum of the currents will exceed the limit set by a fuse or circuit breaker for that circuit and the circuit will be interrupted.

With the above relationships for combining resistances, along with the voltage law, current law, and Ohm's law, most resistive circuits can be analyzed. If both series and parallel combinations are present in the circuit, the analysis may require a step-by-step application of the principles to different parts of the circuit.

Example. A 100 volt battery supplies a 20 ohm resistance which is in series with a parallel combination of a 10 ohm and a 30 ohm resistance. Find the voltage drop across the 20 ohm resistance, the total current, and the current through the 10 ohm resistance.

Solution. The total current must flow through the 20 ohm resistance and then the current divides, with a fraction of it flowing through each of the parallel resistors. The total current can be found from Ohm's law, $I = V/R_{equivalent}$. The equivalent resistance of the parallel combination can be found with the use of equation 13–3.

$$\frac{1}{R_{parallel}} = \frac{1}{10\,\Omega} + \frac{1}{30\,\Omega} = \frac{4}{30\,\Omega}$$

$$R_{parallel} = \frac{30\,\Omega}{4} = 7.5 \text{ ohms.}$$

Now using equation 13–2, the equivalent resistance of this resistance in series with the 20 ohm resistance is

$$R_{equivalent} = 20\Omega + 7.5\Omega = 27.5\Omega.$$

The total current is then

$$I_{total} = \frac{100 \text{ volts}}{27.5 \text{ ohms}} = 3.64 \text{ amperes.}$$

Since all of this current must flow through the 20 ohm resistance, the voltage drop across it is

$V_1 = IR = (3.6 \text{ amps}) (20\ \Omega) = 72.7 \text{ volts},$

leaving $V_2 = 100$ volts $-$ 72.7 volts $=$ 27.3 volts to drop across each member of the parallel combination. The current through the 10 Ω resistor is

$$I = \frac{27.3 \text{ v}}{10} = 2.73 \text{ amperes},$$

leaving 3.64 amps $-$ 2.73 amps $=$ 0.91 ampere to flow through the 30 ohm resistor.

ELECTRICAL ENERGY AND POWER

Most of the uses of electricity involve the conversion of electrical energy into some other form of energy. For example, an electric motor converts electrical energy to mechanical energy; a heater, toaster, or oven produces heat energy. The electric power supplied at a given instant is equal to the product of the voltage and the current,

$$P = VI. \qquad \textbf{13–4}$$

An examination of the units of these quantities will show that the product of volts × amperes gives watts. A summary of the commonly used electrical quantities and their units is given in Table 13–1. Since a volt is by definition a joule per coulomb and an ampere is a coulomb per second:

$$\text{volt} \times \text{ampere} = \frac{\text{joule}}{\text{coulomb}} \times \frac{\text{coulomb}}{\text{second}}$$

$$= \frac{\text{joule}}{\text{second}} = \text{watt}.$$

This relationship between power, current, and voltage can be useful for determining the current which will flow through a given appliance, since the power consumption in watts must be stated on the chassis of the appliance. The power relationship, equation 13–4, can be applied to AC as well as DC circuits if the AC appliance can be classified as a resistance (as opposed to a capacitor or inductor, as discussed later). The relationship $P = VI$ gives a reasonably good approximation for most common AC electrical devices.

Example. Most household AC circuits supply approximately 120 volts. If a given circuit has a maximum current of 15 amperes, can a 1200 watt electric heater, a 100 watt light bulb, and a grill with electrical resistance of 20 Ω be operated simultaneously?

Solution. These devices would be connected in parallel so that each is supplied by a voltage of 120 volts. Using equation 13–4, the heater will require a current

$$I = \frac{P}{V} = \frac{1200 \text{ watts}}{120 \text{ volts}} = 10 \text{ amperes}.$$

The 100 watt light bulb will require

$$I = \frac{100 \text{ watts}}{120 \text{ volts}} = 0.83 \text{ ampere}.$$

Ohm's law, $V = IR$, is required to find the current passing through the grill:

$$I = \frac{V}{R} = \frac{120 \text{ volts}}{20\ \Omega} = 6 \text{ amperes}.$$

The total current is then 16.83 amperes, which exceeds the current limit and would blow a fuse.

The total power available from a given circuit can be calculated from the relation-

TABLE 13–1 Summary of Electrical Quantities.

ELECTRICAL UNIT	DEFINITION	PHYSICAL QUANTITY MEASURED	USUAL SYMBOL FOR PHYSICAL QUANTITY
volt (v)	joule/coul	electrical potential energy per coulomb of charge	V
ampere (amp)	coulomb/sec	electric current	I
ohm (Ω)	volt/amp	resistance to charge flow	R
watt (W)	joule/sec	power	P

ship $P = IV$. For a 120 volt circuit which is fused at 15 amps, the maximum power from that circuit is

$$P = IV = (15 \text{ amps})(120 \text{ volts}) = 1800 \text{ watts.}$$

Overloading a circuit and blowing a fuse could have serious consequences if a patient is being supported by a respirator or other device which is connected to the circuit. A reasonable precaution would be to add the power ratings of all appliances to be connected to see if they total more than 1800 watts, if the fuse or breaker limits the circuit to 15 amps.

Some of the electrical energy in a circuit is converted into internal energy. Moving electrons carry the charge through the conducting wires. These electrons undergo collisions with atoms in the wire and give up kinetic energy to them. As we have seen, internal energy is associated with the random motion of atoms and molecules, so this kinetic energy appears as internal energy as indicated by a rise in temperature. It is not implied that specific electrons must travel all the way through the wire to conduct a current; they impart motion to nearby electrons in a chain process which transmits electric current through a wire at a speed near the speed of light (186,000 mi/sec). The actual net drift velocity of the electrons down a wire is only a few cm/sec, but numerous collisions produce a significant rise in temperature.

Sometimes the electrical heating is the desired effect, but in other applications it represents a loss of energy which might otherwise have accomplished useful work. An electric heater, for example, is just a resistor which converts electrical energy into internal energy. If the heater has a resistance of 12 ohms, then when it is connected to a 120 volt circuit a current of 10 amperes will flow through it. This will produce heat energy at the rate

$$P = IV = (10 \text{ amps})(120 \text{ volts}) = 1200 \text{ watts.}$$

For the calculation of the power dissipated in resistances it is sometimes useful to convert the power relationship to another form using Ohm's law. If either V or I is eliminated from the power equation by the use of the equation $V = IR$, then the power equation can take the alternate forms:

$$P = IV = I^2R = \frac{V^2}{R}. \qquad \textbf{13–5}$$

It has been pointed out that the voltage drop along a power cord is normally quite small; but any wire represents some electrical resistance, and there is some electrical power loss in the power cord. For large appliances this may be noted by the fact that the power cord will be noticeably warmer during operation. A hot power cord may indicate malfunctioning equipment and excessive current. The purpose of an electric motor is to produce mechanical energy by rotating, but it represents some electrical resistance which causes heat production. If the motor conducts a large current to overcome a heavy load, that electric current will also cause a greater rise in temperature.

The electric energy supplied may be expressed in joules, but the joule is often an inconveniently small unit. The kilowatt-hour (kWh) is often used as an energy unit. The equivalent number of joules is

$$1 \text{ kWh} = (1000 \text{ joules/sec})(3600 \text{ sec})$$
$$= 3.6 \times 10^6 \text{ joules.}$$

The economy of electrical energy is apparent since 3.6 million joules, one kilowatt-hour, can be obtained for a few cents.

THE FUSE AND GROUND WIRE

The fuse and ground wire are discussed together because they represent the two basic protection mechanisms necessary for the safe operation of electrical devices. As will be shown, neither is adequate by itself.

Fuses and circuit breakers are current limiting devices. In the fuse there is a small segment of wire which will melt or "burn out" (thus breaking the circuit) if a certain electric current passes through it; the current capacity of the fuse is determined by the size of this wire. Circuit breakers interrupt the current path mechanically when the current limit is reached and have the advantage that they can be reset and used repeatedly. There is normally a main fuse or circuit breaker where an electric power line enters a building, and then the line is branched to several individual paths (circuits), each of which is protected by a smaller fuse. For household wiring, the main fuse capacity may be in the range from 50 to 200 amperes, and the individual circuits are protected by 15 or 20 amp fuses or circuit breakers. Larger fuses are used for stoves, air-conditioners, and other large appliances. These individual circuit fuses may be inadequate to protect an electronic device; 15 amperes might do serious damage to delicate circuit elements. Therefore, most electronic devices have one or more individual fuses of smaller capacity to protect delicate components. These fuses may have capacities as small as a few milliamperes.

Consideration of an ordinary electrical appliance with a metal case will show that the fuse alone does not offer adequate protection against electrical shock. Figure 13–4(a) shows a diagram of an electrical device with no ground wire. Normal two-wire connections consist of one wire to which the high voltage is supplied (commonly called the "hot" wire) and a return wire called the "neutral" to complete the conducting circuit. The usual color code when wiring AC receptacles is that the black wire is "hot," the white wire is "neutral," and the green wire is the ground wire. This color code cannot be depended upon when a possible shock hazard exists, however, and of course either prong of a two-prong plug can be plugged into the "hot" side of the receptacle. The "hot" wire has an effective voltage between 110 and 120 volts, and it is this wire which has the fuse in it so that no voltage reaches the appliance after the fuse is blown. Both the wires are normally insulated from the metal frame of the appliance. The appliance motor or other electrical apparatus forms the electrical resistance in which the electrical energy is used. Although the neutral carries all of the current to complete the circuit, it has a voltage close to zero volts since it has very little resistance compared to the appliance.

If the insulation on the power cord fails so that both of the wires touch the metal frame, this forms a short circuit and will cause the fuse to blow ($I = V/R$; small R, large current). If, however, only the "hot" wire makes contact with the frame, the fuse will not blow if the appliance is not grounded. The metal frame will not constitute a completed circuit in which current can flow, but the entire metal frame will now have 120 volts of electric voltage on it. If a person touches the frame, he may provide a conducting path to ground, causing current to flow through him. An appliance under these conditions may represent a lethal shock hazard, yet continue to operate normally.

The ground wire is a separate wire attached to the metal frame of the appliance. The other end of the ground wire is normally connected to a cold water pipe or some other metal object which goes directly into the ground. The earth represents an essentially infinite reservoir of charge, and any electric voltage on the metal frame will cause current to flow to ground to remove the excess charge. The ground wire will remove any static charge buildup, as mentioned earlier, because the mutual repulsion of like charges will cause them to flow to the earth in order to get further apart. Once the static charge is removed, there will be no current flowing in the ground wire unless there is some malfunction; a current due to a malfunction is called a "fault current."

If the malfunction described above occurs in an appliance in which there is a ground wire but no fuse, as illustrated in Figure 13–4(b), the ground wire will constitute a short circuit to ground and a large fault current will flow. This high current will overheat the wires and very likely cause a fire. Probably the majority of electrical fires are caused by improperly fused circuits or curcuits in which the fuses have been bypassed (for example, by putting a penny behind the fuse). The ground wire in this case removes the shock hazard but at the expense of producing a fire hazard.

If the appliance circuit is protected by both a fuse and a ground wire as shown in Figure 13–4(c), both the shock hazard and the fire hazard are removed. If the "hot" wire makes contact with the metal frame, the ground wire will force the fuse to blow by conducting a large fault current to ground. The ground wire will prevent charge buildup on the frame, and thus the shock hazard is effectively removed.

ALTERNATING CURRENT

With the exception of automobile electrical systems and other battery operated systems, essentially all electrical energy is supplied in the form of alternating current.

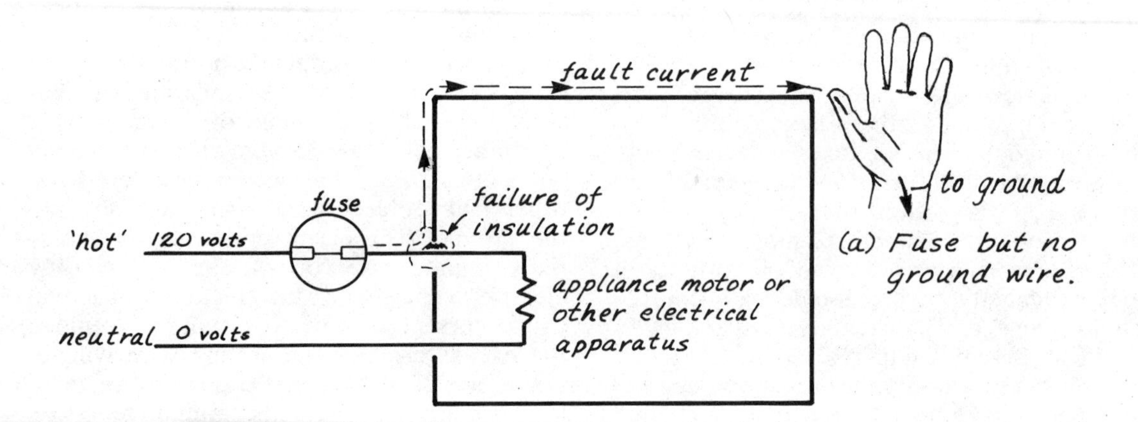

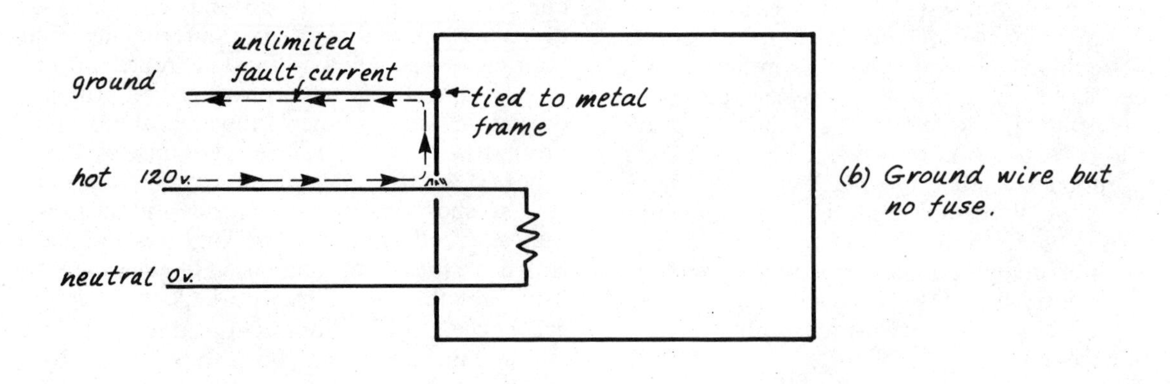

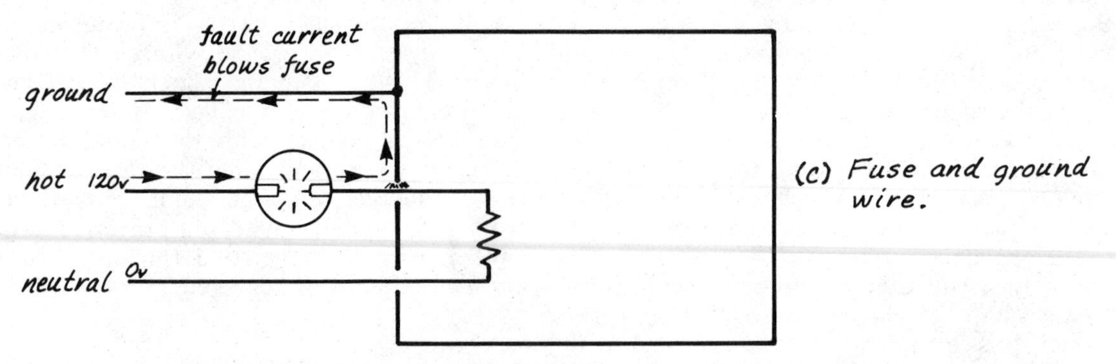

Figure 13–4 Both the fuse and the ground wire are required for electrical safety.

The basic reason is the ease of generating alternating current, but it has other advantages. The voltage supply in the United States is a 60 cycle/second sinusoidally varying voltage like that illustrated in Figure 13–5. The most common voltage supplied is in the range from 110 to 120 volts. This quoted value of 120 volts does not mean that the *peak* voltage is 120 volts. This 120 volts is the *effective* value for the AC voltage. The voltage varies between positive and negative voltage peaks which are $\sqrt{2}$ times the effective voltage. As shown in Figure 13–5, the common 120 volt AC voltage waveform varies between 170 and −170 volts at a frequency of 60 cycles/sec (the frequency unit of cycles per second has been given the name hertz, abbreviated Hz). When an AC current of 10 amperes is quoted, this also refers to the *effective* value of the current. The current in an AC circuit reverses directions, following the 60 Hz variation of the voltage. For a 10 amp AC current, the instantaneous current will reach $\sqrt{2} \times 10 = 14.1$ amperes in one direction, then reverse and reach a current of 14.1 amperes in the other direction one half-cycle later. If the average power used in a resistance is calculated, it is found to be

$$P_{avg} = \frac{V_{peak} I_{peak}}{2} = \left(\frac{V_{peak}}{\sqrt{2}}\right)\left(\frac{I_{peak}}{\sqrt{2}}\right) = V_{eff} I_{eff}.$$

This illustrates the fact that the AC power supplied to a resistor takes the same form as equation 13–4, $P = VI$. Similarly,

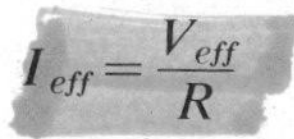

$$I_{eff} = \frac{V_{eff}}{R}$$

in an AC circuit; therefore, Ohm's law applies if the effective current and voltage values are used. It is universal practice to quote AC currents and voltages as the effective values unless the details of the waveform are being considered. Unless otherwise noted, AC voltages and currents may be assumed to be the effective values.

In addition to electrical resistance, other types of circuit elements must be considered when dealing with alternating current. Capacitors and inductors are the other basic types of circuit elements for AC circuits. The capacitor may be considered to be a charge and energy storage device, as described in Chapter 12 in terms of the parallel plates, and it is represented by the symbol C. An inductor is made by winding a coil of wire, often around a magnetic core. An inductor is represented by the symbol L and may be considered to provide electrical "inertia." That is, it tends to oppose any change in electric current through it by generating an opposing voltage, and this opposition to current change might be called electrical inertia.

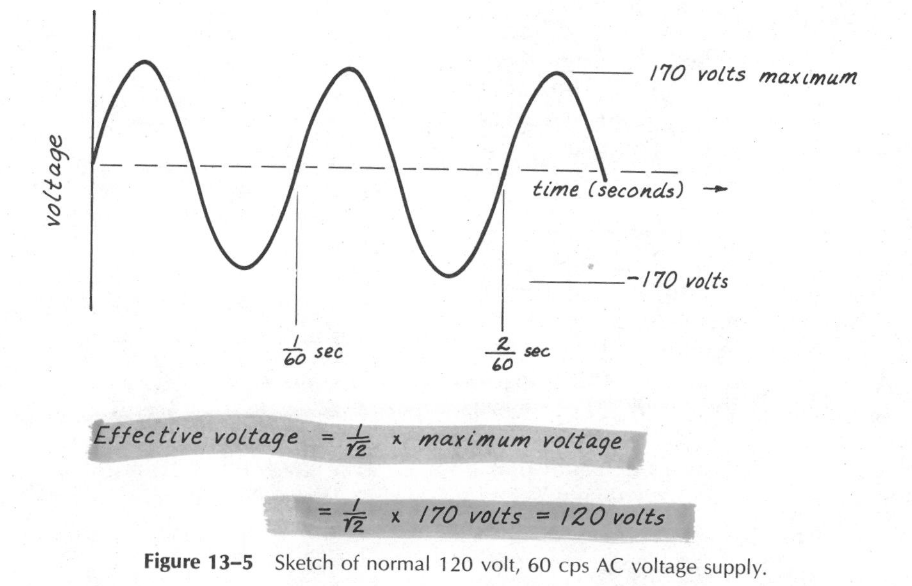

Figure 13–5 Sketch of normal 120 volt, 60 cps AC voltage supply.

Both capacitors and inductors impede the flow of electric current. The impedance offered is dependent upon the frequency of the AC voltage applied. For a capacitor the impedance is

$$\textbf{Capacitor: } Z_C = \frac{1}{2\pi fC} \qquad \textbf{13–6}$$

where f is the frequency in hertz. Thus, an ideal capacitor is an open circuit for $f = 0$ (DC), and presents a low impedance at high frequencies. For an inductor

$$\textbf{Inductor: } Z_L = 2\pi fL, \qquad \textbf{13–7}$$

so an ideal inductor represents no impedance (short circuit) to DC but a high impedance at high frequencies. The effects of resistors, capacitors, and inductors upon current flow do not add directly. If a resistor R, a capacitor C, and an inductor L are placed in series, the resulting impedance is given by

$$Z = \sqrt{R^2 + \left(2\pi fL - \frac{1}{2\pi fC}\right)^2}, \qquad \textbf{13–8}$$

and the current and voltage are related by a modified Ohm's law

$$I_{eff} = \frac{V_{eff}}{Z} \qquad \textbf{13–9}$$

It is quite beyond the scope of this text to develop the principles of AC circuit theory. These relationships are presented to show that the simpler principles of DC circuit theory can be applied directly when only resistances are encountered, but that more complicated relationships are necessary when capacitors and inductors are present. Capacitors and inductors are crucial elements in electronics, since the frequency dependence of their impedances can be utilized to filter out unwanted signals and otherwise contour the AC electrical signals of interest. If the impedance represented by equation 13–8 is examined while the frequency, f, is varied, it is found that the impedance goes through a sharp minimum at a certain frequency where the two terms involving the frequency cancel each other. This phenomenon is called "resonance" and occurs at the frequency

$$f = \frac{1}{2\pi\sqrt{LC}}. \qquad \textbf{13–10}$$

At this resonant frequency, a much larger current would flow than at any other frequency with the same voltage applied. Resonance in AC circuits is a crucially important idea in electronics, since it makes possible the enhancement of a desired electric signal while signals of all other frequencies are discriminated against. For example, radio signals of all frequencies strike the antenna of your radio, but by tuning the resonant frequency of your radio to the desired station, all other stations can be effectively eliminated. Amplifiers, filters, and other devices use similar principles to enhance the desired signal when biomedical electrical signals are measured. For further development of the principles of AC circuits, see reference 1.

HOUSEHOLD ELECTRICAL SUPPLY

Since the necessary background has been established, it seems appropriate to discuss some practical aspects of normal household wiring. A portion of a typical supply system is sketched in Figure 13–6. Electrical supply systems for business or commercial application are similar.

The electrical distribution system takes charge from the earth at the electric "power plant" and does work on it to raise its energy (voltage = electrical energy per unit of charge). This energy may come from burning coal or oil, from the gravitational energy of water behind a dam, or from a nuclear reactor. The distribution system supplies the charge to a house at a high voltage. When electrical energy is being used, and electric current flows from the high voltage supply through the household supply system, doing an amount of work equal to the charge times the drop in voltage ($\mathcal{W} = qV$, equation 12–5). This current finally returns to the earth at some point, completing the cycle:

Electrical energy cycle

1. **Charge taken from the earth.**
2. **Work done on charge to raise its voltage.**

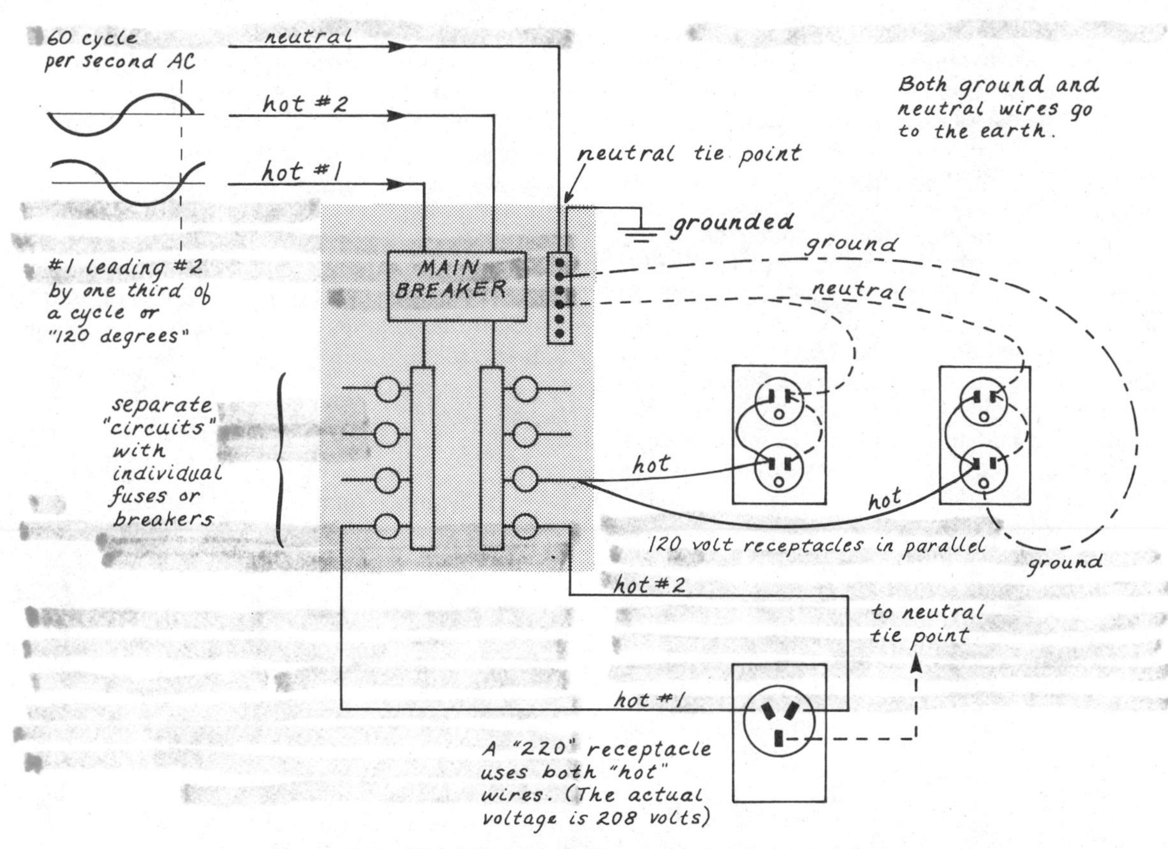

Figure 13–6 Partial diagram of a household electrical supply system.

3. **Charge does work at a distant location as its voltage drops.**
4. **Charge is returned to the earth.**

A typical house in the United States has three wires entering it from the electrical distribution system. One of these wires is a return wire usually referred to as the "neutral" wire. It is tied to ground at the house and also at the supply point from the distribution system (e.g., the utility pole outside the house) as a backup ground. The other two wires are the high voltage supply lines and are popularly referred to as the "hot" wires. Both these hot wires carry a nominal effective voltage of 120 volts at a frequency of 60 cycles per second, but they are not identical. One of them will reach its peak voltage a third of a cycle ahead of the other, and, based on 360 degrees as a full cycle, is said to "lead in phase" by 120 degrees. The dual supply adds to the versatility of the electrical system, as will be shown.

If a 120 volt appliance is to be operated, it can be attached to one of the hot wires and the neutral. But if a higher effective voltage is desired, such as for an electric range or air conditioner, the appliance can be connected to both hot wires. Since the two hot wires do not reach their maximum voltage at the same time, one may have a negative voltage while the other has a positive voltage and, therefore, the effective voltage between them is greater than 120 volts. For the 120 degree phase difference, two 120 volt supply wires give an effective voltage of 208 volts. This process is rather like subtracting two vectors of length 120 which have an angle of 120 degrees between them; the length of the resulting vector would be 208. Though the effective voltage is 208 volts, a circuit which used both hot wires is popularly called a 220 volt circuit.

As indicated in Figure 13–6, as the two hot wires enter a house, they pass through a main breaker which limits the total amount of current that can flow into the house. Then hot wires to a number of individual circuits branch from each main hot wire, having smaller fuses or breakers which typically limit the circuit currents to 15 or 20 amperes. These individual circuit hot wires may lead to receptacles as illustrated, with a neutral wire returning from the other side of the receptacle to the main neutral tie point, which in turn goes to the earth. Most U.S. electrical codes now require three-prong receptacles, in which the third prong (the round prong) is a separate ground connection. Note that both the ground wire and the neutral are tied to ground at the main breaker box, but they have separate functions, as described in the earlier section on the ground wire. The neutral wire is the normal return path to the earth for the current. The ground wire is tied to the metal frame of the appliance and does not normally carry a current, but will do so as a protective measure if an electrical malfunction causes a voltage to be supplied to the frame of the appliance.

TRANSFORMERS

One of the advantages of alternating current is that the voltage can be easily altered with the aid of a transformer, while DC sources such as batteries have fixed voltages. A transformer is basically a set of two coils which are usually connected by a magnetic iron core, as shown schematically in Figure 13–6. When an AC voltage V_1 is applied to the primary coil as shown, it generates a magnetic field in the iron core (as discussed in Chapter 12 in connection with electromagnets). This magnetic field penetrates the secondary coil and induces an AC voltage in it. Any changing magnetic field will generate a voltage in nearby conductors. If the number of turns in the secondary coil, N_2, is greater than the number in the primary coil, N_1, the device constitutes a step-up transformer, producing voltage out of the secondary coil, V_2, which is higher than the primary voltage V_1. The ratio of the voltages is equal to the ratio of the numbers of turns in the coils if none of the magnetic field is lost from the core. The efficiency of commercially available transformers is so high that the relationship

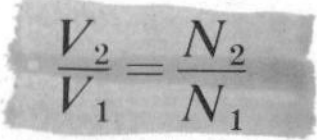

$$\frac{V_2}{V_1} = \frac{N_2}{N_1}$$

is a good approximation. In Figure 13–6, N_2 is twice as large as N_1; therefore, a 110 volt input would produce a 220 volt output. If the transformer is reversed so that N_2 is ½ N_1, the result is a step-down transformer which would yield 55 volts when supplied by 110 volts on the primary. Transformers with very high ratios of turns between the secondary and primary are used to generate the voltages (on the order of 20,000 to 100,000 volts) needed to produce x-rays. Many other applications of high voltages utilize transformers to multiply the AC supply voltage.

It is important to note that the transformer does not multiply energy; that would violate the principle of conservation of energy. Suppose a transformer is supplied with 2 amperes of current at 100 volts. This corresponds to a power of 200 watts; therefore, the output power from the transformer is limited to 200 watts. If the transformer has an output voltage of 400 volts, the maximum current which could be supplied would be

$$I = \frac{P}{V} = \frac{200 \text{ watts}}{400 \text{ volts}} = \frac{1}{2} \text{ ampere.}$$

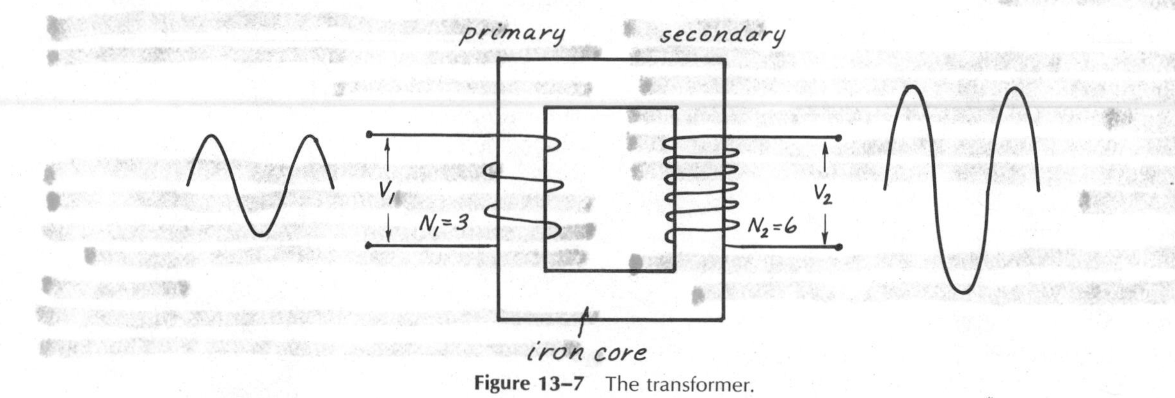

Figure 13–7 The transformer.

Some loss of electrical energy to internal energy always occurs (the temperature of the wires and transformer core will rise slightly), so the output current would be slightly less than ½ ampere.

Note in Figure 13–7 that there is no direct electrical connection between the primary and secondary coils of a transformer. This is often used to advantage in increasing the safety of electrical devices. If the electrical energy for an electrical appliance comes from the secondary of a transformer, the appliance is said to be "transformer isolated" since it has no direct connection to the electrical outlet. The safety implications of this isolation will be discussed in Chapter 14.

SUMMARY

The understanding of practical electric circuits can be facilitated by the analogy between an electric circuit and a fluid circuit. By analogy, voltage may be identified with pressure, electric current in amperes with volume flow rate, and electrical resistance with resistance to flow in a fluid circuit. Ohm's law, voltage = current × resistance, is analogous to Poiseuille's law for fluid flow. For electric circuits, the sum of the voltage drops around any closed circuit must be equal to zero, and the electric current into any junction must equal the current out. These two statements are referred to as the voltage law and the current law. Along with Ohm's law, they provide the basis for calculations involving direct current (DC) circuits. The electric power used in any element in a DC circuit is equal to the current in amperes times the voltage drop across it.

The fuse and the ground wire are the basic protection devices which make possible the safe use of electrical devices. The fuse is a current limiter which interrupts the circuit when excessive current flows, thus preventing fires due to overheating and warning of overload conditions. The ground wire conducts excess charge to the earth, preventing shock hazards due to charge buildup and forcing the fuse to blow and interrupt the circuit in case of major malfunction. Neither is adequate alone; both should be used for electrical safety.

Alternating current (AC) voltages and currents are usually measured in terms of effective values so that Ohm's law and the power relationship can be used for many practical calculations. The presence of other circuit elements, capacitors, and inductors, must be considered in calculating the effective electrical resistance. One of the advantages of AC voltages is the fact that they can be raised or lowered to the desired values by use of a transformer.

REVIEW QUESTIONS

1. Compare the behavior of an electrical circuit with that of a water circuit. What are the analogs to voltage, current, and resistance in the water circuit?

2. What is meant by a short circuit and what are the associated electrical hazards?

3. What is meant by an open circuit and what are the associated electrical hazards?

4. Why isn't a bird electrocuted when it lands on a high voltage power line?

5. What is the specific function of a fuse or circuit breaker? What hazards exist when the fuse is bypassed?

6. What is the specific function of a ground wire? What hazards exist if it is omitted or broken? Why are both the ground wire and the fuse required for electrical safety?

7. What is meant by AC and DC currents? Give examples of the use of each.

8. What are some of the advantages of AC electric power as compared with DC?

9. If a transformer increases the voltage from 100 to 1000 volts, by what factor is the current changed? What general statements can be made about the ratio of the output electric power to the input power?

PROBLEMS

Worked Example: A charge of 120 coulombs flows through a wire in one minute. What is the electric current in amperes?

Solution: The ampere is one coulomb per second. Therefore

$$I = \frac{Q}{t} = \frac{120 \text{ coul}}{60 \text{ sec}} = 2 \text{ amperes.}$$

1. How many electrons are flowing past a point in a wire per second if the electric current is one ampere?

Worked Example: A 12 volt automobile battery is simultaneously supplying energy to a 1.0 ohm light system, a 4.8 ohm radio, and a 6 ohm blower. (a) What current flows through each, and what total current flows from the battery? (b) What power is consumed by each, and what toal power is supplied by the battery? (c) What is the equivalent resistance of the three parallel resistances?

Solution: (a) Since the elements will always be connected in parallel, each will have a 12 volt drop across it.

$$\text{Lights: } I = \frac{12\text{V}}{1\Omega} = 12 \text{ amps}$$

$$\text{Radio: } I = \frac{12\text{V}}{4.8\Omega} = 2.5 \text{ amps}$$

$$\text{Blower: } I = \frac{12\text{V}}{6\Omega} = 2.0 \text{ amps}$$

Total Current: I = 16.5 amps from battery.

(b) The electric power is given by $P = IV$.

Lights: P = (12 amps)(12 volts) = 144 watts

Radio: P = (2.5 amps)(12 volts) = 30 watts

Blower: P = (2.0 amps)(12 volts) = 24 watts

Total Power: P = 198 watts.

(c) The equivalent resistance of the three elements can be found by using Ohm's law.

$$R_{equiv} = \frac{V}{I_{total}} = \frac{12\text{ V}}{16.5\text{ amps}} = 0.73\text{ ohm.}$$

Alternatively, it can be found by the formula for adding parallel resistances.

$$\frac{1}{R_{equiv}} = \frac{1}{1.0\ \Omega} + \frac{1}{4.8\ \Omega} + \frac{1}{6.0\ \Omega} = 1.37/\Omega$$

$$R_{equiv} = (1/1.37)\ \Omega = 0.73\text{ ohm.}$$

2. What is the voltage drop across a resistance of 20 ohms when a current of 6 amperes is flowing through it?

3. What is the electrical resistance of an automobile headlight which draws 5 amps from a 12 volt battery?

4. What is the resistance of a heater if it requires 1200 watts of power when supplied by a 120 volt receptacle?

5. A heater requiring 8 amps and a lamp requiring 2 amps are connected in a parallel across a 120 volt line. What is the resistance of each? What power is used in each?

6. If two 10 ohm resistances were connected in parallel to a 120 volt supply, what would be the current through each and the total power dissipated in the two resistances? Calculate the same quantities if the resistances were connected in series.

7. An appliance is rated at 600 watts when supplied by a 120 volt electrical supply.

   a. What current in amperes will flow through it?
   b. What is its resistance in ohms?

8. A 30 ohm resistance and a 10 ohm resistance are connected in parallel to a 120 volt receptacle.

   a. Find the current through each.
   b. What is the total power supplied by the receptacle to these resistances?

9. How many watts of power could be delivered by a 110 volt line fused at 15 amperes?

10. Could a 1200 watt heater and a 500 watt appliance be used simultaneously on a 120 volt line which is fused at 15 amperes?

11. A 5 ohm resistance and a 10 ohm resistance are connected in series across a battery of unknown voltage. A current of 5 amperes is observed to flow in the 5 ohm resistance.

   a. What is the battery voltage?
   b. How much current will flow in the 10 ohm resistance?

12. A 5 ohm resistance and a 10 ohm resistance are connected in parallel across a battery of unknown voltage. A current of 5 amperes is observed to flow in the 5 ohm resistance.

    a. What is the battery voltage?
    b. How much current will flow in the 10 ohm resistance?

13. An electric current of 200 milliamps flows through the bulb of a small flashlight when a voltage of 3 volts is supplied by the batteries.

    a. What is the resistance of the bulb?
    b. What power is being used?

14. In household wiring, the lights of several rooms are usually placed in parallel on a single circuit. How many 120 watt lights could you place on a single circuit which has a current limit of 20 amperes set by a circuit breaker? The voltage is 120 volts.

15. A 120 volt electrical supply is connected to two resistors of resistance 20 ohms and 60 ohms.

    a. Calculate the current and power supplied if the resistances are in series.
    b. Calculate the current and power if the resistances are in parallel.

16. A 120 volt household circuit is fused at 20 amperes. A 600 watt heater and a 60 ohm light are connected to this circuit (in parallel).

    a. What is the current in amperes through the 600 watt heater?
    b. What is the current through the 60 ohm light?
    c. What is the resistance of the 600 watt heater?
    d. What is the minimum resistance R which could be used as a third appliance without blowing the 20 amp fuse?

17. If you have a 1200 watt hairdryer which operates from a 120 volt circuit, how many amperes of current flow through the hairdryer? What is its electrical resistance?

18. A 2 ohm and a 4 ohm resistance are wired in parallel and the combination is then placed in series with a 5 ohm resistance and attached to a battery. The current in the 2 ohm resistance is measured to be 6 amperes.

    a. What is the current in the 4 ohm resistance?
    b. What is the current in the 5 ohm resistance?
    c. What is the battery voltage?

19. Although the electrical resistance of the wires in an electric circuit is often negligible, there are circumstances where it has a noticeable effect. The resistance of the wire is in series with the appliance. Assume that the wire in a given household circuit has a resistance of 0.2 ohms.

    a. A refrigerator has a current of 2 amperes flowing through it when it is running normally. If it is connected to the 120 volt circuit described above, how much of the voltage will actually be applied to the refrigerator?
    b. If the refrigerator uses a current of 15 amperes while starting, what will be the voltage applied to it?

20. A 600 watt toaster was connected to the same 120 volt receptacle as a refrigerator, and the 15 amp circuit fuse blew when the refrigerator motor was turned on. The following calculations will allow you to an-

alyze the problem. (Most electrical codes require a refrigerator receptacle to be separately wired so that it is the only load on that circuit.)

a. What is the current in the 600 watt toaster when it is operating?
b. How much total power can be supplied by the 120 volt circuit, fused at 15 amperes?
c. If the current to the refrigerator is 12 amps when the motor is starting and 2 amps when it is running steadily, what is the power consumption in each case?

21. A light duty extension cord has a resistance of 0.01 ohm per meter of length, considering both wires.

a. What is the resistance of a 50 meter extension cord?
b. If 120 volts is supplied to the cord, how much voltage would be available to operate a 50 ohm appliance at the end of the cord?
c. Suppose a short circuit occurred at the end of the cord so that only the resistance of the cord was presented to the 120 volt supply. If there were no fuse, how much current would flow?

22. A troublesome problem with battery-operated equipment is the fact that the battery may appear to be in good condition when tested with a voltmeter and yet fail to operate the equipment. This problem arises because of a buildup of internal resistance in the battery when it is near failure. This exercise explores this problem.

a. If a 6 volt battery has an internal resistance of 10 ohms, the voltage at the terminals of the battery will be less than 6 volts by the amount of voltage drop across the internal resistance. If the voltage at the terminals is measured by a voltmeter of resistance 20,000 ohms, what voltage will be measured? (The internal resistance and voltmeter resistance are in series.)
b. If this battery is used to supply an instrument which has a resistance of 20 ohms, what voltage will be produced at the battery terminals to operate the instrument?
c. Repeat (a) and (b) for the case where the battery is nearer to complete failure and has an internal resistance of 50 ohms.

23. Motors require more electric current when starting than when running at normal speeds. This is because their actual resistance is very low, but they generate a "back voltage" when running which reduces the effective voltage that is applied to that low resistance.

a. The actual resistance of a motor is 4 ohms. If the motor is prevented from turning when the 120 volts are applied to it so that the entire voltage is applied to the 4 ohm resistance, what current will flow through the motor?
b. When the motor is running at full speed with no "load" applied to it, it generates a back voltage of 112 volts which opposes the 120 volt supply. How much current will now flow through the 4 ohm resistance of the motor?

24. A 3-way bulb for a lamp can operate at powers of 30 watts, 70 watts, or 100 watts. It accomplishes this by having two filaments with resistances R_1 and R_2. The 30 W and 70 W powers are obtained by using the two filaments separately, and the 100 W by using the two in parallel across the 120 volt electrical supply.

a. Find the resistances R_1 and R_2 of the filaments.
b. Find the equivalent resistance when the filaments are connected in parallel.
c. Confirm that the two in parallel will use a power of 100 watts.

25. An electric range is supplied with 208 volts. To obtain a number of power settings for versatility in cooking, it uses three heating elements at one cooking location with resistances $R_1 = 108\Omega$, $R_2 = 72\Omega$, and $R_3 = 36\Omega$.

    a. If these three elements are connected in series, how much electric power would be used?
    b. How much power would be used if they were connected in parallel?
    c. If the elements were used singly to get three more power settings, how much power would be used for each heating element alone?
    d. Three more power settings could be obtained by using parallel combinations of two of the heating elements. What powers could be obtained?

26. If an automobile starter has an effective resistance of 0.1 ohm, how much current will flow through the starter when it is connected to a 12 volt battery? How much current will flow if the battery terminals are corroded so that they provide a resistance of 0.2 ohm in series with the starter resistance? Explain why such corrosion resistance will affect engine starting more than it will affect the lights, horn, and other devices which have higher resistances than the starter.

27. What would be the cost of operating an 80 watt radio for 4 hours a day for 30 days if the cost of electrical energy is 3¢ per kilowatt-hour?

28. An x-ray machine operates at 200,000 volts with a current of 10 milliamps. This power must be produced by a transformer which raises the voltage from 220 volts to 200,000 volts. Assuming 100% efficiency, how much current must flow in the 220 volt line to operate the x-ray machine?

REFERENCE

1. Blackwood, O. H., Kelly, W. C., and Bell, R. M. *General Physics, 4th ed.* New York: Wiley, 1973.

CHAPTER FOURTEEN

Electrical Safety in the Hospital

INSTRUCTIONAL OBJECTIVES

After studying this chapter, the student should be able to:

1. Outline the levels of the physiological effects of electric current and explain the usual cause of death from electric shock.
2. State the conditions under which a patient will be "microshock sensitive."
3. Explain why the ground connection is so critical when a patient is microshock sensitive.
4. List the precautions to be taken with a microshock sensitive patient.

With the proliferation of electrical and electronic equipment in hospitals, the problems of electrical safety have multiplied. The problems have been complicated by the fact that internal catheters and pacemaker wires make the patient much more susceptible to electric shock. Health care personnel are required to operate electrical devices of increasing complexity. Therefore, it is important to understand some of the fundamentals of electrical safety.

THE PHYSIOLOGICAL EFFECTS OF ELECTRIC CURRENT

The harmful effects of electricity are dependent mainly upon the amount of electric current which flows through the body, the duration of the current, and the path it follows through the body. The current is the primary variable which determines the seriousness of a shock, and a shock depends on voltage and resistance only in the sense that they together determine how much current will flow. The question often asked is, "How much voltage does it take to be dangerous?" It is not possible to give a general answer to this question, since the physiological effects are not directly dependent upon the voltage. It is possible that a person receiving a 120 volt shock will hardly feel it, if he is insulated by normal clothing and shoes and is in a dry environment. However, if he were standing barefoot on a wet floor, the shock would be severe and very possibly fatal. The electrical resistance of the body varies

TABLE 14–1 The Physiological Effects of 60 Hz A.C. Current Through Intact Skin into the Body Trunk.

CURRENT (1 SECOND CONTACT)	PHYSIOLOGICAL EFFECT	VOLTAGE REQUIRED TO PRODUCE THE CURRENT WITH ASSUMED BODY RESISTANCE:	
		10,000 ohms	*1000 ohms*
1 milliampere	Threshold of feeling	10 V	1 V
5 milliamperes	Accepted as maximum harmless current	50 V	5 V
10–20 milliamperes	Beginning of sustained muscular contraction ("can't let go" current)	100–200 V	10–20 V
50 milliamperes	Pain, possible fainting and exhaustion. Heart and respiratory functions continue	500 V	50 V
100–300 milliamperes	Ventricular fibrillation, fatal if continued. Respiratory function continues	1000–3000 V	100–300 V
6 amperes	Sustained ventricular contraction followed by normal heart rhythm (defibrillator). Temporary respiratory paralysis and possibly burns	60,000 V	6000 V

widely; therefore, the voltage required to produce a dangerous current level will vary widely. The skin normally offers a large resistance to electric current. This resistance is strongly affected by moisture. The electrical resistance measured from one hand to the other can vary from over 1,000,000 ohms for very dry skin to less than 1000 ohms for wet skin.

A considerable amount of research has been done to determine the levels of 60 Hz AC current associated with given physiological effects. Table 14–1 lists representative currents that flow when electric shock occurs through intact skin to the body trunk. Most of the data here are taken from references 1 and 2. The current listed in the left column of the table is the primary variable. The voltages required to produce those currents are listed in the last two columns for assumed body resistances of 10,000 and 1000 ohms. Though these are smaller than the normal resistance, they are within the range of measured values for body resistance.

Note that sustained currents above 100 milliamps through the body trunk may be fatal. Current medical opinion is that the usual cause of death due to shock is ventricular fibrillation: random chaotic activity of the heart muscle rather than the periodic, coordinated ventricular contractions necessary to pump the blood. This is in contrast with earlier opinion that respiratory paralysis was the main cause of death. The heart is particularly sensitive to electric shock. It is basically an electrically controlled pump; its coordinated action is stimulated by internally generated electrical impulses which travel in a specified way through the heart (see Chapter 16). Periodic electrical impulses from outside the body can upset this coordinated action. If the cells in a small area of the heart are disturbed, this disturbance tends to propagate to neighboring areas and produce the random activity associated with ventricular fibrillation.

The other muscles of the body are also sensitive to electric shock. They can, like the heart, be considered to be electrically controlled since the activating nerve impulses travel to the muscles in the form of electrical pulses. As will be noted from Table 14–1, externally generated 60 Hz currents above 20 milliamps tend to cause sustained contraction of the muscles. This adds to the hazard of electric shock, because a person holding a wire or other object cannot release his grip if the current is above this threshold.

The body is more sensitive to periodic currents in the range around 60 Hz than to lower or higher frequencies, presumably because the normal nerve impulses are repetitive with similar rates. Figure 14–1 shows the variation of the sensitivity threshold and "can't let go" current with the frequency (3). The sensitivity peaks near 60 Hz and drops greatly for high frequencies. The drop in sensitivity at very high frequencies makes practical the use of electrocautery devices at around 2 million Hz (2 MHz). Very high frequencies tend to produce burns rather than muscle contractions. Other high frequency electrical devices are used to produce internal heating in the body for therapeutic purposes; at these frequencies there is little tendency toward muscle contractions or convulsions, and little shock hazard in the usual sense.

Electric currents above the 100 milliamp level are sometimes used for therapeutic purposes. As shown in Table 14–1, large currents cause a sustained ventricular contraction. If the heart of a patient is in a state of ventricular fibrillation, a large current flow for a brief time will stop the fibrillation, and often the heart will resume a more normal pumping action after the current is released. This is the principle of operation of the defibrillator, which will be discussed further in Chapter 15. A single voltage pulse of up to 10,000 volts is produced by the defibrillator. This causes a high current flow. Therefore, large electrodes are used in contact with the patient to spread the current over a larger area and reduce the likelihood of burns.

In electric shock therapy, a current on the order of 0.5 to 1.5 amperes is conducted between the temples of the patient. The current produces generalized convulsions and unconsciousness. The therapeutic value of these treatments has been demonstrated in the care of patients with certain emotional disorders. Care must be taken to limit the current to the desired path through the head. This is done by placing metal electrodes of about 2 in² area on the temples after covering the area with a conducting jelly. A typical voltage application would be 90 volts between the two electrodes for a period of about 1/10 sec (4).

MICROSHOCK HAZARDS

The preceding discussion of current levels in the body assumed that the current en-

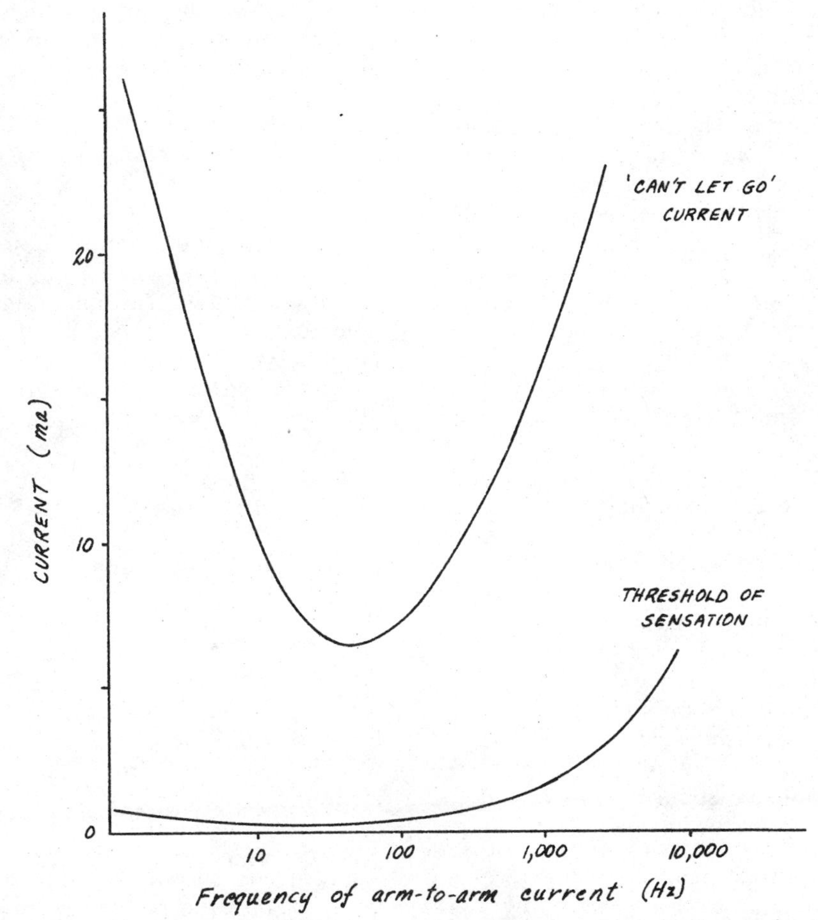

Figure 14–1 Approximate currents for threshold of sensation and sustained muscular contraction as a function of frequency (3).

tered through the skin and was distributed through the body. When an internal electrical path to the heart exists, the dangerous current for shock may be less by a factor of about a thousand. Experiments with dogs have produced ventricular fibrillation with currents as low as 20 microamperes when the current was introduced directly into the heart (5). This 20 μa current is actually far below the threshold of perception when conducted through the skin. This is a factor of 5000 smaller than the 100 ma current required to initiate ventricular fibrillation when the current enters through the skin. The voltage required to produce this current, of course, depends upon the resistance of the current path. If the resistance is assumed to be 1000 ohms, then the voltage required to produce a hazardous current may be as low as

$$V = IR = (20\mu\text{a})(1000\ \Omega) = 20 \text{ millivolts.}$$

When conditions exist such that a patient has an internal electrical path to the heart, the patient may be said to be "microshock sensitive."

It is important to be able to recognize the conditions under which a patient may be microshock sensitive. The primary contributing factor is the existence of an electrical conductor which extends inside the body. A pacemaker wire is the obvious example, but most types of catheters also provide a conducting path. Pure water is an insulator, but a saline solution is an electrical conductor and can conduct a current to the heart; therefore, any catheter carrying a conducting fluid into the body produces a low resistance path to the heart. Even urinary catheters and catheters used to drain fluid from the body may represent a conducting path. In coronary care units, patients may have conducting catheters extending directly into the heart for ECG measurements or the measurement of the pressure in the heart chambers. The devices mentioned offer differing conductivities to the heart, but they have in common the fact that they bypass the high electrical resistance of the skin.

The second condition which is necessary to produce a microshock hazard is a ground connection from the patient to complete the electrical circuit so that current can flow in the internal path described above. Even when a voltage is applied to an internal conductor, a current will not flow unless there is a second conductor to complete the circuit.

Two electrical connections to the body are required for a person to be shocked. The flow of electric current is usually from a high voltage point to ground. To complete an electrical circuit you must not only have a point of contact to the high voltage supply but also a contact to ground, like the return wires discussed in Chapter 13 for household circuits. In the language of those circuits, you have a "hot" wire and a "neutral" wire that takes the current to the earth or "ground" voltage. A bird landing on a high voltage wire is not shocked, even though he may be in contact with a voltage of hundreds of volts, because it has no contact with a ground — both of its feet are at the same high voltage. If a high voltage wire falls across your car, you will usually not be shocked because rubber insulating tires prevent the circuit from being completed to ground. If you stepped out of the car and made contact with the ground while still touching the metal car, the results would be disastrous. In a microshock hazard, a catheter may be the point of contact to a high voltage or it may be the contact to ground, but in either case it contributes to completing an electrical circuit that leads to a shock.

The Critical Nature of the Ground Connection

Since a ground connection is part of the electrical circuit which leads to a shock, as discussed above, careful attention must be paid to ground connections when there is a potential shock hazard. A source of voltage will cause current to flow through any available path to ground, analogous to water seeking the lowest point by whatever path available.

DO NOT GROUND THE PATIENT. The purpose of electrical safety measures is to avoid making the patient part of an electrical circuit. If the patient is grounded, then any source of voltage which touches the patient will cause a current to flow to ground through the patient. If there is no electrical path to ground from the patient, then no current would flow: *two* accidental electrical connections would have to be made simultaneously to make the patient part of a current path. Therefore, isolation of the patient from electrical ground offers a considerable measure of safety.

Isolating the patient from ground is not

always a simple matter. Older models of ECG monitors have the lead to the right leg grounded. This has been recognized as a hazard, since many patients on continuous ECG monitors will have other electrical apparatus in contact with the body. Modern ECG monitors are isolated from ground by means of transformers. As mentioned in Chapter 13, if a piece of equipment receives its power from the output of a transformer, it has no direct electrical connection with the electrical outlet and will therefore not be grounded by the electrical supply system. Other devices connected to the patient, including catheters, can provide conducting paths to the ground. It is recommended that all devices which are connected to the patient for extended periods be checked to make sure that they do not provide an electrical path to ground.

Ground Electrical Equipment near the Patient. All electrical equipment, such as lamps, motors for electric beds, and instruments used in patient care, should be connected to suitably grounded electrical outlets with three-wire cords (6). The third wire is a direct connection to ground and will prevent a voltage buildup on the metal frame of the equipment. As discussed in Chapter 13, a fuse alone offers no protection against electric shock in certain circumstances. A short circuit between the high voltage lead and the case could produce 120 volts on the case without blowing a fuse and without apparent malfunction of the equipment. Note that since the ground wire is a protection device and not part of the main circuit, the equipment will continue to operate normally if the ground wire is broken. It is therefore advisable to periodically check the ground wires of electrical equipment used in intensive care units where intracardiac catheters are in use. If a tingling sensation is felt when the metal parts of a piece of equipment are touched, this indicates not only that the equipment is not grounded, but also that a serious leakage current exists.

Modern patient monitoring devices are designed so that their cabinets can be grounded without grounding any of the connections to the patient. This makes possible the grounding of all electrical devices within the reach of the patient while leaving the patient ungrounded. As a final precaution, it is desirable to use equipment with nonmetal cases near microshock sensitive patients, and to cover exposed conducting wires into the patient (e.g., pacemaker wires) with a nonconducting material such as plastic.

THE NATURE OF LEAKAGE CURRENTS

Leakage currents are currents which flow from the high voltage components of an electrical device to its metal frame or case. In the ideal case, the high voltage components would be perfectly isolated from the case. The impedance to current flowing to the case would be infinite and no leakage current would exist. Modern insulating materials are very good and the resistance between the case and high voltage can be made extremely large, so that leakage current would not pose a large problem if only DC voltages were present. But AC voltages can cause currents by capacitive and inductive coupling even when no DC current could flow. Capacitive coupling occurs because any excess charge on interior metal parts will tend to attract the opposite polarity charge on the case. When the charge on the interior parts reverses 60 times per second (60 Hz AC), then the outside induced charge tends to reverse also, causing a 60 Hz charge motion or current in the case.

It is practically impossible to eliminate leakage currents completely. Until recently, a leakage current maximum of 5 ma was allowed by the Underwriters' Laboratories (UL). This was based upon the acceptance of 5 ma as the maximum safe current conducted into the body *through intact skin*. However, this is far above the 20 μa current that could be fatal if conducted directly to the heart by an intracardiac catheter. Normally, the third wire (ground wire) of the equipment would carry essentially all of the leakage current, but if the electrical device used a two-wire cord or if the ground wire were broken, a shock hazard would exist.

Example. An ill-advised situation exists in which the case of an electrical instrument makes electrical contact with a catheter that leads to a patient's heart. The patient is grounded so that the resistance through the patient to ground from this catheter is only 1000 ohms. The instrument has a leakage current of 5 ma (legally acceptable at the time of manufacture). The instrument is protected by a ground wire which has a resistance of 1 ohm to ground. The 1 ohm resistance to ground and the 1000 ohm resistance to ground through the patient are then in parallel. Calculate the current through the patient if the ground wire is intact, and the current if the ground wire becomes disconnected.

Solution. With the ground intact, essentially all the current will flow through the 1 ohm ground wire. The instrument frame will have a voltage

$$V = IR = (5 \text{ ma})\ (1\ \Omega) = 5 \text{ mv}$$

with respect to ground. The current through the patient would then be

$$I = \frac{V}{R} = \frac{5 \text{ mv}}{1000\ \Omega} = 5 \times 10^{-6} \text{ amp or } 5\ \mu\text{a},$$

presumably a safe value. But if the ground wire breaks, the entire 5 ma could flow through the patient and cause a fatal shock.

Two of the advisable safety measures have been violated to create the situation in the above example. First, the patient should have been electrically isolated from the case of the electrical instrument. Second, the patient should have been otherwise isolated from ground so that the 1000 ohm path to ground would not have existed to provide the second connection necessary to complete the electrical circuit. Either precaution would have eliminated the danger in this particular case. As added protection it has been proposed that 10 μa leakage current be set as the maximum allowable leakage current for electrical instruments in the vicinity of microshock-sensitive patients.

As of September 30, 1974, a leakage current maximum of 100 μa to the frame of an instrument had been set by the Underwriter's Laboratories. Ordinary wires to a patient must not have more than 50 μa leakage current, and isolated patient wires, such as ECG wires, must not have more than 10 μa. From the instrument manufacturer's point of view, these are extremely stringent limits, but they must be met to qualify for UL listing.

EXAMPLES OF HAZARDS

The precautions suggested thus far should remove the shock hazard to microschock sensitive patients, but many hospitals have situations in which these precautions are not observed. For example, it would not be extremely uncommon to find a patient grounded at two different points on his body. As long as leakage currents are small and the two ground wires have the same voltage with respect to ground, this may not be hazardous. However, conditions like the following example represent a hazard.

Example. Several electrical instruments must be connected to a patient, and an extension cord is extended into the hall to power one of them. This results in two separate, unconnected ground wires which are in contact with the patient through the instrument cases. Each has a resistance of 1 ohm to ground. Another electrical device connected to the hall electrical circuit malfunctions, causing a short circuit to ground which blows a fuse.

Possible Result. The hall circuit ground wire may carry a fault current of 20 amperes for several seconds before a 20 amp fuse burns out and interrupts the circuit. Twenty amperes flowing through the 1 ohm ground wire would cause a 20 volt potential on the case of the patient-related instrument for the duration of the fault current. The patient, having a separate ground connection to complete the circuit, would be subjected to a 20 volt "shock" for the duration of the fault current. While this would not be extremely serious if both ground connections were to intact skin, it could be fatal if one were attached to an internal catheter.

This example illustrates the advisability of connecting all patient-related electrical equipment to the same circuit, and the importance of having ground wires with a very low resistance to ground. Many modern hospitals are installing modular electrical panels, manufactured to provide all necessary electrical outlets with a common, low resistance ground connection which meets rigid specifications.

Example. An attendant who is changing a surgical dressing on a pacemaker wire has one hand in contact with a bedside metal lamp, which is plugged into an electrical outlet with a two-wire cord, as illustrated in Figure 14–2. The electrical resistance of the attendant's body is 9000 ohms and the resistance through the patient's heart to ground via his electric bed is 1000 ohms, for a total series resistance of 10,000 ohms to ground. The surface of the lamp has a leakage voltage of 5 volts. How much current will flow through the attendant and the patient and what are its likely effects?

Solution. The current which flows through both attendant and patient is

$$I = \frac{V}{R} = \frac{5\ v}{10{,}000\ \Omega} = 0.0005 \text{ amp}$$

$$= 0.5 \text{ milliamp.}$$

This current is below the 1 milliamp threshold of perception for current flowing through intact skin, so the attendant will be unlikely to feel it at all. However, it is above the 0.05 milliamp presumed threshold for the initiation of ventricular fibrillation when conducted directly to the patient's heart via the pacemaker wire. This is an unlikely collection of circumstances, but it illustrates the physical possibility of conducting a potentially fatal current to a patient through your own body without being aware of it.

Further examples of the sometimes subtle shock hazards which may occur in hospitals may be found in an excellent booklet, *Patient Safety* (reference1). Considerable attention has been given to the existence of electrical hazards in hospitals, and modern medical instruments offer some of the means for overcoming these hazards.

RECOMMENDATIONS FOR SAFETY

The discussion in this chapter has dealt mainly with the safety measures necessary to protect microshock sensitive patients with internal conducting catheters or other internal electrical paths to the heart. In the final analysis, the main safety measures recommended are the measures which should be taken to protect all patients from electric shock. The microshock sensitive patient requires certain additional precautions, but the attention given this problem has been taken advantage of by some commercial interests to sell expensive equipment.

For general electrical safety, the following measures are appropriate:

1. All electrical equipment near the patient should be connected to electrical outlets with three-wire cords to provide a ground connection for the equipment.
2. The patient should not be grounded. Measures should be taken to isolate the patient from ground by making sure that electrical wires or metal objects in contact with the patient are not grounded.
3. Bare metal cases of electrical equipment should be kept out of the reach of the patient.
4. If electrically operated measurement or treatment apparatus is used, all points in contact with the patient should be electrically insulated from the metal case of the electrically powered part of the apparatus.

Number 1 is crucial because it is easy to accidentally violate the other three. A routine electrical maintenance program should be car-

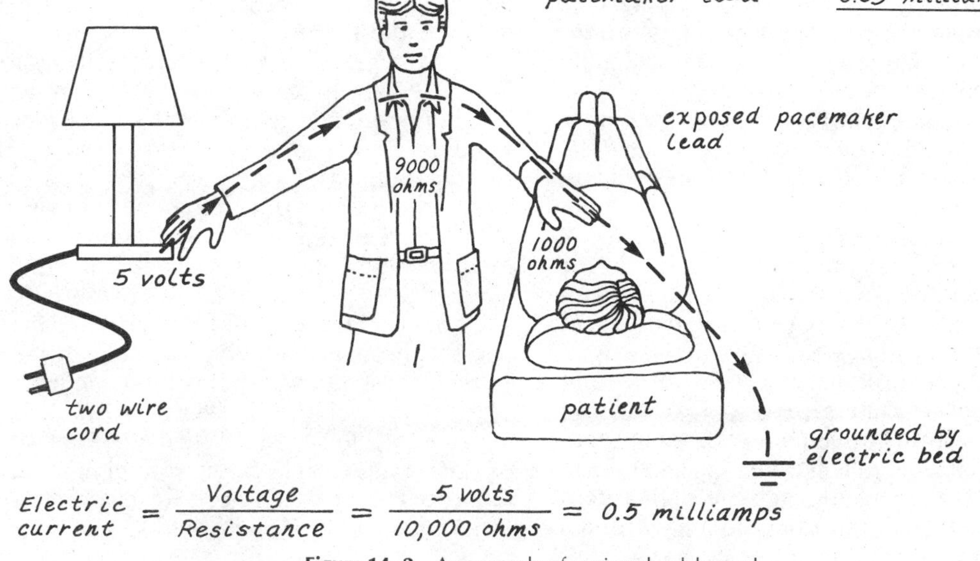

Figure 14–2 An example of a microshock hazard.

ried out in which the ground wires of electrical equipment are periodically checked. It must be kept in mind that equipment will operate normally with a broken ground wire. An electrical resistance check must be made to be sure that it is intact.

Additional precautions are necessary when the patient is microshock sensitive because of the presence of an internal conducting catheter or pacemaker wire:

5. Avoid contact with a bare pacemaker wire or the conducting part of a catheter while touching any metal object with the other hand. This precaution is necessary to avoid becoming part of an electrical circuit which conducts electricity to the patient. It must be kept in mind that a dangerous current could flow through you to a pacemaker wire without causing you to have a sensation of "shock." A 50 μa current in a pacemaker wire could produce a ventricular fibrillation, yet it is far below the threshold of perception when conducted through a person's skin. A sensible precaution would be to tape or otherwise cover exposed pacemaker wires except when it is necessary to manipulate them (7).

6. Connect all electrical equipment associated with a microshock sensitive patient to receptacles which have a common, low resistance ground. In most hospitals, the beds used to treat such patients will have individual electrical panels where all electrical equipment can be connected. Very stringent requirements are being developed for such panels so that they usually provide the optimum in safe grounding for equipment. The point of this recommendation is that if two pieces of electrical equipment have different grounds, a malfunction in one could cause a fault current and produce a voltage difference between the two instruments.

Modern electrical equipment has made possible great advances in health care, and the use of sophisticated electronic devices will undoubtedly increase. The positive aspects of such devices can be maintained without shock hazards if proper attention is paid to the grounding of equipment and the electrical isolation of the patient from hazardous current paths to electrical ground. The nurse or other attendant can contribute to this effort by maintaining a close watch on the ground wires of equipment, noting frayed wires or other obvious electrical hazards, and making an effort to abide by the six recommendations above. Further information about this role can be found in reference 8. It is recommended that new equipment be tested for leakage current and for proper grounding, and that a periodic check be made on all electrical devices.

A considerable amount of sophistication is required to satisfy recommendations 2 and 4 when ECG and other electrical measurements are being made. Basic ECG monitors, particularly older models, ground the right leg of the patient. Other electrical devices tend to ground some part of the body. Isolation transformers are required to accomplish the electrical isolation of the patient. Such isolation transformers are a standard part of modern ECG equipment. Further information about patient isolation can be found in references 1 and 6. Since some electrical hazards are difficult to detect, elaborate protection systems are marketed to detect the presence of leakage currents, fault currents, or open ground connections. These systems sound an alarm or interrupt the circuit when a malfunction occurs (1). There are circumstances in which the interruption of electric power to support devices such as respirators may pose a greater threat to the patient than the electrical hazard of a broken ground. Technically trained personnel are required to evaluate these factors before purchasing expensive electrical protection devices. The electrical requirements for hospital electrical wiring systems and equipment are described in references 9 and 10.

STATIC ELECTRICITY

After the emphasis on the danger of small voltages when electrically conducting pathways to the heart exist, it might appear that static voltages would also represent a shock hazard. Quite high voltages can be produced by shuffling one's feet on a dry carpet or by removing a sweater or in other ways. But these examples of static electricity represent very tiny amounts of energy, and though they may produce a momentary sensation of shock, they are not a dangerous shock hazard. The static electricity discharges are essentially instantaneous and will not produce a significant current through the body. The concern about static electricity in hospitals has been because of the danger of a static electricity spark causing an explosion or fire. When ether and other explosive anesthesia gases were in more common use, there was a need for elaborate precautions to prevent such sparks.

SUMMARY

The physiological effects of electric shock are directly dependent upon the amount of electric current which flows, and upon its path through the body. Although the amount of current is dependent upon the voltage applied and the electrical resistance of the path, the current in amperes is the primary parameter for the classification of shock hazards. The physiological effects are dependent upon the frequency of the current. Very high frequencies tend to produce heat rather than muscular contractions and are thus useful for cautery and electrosurgery, but the body is extremely sensitive to electric shock from ordinary 60 Hz AC currents. When conducted through skin, current above 10 milliamps may produce sustained involuntary muscular contraction and currents above 100 milliamps may cause death through ventricular fibrillation. When the high electrical resistance of intact skin is bypassed by a conducting catheter or pacemaker wire to the heart, the patient may be more than a thousand times more sensitive to electric currents, since experiments indicate that a current of 20 microamps flowing directly to the heart might initiate ventricular fibrillation. Such patients are said to be "microshock sensitive." Such patients should be protected from shock by properly grounding all electrical equipment, isolating the patient from ground, keeping electrical appliances out of the patient's reach, and other measures necessary to avoid making the patient part of an electrical circuit.

REVIEW QUESTIONS

1. What is the primary physical variable which determines the seriousness of an electric shock (voltage, current, resistance, etc.)?
2. Why is it not possible to say how many volts represent a shock hazard?
3. When a person is shocked by a wire, he often cannot let go of the wire. Explain.
4. What current is required to produce ventricular fibrillation if conducted through the skin? If conducted through a catheter or pacemaker wire?
5. When high voltages are used for electrosurgery or electrocautery, why is the patient not shocked by these voltages?
6. How does isolating a patient from electrical ground make him less susceptible to shock?
7. What conditions make a patient "microshock sensitive"?
8. What precautions should be taken to protect the microshock sensitive patient?
9. In terms of the operation of an electric device, what indication do you usually get if the ground wire is disconnected or broken?

PROBLEMS

1. If the body's electrical resistance were 20,000 ohms through intact skin, how much voltage would be required to produce a potentially fatal 100 milliamps of current through the body? If there were a direct path inside the body to the heart which had a 20,000 ohm resistance to ground, what voltage would be required to produce a potentially fatal 50 microamperes of current directly through the heart?

2. A person makes contact with a common 120 volt, 60 cycle/sec AC electrical wire.

   a. What will be the current flow and the likely physiological effect if the person has dry shoes and dry skin and a resistance of 1,000,000 ohms to ground?
   b. What will be the current and the likely physiological effect if the person is standing barefoot on a wet concrete floor and has a resistance of only 1000 ohms to ground?

3. If a person grasps an object which is at a high voltage and which causes a current of 20 milliamps or more to flow through the body, the person will not be able to release the object because the muscles in the hand and arm will forcibly contract. If the person has a resistance to ground of 200,000 ohms, what voltage would be required to produce this "can't let go" current?

4. Many hospitals have stringent standards related to the resistance of ground wires since a sizable resistance in the grounding circuit can lead to unexpected voltages being applied to a patient through monitoring equipment.

   a. Suppose the resistance of a ground wire is 0.5 ohm and a piece of equipment is malfunctioning so that a fault current just under 15 amperes is flowing through the ground wire. The appliance is apparently operating normally and the fuse doesn't blow. How much voltage could this ground wire apply to a patient?
   b. How low would the ground wire resistance have to be to limit the voltage to 0.1 volt?

REFERENCES

1. Hewlett-Packard Co. *Patient Safety*. Application Note AN 718. Medical Electronics Division, 175 Wyman St., Waltham, Mass., 1971.
2. Bruner, J. *Hazards of Electrical Apparatus*. Anesthesiology, Vol. 28, March–April, 1967.
3. Strong, P. *Biophysical Measurements*. Tektronics, Inc., Beaverton, Oregon 97005, 1970, Chapter 17.
4. Flitter, H. H. *An Introduction to Physics in Nursing,* 7th ed. St. Louis: C. V. Mosby Co., 1976.
5. Starmer, C. F., Whalen, R. E., and McIntosh, H. D. *Hazards of Electric Shock in Cardiology*. American Journal of Cardiology, Vol. 14, 1964, pp. 537–546.
6. Walter, C. W. *Electrical Hazards in Hospitals and that Green Wire*. Fire Journal FJ71–10, Vol. 65, Nov. 1971.
7. Hewlett-Packard Co. *Using Electrically Operated Equipment Safely With the Monitored Cardiac Patient.* Waltham, Mass., 1970.
8. Rockwell, S. M. *Electricity — It Doesn't Need to Be a Problem*. RN Magazine, 35, June 1972.
9. The following safety standards may be obtained from the National Fire Protection Association, 60 Batterymarch St., Boston, Mass. 02110:

   NFPA 56A Inhalation Anesthetics
   NFPA 76A Essential Electrical Systems for Hospitals
   NFPA 76BM Safe Use of Electricity in Hospitals
   NFPA 76CM High Frequency Electrical Equipment in Hospitals
   NFPA 70 National Electrical Code, Article 517
10. Underwriters' Laboratories. *Safety Standards for Medical and Dental Appliances*. Safety Standard No. 544.
11. Dalziel, C. F., and Lee, W. R. *Reevaluation of Lethal Electric Currents*. IEEE Transactions on Industry and General Applications, Vol. IGA-4, Sept/Oct 1968, p. 467, and Vol. IGA-4, Nov/Dec 1968, p. 676.
12. Dalziel, C. F. *Electric Shock Hazard*. IEEE Spectrum, Vol 9, p. 41, February, 1972.

CHAPTER FIFTEEN

Electrical and Electronic Instruments

INSTRUCTIONAL OBJECTIVES

After studying this chapter, the student should be able to:

1. Name and describe briefly several examples of sensors, amplification devices, and display devices used in medical instrumentation.
2. State the advantages and disadvantages of thermocouples, thermistors, and platinum resistance thermometers for temperature monitoring.
3. Name and describe the functions of the basic components of the defibrillator, and describe the physiological effect of the defibrillator.
4. State typical values for the voltage and frequency associated with an electrosurgical instrument and describe the physiological effects of such instruments

With the rapid proliferation of electrically operated instruments for medical use, any attempt to catalog such instruments would be immediately obsolete. However, it is important to establish a framework of basic physical principles to aid in the understanding of new instrumentation. It also seems appropriate to outline the physical principles associated with some of the more important and established instruments. Though the distinction is not crucially important, electrically operated equipment, such as motors and pumps, are distinguished from electronic instruments with active electronic elements such as vacuum tubes and transistors which amplify or generate electrical signals for diagnosis or treatment.

Some of the instruments used for physiological measurements will be considered first. Most such instruments will consist of one or more of the following: (1) sensing element, (2) amplifier, and (3) display device.

SENSING ELEMENTS FOR PHYSIOLOGICAL MEASUREMENTS

A number of physiological measurements are made by using sensors which convert the physiological data into small electrical signals. Such devices are often called "transducers." The advantage of the electrical "image" of the physiological data is that it can be easily transmitted to a remote monitoring point, amplified if necessary, and then displayed in a manner which offers maximum ease of measurement.

THERMOCOUPLES. A thermocouple is a temperature sensor which consists of a junction between two different metals. It is usually in the form of two tiny wires which are spot-welded together at one point. When dissimilar metals are brought together, a contact voltage is generated which is proportional to the temperature of the junction. Fine wires of copper and constantan, an alloy, are often used. The voltages generated near room temperature are on the order of a few millivolts. For a copper-constantan thermocouple, the voltage output changes by 0.04 millivolt for each °C change in temperature (2). Of course, if a complete circuit is made by tying the wires together at both ends, the voltages generated will oppose each other. If both ends are at the same temperature, there will be a zero net voltage. However, if one junction is put in an ice and water mixture at 0°C, there will be a net voltage which is proportional to the temperature difference between the two junctions. One of the wires can then be interrupted and attached to a sensitive voltage measuring device. The voltage can be calibrated and converted to a temperature indication on a meter or digital display.

The advantages of the thermocouple as a temperature sensor are accuracy and small size. Thermocouples can be made small enough to be incorporated in the tip of a hypodermic needle. Thermocouple sensors in the form of probes which can be inserted into the patient's ear are used to record temperatures during surgery. Oral and rectal thermocouple probes are also used.

One disadvantage of the thermocouple is the need for the second junction in an ice and water mixture. This difficulty is circumvented by calibrating the thermocouple with the second junction at the average room temperature, sometimes with compensating circuitry to correct for room temperature variations. The reference junction in an ice and water mixture is necessary for extremely accurate temperature measurements, since there is some loss of accuracy in the room-temperature method. The other main disadvantage of the thermocouple is that its output is a very small DC voltage (millivolts). It must be amplified for convenient display, and the amplification of DC signals is inherently more difficult than the amplification of AC signals.

THE ELECTRICAL RESISTANCE THERMOMETER. Because the thermal agitation of the atoms of a metal increases with temperature, its resistance to electric current flow increases. The fact that the resistance is proportional to temperature is used to advantage by constructing temperature sensors of platinum wire or tape. If a constant, controlled voltage is placed across the sensing element, then the current flowing through the sensor can be calibrated in terms of temperature. This is the most accurate of the electrical temperature sensors used for physiological measurements, but it has the disadvantage that it cannot be miniaturized like the thermocouple and thermistor probes.

THE THERMISTOR. The thermistor temperature sensor is similar to the resistance thermometer in that the electrical resistance changes with temperature. However, with the thermistor the resistance *decreases* with temperature increase, and the percentage change is much larger than that of the metallic resistance element. The resistance of a thermistor may change 4% to 6% per degree centrigrade change in temperature (4) compared to 0.37% per degree centigrade for a platinum resistance thermometer.

The thermistor is composed of a hard, ceramic-like material made up of a compressed mixture of metal oxides. The material can be molded into many shapes and can be miniaturized. The use of the thermistor is increasing, since miniaturization makes it suitable for probing and implantation. It is more sensitive than the thermocouple and more easily calibrated. Though it doesn't have the extreme accuracy of the platinum resistance thermometers, temperature differences as small as 0.01°C are measurable (4).

A thermistor can be incorporated into an AC circuit to obtain an AC temperature signal, which can be amplified more easily and reliably than the DC signal from a thermocouple. This property, along with miniaturization, makes it suitable for use in a telemetering capsule or "radio pill" which can be swallowed or implanted. Temperature data are then transmitted to an external radio receiver. Larger thermistors can be used with larger electrical currents, and they find application as temperature control devices. For example, thermistors are used in many dialysis machines as safety devices to shut off the dialysate flow if the temperature of the dialysate liquid is outside narrow temperature limits.

In addition to the direct measurement of temperatures, temperature sensors can be used as indirect sensors for other physiological variables. For example, a thermistor which is heated by passing a current through it can be used to monitor respiration. If the heated thermistor is placed in the respiratory air stream, it is cooled by both inspiration and expiration. The cooling effect of the air flow changes the resistance and causes a current change which is correlated to respiration.

PRESSURE TRANSDUCERS. A pressure transducer produces an electrical signal proportional to the pressure exerted upon it. Microphones can be classified as pressure transducers, since they form an electrical "image" of the pressure variation in the air caused by sound. The variety of pressure transducers is so great that a survey is quite impractical. The discussion here will be limited to two main types of pressure transducers used in direct blood pressure measurements (1, 2, 3).

Direct blood pressure measurements require access to the circulatory system. For arterial pressure this access is usually in the form of a needle or catheter inserted into the brachial artery at the elbow. Once the needle or catheter is inserted, the pressure transducer can be placed on the external end of the catheter. The catheter is usually filled with a saline solution. Since the pressure in the fluid is transmitted undiminished through the catheter (Pascal's law, Chapter 5), the pressure at the outer end gives a measurement of the arterial pressure. The type of transducer usually used to measure this pressure is referred to as the "strain-gauge" type of transducer (4) because of its similarity to industrial strain-gauge techniques used for measurement of strains in metals. The pressure exerted upon a diaphragm stretches one or more small sensing wires. This stretching changes the electrical resistance of the wires and therefore alters the current through them.

The measurement of the pressure in the cavities of the heart requires the use of extremely small catheters called "drift catheters," which are inserted into the venous system and are allowed to drift with the blood flow into the heart (3). The catheter sizes are so small that the outside pressure transducer is unsatisfactory; since the tube is so small, the pressure pattern corresponding to the cycle of the heart is distorted. To overcome this distortion, very small strain-gauge type transducers have been developed which fit on the inserted end of the catheter. Since a current must be passed through this strain-gauge type transducer, this requires a fairly low resistance electrical conductor into the heart, with the consequent problems of electrical safety.

An alternative type of transducer for these small catheters is the piezoelectric transducer. When a pressure is exerted upon a piezoelectric crystal, a small voltage is generated between the surfaces of the crystal. This voltage is proportional to the pressure and can be calibrated to measure the blood pressure. Small piezoelectric transducers can be mounted on the ends of small catheters and inserted into the heart cavities. They produce small voltages which are measured by sensitive external voltage measurement devices. These transducers have the advantage that a voltage need not be applied to them, but they must nevertheless be used with caution since the measurement wires represent an electrical conducting path to the heart and make the patient microshock sensitive. Other piezoelectric transducers are used for microphone purposes in phonocardiography.

THE OXIMETER. As an example of a sensing element which uses light, one of the methods for determining the oxygen content of the blood will be considered. The oximeter is used to analyze a blood sample to determine the percentage of oxygen saturation of the hemoglobin (1, 5). A small sample of blood is illuminated with red light, and the reflected light from the blood enters a photoelectric cell. The amount of light reflected at that particular color is found to be dependent upon the oxygen saturation of the hemoglobin. Remarkably, the reflection is essentially independent of the total hemoglobin concentration and is determined by the ratio of the concentrations of oxyhemoglobin and hemoglobin. The photoelectric tube produces a small DC voltage which is proportional to the light intensity which reaches it. This voltage can be amplified, calibrated, and used to display the oxygen saturation percentage on a meter or recorder.

ELECTRODES FOR pH, P_{CO_2} AND P_{O_2}. Three of the most important parameters related to blood chemistry are the partial pressures of oxygen and carbon dioxide, P_{O_2} and P_{CO_2}, and the acidity or alkalinity of the blood, indicated by the pH. The direct determination of these quantities requires rather sophisticated techniques of physical-chemistry, but sensing elements have been developed which can provide electrical signals from which approximate measurements can be made.

A common pH electrode consists of a small glass bulb which is selectively perme-

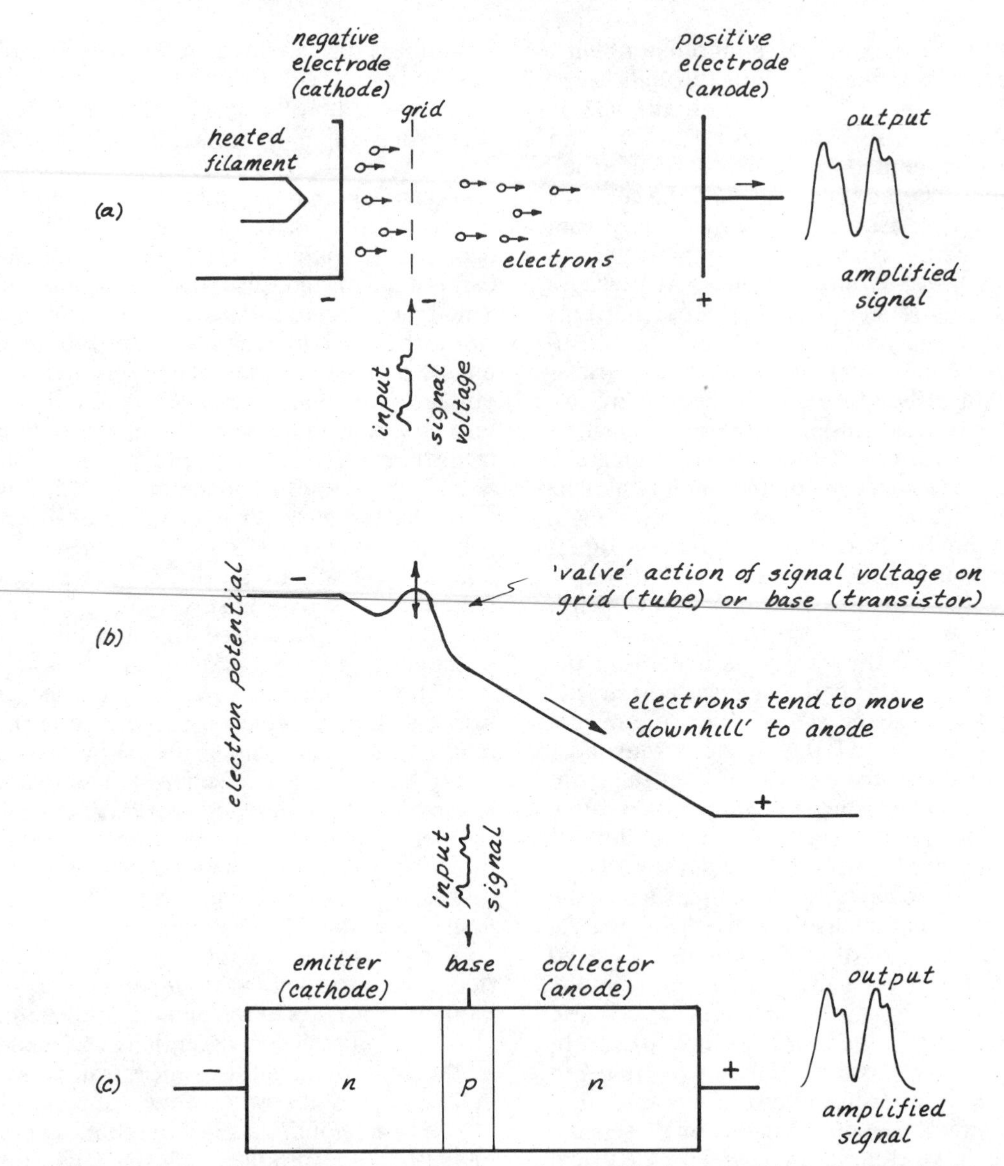

Figure 15–1 Amplification by a triode vacuum tube. (a) Schematic of tube components, showing sketch of amplified signal. (b) Illustration of "valve" action of the signal voltage on the grid. (c) Transistor.

able to hydrogen ions. Inside the bulb is a silver electrode which is coated with silver chloride and immersed in a buffer solution with pH = 1. When the electrode is placed in a liquid of unknown pH, the diffusion process across the permeable glass "membrane" produces a small voltage on the electrode. This voltage can be calibrated and used to determine pH's in the range from 0 to 8 with very small errors, and extended to a pH of 14 with special techniques. The pH of the blood depends upon the partial pressure of CO_2, and a modified pH electrode can be used to measure the P_{CO_2}. Alternatively, the ordinary pH electrode can be used to obtain the P_{CO_2}. This is done by first determining the pH of the blood, and then making two additional pH determinations on the blood sample after it has been allowed to come to equilibrium with gases of known P_{CO_2}, for example 30 mm Hg and 60 mm Hg. Since the P_{CO_2} of the blood will normally lie between these partial pressures, it can be determined graphically by interpolation between the known P_{CO_2} values. The partial pressure of oxygen can be measured by means of a platinum electrode embedded in glass, with one end exposed to the solution being analyzed. When a negative voltage of about -0.6 to -0.9 volt is applied to the electrode, oxygen is reduced by the electrode (4). The rate of reduction and the resultant current in the electrode are proportional to the oxygen partial pressure, Po_2. These processes are described in detail in Reference 4.

This discussion of sensing elements for physiological measurements has been of necessity somewhat superficial and incomplete. The variety and range of sophistication of the devices used are great. The point of this survey is to give some perspective concerning the types of devices used, and to emphasize that a wide variety of physiological parameters can be measured by producing proportional electrical "images" of the parameters. Blood flow, pulse rate, muscle contraction, and various measurements of body motion are examples of other physiological variables which can be measured with the aid of such electrical sensing devices or transducers. The references at the back of the chapter, particularly Reference 4, should be adequate for further study in the broad field of biomedical transducers.

AMPLIFIERS

A considerable amount of electric power is required to operate the meters and other display devices used to evaluate physiological data. Since the electrical output of most sensing elements (transducers) is much too small to be displayed directly, an amplifier is required. In addition to providing a power level adequate to operate the display device, the amplifier must produce a distortion-free replica of the original signal so that none of the information contained in the original is lost.

TRIODES AND TRANSISTORS. Transistors and triode vacuum tubes are the basic active elements in most amplifiers. Transistors and other solid state circuit components are usually smaller, cheaper, more reliable, and mechanically sturdier than their vacuum tube equivalents; therefore, tubes are now limited to applications in which they offer specific advantages. Nevertheless, the vacuum tube triode will be considered first since the basic aspects of amplification can be more easily explained with reference to vacuum tubes.

The triode vacuum tube contains a negative electrode (cathode), a positive electrode (anode), and a grid (Figure 15–1(a)). The cathode is heated to temperatures at which it gives off electrons by a process called thermionic emission. As the electrons are given off, they are repelled by the negative electrode and attracted by the positive electrode. Therefore, the electrons tend to move "downhill" to the positive electrode, causing a current to flow through the tube. If the grid is given a negative voltage, however, it forms a repulsive barrier which the electrons must surmount before they can move to the positive electrode (Figure 15–1(b)). Since the grid is close to the cathode, a small negative voltage can stop the flow of electrons through the tube. Within the range of grid voltages at which electrons are flowing through the tube, the relatively large flow of electrons through the tube can be varied over a wide range by a small control voltage on the grid. The grid then acts as the control element in an electrical "valve" which controls the large flow of charge through the triode. The current through the tube is supplied by a battery or a DC power supply operated from the AC power line and thus can be very large. The small signal from the physiological data transducer is placed on the grid to control this larger flow. The voltage output from the tube forms a large electrical replica of the small control voltage on the grid, as illustrated in Figure 15–1(a). This valve action accomplishes the desired amplification of the input signal. Several tubes can be placed in sequence so that the output of the one tube is applied to the grid of the successive tube. By successive stages of amplification, almost any desired amplification can be achieved, within the voltage limits of the tubes employed. Other practical limits are set by electrical "noise" and distortion, but amplification of signals by factors of several million is not uncommon.

The transistor is the solid-state analog to the triode vacuum tube. It also accomplishes the amplification of a small electrical signal by a valve type control of a large current from its power source. However, the physical mechanism is quite different. The electronically useful characteristics of the semiconductors silicon and germanium depend upon a small percentage of "impurity" elements such as antimony and indium which are introduced into the solid semiconductor crystals by controlled diffusion during the manufacturing process. An impurity such as antimony contributes an excess electron to the semiconductor crystal. A semiconductor crystal with a concentration of antimony is called a negative or n-type semiconductor. Indium atoms create a deficiency of electrons, and the resulting semiconductor is a positive or p-type semiconductor. If a thin layer of one type of semiconductor is sandwiched between two layers of the opposite type, the result is a transistor. These three elements of the transistor are called the emitter (cathode), base (grid), and collector (anode), as shown in Figure 15–1(c). If the collector is given a positive voltage with respect to the emitter, electrons tend to move across the thin base region to the collector. A small controlling voltage on the base region can cause large variations in the emitter-to-collector current. This valve action accomplishes the amplification of the signal as described above.

Other types of solid-state electronic devices make use of various combinations of n- and p-type semiconductors. A discussion of some other types of circuit elements can be found in Reference 6. Solid-state circuit elements can be formed by vacuum deposition of the materials on tiny substrates called "chips." Chips which have areas of a few square millimeters may contain thousands of transistors. This miniaturization capability may be looked upon as a result of the space exploration program. It has produced a continuing revolution in miniature electronics (8).

The amplification elements described above can amplify either a DC or AC signal, but the amplification of DC signals is more difficult. When a changing signal (AC) is applied to the control element of a tube or transistor, the entire change in the output voltage can be attributed to the signal. But when constant (DC) or slowly varying signals are applied, it is difficult to separate the part of the output which is due to the signal from that caused by stray voltages or changes in the power supply.

The design and application of amplifiers is a highly developed segment of our technology. The further development of the principles of amplifiers for biomedical application may be found in the references (1, 4, 6). As a component in an instrument package, the function of the amplifier is that of producing an enlarged, distortion-free replica of the output of the transducer. It must then couple that replica to a display device so that an accurate physiological measurement may be made with a reasonable expenditure of time and effort.

DISPLAY DEVICES

The numerical values in physiological data are usually taken from a display device such as an oscilloscope, a meter, a chart recorder, or a digital display device. The ease of reading such devices tends to deemphasize the fact that often intricate calibration procedures are required to produce the display. It must be kept in mind that in the sensing element, amplifier, and display device there are possible sources of error. Therefore, electronic instruments must be routinely calibrated to maintain their accuracy. It is the responsibility of the person making the measurement to note whether a numerical measurement is reasonable in the light of normal values for the parameter being measured. Extraordinary values should be verified by recalibration of the instrument or remeasurement with another instrument before drastic action is taken.

THE OSCILLOSCOPE. The cathode-ray oscilloscope is an extremely versatile display device for monitoring physiological data. The principles of operation of the oscilloscope have been described in Chapter 12. The oscilloscope can be used to display voltages from DC to very high frequency AC. Its particular advantages stem from the visual display of the voltage as it varies with time, rather than just a measurement of the average voltage. This is particularly valuable for use as a cardiac monitor, since the repetition rate and the detailed shape of the ECG voltage pattern are of interest. Further comments on the use of the oscilloscope as an ECG monitor will be made in Chapter 16.

Multiple displays on a single oscilloscope screen may be obtained either by using multiple electron guns at the back of the tube or by electronically switching a single beam at a very high rate so that a single beam produces several separately controlled displays. In this way a single oscilloscope can be used to monitor several physiological variables simultaneously. The monitor in an operating room may simultaneously display systolic and diastolic pressure, pulse rate, respiration rate, the ECG signal, and perhaps others. Display units which are capable of 10 simultaneous displays are commercially available.

METERS AND CHART RECORDERS. Probably the most common electrical display device is the moving coil meter. The principle of operation of such meters is based upon the fact that a current-carrying wire will experience a force if placed in a magnetic field. If a wire is placed between the poles of a magnet as shown in Figure 15–2(a) and a current I passes through it, it will experience a force as shown. The force is proportional to the current. If a coil of wire is placed in the magnetic field as shown in Figure 15–2(b), a force will be exerted on both sides of the coil. This will cause a torque which tends to rotate the coil. The torque is proportional to the current through the coil. If the coil is restrained by a spring, the amount of rotation of the coil will be proportional to the torque and therefore proportional to the current. If an indicating "needle" or pointer is attached to the coil, the position of the pointer can be calibrated in terms of the electric current. This basic arrangement is called a galvanometer, and it is the basis for all types of moving coil meters.

The moving coil meter is basically a current measuring device, but since both voltage and resistance are related to current by Ohm's law, $I = V/R$, moving coil meters can be adapted to measure voltage or resistances. If a large resistor is placed in series with the coil, the combination can be used as a voltmeter, since the current flowing through the coil will be proportional to the voltage across the combination. If a battery of known voltage is connected in series with the coil, then it can be used as a resistance meter (ohmmeter). In this case a high current corresponds to a low resistance and vice-versa, since they are inversely proportional. By including a battery and a series of calibrated resistors, a single moving coil can be used for multiple functions. The same coil can be used for AC and DC measurements, but the internal connections are different. With external switches, a given moving coil instrument can be used to measure AC or DC voltages and currents. This versatility and the relatively low cost of such a multimeter make the moving coil meter the display instrument of choice in many applications.

When a permanent record of an electrical signal as a function of time is required, the chart recorder is the logical display device. Paper from a roll is moved at a constant linear speed while an indicator pen makes a continuous record of an electrical signal. The motion of the pen perpendicular to the paper travel is proportional to the applied signal voltage or current. Since a given length of paper represents a time interval, very accurate time measurements between events can be made. One of the limitations of the chart recorder is its slowness of response. The oscilloscope display may be used to accomplish the voltage versus time display for events which happen too rapidly to be followed by the chart recorder pen.

DIGITAL DISPLAYS. With the development of miniaturized solid-state electronics, digital displays have become more feasible. Although several types of digital displays are available, the various types of light-emitting diode displays seem to be most practical at present for application to biomedical measuring instruments. They are small, rugged, and dependable and are easily interfaced to amplification circuits. With the use of solid-state digital circuitry, a signal voltage can be converted to a set of small voltages which are applied to the appropriate electrodes in the light-emitting diode arrays to produce the visi-

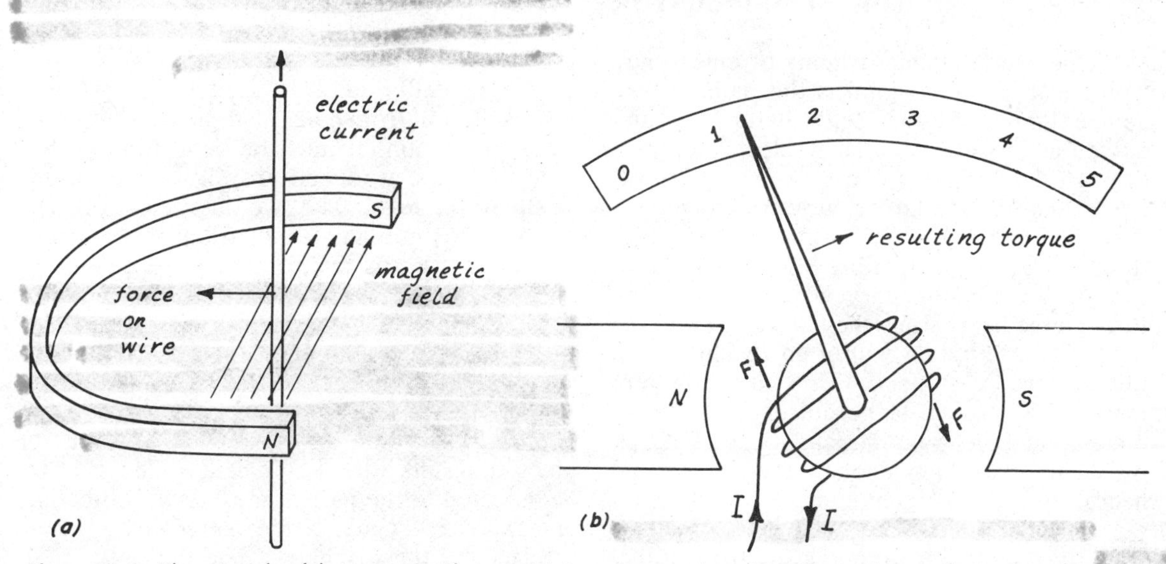

Figure 15–2 The principle of the moving-coil meter. (a) A current-carrying wire will experience a force when placed in a magnetic field. (b) A coil placed in a magnetic field will experience a torque which is proportional to the current flowing in the coil.

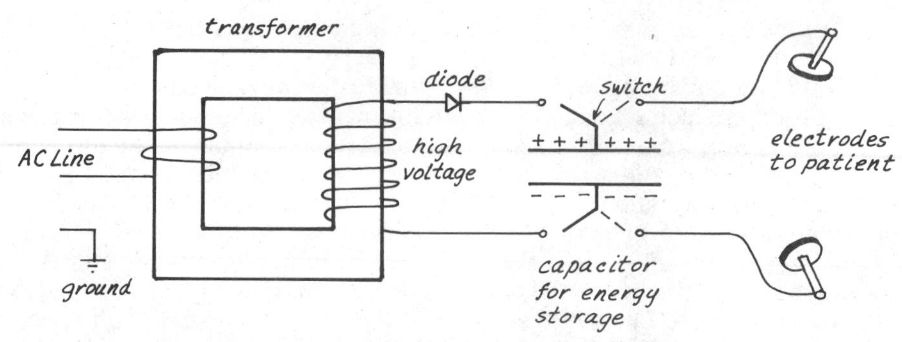

Figure 15–3 Simplified elements of a defibrillator.

ble arabic numerals corresponding to the voltage. The circuitry can sample the signal voltage at short intervals, producing a sequential digital display of the voltage which is practically equivalent to the continuous voltage display on a moving coil meter. Once the electrical signal is converted to digital form, it is easier to add logic or computing circuitry to perform other useful functions. For example, when the temperature probes of certain digital thermometers are placed in the mouth, a sequential sampling of the temperature is displayed. Associated circuitry senses when the temperature of the probe has reached its equilibrium value by the fact that the temperature is no longer changing. Some such thermometers then give a signal to indicate that the temperature has reached a stable value and can be recorded.

THE DEFIBRILLATOR

One of the standard items of emergency equipment in the hospital is the high voltage device known as the defibrillator. As discussed in Chapter 14, a small electric current through the heart can cause ventricular fibrillation, but a much larger current can cause a sustained ventricular contraction. When life-threatening ventricular fibrillation occurs from any cause, a momentary large current flow through the heart stops the fibrillation and the normal heart rhythm will often resume when the current is stopped. This rather severe method has proved to be quite successful in interrupting the random, chaotic activity of the heart cells long enough for the normal heart impulses to regain control.

The defibrillator is basically a large capacitor. It is similar in principle to the parallel plate configuration discussed in Chapter 12. Charge is stored on the plates of the capacitor by a high voltage, usually around 7500 volts but sometimes as high as 10,000 volts (3). This charged capacitor has stored energy due to the charge, typically up to 400 joules (often denoted as 400 watt-seconds on the instruments). The high DC voltage on the capacitor is obtained from the AC line voltage by means of a transformer and a diode as shown in Figure 15–3. The transformer produces a high AC voltage, but the diode permits current flow only to the right in Figure 15–3 and therefore converts it to DC voltage. This contributes positive charge to the upper plate and negative charge to the bottom plate of the capacitor. Large electrostatic forces tend to make the capacitor discharge, but this cannot occur until some conducting path is provided between the electrodes. When the electrodes are placed on the patient's chest and the switch is closed, the capacitor discharges through the patient's body. The duration of the discharge current is typically a few milliseconds. Then the capacitor must recharge for several seconds before it can be used again.

The defibrillator is used for emergency resuscitation from a heart stoppage or ventricular fibrillation. It must be kept in mind that the currents produced are above the potentially lethal level. The two large electrodes (3 to 5 inches in diameter) distribute the current to prevent burns due to high current density. Both electrodes are isolated from ground during the discharge; therefore, any nearby grounded object could conduct a current. It is thus important that the patient not be in contact with a grounded metal bed frame and that no one touch him during the discharge. ECG monitors and other equipment which may be connected to the patient must have protection mechanisms against the surge of current caused by the discharge. The discharge may be triggered manually or synchronized with the ECG signal as discussed in Chapter 16.

ELECTROCAUTERY AND ELECTROSURGERY

Electrosurgical instruments are used extensively for either cutting tissue or welding tissue together. These instruments generate high frequency currents at voltages up to 15,000 volts (3). As discussed in Chapter 14, the physiological effects of electric currents are strongly dependent upon the frequency. The currents involved in electrocautery and electrosurgery could be lethal if they were 60 Hz currents, depending upon the path through the body. But the frequencies are too high to cause ventricular fibrillation in normal usage, and the possibility of burns is the main electrical danger. In terms of muscle contraction, and presumably the tendency toward ventricular fibrillation, the body is most susceptible to shock in the region from 10 to 100 Hz, and the sensitivity drops off quite rapidly for high frequencies (see Figure 14–1).

When the high frequency voltage is applied to a sharp metal probe or blade, it becomes an effective cutting tool. There is some tendency toward muscle contraction, but the main effect is to produce a local burn. Frequencies on the order of 500 to 600 KHz have been found to be effective for cutting operations, and higher frequencies in the range from 2 to 4 MHz are more effective for coagulation and cautery (7). The commercially available instruments usually supply frequencies in each of the ranges. The high current density caused by several amperes of current entering the body through the small area of the probe accounts for the cutting and cauterizing action. Precautions must be taken to ensure that the current doesn't leave the body through a small area, since that would cause severe burns at the exit point. A large conducting buttock plate is usually placed under the patient to provide a large area for the exit current. Suitable precautions must be taken to prevent current flow through small ECG leads, for example, which could cause burns at those points.

SUMMARY

Electronic instruments for physiological measurements usually consist of a sensing element which produces an electrical image of the physiological variable, an amplifier to increase the size of the electrical signal, and a calibrated display device from which the measurements can be taken. In the case of measurements such as the ECG and EEG, a voltage produced by the body is measured, but with most other measurements a nonelectrical physiological variable is converted to an electrical signal by a sensor which is often called a "transducer." For the measurement of body temperature, thermocouples, platinum resistance elements, or thermistors are usually used as sensors. Various types of pressure transducers are used for the measurement of blood pressure. Sophisticated electrodes have been developed for the measurement of the pH, P_{CO_2} and P_{O_2} of the blood.

Amplifiers make use of the energy available from an external power supply to produce a faithful, enlarged replica of the original electrical signal. This is basically accomplished by a "valving" process whereby the small electrical signal controls a larger electric current flow through an active electronic device like a vacuum tube or transistor. Amplifications by factors of several thousand are common, but such amplifications are much easier for AC signals than for DC signals. The amplified signals may be displayed on an oscilloscope, a meter, a chart recorder, or a digital display device, or they may be interfaced to a computer for further refinement of the data.

Electronic devices have become important treatment tools in the hospital. The defibrillator has become a standard operating room accessory for the treatment of ventricular fibrillation. It consists of a high voltage transformer and a rectifier which store energy on a large capacitor. When this capacitor is discharged through the patient, it causes a generalized contraction of the heart which often allows the normal heart cycle to be restored. Also in common use is a high voltage, high frequency generator which supplies currents for electrocautery and electrosurgery.

REVIEW QUESTIONS

1. What are the advantages of thermistor probes over other types of temperature sensors?

2. Explain how an amplifier is similar to a valve.

REFERENCES

1. Camishion, R. C. *Basic Medical Electronics*. Boston: Little, Brown and Co., 1964.
2. Stacy, R. W. *Biological and Medical Electronics*. New York: McGraw-Hill, 1960.
3. Strong, Peter. *Biophysical Measurements*. Tektronix, Inc., Beaverton, Oregon 97005, 1970.
4. Geddes, L. A., and Baker, L. E. *Principles of Applied Biomedical Instrumentation*, 2nd ed. New York: Wiley, 1975.
5. American Optical Corporation. *Instruction Manual for the Reflection Oximeter*. Medical Division, Grosby Drive, Bedford, Mass. 01730.
6. Diefenderfer, A. J. *Principles of Electronic Instrumentation*. Philadelphia: W. B. Saunders Co., 1972.
7. Dummer, G. W. A., and Robertson, J. M., eds. *Medical Electronic Equipment, 1969–70*. New York: Pergamon Press, 1970.
8. Hittinger, W. C. *Metal-Oxide-Semiconductor Technology*. Scientific American, August 1973, p. 48.

CHAPTER SIXTEEN

Bioelectricity

INSTRUCTIONAL OBJECTIVES

After studying this chapter, the student should be able to:

1. State and describe the three basic influences which determine the relative ion concentrations inside and outside the membranes of living cells.
2. Sketch a typical action potential for a cell membrane and describe what is happening to the sodium, potassium, and chlorine ions during each stage of the action potential.
3. Identify the basic features of the electrocardiogram in terms of the action potentials of the heart.
4. Describe the action of the electronic pacemaker in terms of its effects upon the action potentials of the heart, and compare its effects with those of the natural pacemaker.

Bioelectricity refers to electrical phenomena in living tissues. It has long been known that small electric currents are generated in tissue and that the conduction of electric impulses is important in the functioning of nerves. The external measurement of electric signals generated by the body is the basis for several diagnostic techniques, such as the ECG and EEG. The pacemaker is an example of an instrument which simulates normal bioelectric potentials when the body is not capable of producing them.

THE LIVING CELL AS AN ELECTRIC SOURCE

The use of the same word, "cell," to refer to a small battery and to the functioning unit of living tissue is somewhat appropriate, since the living cell can generate a small voltage. To be more specific, the membrane of the living cell can maintain a voltage difference between the inside and outside of the cell. This is analogous to the charged parallel plate arrangement discussed

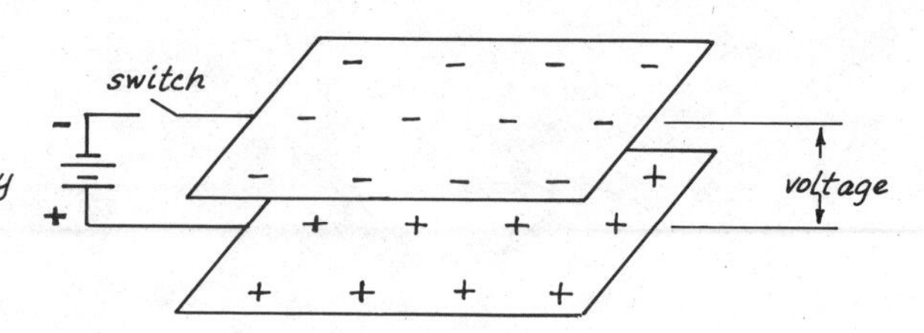

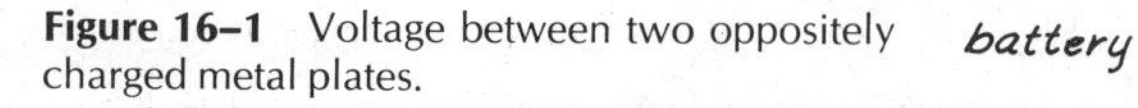

Figure 16–1 Voltage between two oppositely charged metal plates.

in Chapter 12. If unlike charges are placed on parallel conducting plates as shown in Figure 16–1, a voltage will exist between the plates. Energy is required from a battery or other source to establish this unequal charge distribution. If the switch in Figure 16–1 is opened after charging the plates, the charge will remain at rest on the plates because it has no available discharge path. However, because of the mutual repulsion of the like charges and the attraction toward the unlike charge on the opposite plate, the charge has potential energy and would move if a conducting path between the plates were available. As discussed in Chapter 12, this electric potential energy per unit charge is measured in volts (joules/coulomb).

When a living cell is in its normal or "rest" state, it maintains a voltage of about 70 to 90 millivolts between the inside and outside of the cell (1,2). The inside of the cell is negative with respect to the outside. This voltage across the cell membrane is referred to as the "membrane potential" or "rest potential" of the cell. The origin of this voltage is considerably more complicated than that of the parallel plate arrangement of Figure 16–1, but the analogy may be helpful.

The behavior of the electrolytes potassium, sodium, and chloride is one of the major keys to the origin of bioelectricity. The salts potassium chloride (KCl) and sodium chloride (NaCl) are dissociated in solution, and the charged ions K^+, Na^+, and Cl^- constitute mobile charge carriers. These electrolytes are present in varying concentrations inside and outside the cells. Their movements through the cell membranes are governed by three main influences: (1) the tendency toward diffusion from a higher to a lower concentration (Chapter 9), (2) the tendency to move away from like charges and toward unlike charges, and (3) the permeability of the membrane to the particular ion.

The resting state of the cell membrane is a result of the balancing of opposing influences, as illustrated in Figure 16–2. Large concentration gradients exist across the membrane for each of the three electrolytes. Potassium ions are more concentrated inside the cell by a factor of roughly 30 to 1. The sodium and chloride ions are more concentrated outside, with concentration ratios of roughly 10:1 for Na^+ and 20:1 for Cl^-. When the negative voltage (rest potential) is established across the membrane, the electrical forces oppose the migration of the K^+ and

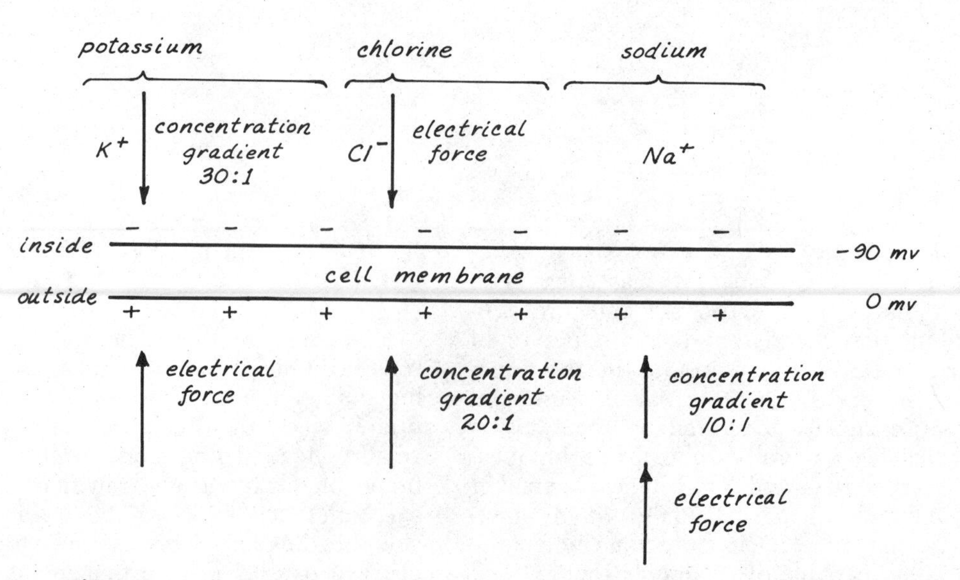

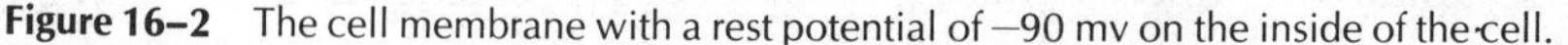

Figure 16–2 The cell membrane with a rest potential of −90 mv on the inside of the cell.

Cl^- ions down their concentration gradients. Although the membranes are permeable to these ions, they reach an equilibrium state in which a large concentration gradient exists. Both the concentration gradient and the electrical force tend to move sodium ions into the cell (Figure 16–2), but the permeability of the membrane to Na^+ is very low during the rest state. Despite the low permeability to Na^+, some of these ions must penetrate the membrane. An active transport mechanism referred to as the "sodium pump" is postulated (1) to move an equivalent number of Na^+ ions outward against the electric and diffusion influences to maintain the rest potential.

Energy is required to establish the electric potential energy associated with the rest state of the cell membrane. This is directly analogous to the charging of the parallel plates in Figure 16–1. The availability of this energy is a characteristic of living tissue (3); the potential cannot be generated when the cell is dead. Once the potential is established, it will discharge if a suitable electrical conduction path is made available.

THE ACTION POTENTIAL. When a cell is sufficiently stimulated, it "fires" or releases some of the stored energy. The interior potential of the cell quickly rises from about −90 millivolts to about +20 to +30 millivolts. This process is called depolarization. The repolarization process begins immediately and builds the interior potential back to the −90 millivolt rest potential. The voltage pulse produced by the depolarization-repolarization process is referred to as the action potential (Figure 16–3). The depolarization process is closely related to the conduction of sodium ions into the cell. During depolarization the cell membrane becomes permeable to Na^+ and the sodium rushes into the cell, driving the potential positive. The threshold requirement for stimulating a cell involves increasing the permeability to Na^+ sufficiently for the influx to start. As indicated in Figure 16–2, both the electrical force and the concentration gradient tend to make the system unstable toward inward Na^+ diffusion; once the influx starts, it is self-sustaining until it is counteracted by the outward migration of potassium ions. The lowered electrical force allows potassium ions to migrate outward. With energy supplied by the cell, the potassium diffusion process takes control for the repolarization or "recharging" of the cell membrane.

The sequence of cell membrane events during the production of an action potential is shown in Figure 16–4. This sequence must be completed, with the membrane restored to its high energy polarized state, before another action potential can be generated. In response to a strong, continued stimulus, several hundred action potentials per second can be generated.

In the case of large cells, such as nerve cells with long extensions called axons, an action potential can be generated in one part of a cell and transmitted to other parts. This involves the successive stimulation of neighboring membrane areas so that the depolarization-repolarization pulse propagates through the cell. The electrical impulse can also be propagated across tissue as one cell stimulates neighboring cells. It must be

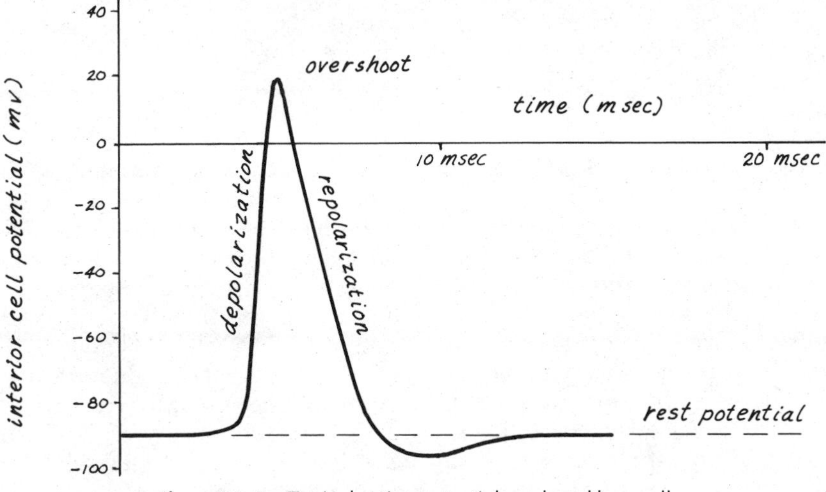

Figure 16–3 Typical action potential produced by a cell.

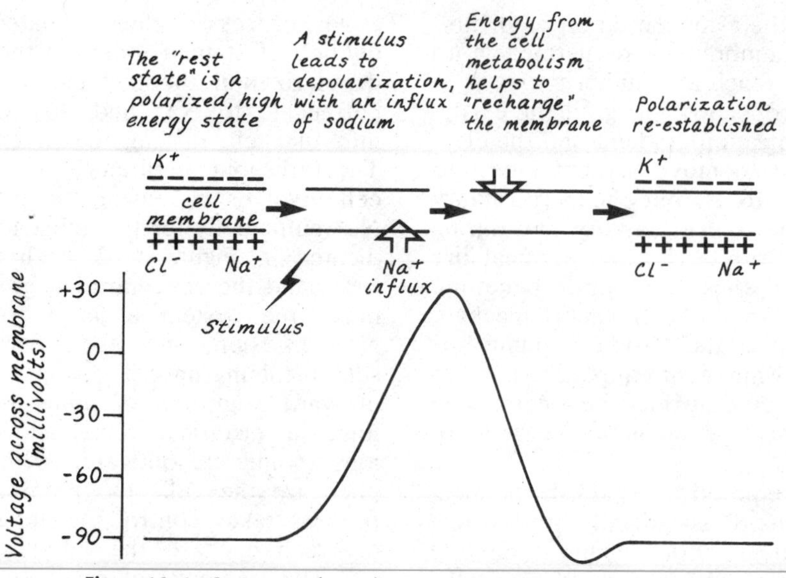

Figure 16–4 Sequence of membrane events during an action potential.

noted that this electrical conduction process is quite different from the conduction of electric currents in wires. The electron conduction in wires results in currents which travel at essentially the speed of light (186,000 miles/sec), but the conduction of the action potential pulses along nerves varies speed from about 1 to 100 meters/sec (100 m/sec = 224 miles/hr) (see Reference 1). The impulses travel most rapidly in the large neurons, such as those in the spinal column; the speed is reduced by about a factor of 100 in the smallest nerve fibers.

The propagation of the action potential pulses by nerve fibers and other cells produces measurable electrical voltages at all points in the body. The fact that these action potentials can be measured at the surface of the skin provides the basis for many bioelectric measurements, such as the ECG and EEG.

THE ELECTROCARDIOGRAM

The most widely used of the bioelectric measurements is the electrocardiogram (ECG). It is a direct measurement of voltages produced by the body and therefore does not involve a transducer. The action potentials produced in the heart result in measureable voltages at the skin which can be monitored by external electrodes. These are the largest action potentials measured on the body, producing voltages on the order of 1 mv between ECG leads. Therefore, the ECG signal is much easier to record and measure than the much smaller signals associated with the electroencephalogram (EEG) and other bioelectric measurements. The details of the collection and interpretation of ECG data are discussed thoroughly in the references at the end of the chapter and in other medical literature. The treatment here will be limited to a brief survey.

The pumping cycle of the heart is initiated by electrical impulses generated in a small specialized spot of tissue in the right atrium known as the sinoatrial node (SA node). This SA node constitutes the natural "pacemaker" of the heart. Data from parts of the nervous system external to the heart can cause the SA node to respond to increased or decreased demand for blood, but in the absence of external information, the SA node has its own rhythm which normally controls the heart rate. This series of electrical impulses from the SA node can be thought of as self-stimulated action potentials. When an action potential of the SA node has been completed and its polarization has been restored to a certain threshold state, the SA node spontaneously "fires" again and repeats the sequence. The action potential from the SA node is the first step in an electrical conduction process which controls the pumping action of the heart. This conduction process is illustrated in Figure 16–5. The SA node stimulates atrial contraction, and the impulses travel to the atrioventricular node (AV node). The consequent depolarization of the AV node causes electrical impulses to travel to the myocardium (heart muscle) via a special conducting system. This conducting system is composed of a bundle of fibers known as the bundle of His, which branches to go to the individual ventricle via smaller conducting systems known as the Purkinje fibers. This conducting system provides for almost simultaneous stimulation of all parts of the ventricles so that the ventricles contract sharply for effective pumping action.

The normal heart action is thus dependent upon the generation and conduction of electrical impulses over specified paths within a small time period. Since these electrical impulses are also conducted to the surface of the skin, the ECG offers a means for monitoring the action of the heart. The size, shape, and time sequence of the impulses reaching the skin provide a large amount of diagnostic data. Figure 16–6 shows a typical ECG trace and its correlation to the arterial blood pressure. The various components of the ECG pattern are labeled P, Q, R, S, and T as indicated in Figure 16–6. The P wave is associated with the depolarization of the SA node and marks the initiation of the heart action. The resultant action potential normally requires from 120 to 220 milliseconds to travel to the AV node (2). The subsequent depolarization of the AV node and conduction to the ventricles produces the QRS complex. The repolarization of the AV node and the ventricular conduction system produces the T wave. The identification of the features of the ECG pattern with specific cardiac events makes it an extremely valuable diagnostic tool. The discussion of the interpretation of abnormal ECG's is outside the scope of this book, but numerous review articles are available (4).

The ECG signal can often be used in conjunction with a defibrillator to correct certain cardiac disorders. When a defibrillator is used in the synchronized mode, it is coupled to the ECG monitor. The defibrillator discharge doesn't occur immediately when the discharge switch is closed, but is delayed to occur during the descending part of the R wave. This synchronized use of the

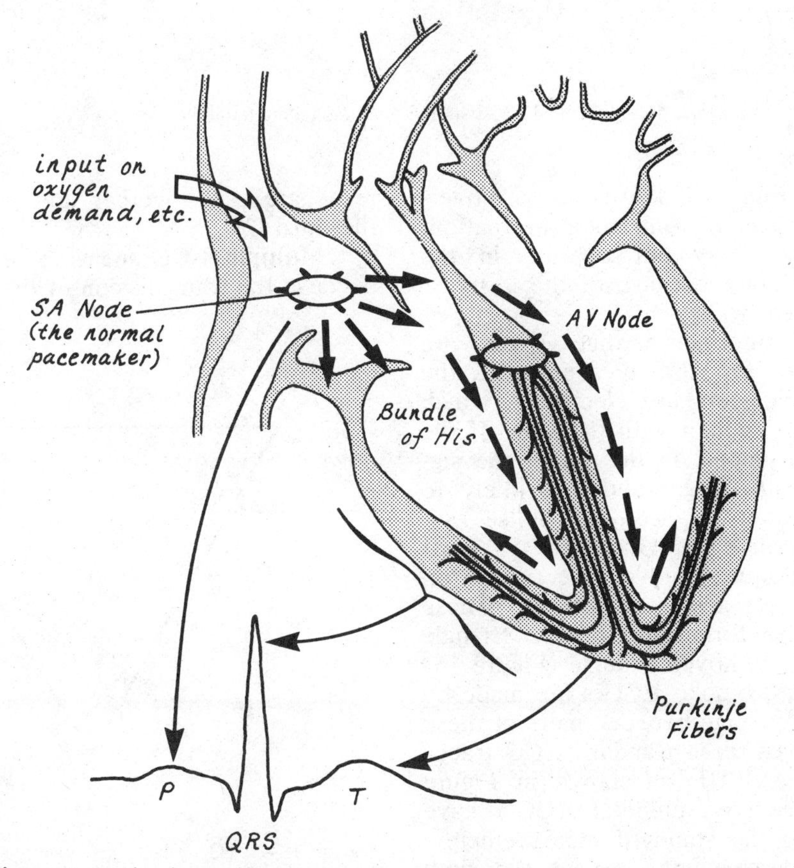

Figure 16–5 The electrical conduction process which controls the heart's pumping cycle.

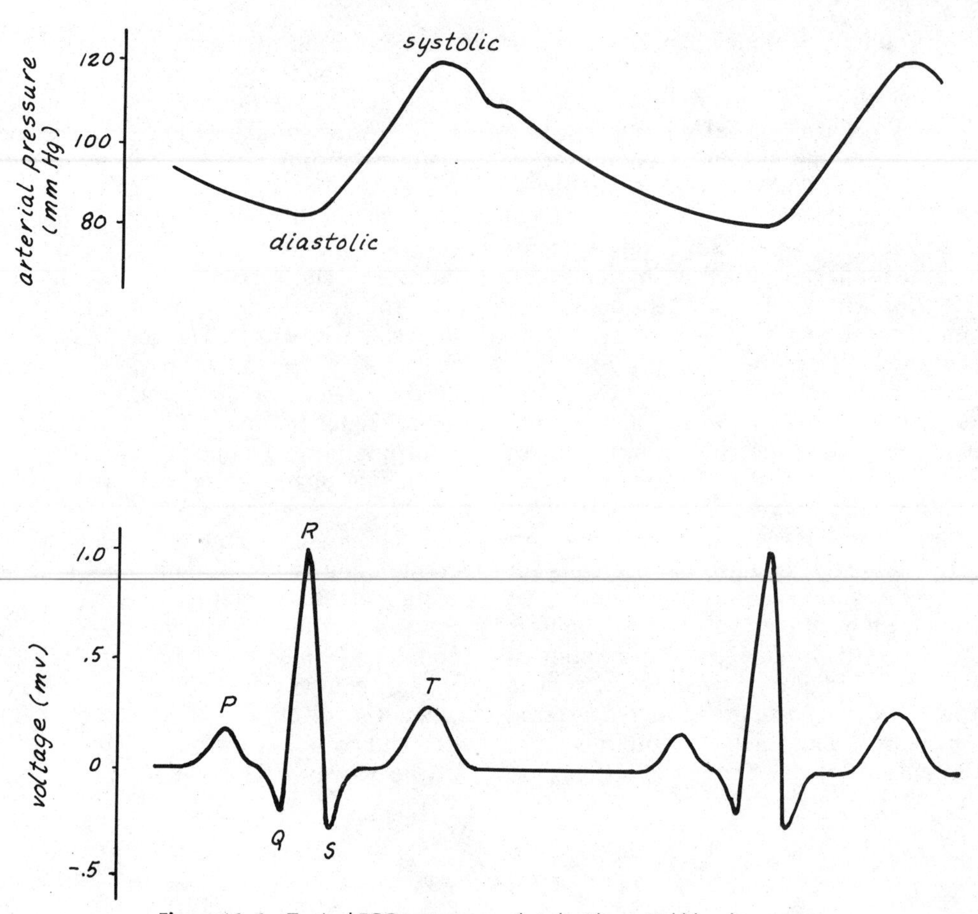

Figure 16–6 Typical ECG pattern correlated with arterial blood pressure.

ECG and defibrillator is known as cardioversion. In the case of ventricular fibrillation, the defibrillator may not operate in the synchronized mode because of the absence of a detectable R wave.

Although internal cardiac wires are sometimes used for ECG measurement, the usual diagnostic tests use electrodes which make electrical contact with the skin. If the electrodes are placed on the chest, the signals obtained are larger and less likely to have large amounts of electrical "noise" superimposed upon them. However, the most common technique involves electrodes on the right and left arms and left leg. These three electrodes form an effective triangle known as the Einthoven triangle (Figure 16–7). The standard bipolar ECG's are made by measuring the voltage between pairs of these leads. This gives three standard ECG traces labeled I, II and III, as shown in Figure 16–7. More recently, unipolar ECG's have been added to the standard measurements; for these measurements two of the limb leads are tied together and compared with the third (2).

Multiple ECG measurements are made because the transmission of the action poten-

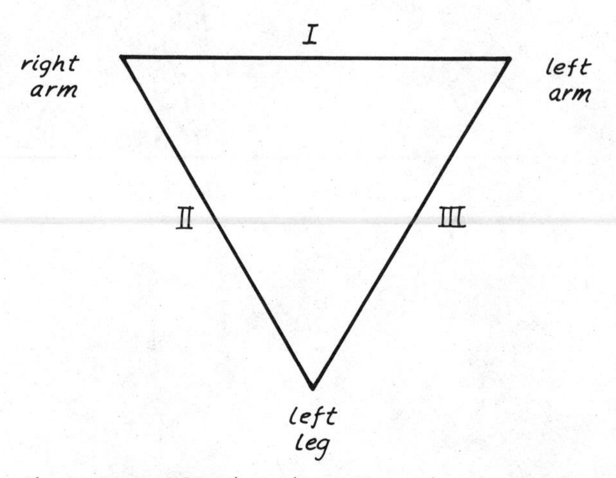

Figure 16–7 ECG electrode positions; the triangle of Einthoven.

tials in the heart is a directional or vector process. The action potential is conducted along a preferred direction. The cardiac vector is associated with the electric field vector produced by the instantaneous charge distribution in the heart during this conduction process. The measurement of the external manifestation of the action potential with a given pair of electrodes gives an indication of one component of the vector. Two independent measurements yield the components of the vector along two directions, and the orientation and relative magnitude of the vector can be deduced. Early attempts at determining the cardiac vector made use of ECG's from electrode pairs on paths which differed in direction by 90° (Figure 16–8(a)). In this way the vector magnitude and direction could be obtained from the Pythagorean relation and simple geometry. With the use of the triangular lead configuration as in Figure 16–8(b), either on the chest or on the extremities, the two directions used will differ by approximately 60°. The resultant vector can be obtained by the vector addition methods discussed in Chapter 3. The resulting "frontal plane cardiac vector" is not a faithful representation of the action potential conduction direction because of distortions introduced by conduction through the body tissue, but it has proved to be useful as a diagnostic aid. The present discussion has been limited to the frontal plane, but other types of vector ECG's are used for research and diagnostic purposes (2,5).

THE ELECTROENCEPHALOGRAM

The electroencephalogram (EEG) is a recording of electrical signals produced by the brain. Usually the measurements are made from electrodes placed on the scalp. The signal strengths involved are much lower than those involved in the ECG. The voltages measured are on the order of 50 microvolts, compared to about a millivolt for the ECG. Consequently, the signals must be amplified by factors of several thousand to be recorded. This, of course makes the problems of stray signals and electrical noise much more severe than with the ECG.

The EEG signals cannot be correlated with specific brain activity as precisely as the ECG pulses can be related to the heart cycle, but many years of experimentation have produced definite correlations with the physiological state. The signals are composed of roughly periodic oscillations of varying frequencies. Much of the available information is contained in the frequencies of the waves. To facilitate analysis, the various frequencies have been divided into bands as indicated in Table 16–1. The alpha waves are associated with the relaxed but alert state. During sleep the alpha waves decrease, and the presence of delta waves indicates deep sleep. The delta waves are strong and dominant when the patient is in a comatose state. Further interpretation can be made in terms of the size and shape of the waveforms. The EEG has been of consider-

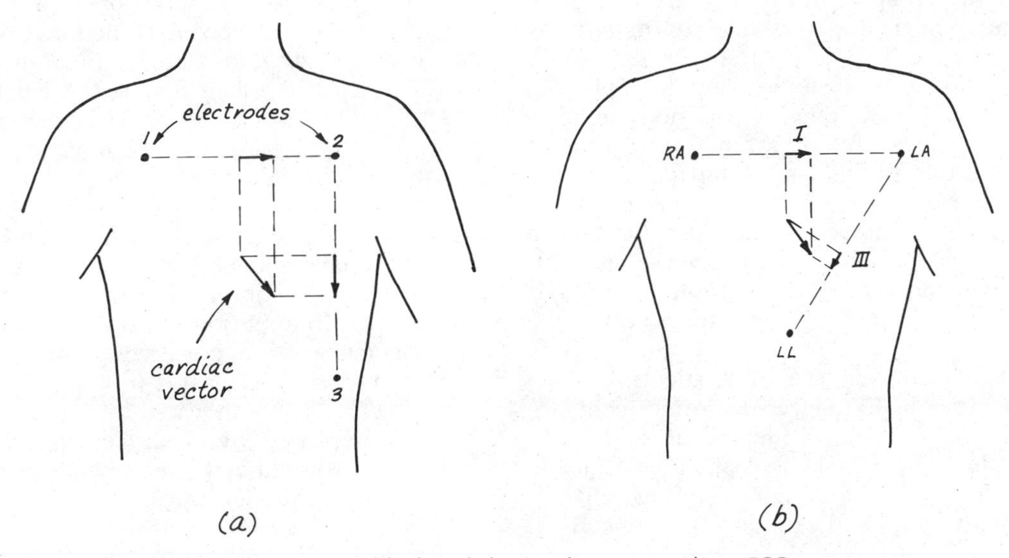

Figure 16–8 Determination of the frontal plane cardiac vector with two ECG measurements.

TABLE 16–1 Electroencephalogram Frequency Bands.

Band Designation	Frequency (Hz)
delta (δ)	0.5–4
theta (θ)	4–8
alpha (α)	8–13
beta (β)	13–22
gamma (γ)	22–30

able value in diagnosing epilepsy, tumors, certain forms of drug addiction, and various other brain-related diseases. Since there is normally a continuous activity in the brain as indicated by the EEG, even during deep sleep, the absence of any EEG signal over a long period of time has been used as one criterion for death.

OTHER BIOELECTRIC MEASUREMENTS

A measurement of electric potentials produced by muscle fibers is referred to as an electromyogram (EMG). The measurement may be made from small needle electrodes inserted into the muscle or from surface contact electrodes. A voluntary contraction of a normal muscle will produce a series of electrical impulses on the order of 1 millivolt from the motor units (6). Muscles in which the nerves have been damaged or destroyed may exhibit a fibrillation pattern of smaller pulses of shorter duration. The technique is of value in diagnosing certain neuromuscular disorders and in monitoring the nerve recovery in injured muscle tissue.

If a high frequency voltage is applied to two or more electrodes on the body and the current is measured as a function of time, it is found that the electrical impedance of the body changes in response to many physiological events. Impedance changes have been used to record data on such diverse phenomena as endocrine activity, autonomic nervous system activity, and respiration rate (5).

The increased use of bioelectric measurements is likely, since electronic technology has reached a high degree of sophistication. Such measurements can now be made with ease and with a high degree of reliability. With proper precautions for electrical safety, such measurements offer increasing advantages as diagnostic aids. It is also possible to use the action potentials from certain controlling nerves to activate motors in prosthetic devices, thus permitting the direct control of such devices by impulses from the brain.

THE ELECTRONIC PACEMAKER

Since many physiological events are stimulated by biologically generated electric impulses, it seems logical that those events could be stimulated by equivalent but externally generated impulses. The most familiar example of the success of artificially generated impulses is the operation of the cardiac pacemaker. The SA node is the natural pacemaker, but when it is disabled or if the conduction system which carries the stimulating impulses to the ventricles is blocked, the synchronous action of the heart is destroyed. The more common cause is a blockage of the impulses. Then the atria contract in response to the SA node but the ventricles contract in a slower, independent rhythm on the order of 30 to 40 beats/minute. The artificial pacemaker is a battery-powered device which generates electrical stimuli at a predetermined rate. Typically these pulses will be on the order of 10 volts with a duration of a few milliseconds and a repetition rate of 60 to 70 per minute. The electrodes make contact with the myocardium, and the arrival of an electrical impulse of sufficient size will cause the entire heart to contract (7). More sophisticated pacemakers operate on a standby basis and become functional only if the normal rhythm falls below a certain rate.

The cardiac pacemaker is the most successful example of an artificial implanted device. The pacemaker is implanted subcutaneously in the abdominal region, and the leads connect it to the heart. The electronic package is easily miniaturized and will operate almost indefinitely. The major problem is the provision of a power supply which will last for long periods of time. Most present pacemakers are powered by small mercury batteries which last 2 to 4 years. When the batteries are exhausted, the pacemaker must be replaced by a surgical procedure. The surgery is minor since it is not necessary to replace the electrodes, but it would be desirable to have a power source which would last 10 to 20 years. Active research is being carried out to develop a suitable long-lived power source.

The types of power sources which show promise as alternatives to the mercury batteries are nuclear fuel sources, piezoelectric sources, bio-galvanic sources, and radio frequency power transmitters (4). The nuclear fuel element produces heat at the junctions of a large number of thermocouples which produce the needed voltage. The piezoelectric sources are powered by piezoelectric crystals, which produce a voltage when subjected to deformation forces. These devices are implanted in positions such that the expansion of the aorta or of the heart provides the necessary deformation. The bio-galvanic sources make use of the fact that body fluids are electrolytes and can form part of a "battery" if suitable electrodes are supplied. In this way the body can supply the necessary energy to operate the source as long as the electrodes of the bio-galvanic source remain intact. An alternate approach to the power source problem is offered by the radio frequency (RF) transmitter method. In this case the battery and electronic package are outside the body. The necessary impulses are generated in the form of high frequency signals which are transmitted to an implanted receiver attached to the cardiac electrodes. This receiver detects the transmitted pulse and applies it to the heart. Alternatively, the transmitted power can be used to recharge a battery for operation of the pacemaker.

The electrical nature of many of the controlling influences in the body offers a large range of possibilities for electrical stimulation or control by means of externally generated impulses. Various types of electrostimulation are now employed, and the variety of such devices is increasing.

SUMMARY

The membranes of living cells store electrical energy by maintaining the insides of the membranes at a small negative voltage with respect to the outside. This rest potential is maintained basically by controlling the concentration gradients of potassium, sodium, and chlorine ions. The membrane normally withstands a strong electrochemical gradient tending to push Na^+ into the cell, but an appropriate stimulus renders the membrane permeable to sodium. The influx of sodium causes an electrical change known as depolarization and the reinstatement of the rest potential (recharging) is known as repolarization. This entire process takes a few milliseconds and produces an electrical pulse known as the "action potential" of the cell. The action potentials of nerve cells and other types of cells are basic to the communication and control systems of the body. These action potentials propagate with greatest efficiency along the nerve networks, but also propagate through ordinary tissues. The measurement of heart action potentials which have propagated to the skin (electrocardiogram) is extremely useful as a diagnostic aid. Electroencephalograms, electromyograms, and other such measurements have proven to be medically useful.

The possibilities for stimulating parts of the body with artificially generated electrical impulses are being actively explored. The electronic pacemaker for the heart has been the most widely used artificial stimulation device. A regularly produced electrical impulse applied to the myocardium can cause the ventricles to contract when the normal biological pacemaker signal from the SA node is absent or blocked.

REVIEW QUESTIONS

1. How does the electrochemical gradient for sodium ions across the resting cell membrane differ from that for the potassium and chlorine ions? How can you account for this difference?

2. Once a cell membrane has been stimulated and the sodium influx has started, what causes the potassium ions to start diffusing outward?

3. Explain the terms depolarization, repolarization, and action potential.

4. After a nerve cell has "fired" or produced an action potential, why is there a waiting period before it can fire again?

5. How do the action potentials of the heart get to the surface of the skin to produce the electrocardiogram signals?

REFERENCES

1. Langley, L. L. *Physiology of Man*, 4th ed. New York: Van Nostrand-Reinhold, 1971.
2. Strong, P. *Biophysical Measurements*. Tektronix, Inc., Beaverton, Oregon 97005.
3. Solomon, A. K. *Pumps in the Living Cell*. Scientific American, Aug 1962, p. 100.
4. Butler, H. H. *How to Read an ECG*. Part I: RN Magazine 36, No. 1, Jan 1973, Part II: No. 2, Feb 1973, Part III: No. 3, Mar 1973.
5. Geddes, L. A., and Baker, L. E. *Principles of Applied Biomedical Instrumentation*, 2nd ed. New York: Wiley, 1975.
6. Bell, G. H., Davidson, J. N., and Scarborough, H. *Textbook of Physiology and Biochemistry*, 9th ed. Baltimore: Williams and Wilkins, 1976.
7. Nose, Y., and Levine, S. N., eds. *Cardiac Engineering*, Advances in Biomedical Engineering and Medical Physics, Volume 3. New York: Wiley-Interscience, 1970.

CHAPTER SEVENTEEN

Elasticity and Wave Motion

INSTRUCTIONAL OBJECTIVES

After studying this chapter, the student should be able to:

1. Define elasticity and describe how elasticity leads to sustained periodic motion and traveling waves in elastic media.
2. Obtain values for the amplitude, frequency, period, and wavelength of a traveling wave, given a plot of displacement as a function of time or distance and appropriate numerical data.
3. State the distinguishing characteristics of longitudinal and transverse waves and state the appropriate category for sound and light waves.
4. Calculate the intensity in watts/cm^2 produced at a point in space by a localized wave source, using the inverse square law for intensity.
5. Define interference and resonance.
6. Describe the Doppler effect and give an example of its use in SONAR type detection systems.

Sound and light are vastly different physical phenomena, but they have in common the fact that both involve traveling waves. The purpose of this chapter is to describe periodic motion and traveling waves, including a discussion of the forces which produce them. The vocabulary necessary for the more detailed description of sound and light is developed in the process.

ELASTICITY

When a spring or a rubber band is stretched, it tends to return to its original length when released, thereby demonstrating elasticity. Elasticity is the property of an object which tends to restore it to its original dimensions after the distorting forces are removed. Most solids exhibit "restoring forces" and tend to regain their shape after small distortions. If the distorting forces exceed the elastic limit, then a permanent deformation will occur. One substance is said to be more elastic than another if it returns to its original shape with greater precision; elasticity is not necessarily correlated with ease of deformation. For example, glass is one of the most elastic of common substances within its elastic limit. Though brittle, glass has the ability to withstand hundreds of deformations and retain its original shape.

Within the elastic limit, it is observed that the deformation of an elastic solid is proportional to the deforming force. In other words, if the stretching force is doubled, the object will stretch twice as far. This linear relationship is called Hooke's law after the seventeenth century English physicist, Robert Hooke. For a one-dimensional motion like the stretching of a spring, Hooke's law can be written

$$F = kx \qquad \textbf{17–1}$$

where F is the deforming force, x is the amount of stretch, and k is a numerical proportionality constant.

PERIODIC MOTION

If a spring is hung vertically and a mass is suspended from the end of it, the spring will stretch until the elastic restoring force is great enough to support the weight of the object. If the mass is now pulled downward and released, it will oscillate up and down in "simple harmonic motion." It cannot settle immediately to rest at the equilibrium point because of the energy which has been given to it. It will move up and down periodically until all that energy has been dissipated into internal energy (disordered molecular motion in the spring and surrounding air). This periodic motion is characteristic of elastic objects moving under the influence of Hooke's law type forces. Since almost all solids obey Hooke's law when they are deformed slightly from their equilibrium configurations, vibrations similar to those of the mass on the spring are very common in nature. Even on the molecular level, the vibrations of atoms in solids vibrate in an approximation of simple harmonic motion. This is the basis for the spring type solid model in Figure 10–1.

Periodic motion is defined as motion which repeats itself at regular time intervals. The *period* of the motion is the time required to complete one full cycle. The *frequency* of the motion is the number of cycles completed per second:

$$f(\text{cycles/sec}) = \frac{1}{T\,(\text{seconds})}$$

where f = frequency and T = period. The unit for frequency is 1/sec. It is often written cycles/sec as a reminder that you are dealing with the number of full cycles or repetitions of the motion, but "cycle" is not a physical unit and can be dropped in calculations. The *amplitude* of the motion is the maximum distance moved from the equilibrium point.

Example. A mass on a spring is pulled down 5 cm and is released. It moves up and down, 5 cm above and below the equilibrium point, at such a rate that it completes five full cycles in 2 seconds. What are the period, frequency, and amplitude of the motion?

Solution. The amplitude of the motion is 5 cm, that being the greatest distance from the equilibrium point. It is important to note that the amplitude is not the full distance of swing (10 cm). The frequency is

$$f = \frac{5\text{ cycles}}{2\text{ sec}} = 2.5\text{ cycles/sec}$$

and the period is just the reciprocal of the frequency,

$$T = \frac{1}{f} = \frac{1}{2.5}\,\frac{\text{sec}}{\text{cycle}} = 0.4\text{ sec}$$

TRAVELING WAVES

If any part of an elastic object is disturbed, the disturbance will tend to propagate to all parts of the object. For example, if one end of a stretched wire is struck or plucked, the resulting vibration will travel to all parts of the wire. Any collection of matter which has a definite equilibrium state will have a tendency toward periodic motion, and in an extended body that periodic motion will take the form of a *traveling wave*. One of the most common examples is the behavior of the surface of a quiet pond when a pebble is dropped into it. Waves will move out from the disturbance in widening circles.

Transverse Waves. The waves on the pond surface are a good example of transverse waves. If a cork is placed on the water surface, it will be noted that the cork will bob up and down when the wave passes but will not move forward with the wave. The surface of the water is moving up and down as the wave train propagates across the water surface. A transverse wave is defined as a wave in which the particles of the medium are oscillating back and forth *perpendicular* to the direction of propagation of the wave. Waves in a string or rope are examples of transverse waves. A sketch of a simple transverse wave is included in Figure 17–1(a). Although light waves do not involve the movement of a material medium to propagate them, it will be seen that light waves as well as radio waves, x-rays, and other types of electromagnetic waves are also transverse waves (Chapter 20).

Longitudinal Waves. Waves in which the periodic motion of the particles is *parallel* to the propagation direction are called longitudinal waves. Figure 17–1(b) illustrates the case in which a long coiled spring is given a displacement at one end. The wave motion propagates as a series of contractions and expansions of the spring. Although it is not so obvious in this case, the individual particles of the spring are undergoing simple harmonic motion about their equilibrium points, the points where they would be if the spring were at rest. Sound waves in air are longitudinal waves, since they propagate by the periodic motion of air molecules parallel to the direction of propagation.

An important characteristic of wave motion is that the speed of propagation is determined by the medium in which the wave is traveling and usually does not depend upon the frequency or amplitude of the wave. (The exceptions in the case of light will be discussed in Chapter 19.) The other parameter needed to describe traveling wave motion is the *wavelength*, represented by the Greek letter λ. The wavelength is given by

$$\lambda = \frac{v}{f} = vT$$

where v = propagation speed, f = frequency, and T = period of the motion. This relationship is fundamental to all traveling wave motion and is usually written in the form

$$v = f\lambda. \qquad \textbf{17–2}$$

Since the speed remains constant for a given type of wave motion, an increase in frequency will result in a shorter wavelength. The wavelength is the shortest repeat distance for the periodic wave, such as the distance between crests in Figure 17–1(a) or the distance between compressions of the spring in Figure 17–1(b). The distance between any two corresponding points on successive waves, such as the distance between successive low points of the waves, will give the same distance λ.

Example. If waves traveling across a water surface with a speed of 3 ft/sec cause

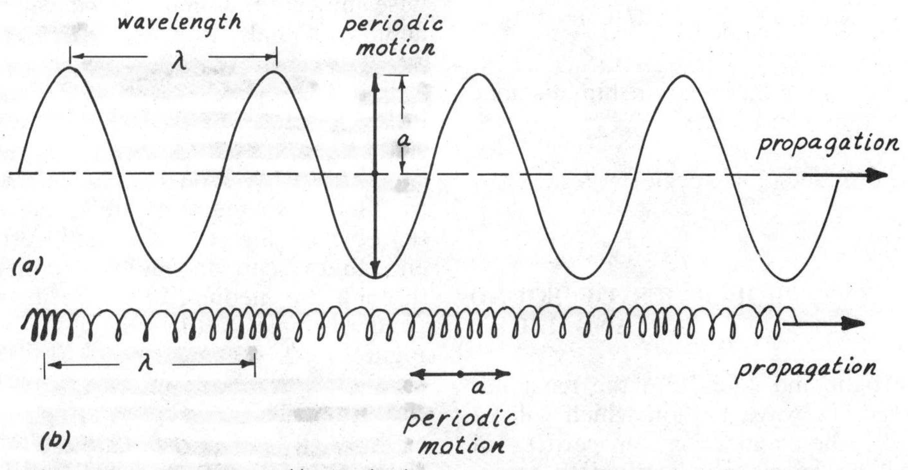

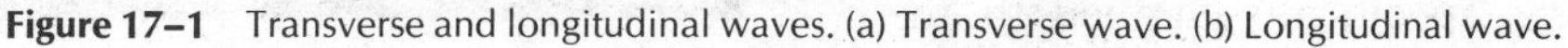
Figure 17–1 Transverse and longitudinal waves. (a) Transverse wave. (b) Longitudinal wave.

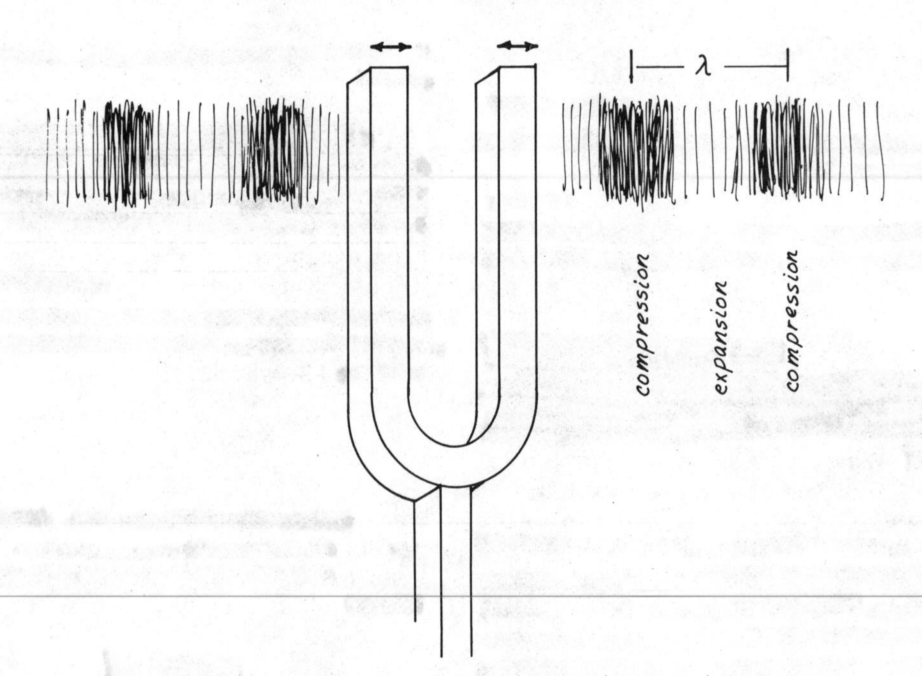

Figure 17–2 The vibration of an elastic object produces sound waves.

a cork to bob up and down with 2 seconds between peaks, what are the frequency and wavelength of the wave motion?

Solution. If the period is 2 seconds then

$$f = \frac{1}{T} = \tfrac{1}{2} \text{ cycle/sec.}$$

Applying equation 17–3,

$$v = f\lambda$$

$$3 \text{ ft/sec} = (\tfrac{1}{2} \text{ cycle/sec})\lambda$$

$$\lambda = \left(2 \frac{\text{sec}}{\text{cycle}}\right) (3 \text{ ft/sec}) = 6 \text{ ft.}$$

Alternatively, just the relationship distance = speed × time can be used:

$$\lambda = (3 \text{ ft/sec})(2 \text{ sec/cycle}) = 6 \text{ ft.}$$

WAVE PROPERTIES OF SOUND AND LIGHT

Since light and sound are the most important types of wave motion which will be considered, the remaining properties of waves will be specifically applied to them. Further descriptions and applications of sound and light are contained in Chapters 18 and 19 respectively.

Sound is a traveling pressure wave which may be propagated through the air or through solid or liquid materials. It cannot travel through a vacuum. This traveling wave phenomenon is said to be in the "audible" range when its frequency is between 20 and 20,000 cycles/sec, though the range of human hearing varies widely. Waves with frequencies above the range of human hearing are classified as "ultrasonic" sound, and those with lower than audible frequencies are said to be "infrasonic." Unless otherwise indicated, sound will be taken to mean audible sound, that is, periodic pressure variations with frequencies in the audible range. The phenomenon of pitch perception is basically one of measuring the frequency of an incoming sound wave, a higher frequency being perceived as a higher pitch.

Sound waves originate when some elastic object vibrates back and forth rapidly enough to send an audible frequency wave through the medium in which it vibrates. A good example is the tuning fork shown in Figure 17–2. If the elastic fork is struck, it continues to vibrate at its natural frequency for a time. The moving prongs strike air molecules, causing a compression of the air, and then move in the other direction, producing a partial vacuum or expansion. After the tuning fork strikes the air molecules to create the compression, those molecules in turn strike neighboring molecules. The series of successive molecular collisions cause the compression of the air to move outward from the fork. After a time interval equal to the period of the tuning fork, another compression follows. Thus a succession of compressions travels through the air, creating a traveling pressure wave with a frequency equal to that of the periodic motion of the tuning fork.

After being struck by the tuning fork and then colliding with neighboring molecules, the air molecules near the tuning fork will rebound backward toward the fork. The average motion of the molecules is like simple harmonic motion back and forth in the direction of wave motion. Therefore, the sound wave is a longitudinal wave as mentioned above. This regular periodic motion is superimposed upon the random thermal motion of the molecules due to the internal energy of the air. Since there are about 3×10^{19} molecules per cubic centimeter of air under standard conditions, then for all practical purposes, the second wave can be visualized as a wave motion in a continuous fluid.

The speed of sound is determined by the medium through which it is traveling. The speed of sound in air at 0°C is 331.5 m/sec (1087 ft/sec) and it increases about 0.6 m/sec for each degree Celsius above 0°C. The speed of sound is independent of the atmospheric pressure and independent of the frequency and amplitude of the sound wave. The speed of sound in solids and liquids is considerably higher than the speed in air, but the frequency or pitch will remain the same since it is determined by the sound source.

Example. If the audible frequency range of sound is taken to be 20 to 20,000 cycles/sec, what is the range of wavelengths involved at 0°C?

Solution. Given v = 1087 ft/sec as the velocity of sound at 0°C, the wavelengths can be calculated with the use of equation 17–2, $v = f\lambda$. At 20 cycles/sec

$$1087 \text{ ft/sec} = (20 \text{ cycles/sec})(\lambda)$$

$$\lambda = 54.4 \text{ ft.}$$

At 20,000 cycles/sec the wavelength is 1/1000 as great or 0.05 ft, about 0.7 inch.

Light is a traveling wave phenomenon consisting of propagating electric and magnetic fields. The further description of its basic nature is the subject of Chapter 20. At this point it is sufficient to say that it is a transverse wave phenomena which travels through a vacuum at the enormous velocity of 3×10^8 m/sec (186,000 miles per second), traversing the 93 million mile distance between the sun and the earth in just over 8 minutes. The perceived color of light is basically a measurement of the frequency of the light. Red light corresponds to the lowest visible frequency, and increasing frequencies correspond to the successive colors of the rainbow spectrum until the highest visible frequency, violet, is reached. The color of light is analogous to the pitch of sound, in that both are determined by frequency. Waves with frequencies just below the visible are denoted "infrared" and frequencies just above the visible are "ultraviolet."

ENERGY IN WAVES

All traveling waves carry with them a certain amount of energy. The rock thrown into a pond gives energy to the water which is carried outward in the widening ripples. The energy travels with the waves in their ordered motion until it is finally dissipated into disordered motion of the water molecules (internal energy).

Consider a small sound source in air which radiates sound in all directions. The sound source produces a certain amount of power which may be expressed in watts. Since the power at a given distance from the source is spread over a large area, it is convenient to use the power per unit area rather than the total power (see Figure 17–3). The power per unit area is the *intensity* of the radiation.

$$\text{intensity} = I = \frac{P}{A}. \qquad \textbf{17–3}$$

The intensity is commonly measured in watts per square centimeter. If the source is very small and emits its energy uniformly in all directions, it spreads its energy in a spherical distribution. The intensity meas-

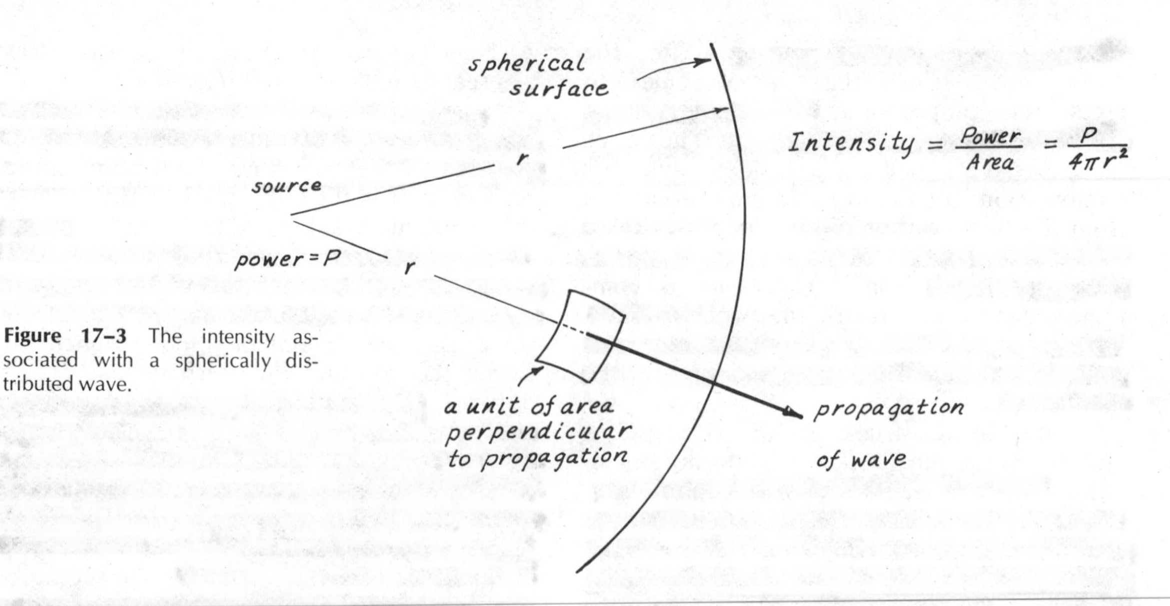

Figure 17–3 The intensity associated with a spherically distributed wave.

ured at a distance r from the source will then be

$$I = \frac{P}{4\pi r^2} \qquad \textbf{17–4}$$

where P is the total power of the source and $4\pi r^2$ is the area of the sphere of radius r. This intensity relationship is known as the *inverse square law*. If radiation follows the inverse square law, the intensity will drop off by a factor of four when the distance r is doubled, by a factor of nine when the distance is tripled, and so forth. Even if the source radiates uniformly, the intensity distribution will depart from the inverse square law if there are reflections from nearby surfaces, such as the reflection of sound from walls. The intensity from a point source of light in an open area will obey the inverse square law if reflections are negligible. In any case, the defining relationship equation 17–3 can be used to determine intensity.

Example. It is assumed that the light from a small lamp follows the inverse square law. By what factor would the light intensity (brightness) increase if the lamp is moved from a distance of six feet from a desk to a distance of three feet?

Solution. If I_1 and I_2 represent the intensities at 6 ft and 3 ft respectively, then

$$\frac{I_2}{I_1} = \frac{P/4\pi r_2^2}{P/4\pi r_1^2} = \frac{r_1^2}{r_2^2} = \frac{(6\text{ ft})^2}{(3\text{ ft})^2} = 4$$

and therefore I_2 (at 3 ft) $= 4 \times I_1$ (at 6 ft).

$I_1 r_1^2 = I_2 r_2^2$

Cutting the distance to the source in half raises the light intensity by a factor of four. This inverse square law also applies to other types of wave motion such as gamma rays and x-rays when they are spherically distributed (see Chapter 21). It is important for a person routinely exposed to gamma radiation to realize that doubling the distance from the source will cut the radiation by a factor of four.

INTERFERENCE AND STANDING WAVES

Interference refers to the addition of two or more waves which pass the same point in space. If water waves from two different sources meet on the surface of a lake, there will be places where the crest from one set of waves will coincide with a trough from the other. If the amplitudes of the waves are equal, they will completely cancel at this point, leaving a momentary flat spot on the lake surface. This is an example of *destructive interference*. *Constructive interference* occurs where two wave crests coincide and add to produce a larger crest (see Figure 17–4).

Interference between sound waves may be demonstrated by striking a tuning fork and then rotating it by your ear. The sound intensity will rise and fall as areas of constructive and destructive interference coincide with the ear. The two prongs of the tuning fork constitute two wave sources. At some angles these waves will add and at other angles they will subtract from each other. Another example of the interference of sound waves is the phenomenon of "beats" when two musical tones near the same frequency are sounded. As the two waves of different wavelengths reach the ear, their crests will coincide at one instant and then oppose each other a short time later. The result is a periodic rising and falling of the sound intensity, which occurs at a rate equal to the difference between the two frequencies. For example, if the frequencies 440 cycles/sec and 442 cycles/sec are sounded simultaneously, the sound intensity will rise and fall twice per second. If the frequencies differ by 5 cycles/sec, then 5 beats per second will be heard. These beats are the basis for the tuning of musical instruments to the same pitch and are the fundamental phenomena underlying the concepts of consonance and dissonance in music.

Interference between light waves is responsible for the rainbow-like colors seen in soap bubbles and thin oil films. A portion of the light striking a thin film of oil will reflect from the top surface, and a portion will enter the oil and reflect from the bottom surface. Depending on the wavelength of the light (color), the two reflected waves may interfere constructively or destructively. Since white light is a mixture of all wavelengths (colors), the removal of one wavelength by destructive interference will remove one color from the white light, leaving the complementary color as the apparent color of the oil film at this point. (See Chapter 19 for further discussion of color vision).

When a wave is produced in an elastic medium like a stretched guitar string, the wave will reflect at the end of the string and return. The reflected wave may interfere constructively or destructively with the incident wave. When the length of the string corresponds to a half-wavelength or an integer multiple of half-wavelengths ($\lambda/2$, $3\lambda/2$, 2λ, $5\lambda/2$, etc.) then the interference between reflected waves will cause the condition known as a *standing wave*. It is called a standing wave because the string will appear to vibrate up and down in fixed segments as illustrated in Figure 17–5. Wavelengths satisfying the standing wave condition will tend to be sustained by the string while other wavelengths will quickly die away. The frequencies corresponding to the standing waves are said to be natural frequencies or "resonant" frequencies, similar in principle to the natural frequency of a mass on a spring discussed earlier.

A detailed discussion of the production of standing waves is beyond the scope of this book, but the "resonant" nature of such waves is important in the production of sound waves. All objects which produce a sustained, single frequency sound do so by means of some sort of standing wave or resonant phenomenon. In the case of a trumpet, the resonant vibration of the air column produces the sustained sound. The resonant vibration of a drum head moves the air and

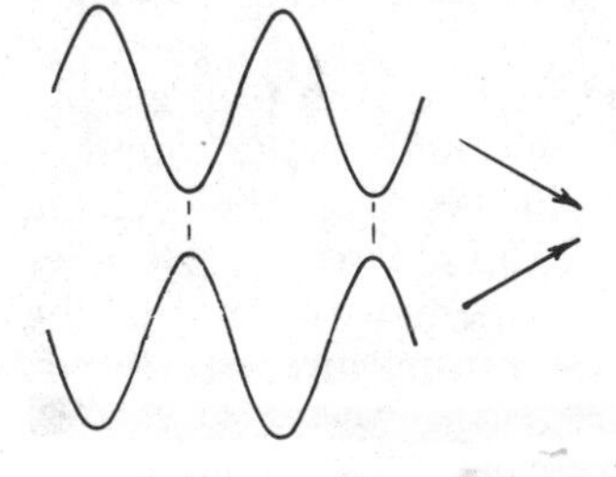

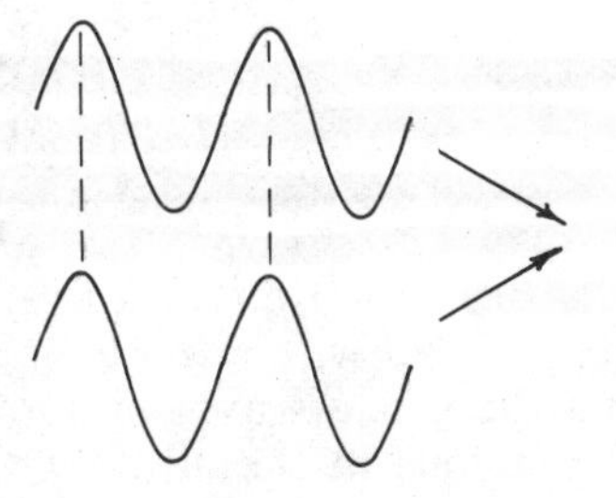

Figure 17–4 Constructive and destructive interference of waves.

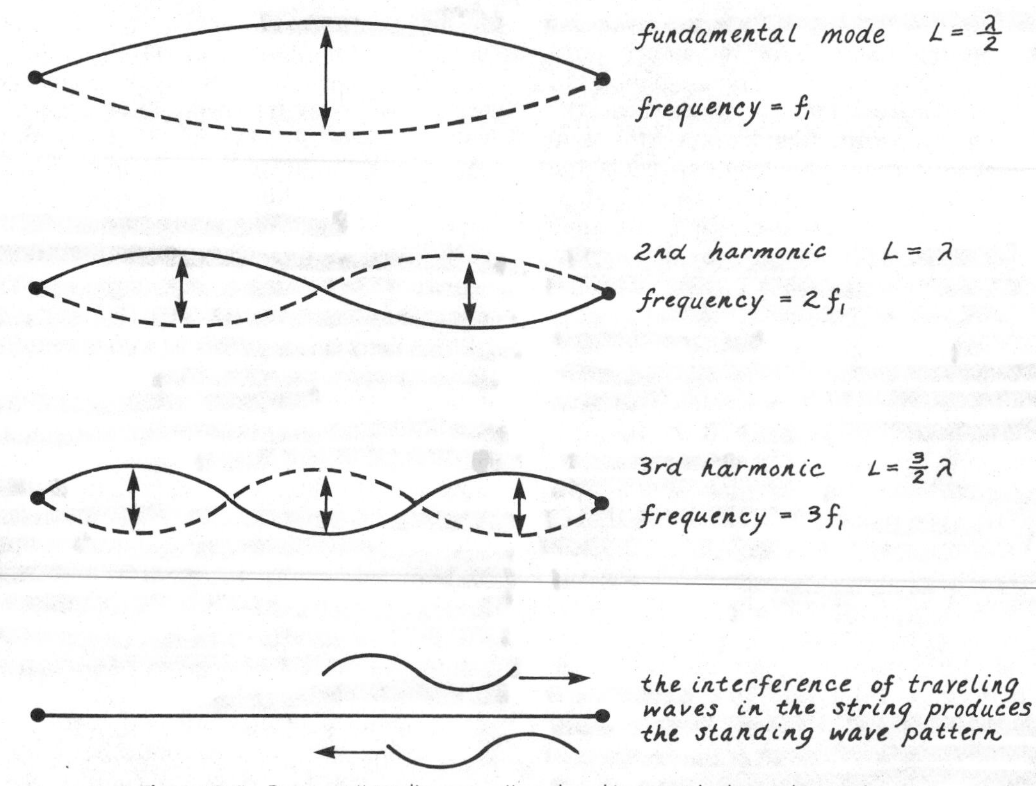

Figure 17–5 Resonant "standing waves" produced in a stretched string by interference.

produces a sound wave in the air. When a guitar string is plucked, it produces not only the fundamental frequency corresponding to the fundamental standing wave mode illustrated in Figure 17–5 but also the higher resonant frequencies. These higher "overtones" are integer multiples or "harmonics" of the fundamental frequency. The fundamental vibration frequency of the string can be raised by increasing the tension in the string. A heavier string with the same tension and length will have a lower fundamental frequency (e.g., the bass strings on a guitar).

The sound of the human voice is produced by forcing air through the larynx. This air flow sets the vocal cords into vibration, producing a periodic pressure variation in the air (sound). Although the vocal cords are more like membranes than strings, the analogy with the guitar string can be used to some extent. When the tension of the vocal cords is increased, the frequency of the sound is increased. Male voices are usually lower than female voices partly because the vocal cords are more massive. The vocal cords alone would not produce much sound. In the process of opening and closing they create periodic puffs of air which excite the resonant frequencies of the air spaces above them in the throat, nasal cavity, oral cavity, and sinuses, which are more efficient sound producers.

THE DOPPLER EFFECT

When a vehicle with a siren passes you, a noticeable drop in the pitch of the sound of the siren will be observed as the vehicle passes. This is an example of the Doppler effect. The effect is evident with any moving sound source such as a train, aircraft, or car. It also occurs if the observer is moving and the sound source is fixed. The clanging of a warning bell at a railroad crossing will drop in pitch as you pass it on a rapidly moving train.

The explanation of the Doppler effect is based on the fact that a sound wave consists of a series of high and low pressure areas in the air which propagate with a speed of approximately 1100 ft/sec. Suppose that a fixed sound source emits a sound with a frequency of 1100 Hz, so that there is a wavelength of 1 ft between compressions in the air. Then, as indicated in Figure 17–6, the observer would perceive the frequency as 1100 Hz. Now suppose the sound source is moving toward the observer with a speed of 550 ft/sec, one half the speed of sound. During one full period of the sound source, an air compression would have moved outward 1 foot, but the sound source would have moved 6 inches. It would then cause the next compression to be only 6 inches from the previous one, giving an effective wavelength of 6 inches instead of 1 foot. Since the sound speed, 1100 ft/sec, depends only upon the medium (air), the observer will receive the frequency (from equation 12–3):

$$f_0 = \frac{v}{\lambda} = \frac{1100 \text{ ft/sec}}{0.5 \text{ ft}} = 2200 \text{ Hz.}$$

Now if the observer were traveling in the opposite direction at half the speed of sound, it would increase the wavelength by 6 inches as illustrated in Figure 12–6(c). The observed frequency would be

$$f_0 = \frac{v}{\lambda} = \frac{1100 \text{ ft/sec}}{1.5 \text{ ft}} = 733 \text{ Hz.}$$

If such a sound source passed you at this speed, the frequency you observed would drop from 2200 Hz to 733 Hz as it passed, a rather large drop. The amount of pitch drop is proportional to the speed of the sound source and could be used to measure its speed. This speed-dependent pitch drop explains why the sound of a fast aircarft changes from a high pitched whine to a low, thundering roar as it passes overhead. The Doppler effect for a moving sound source can be summarized by the equation

$$\frac{f_{\text{observed}}}{f_{\text{source}}} = \frac{v}{v \pm v_s} \qquad \textbf{17–5}$$

where v is the speed of sound and v_s is the source speed. The minus sign is used for an approaching sound source and the positive sign for a receding source.

ULTRASONIC SOUND

Ultrasonic sound consists of the frequencies above the range of human hearing, beginning about 20,000 Hz and going upward

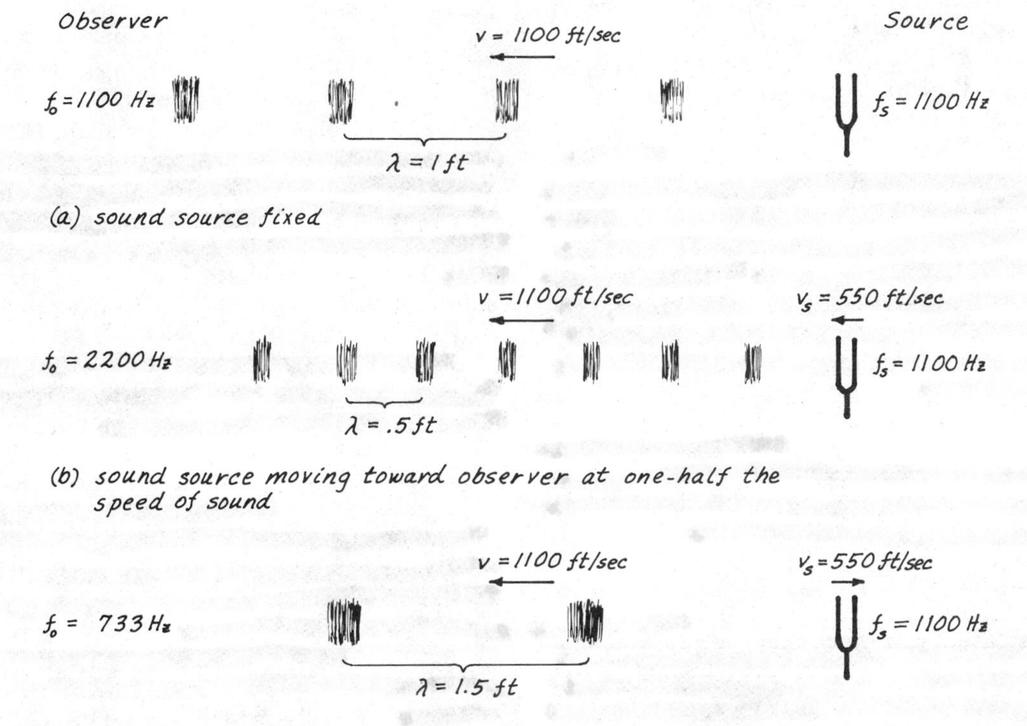

Figure 17–6 Illustration of the Doppler effect.

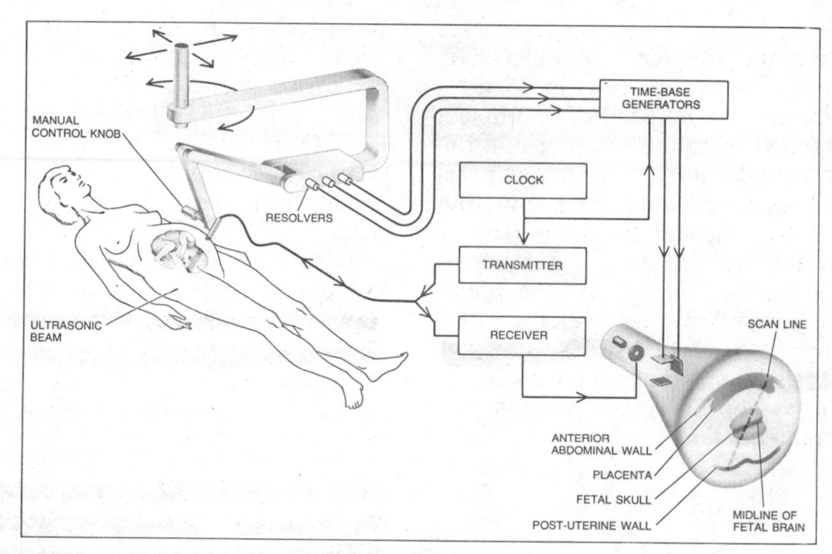

Figure 17–7 Two-dimensional ultrasonic scan of the abdomen for the examination of the fetus. (From Devey, G. B., and Wells, P. N. T. Ultrasound in Medical Diagnosis. Scientific American, Vol. 238, May 1978. Copyright © 1978 by Scientific American, Inc. All rights reserved.)

without a definite upper limit. Frequencies as high as 15 million cycles/sec (15 MHz) are routinely used in medical applications.

Most of the diagnostic uses of ultrasonic sound make use of echo techniques analogous to the SONAR echo-location of underwater objects by submarines and fishing vessels. An echo is a reflection of a sound wave from some interface where the nature of the medium in which the sound travels undergoes a significant change. The common example is the echo caused by sound reflection from a mountain or tall building. In water, any solid object will cause an echo, and SONAR can be used to measure the distance to the bottom or to detect moving objects such as fish or submarines. Measurement of the time elapsed before the echo is received and a knowledge of the speed of sound in the medium make possible an accurate range determination. Note that such range determinations use the propagation of very short pulses of sound energy. The emphasis in this chapter has been upon sustained wave motion, but brief pulses of sound have the same propagation speeds and properties.

When ultrasonic waves are directed into the body, reflections occur at interfaces between different tissues or fluids. The pattern of the reflections produces a visualization of interior body tissue structures. The minimum size of a resolved object depends upon the wavelength of the sound; the sound wavelength should be considerably smaller than the object in order to visualize it clearly. The same principle applies to vision. We see ordinary objects with clear detail because the light wavelengths which are reflected to our eyes are much smaller than the objects seen. When 15 MHz sound waves are used, the wavelength is about 0.1 millimeter, so the sound wavelength is then not a limiting factor. However, the higher the frequency the more the transmission of the ultrasound is attenuated, so a compromise must be made between resolution of the image and the depth of penetration of the scan. For the examination of the eye, frequencies up to 20 MHz can be used, but for abdominal examinations frequencies of 1 to 3 MHz are used to obtain the necessary depths of penetration. Present techniques involve a series of short sound pulses, with the echoes received and displayed on an oscilloscope (3).

Cross-sectional views of the abdomen may be obtained with two-dimensional scanners, such as the one outlined in Figure 17–7. In such scanners the ultrasonic beam is moved back and forth across the abdomen in a given plane, while the position and orientation of the probe is registered by the equipment and stored along with the ultrasonic information so that a two-dimensional image can be constructed. At each position of the probe the intensity and time of arrival of each echo is measured and stored in the scanner. The individual echo patterns are combined to form a two-dimensional image such as the image of a normal fetus shown in Figure 17–8. Such images can be constructed

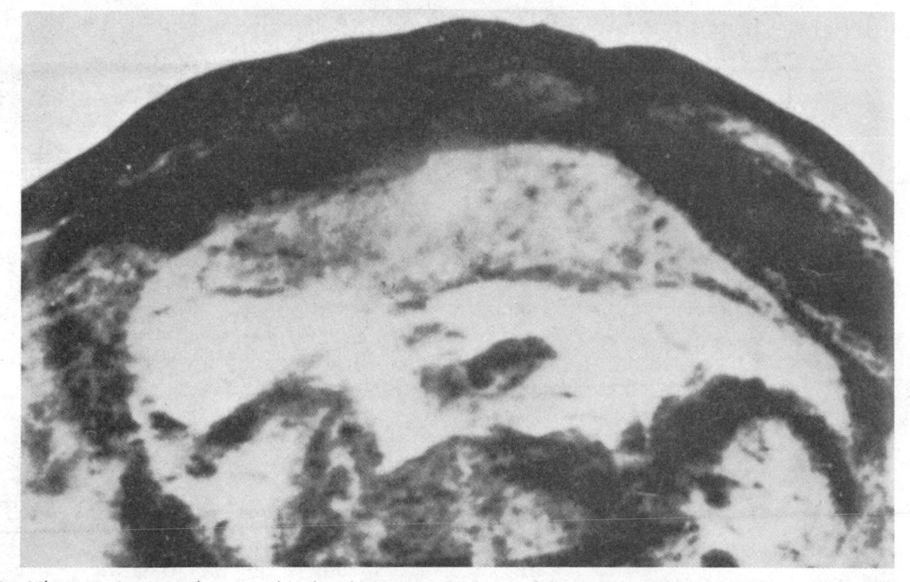

Figure 17–8 Ultransonic scan showing the development of a normal fetus. (From Devey, G. B., and Wells, P. N. T. Ultrasound in Medical Diagnosis. Scientific American, Vol. 238, May 1978. Copyright © 1978 by Scientific American, Inc. All rights reserved.)

in times on the order of fifteen seconds and displayed on a television type display screen (5).

The echo displays can be used to study either stationary or moving objects. They are particularly advantageous for use in obstetrics (5), since they share none of the dangers characteristic of x-rays. Ultrasonic scanners making use of the Doppler effect have been used to study the motion and function of the heart, the blood flow through major arteries, and for the early detection of fetal heart tones. Ultrasonic techniques have been used to locate tumors, cardiovascular disorders, and defects in the eye.

Lower frequency ultrasonic sound in the range from 20,000 to 40,000 Hz is used in ultrasonic cleaning devices. Ultrasound in water produces large numbers of tiny bubbles in a process called "cavitation." This cavitation is very effective for cleaning surgical instruments. The cavitation effect is also utilized in ultrasonic nebulizers. The agitation of the water produces an aerosol of water and/or medication. The frequencies employed in commercial ultrasonic nebulizers are usually on the order of 1.4 MHz.

Other properties of waves, such as absorption, refraction, and diffraction, will be discussed as they apply directly to sound and light in Chapters 18 and 19 respectively.

SUMMARY

Elasticity is the property of an object which tends to return it to its original configuration after being deformed. A simple elastic system such as a mass on a spring will undergo simple harmonic motion about its equilibrium point after a disturbance. Its restoring force always pulls it toward its equilibrium configuration, but it cannot come to rest there until all of its energy is dissipated. In the case of an extended elastic medium, a disturbance will propagate outward from the source in the form of traveling waves. Such waves are characterized by amplitude, velocity, frequency, and wavelength, the last three being related by the basic equation $v = f\lambda$. Waves are classified as either transverse or longitudinal waves. Sound in air is a longitudinal wave. While light is not associated with elasticity, it can be considered to be a traveling wave phenomenon consisting of transverse waves. Traveling waves carry energy and, like the mass on a spring, the associated periodic motion will not cease until the energy imparted by the generation of the wave has been dissipated.

Sound, light and other types of waves demonstrate the phenomena of reflection, interference, and diffraction and refraction. As an example of wave interference, two sound waves of differing frequencies may periodically interfere constructively and then destructively, producing a pulsating "beat" phenomenon. Sound interference in a given air cavity can lead to enhancement of sound production at a particular frequency (resonance).

The frequency of a wave is basically determined by its source, but if a sound source is moving with respect to the observer, the observer will receive an altered frequency (Doppler effect). An echo or reflection from a moving object will have a shifted frequency, depending upon the speed and direction of the object's motion. The time of arrival and frequency of echoes can be used to measure the distance to an object and its velocity. Biomedical applications of ultrasonic sound are being developed along these lines.

REVIEW QUESTIONS

1. A characteristic of elastic media is the fact that they have a definite equilibrium configuration at which they eventually come to rest after being disturbed. Why do they not immediately come to rest after being disturbed instead of oscillating back and forth through the equilibrium point?

2. Give some examples of interference of waves.

3. Can sound waves travel through a vacuum?

4. Does changing the frequency of a traveling wave also change its speed of propagation?

5. What is the difference between a longitudinal and a transverse wave? Name some examples of each.

6. What part do interference and resonance play in the production of sustained sounds?

PROBLEMS

1. What is the wavelength of a 440 Hz sound if the speed of sound is 1100 ft/sec?

2. A water wave passing a point causes a cork to bob up and down through a total distance of 1 foot, completing one complete cycle every 1.5 sec. It is noted that the wave crests are 4 ft apart. What are the amplitude, frequency, wavelength, and propagation speed associated with the wave?

3. A 100 watt incandescent light bulb actually radiates only about 10 watts of power in the form of visible light. Assuming that it is radiated equally in all directions, what is the light intensity in watts per square centimeter at a distance of 1 meter from the light bulb? At a distance of 2 m? 3 m?

4. A very intense sound source might produce 1 watt of acoustic power. What would be the sound intensity in watts/cm^2 at a distance of 4 meters from this sound source, assuming it radiates equally in all directions? What would be the sound intensity at a distance of 8 meters?

5. If the musical notes A_4 (440 cycles/sec), C_4 (262 cycles/sec), and E_4 (330 cycles/sec) are sounded simultaneously, what other frequencies will be produced by their interference?

6. A distant pile driver is heard 3 seconds after the drive is observed striking the pile. How far away is the pile driver, assuming a sound speed of 1100 ft/sec? Should the time required for the light to reach your eye be included to get an accurate answer?

7. To determine the distance to a cliff, a gun is fired and the time of arrival of the first echo is recorded. At 20°C the speed of sound is 344 m/sec and the echo time is 4.2 seconds. What is the distance to the cliff?

8. Some pulse detectors are based upon the Doppler shift experienced by ultrasonic sound when it bounces off the moving blood in an artery. If the ultrasound frequency is 2 MHz, the sound speed in the body is 1450 m/sec, and the blood flow velocity is 5 cm/sec toward the sound source, what would be the frequency of the echo from the blood? If this echo were mixed with the original frequency, what beat frequency would be produced? If this frequency were amplified and converted into sound by a speaker, would it give an audible blood flow signal?

9. The smallest object which can be resolved with a given frequency of ultrasound in a diagnostic scanner is on the order of the wavelength of the ultrasound used. If 20 MHz ultrasound is used for a scan of the eye, and the speed of the wave is 1500 m/sec, how small an object could be detected in the eye?

10. The practical depth of penetration of ultrasound for diagnostic scans is about 200 times the wavelength of the sound. If the speed of the sound in tissue is 1500 m/sec, what maximum frequency could be used for an abdominal scan where a depth of penetration of 30 centimeters is required. If the smallest object discernible in the scan has a dimension equal to the wavelength of the sound, what is the smallest detail observable?

REFERENCES

1. Evans, F. G. *Stress and Strain in Bones: Their Relation to Fractures and Osteogenesis*. Springfield, Illinois: Charles C Thomas, 1957.
2. Cromer, Alan H. *Physics for the Life Sciences*. 2nd ed. New York: McGraw-Hill, 1977.
3. Bartrum, R., and Crow, H. C. *Gray-Scale Ultrasound.* Philadelphia: W. B. Saunders, 1977.
4. Dummer, G. W. A., and Robertson, J. M., eds. *Medical Electronic Equipment, 1969–1970*. New York: Pergamon Press, 1970.
5. Devey, G. B., and Wells, P. N. T. *Ultrasound in Medical Diagnosis*. Scientific American, Vol. 238, p. 98, May 1978.

CHAPTER EIGHTEEN

The Physics of Hearing

INSTRUCTIONAL OBJECTIVES

After studying this chapter, the student should be able to:

1. Outline the steps of the hearing process in terms of the parts of the ear and their specific functions.

2. Describe the basic mechanism for detecting the pitch of a sound.

3. Express the ratio of two sound intensities in decibels, given the numerical value for the intensity ratio.

4. Express a sound intensity in decibels, given its intensity in watts/cm^2 and the standard value in watts/cm^2 for the threshold of hearing at 1000 Hz.

5. Define the terms intensity and loudness for sound, pointing out the distinction between them.

6. Explain why hearing tests must be made over a wide frequency range to accurately assess difficulty in understanding speech sounds.

As discussed in the previous chapter, sound in air is a traveling wave phenomenon in which the pressure at any point is periodically increased and decreased with respect to the ambient atmospheric pressure. These periodic pressure variations are received by the ear and passed on to the brain in the form of information about the pitch, loudness, and quality of the sound.

THE MECHANISM OF THE EAR

Figure 18–1 is a simplified sketch of the human ear with the cochlea unrolled from its normal spiral configuration. The acoustical function of the pinna is based upon the fact that the sound energy received is proportional to the area of the wavefront intercepted. Although the effectiveness of the human pinna is minimal, it tends to concentrate more sound energy into the auditory canal. In the optimum frequency range between 2000 and 5500 Hz, this focusing effect plus the resonance of the auditory canal achieves an amplification of about two (1). That is, the sound pressure at the eardrum is about twice what it would be without those effects. The oval window could presumably receive sound energy directly in the absence of the other structures, but the tympanic membrane and ossicles act as amplifiers to increase the effectiveness of this reception. Since the eardrum is 15 to 30 times as large as the oval window (1), it receives 15 to 30 times as much sound energy as would the oval window alone. A large fraction of this energy is passed to the oval window by the set of three small bones known as the ossicles (hammer, anvil, and stirrup). These three bones constitute a compound lever system which multiplies the sound force exerted on the eardrum.

When very soft sounds are received, this level system is thought to have a mechanical advantage of up to three. To reduce the sensitivity and protect against very loud sounds, this multiplication of force is reduced as the sound intensity increases, and the stirrup is actually pulled away from the oval window if the sound is sufficiently loud. Very loud sounds also trigger a set of muscles which tighten the eardrum and lessen its responsiveness to the sound. Acting together under optimum conditions, the three amplification mechanisms may produce an amplification to almost 180 times the sound pressure level in the air outside the ear (a factor of two from the pinna and auditory canal, a factor of 30 from the eardrum, and a factor of 3 from the ossicles).

The amplified mechanical force transmitted to the oval window by the ossicles results in a hydraulic pressure in the cochlear fluid. This pressure is transmitted throughout the fluid, creating a wavelike ripple in the basilar membrane. The behavior of this wave as it travels through the cochlea is apparently the key to our ability to distinguish different frequencies (pitches) of sound. A high frequency wave will peak near the oval window as illustrated in Figure 18–1, and excite the basilar membrane in that area. When nerve cells in that area of the basilar membrane relay a signal to the brain, it is perceived as a high pitch. A low frequency wave will peak near the end of the cochlea. Signals from that area of the basilar membrane are perceived as low pitches.

The sketch of the waves in the cochlea in Figure 18–1 is not to scale, since most of the wavelengths of audible sound are much longer than the entire cochlea. As indicated in Chapter 17, the lowest audible frequencies have wavelengths in air in excess of 50 feet, whereas the entire cochlea, when coiled, has a maximum dimension on the order of one centimeter. The pressure waves in the cochlea are converted to electrical impulses in the delicate organ of Corti and are transmitted to the brain. Our understanding of these mechanisms is primarily due to the painstaking research of Georg von Bekesy (7) who received the Nobel Prize in 1961 for his work. For further descriptions of the hearing mechanism, Reference 1 is recommended for an excellent pictorial review, and References 8 and 9 provide further details.

THE RANGE AND SENSITIVITY OF HUMAN HEARING

The remarkable sophistication of the human ear may be more fully appreciated after examination of some of the data about

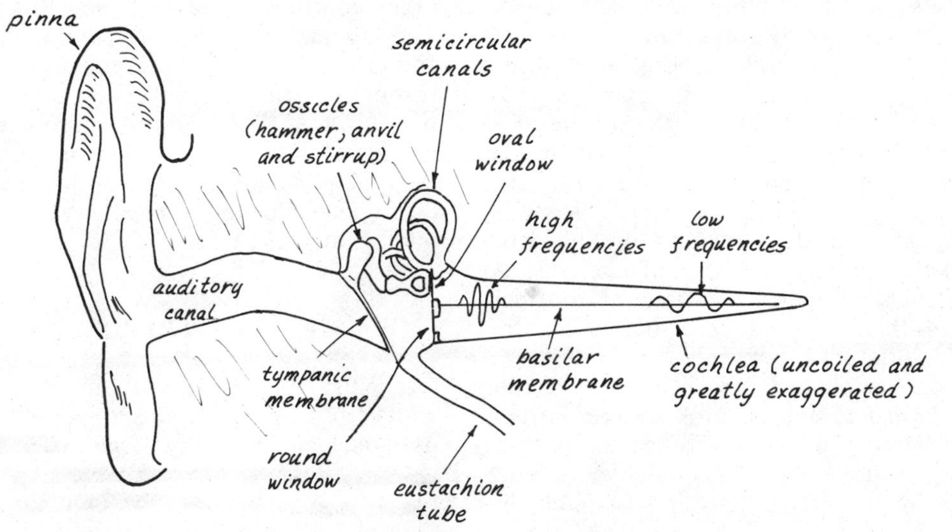

Figure 18–1 Sketch of human ear with cochlea uncoiled.

its sensitivity. The threshold of human hearing occurs at a sound intensity of about 10^{-16} watts/cm^2 at 1000 Hz. This corresponds to periodic pressure changes in the air which are less than one-billionth (10^{-9}) of the surrounding atmospheric pressure. Yet sound intensities up to 10^{13} times as great as this threshold intensity will not damage the hearing mechanism when exposure is brief. Since the sound intensity is proportional to the square of the amplitude of pressure variation above and below atmospheric pressure, an intensity level 10^{12} times the threshold intensity would correspond to 10^6 times the pressure variation at threshold. Referring to the pressure variation at threshold of about 10^{-9} atmosphere, it is seen that the air pressure variation associated with very loud sounds is usually less than one thousandth (10^{-3}) of atmospheric pressure.

The quoted frequency range for human hearing varies considerably, but the range from 20 to 20,000 cycles/sec is a convenient inclusive range. (This is usually written as 20 to 20,000 Hz in accordance with the international adoption of the Hertz as the unit for frequency). The upper frequency limit for hearing depends, of course, upon the intensity of the sound. At a given intensity the upper frequency limit is usually higher for women than for men. This limit usually decreases with increasing age. Loss of the high frequency sensitivity with age is called presbycusis.

The human ear is quite sensitive to frequency differences between two sounds when sounded together or separately. When the tones are sounded simultaneously, the frequency difference is detected by means of the beats caused by interference between the sound waves, as discussed in Chapter 17. The ability to discriminate between different frequencies when sounded separately depends upon the fact that different frequencies excite different areas on the basilar membrane as discussed above, but the "peaking" of the waves in the cochlea does not appear to be sharp enough to explain the remarkable pitch sensitivity of the ear. At low pitches, a frequency change of considerably less than 1 Hz can be detected. The frequency difference between the lowest two notes on the piano is only 1.6 Hz. At 1000 Hz a frequency change of 3 Hz can be perceived by most people (2). Since the peaking of the pressure waves in the cochlea is much too broad to explain this kind of pitch discrimination, there must be other mechanisms to aid this basic process. It is thought that there must be a "coding" process involved in the transmission of nerve impulses to the brain which enhances the frequency sensitivity.

Although the pitch difference that is just noticeable varies in terms of the number of cycles/sec, it represents about the same frequency ratio or "interval" at all frequencies. That is, if it takes a 3 Hz change to be heard at 1000 Hz, it takes only about 0.3 Hz to be heard at 100 Hz. This illustrates the basic fact that the ear is sensitive to frequency ratios rather than to absolute frequency changes. This is a fundamental fact in music. When the entire frequency level of a musical composition is changed by changing the "key," the composition sounds basically the same because the frequency ratios between successive notes are maintained. The actual number of cycles per second between successive notes may have changed considerably.

Besides the ability to distinguish between sounds of different pitch and intensity, the ear can distinguish between sounds of different "quality" even though they may have the same pitch and intensity. The ear has no trouble distinguishing between a trumpet and a clarinet, even though they may be playing the same note at the same loudness. Different types of vibrating objects will produce different "overtones" or higher frequencies above the fundamental frequency, as discussed in Chapter 17. The varying frequencies and loudness of these overtones is the basic difference between the sound qualities of different sound sources. (For other contributing factors, see Reference 2). The human voice can vary the quality of the sound by changing the configurations of the resonant cavities involved in sound production so that the overtone frequencies and amplitudes are changed. This is an important factor contributing to the ability to produce distinguishable and reproducible sounds for intelligent speech.

THE DECIBEL SCALE

When direct intensity measurements in watts/cm^2 are used, the range of numbers used is inconveniently large. The sound intensity at the pain threshold is 10^{12} or 10^{13} times as intense as the threshold of hearing intensity. The range of intensities used for normal communication and music covers an intensity ratio of about a million to one from the softest to the loudest. To obtain a more manageable set of numbers it is convenient to use a logarithmic scale of intensities, the decibel scale. The intensity in decibels is defined as

$$\text{intensity (decibels)} = 10 \log_{10}\left(\frac{I}{I_0}\right) \qquad \textbf{18–1}$$

where I = intensity of sound in watts/cm² and I_0 = the threshold of hearing intensity at 1000 Hz (10^{-16} watts/cm²). The logarithm to the base 10 is defined as:

$\log_{10}(x)$

= power to which you must raise 10 to get the number x.

Examples of logarithms are:

$$\log_{10}(1) = 0 \text{ since } 10^0 = 1$$

$$\log_{10}(10) = 1 \text{ since } 10^1 = 10$$

$$\log_{10}(100) = 2 \text{ since } 10^2 = 100$$

$$\log_{10}(1000) = 3 \text{ since } 10^3 = 1000$$

$$\log_{10}(1{,}000{,}000) = 6 \text{ since } 10^6 = 1{,}000{,}000$$

$$\log_{10}(40) = 1.6 \text{ since } 10^{1.6} = 40$$

(must use log tables).

For most numbers a table of logarithms is needed, but for the powers of ten the exponent is the logarithm, or the logarithm is just the number of zeros if the number is written out.

If an intensity is stated as a multiple of the threshold intensity, I_0, then the intensity in decibels can be calculated from equation 18–1.

Example. Find the intensity in decibels if the sound intensity is equal to the threshold of hearing intensity. Then calculate the corresponding decibel levels associated with intensities 100 times, 1,000,000 times, and 40 times the threshold intensity.

Solution. For the threshold level

$$I(\text{db}) = 10 \log_{10}\left(\frac{I_0}{I_0}\right) = 0 \text{ decibels}$$

since $\log_{10}(1) = 0$. For the other levels,

$$I = 100\, I_0,\; I(\text{db}) = 10 \log_{10}\left(\frac{100\, I_0}{I_0}\right) = 20 \text{ db}$$

For $I = 1{,}000{,}000\, I_0$

$$I(\text{db}) = 10 \log_{10}\left(\frac{1{,}000{,}000\, I_0}{I_0}\right) = 60 \text{ db}$$

For $I = 40\, I_0$,

$$I(\text{db}) = 10 \log_{10}\left(\frac{40\, I_0}{I_0}\right) = 16 \text{ db.}$$

Zero decibels is the threshold of hearing level at 1000 Hz, which is accepted to be 10^{-16} watts/cm² for standardization. The pain level is 10^{12} to 10^{13} times the threshold level and therefore corresponds to 120 to 130 decibels. The useful intensity range for communication is from about 40 to 100 decibels (from $10^4\, I_0$ to $10^{10}\, I_0$). The approximate decibel levels for some ordinary sounds are given in Table 18–1. Several of those values were taken from Ackerman (3).

The ratio of the intensities of two sounds can be expressed in decibels. Two sounds A and B with intensities I_A and I_B can be compared by the relationship:

$$\text{intensity difference (db)} = 10 \log_{10}\left(\frac{I_A}{I_B}\right). \qquad \textbf{18–2}$$

If $I_A = 100\, I_B$, then

intensity difference (db)

$$= 10 \log_{10}\left(\frac{100\, I_B}{I_B}\right) = 20 \text{ decibels.}$$

Then sound A is said to have an intensity 20 decibels higher than sound B. This type of comparison in terms of decibels is used in hearing tests, as will be seen.

TABLE 18–1 Typical Decibel Levels for Normal Sounds.

Decibel Level at 1000 Hz		Multiple of Threshold Intensity
160	Bursting of eardrum	10^{16}
140	Severe pain	10^{14}
120	Pain threshold	10^{12}
100	Damage to hearing after prolonged exposure; average factory, loudest passages of orchestra for close observer (*fff*)	10^{10}
80	Class lecture, loud radio	10^8
60	Conversational speech	10^6
40	Very soft music (*ppp*), typical living room	10^4
20	Very quiet room	10^2
0	Threshold of hearing	1

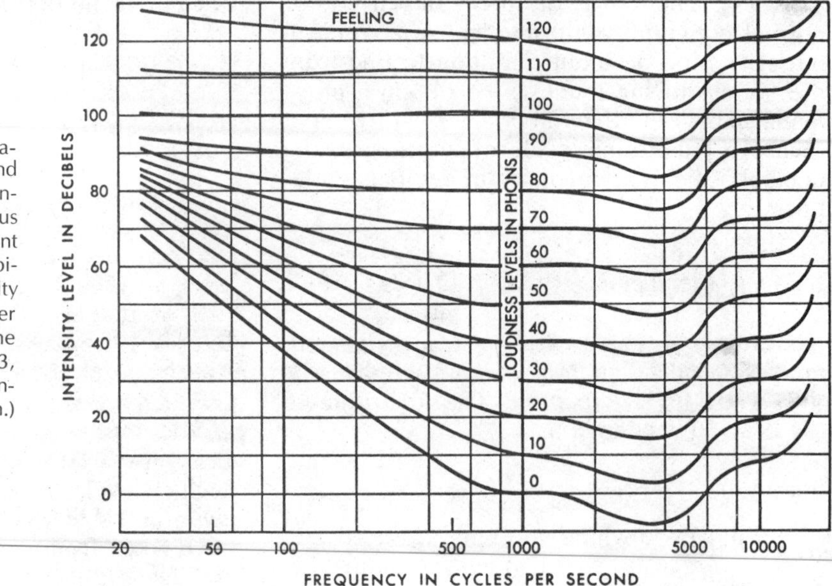

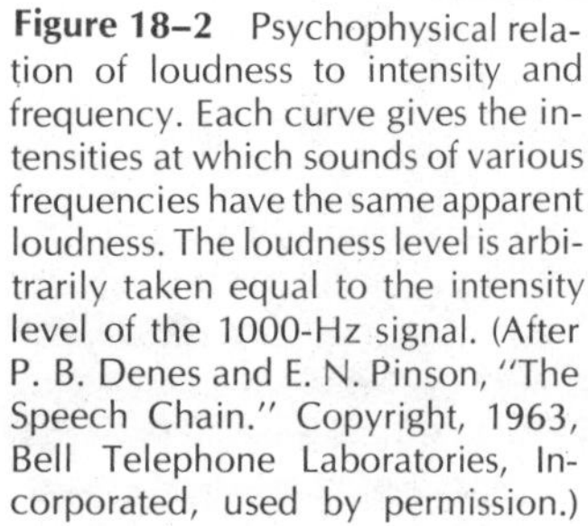
Figure 18–2 Psychophysical relation of loudness to intensity and frequency. Each curve gives the intensities at which sounds of various frequencies have the same apparent loudness. The loudness level is arbitrarily taken equal to the intensity level of the 1000-Hz signal. (After P. B. Denes and E. N. Pinson, "The Speech Chain." Copyright, 1963, Bell Telephone Laboratories, Incorporated, used by permission.)

The just noticeable difference in intensity between two sounds is in the neighborhood of 1 decibel. Studies of the just noticeable difference indicate that it ranges up to about 1.5 db for soft sounds at low frequencies and down to 0.5 db or lower for loud sounds at high frequencies (2).

THE DISTINCTION BETWEEN LOUDNESS AND INTENSITY

Sound intensity is defined as the acoustic power per unit area, and it is therefore an objective physical measurement which is independent of the frequency of the sound. Loudness could be defined as the physiological perception of the sound intensity. At a given frequency, a more intense sound will be perceived as louder. Therefore, loudness could be said to be proportional to intensity at a given frequency. However, since the ear's sensitivity varies with frequency, two sounds of the same intensity will in general not be perceived to have the same loudness unless they have the same frequency. For example, the lower pitched instruments in an orchestra must produce sounds of greater intensity because the ear is less sensitive at those frequencies. No matter how intense a 40,000 Hz sound is made, it will not be perceived as "loud" since the human ear does not respond to that frequency, except that it might cause pain.

Figure 18–2 displays a collection of equal-loudness curves based upon the loudness of a 1000 Hz reference tone. The plot is in decibels as a function of the frequency of the sound, and supposedly represents average values for persons with normal hearing. However, there is some disagreement about the exact shape of the curves. For example, to plot the curve marked 60 phons, the subject was exposed to a 1000 Hz tone at a level of 60 decibels. Then at other frequencies he was asked to raise or lower the sound intensity until it sounded the same *loudness* as the 1000 Hz tone. At 60 Hz the intensity must be raised to about 76 decibels to sound as loud as the 1000 Hz tone at 60 db, since the ear is much less sensitive at 60 Hz. These curves form the basis for the *phon* scale for measuring loudness. Sound of any frequency which is perceived to be as loud as a 1000 Hz tone at intensity 60 db is said to have a loudness of 60 phons. Therefore, at 1000 Hz the loundess in phons and the intensity in decibels are always numerically equal.

Using the equal loudness designation, phon, the threshold of hearing could be said to be zero phons, regardless of the frequency. It is not correct to refer to zero decibels as the threshold of hearing unless the sound has the frequency 1000 Hz. From Figure 18–2 it can be seen that the threshold of hearing (0 phons) at 60 Hz is at about 48 decibels, or almost 100,000 times as intense as the zero decibel level. The hearing sensitivity has dropped by 48 decibels at this frequency. Note that the sensitivity curve is flatter at high intensities. The ear's sensitivity is not so frequency dependent at high sound levels.

The variation of the sensitivity of the ear with frequency is illustrated in Figure 18–3. If a tone of constant intensity (60 decibels in this case) is started at a very low frequency and swept through the audible range, the loudness will vary with the sensitivity of the ear. The plot of loudness in phons in Figure 18–3 is then a plot of the ear's sensitivity as a function of frequency. Note that the range of maximum sensitivity is the range from 2000 to 5000 Hz. It is found that this frequency range is the most important for the understanding of speech.

Note that the inaudible regions in Figure 18–3 are not indications of the ultimate limits of human hearing, but these sounds are inaudible at an intensity of 60 decibels. The range of hearing can be extended by raising the intensity, but practical limits occur when the intensity must be raised to the physical pain threshold in order to be audible.

HEARING TESTS

Testing for hearing loss is generally referred to as audiometry. The two basic approaches to testing are pure tone audiometry and speech audiometry. A "pure tone" is a tone of single frequency. Pure tone audiometry is the simplest and most common form of testing. Speech audiometry has some obvious advantages since speech sounds are the most important sounds which we must interpret, but the tests are more complex and difficult to standardize. The discussion here will be limited to pure tone audiometry, since it includes all the relevant physical principles. Reference 5, page 207, is recommended for a discussion of speech audiometry.

The fundamental frequency range of the human voice is from about 85 to 1100 Hz (1), yet a person with normal hearing in this range and severe hearing loss above 1100 Hz cannot understand speech sounds at normal levels. The physical reason for this may be understood from the examination of some simple speech sounds. Figure 18–4 contains reproductions of oscilloscope displays of three sustained vowel sounds. These displays were obtained by using a microphone, which produces an electrical "image" of the sound by producing voltages which parallel the changes in air pressure accompanying the sound waves. The output of the microphone was used as the input to an oscilloscope to alter the vertical deflection of the oscilloscope beam while the beam main-

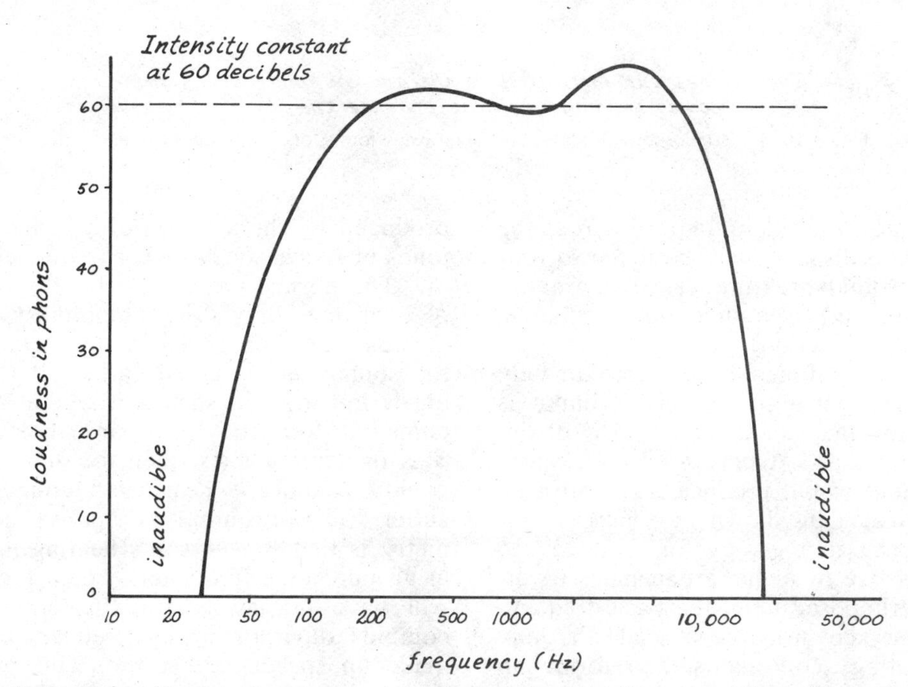

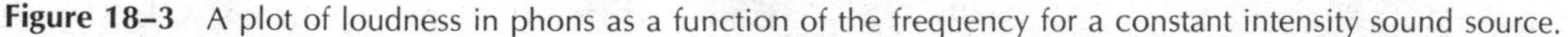
Figure 18–3 A plot of loudness in phons as a function of the frequency for a constant intensity sound source.

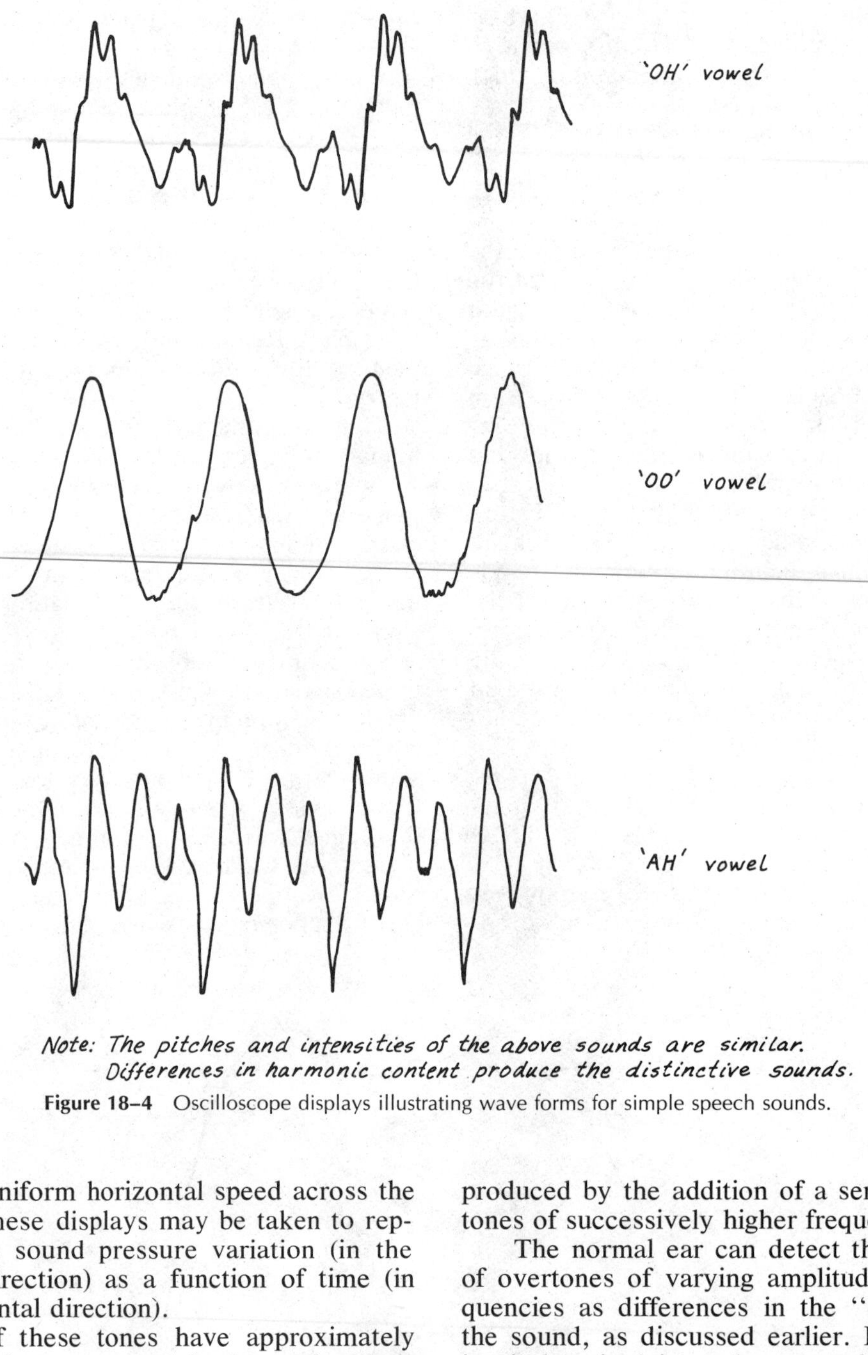

'OH' vowel

'OO' vowel

'AH' vowel

Note: The pitches and intensities of the above sounds are similar. Differences in harmonic content produce the distinctive sounds.

Figure 18–4 Oscilloscope displays illustrating wave forms for simple speech sounds.

tained a uniform horizontal speed across the screen. These displays may be taken to represent the sound pressure variation (in the vertical direction) as a function of time (in the horizontal direction).

All of these tones have approximately the same frequency and loudness, but it is important for the ear to be able to distinguish these sounds. A pure tone (i.e., a single frequency) would produce a smooth sine wave, such as that shown in Figure 17–1. The difference in the waveforms is caused by the presence of higher frequencies (overtones) superimposed upon the basic frequency. It can be shown mathematically that any such repeating, continuous waveform, no matter how complicated in form, can be reproduced by the addition of a series of pure tones of successively higher frequencies.

The normal ear can detect the presence of overtones of varying amplitudes and frequencies as differences in the "quality" of the sound, as discussed earlier. If the hearing is impaired in such a way that the fundamental frequency is heard, but not the overtone frequencies, then the differences in quality cannot be heard and understanding suffers. The presumption of pure tone audiometry is that if a person's hearing is normal at all pure tone frequencies, then the person will have normal hearing in terms of the complex superpositions of frequencies which make up speech and music. This presumption is difficult to prove directly, but the evidence of experience shows that it is reasonably valid.

A frequency analysis of speech indicates that most of the acoustic energy falls within the range from 300 to 3000 Hz (Reference 5, page 168). In most hearing tests, the threshold of hearing is measured over the inclusive range from 125 to 8000 Hz. It is normally measured at octave intervals (125, 250, 500, 1000, 2000, 4000, and 8000 Hz). As pointed out earlier, the human ear is sensitive to frequency ratios rather than to absolute numbers of cycles/sec, so the above series represents measurement over six equal "intervals" or ratios. Sometimes the frequencies 1500, 3000, and 6000 Hz are added to the sequence. Such a sequence not only gives an indication of the ability to understand speech sounds, but also gives some evidence about the nature of the difficulty if there is hearing loss. Since a given frequency corresponds to a certain area on the basilar membrane and a given set of nerve "channels," sensitivity as a function of frequency may suggest the location of the difficulty. More important, as indicated below, are indications about difficulties associated with the middle and outer ear, since the inner ear is not easily accessible for corrective measures.

Hearing tests generally include the evaluation of the threshold of hearing at the series of frequencies given above. The instrument used is an *audiometer,* which includes an electronic oscillator to produce pure tones at the specified frequencies, an attenuator to control the loudness, and compatible earphones. The methods for threshold testing will not be discussed here (see References 4, 5, and 7). Since the sensitivity of the ear varies widely over this frequency range, the instrument must be calibrated according to the normal hearing curve. Therefore, the threshold of hearing measurement is not an absolute intensity measurement, but a measurement of the ratio of the intensity required for audibility to the normal threshold intensity agreed upon as an international standard. This ratio is measured in decibels, so a 30 decibel hearing loss means that the sound intensity had to be increased 30 decibels above the standard threshold level for that frequency to be heard.

The measured threshold intensities in decibels are plotted on an audiogram such as that shown in Figure 18–5. The plot in Figure 18–5 illustrates presbycusis, or the loss of high frequency sensitivity which occurs with advancing age. The zero level represents the normal threshold, which is sometimes labeled zero decibels Hearing Level (0 db HL). The actual sound intensity or sound pressure level in decibels is sometimes referred to as decibels of Sound Pressure Level (db SPL) in medical literature. Figure 18–6 illustrates audiograms for some of the more common pathological causes of hearing loss. In actual tests there would be two curves for each case to represent the two ears, and bone conduction hearing tests are also included as a part of a complete audiological assessment (5).

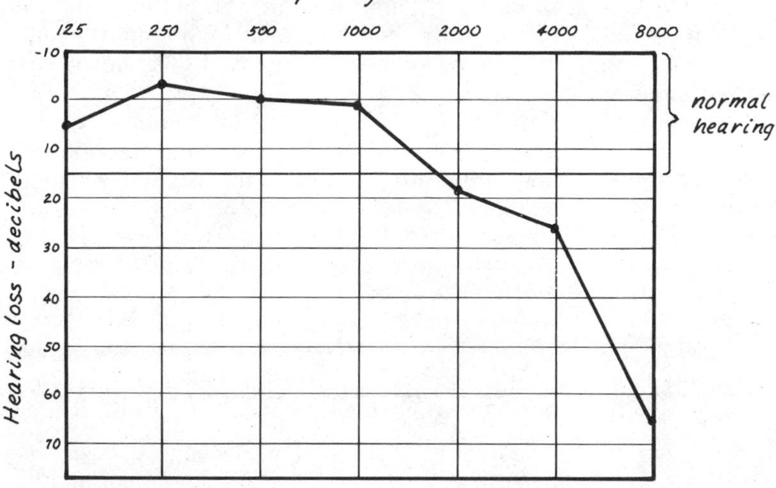

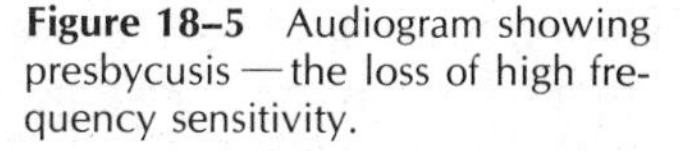

Figure 18–5 Audiogram showing presbycusis — the loss of high frequency sensitivity.

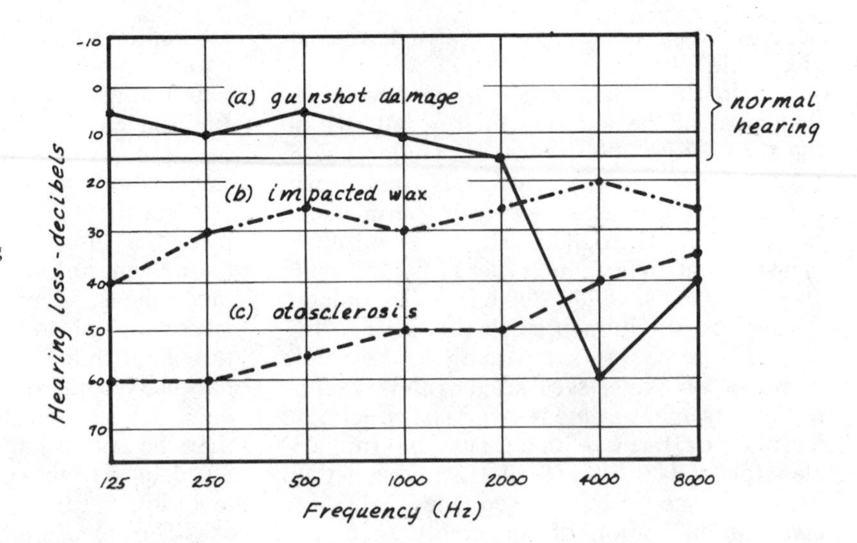

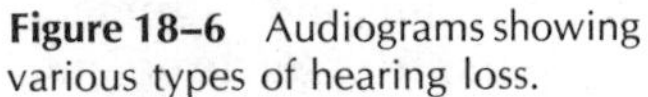
Figure 18–6 Audiograms showing various types of hearing loss.

STANDARDS FOR ENVIRONMENTAL NOISE

State and federal laws protect the industrial worker from excessive noise levels which can produce permanent hearing loss. Numerous local ordinances regulate the general environmental noise level, and national standards are being formulated. The American Academy of Ophthalmology and Otolaryngology has recommended that no worker be exposed to a continuous sound level of 85 decibels for more than 5 hours a day without protective devices.

The sound levels referred to in these recommendations as decibels should more appropriately be expressed in phons or db HL (Hearing Level). It is misleading to measure environmental sounds on a strict decibel intensity scale since that gives equal weighting to all sounds in the audible frequency range, and the ear is not equally sensitive to all frequencies. As previously described, the ear discriminates against both very low and very high audible frequencies. Since it is impractical to precisely measure phons or dbHL when measuring complex environmental sounds, some compromise must be made. The usual approach is to use a sound-level meter with a standard contour filter which discriminates against low and high frequency sounds in a way that approximates the human ear's response.

There are three internationally accepted standard contour filters, known as the A, B, and C contours. The A contour discriminates most strongly against very low and very high frequencies and is the best approximation to the human ear for sounds of soft and medium intensities. If a meter with such a contour gives a reading of 80, it is recorded as 80 dbA to indicate the contour. For routine sound surveys the dbA readings are generally preferred. Straight decibel readings can be very misleading in buildings where the meter may pick up a lot of low frequency sound from air circulation units that may be nearly inaudible to the ear. The B and C contours (recorded dbB and dbC) discriminate against low and high frequencies by lesser amounts and thus have "flatter" response curves, to more nearly approximate the flatter response curve of the human ear at high sound levels. As can be seen from Figure 18–2, the curves at 100 and 110 phons show much less variation in sensitivity with frequency than at lower sound levels. The B and C contours are sometimes useful for surveys of loud traffic noise and industrial noise levels.

SUMMARY

The ear is the most delicate and sophisticated mechanical device in the human body. The tympanic membrane and the ossicles of the middle ear serve to amplify sound vibrations and to offer some protection against very intense sounds. The sound energy transmitted to the fluid of the inner ear (cochlea) triggers nerve endings along the basilar membrane, the perceived pitch of the sound being determined by the area of maximum excitation along this membrane.

Perceived sound may be characterized by pitch, loudness, and quality. Pitch may be considered synonymous with frequency. Sound intensity may be objectively defined as the acoustic power per unit area, regardless of the frequency, but the ear has a limited frequency range and its sensitivity varies widely within that range. The term loudness is used to characterize the ear's response to sound. It is proportional to intensity at a given frequency, but dependent upon the ear's sensitivity to that frequency. Sounds of equal intensity at differing frequencies may be perceived to have differing loudness. It is convenient to express sound intensity in decibels. The decibel scale actually represents the ratio of a sound intensity to a standard threshold intensity. Hearing tests expressed in decibels usually use the decibel ratio between the sound intensity and the normal threshold of hearing intensity for that frequency, often called "decibels hearing level" or db HL. Many sounds which have the same basic pitch and loudness can be distinguished by the ear because of differences in quality. Sound quality is primarily determined by the number and intensity of higher frequencies present in the sound. Quality discrimination is important in the understanding of speech, since the intelligible differences between many speech sounds are quality differences rather than loudness or pitch differences.

Testing for hearing loss, or threshold shift, is referred to as audiometry. Hearing tests are usually made at octave intervals throughout the most sensitive range of human hearing. Audiograms of hearing loss as a function of frequency can help to diagnose diseases, monitor hearing loss from occupational noise, and make possible the contouring of the frequency response of hearing aids to best fit the individual's needs.

REVIEW QUESTIONS

1. Define sound. How does a tuning fork produce sound?

2. Describe the mechanisms in the ear which amplify the sound which reaches the ear. Does the amount of amplification depend upon the intensity of the sound?

3. Describe the basic pitch sensing mechanism of the ear.

4. To support the idea that all sound frequencies travel at the same speed, discuss the effects which would be observed with musical sound if different frequencies traveled at different speeds.

5. Why must sound loudness be distinguished from sound intensity?

6. What are the advantages of expressing sound intensities in decibels rather than in watts/cm^2?

7. Under what conditions could you have very intense sounds which strike a normal ear but cannot be heard?

8. In hearing tests the sequence of frequencies chosen (125, 150, 500 Hz, and so on) is such that the frequency is doubled between test points. Why is this more desirable than regular sets of frequencies separated by a constant number of Hz (e.g., 500, 1000, 1500)?

9. How can you account for the fact that many people with hearing loss can hear speech sound with apparently sufficient loudness, but still have great difficulty understanding speech?

10. If hearing by bone conduction in a given patient was found to be nearly normal but severe hearing loss was indicated with airborne sounds, what types of causes are indicated for the hearing loss?

PROBLEMS

1. If a sound has an intensity 100,000 times the threshold of hearing at 1000 Hz, what is the intensity in decibels?

2. If a quiet room has a sound intensity level of 40 decibels, the intensity is how many times as intense as the 1000 Hz threshold of hearing intensity I_0?

3. If an audiologist has to turn a test tone intensity up to an intensity 1000 times the normal threshold of hearing intensity for a given patient to hear it, what is the patient's hearing loss in decibels?

4. If the normal dynamic range for orchestral music is from about 40 db to 100 db, find the numerical ratio of intensities for the loudest sound compared to the softest sound.

5. What would be the sound intensity in watts/cm^2 at a distance of one meter from a one watt sound source? Rounding this off to the nearest power of ten, find the approximate sound intensity in decibels above the standard threshold of 10^{-16} watts/cm^2.

6. If a jet aircraft engine produced a sound intensity of 130 decibels at a distance of 50 ft and you could hear the sound at a level of 30 db, from what maximum distance could you hear the aircraft, assuming inverse square law reduction of the intensity (see Chapter 17)? (Air absorption of the sound energy helpfully reduces this audible distance considerably).

7. A sound source produces an intensity I_A = 80 db at a distance of 10 feet from the source, and it is assumed to drop off according to the inverse square law.

   a. Find the intensity I_B at a distance of 100 ft from the source as a fraction of I_A.
   b. What would be the intensity I_B in decibels?
   c. If the sound source above were speech and you could understand it at a level as low as 40 db, how far away from the source could you understand the speech?

8. The sound intensity at a distance of 10 m from a spherically radiating sound source is found to be 80 db. What is the acoustic power in watts of the sound source?

9. The human ear is usually most sensitive in the range from 2000 to 5000 Hz. If the speed of sound is 1120 ft/sec, calculate the wavelength range for maximum sensitivity.

10. A cylindrical pipe with one end closed will vibrate most readily at its fundamental resonant frequency. At this frequency the sound wavelength is four times the length of the pipe. If the auditory canal of the ear has an effective length of 2.5 cm and the speed of sound is 344 m/sec at 20°C, what is the resonant frequency of this entranceway to the ear?

11. If 0 decibels represents a sound intensity of 10^{-16} watts/cm^2, what would be the decibel level corresponding to an intensity of 10^{-9} watts/cm^2?

12. Measured in terms of watts/cm^2, it is found that the sound intensity must be increased by a factor of 100 above the normal threshold level for a certain person to hear it. What is the hearing loss in decibels?

13. If a person has a 40 db hearing loss, by what factor must the sound intensity be increased above the normal hearing threshold for that person to hear it?

14. A given sound of frequency 1000 Hz has an intensity of 40 db. What must be the intensity of a 40 Hz tone for it to sound as loud to the normal ear (see Figure 18–2)? What is the intensity in decibels of a 60 Hz tone which has a loudness of 40 phons?

REFERENCES

1. Time-Life Books. *Sound and Hearing,* Life Science Library. New York: Time-Life Books, 1969.
2. Backus, J. *The Acoustical Foundations of Music,* 2nd ed., New York: W. W. Norton, 1977.
3. Ackerman, E. *Biophysical Science*. Englewood Cliffs, N.J.: Prentice-Hall, 1979.
4. Watson, L. A., and Tolan, T. *Hearing Tests and Hearing Instruments*. New York: Hafner Publishing Co., 1967.
5. Rose, D., ed. *Audiological Assessment.* Englewood Cliffs, N.J.: Prentice-Hall, 1971.
6. Stearns, H. O. *Elementary Medical Physics*. New York: Macmillan, 1947.
7. von Bekesy, G. *Experiments in Hearing*. New York: McGraw-Hill, 1960.
8. Stevens, S. S., and Hallowell, D. *Hearing, Its Physchology and Physiology*. New York: Wiley, 1963.
9. Wever, E. G. *Theory of Hearing*. New York: Wiley, 1961
10. Peterson, A. P. G. *Handbook of Noise Measurement*. Concord, Massachusetts: General Radio Company, 1974.

CHAPTER NINETEEN

The Physics of Vision

INSTRUCTIONAL OBJECTIVES

After studying this chapter, the student should be able to:

1. Define refraction and show that refraction can be used to focus the light which passes through a lens.
2. Define the term "focal length" for a converging and a diverging lens and calculate lens strengths in diopters, given numerical focal length values.
3. Determine the size and location of the image formed by a lens, given the focal length of the lens, the size of the object, and the distance from the object to the lens.
4. Describe the components involved in image formation by the eye and describe the nature of the defects which cause nearsightedness, farsightedness, and astigmatism.
5. Describe the process by which the eye changes its focus from a distant object to a close object (accommodation).
6. Calculate the lens strength needed to correct myopic or hyperopic vision, given the relaxed focal length of the eye's lens and the effective lens-to-retina distance.

The human eye is sensitive to electromagnetic waves in a certain narrow frequency range. Such waves are called light or "visible light" to distinguish them from the wide range of electromagnetic waves to be discussed in Chapter 20. The present chapter is devoted to the discussion of visible light and the vision process. The use of lenses for vision correction and the principles of simple optical instruments are considered.

REFRACTION AND LENSES

In free space, light travels in straight lines at the speed $c = 3 \times 10^8$ m/sec. When it enters a transparent medium such as glass, or even air, it travels at a slower speed. The propagation speed is a property of the medium in which it is traveling, as in the case of the other types of traveling waves discussed in Chapter 17. Therefore, light may either accelerate or decelerate as it crosses an interface between two different media. If the propagation speed is perpendicular to the boundary, no visible effect accompanies the change in speed. However, if it strikes the boundary at an angle other than 90 degrees, the direction of a ray of light will be changed as it passes from one medium to the other. This "bending" of light rays at interfaces where the speed of light changes is referred to as *refraction*.

As a model for the phenomenon of refraction, consider a column of marchers moving from a hard parade ground onto softer ground where their marching speed would tend to be slower. If the column is moving onto the slower medium at an angle other than 90 degrees, as in the sketch in Figure 19–1, one side of the column will cross the boundary sooner than the other. Therefore, while crossing the boundary, part of the column will be traveling at the slower speed characteristic of the new medium while the other part will be traveling at the original speed. At cross-section B in Figure 19–1, the two marchers on the right are traveling at a slower speed than the other two. This tends to rotate the column and change the direction of march as illustrated.

As light moves from air into glass, its speed must decrease. As shown in Figure 19–2, its direction of propagation is deflected toward the perpendicular line PP' as it moves into the glass. The path traveled by the light is reversible; therefore, light traveling from glass into air would be deflected away from the perpendicular line. The amount of deflection or "bending" of the ray depends upon the change in speed which occurs. Instead of using the light speed directly, the optical properties of substances are usually specified in terms of the index of refraction, n, which is defined by

$$n = \frac{\text{speed of light in space}}{\text{speed in the medium}}.$$

Since the speed of light in all material media is less than the free space speed, the index of refraction is always greater than or equal to one. The index of refraction for air is about 1.00029, which is so close to the free space value that $n = 1$ is used for practical calculations. The index of refraction for water is about 1.33, and the range for common types of glass is about 1.5 to 1.9.

As an example of the effects of refraction, consider the bending of light rays as they pass from water into air. Since light travels faster in air than in water, the light is bent away from the vertical as it leaves the water. If light reflected from a fish leaves the water and reaches your eye, as illustrated in Figure 19–3, the refraction at the surface changes the apparent position of the fish. The eye presumes that the direction to the object observed is along the direction of the light rays reaching the eye, so the fish appears to be closer to the surface as shown. This refraction makes deep,

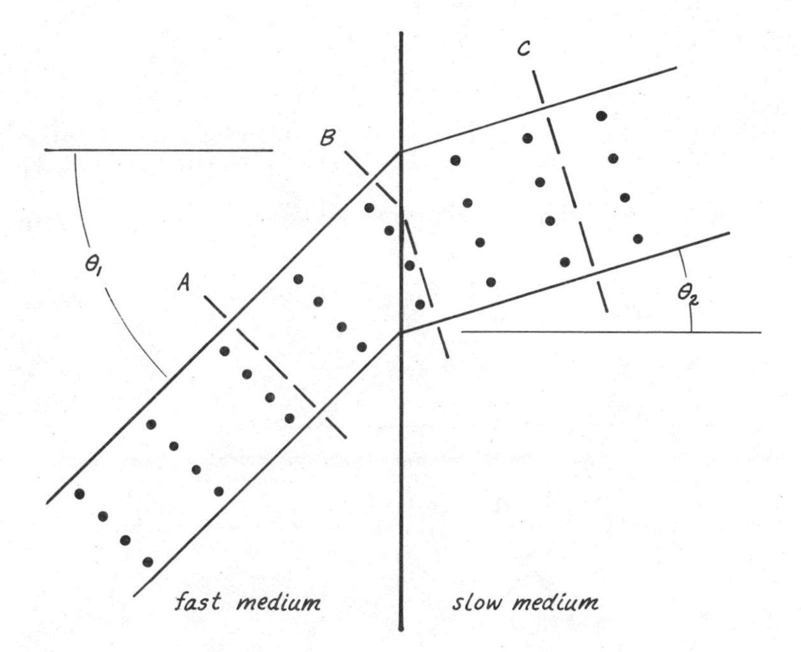

Figure 19–1 Refraction at a boundary where the velocity changes.

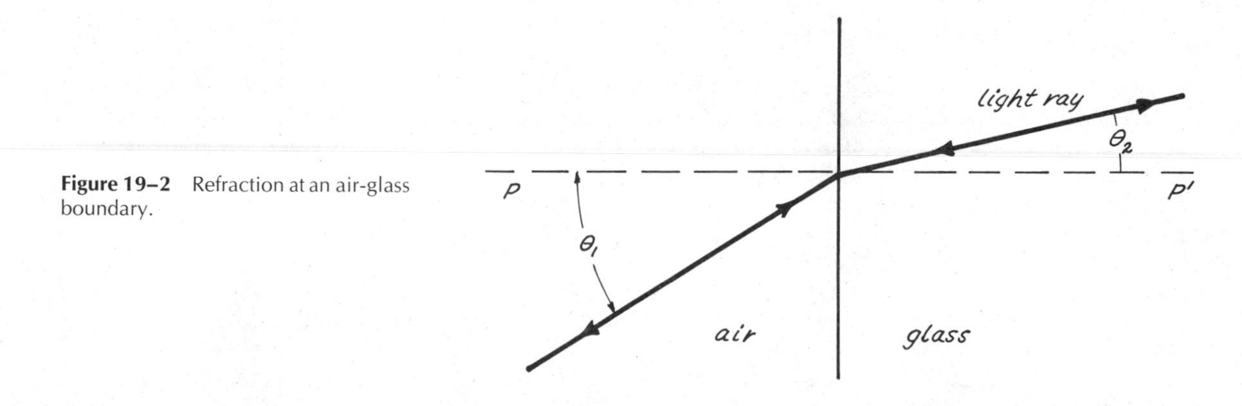

Figure 19–2 Refraction at an air-glass boundary.

clear bodies of water appear deceptively shallow. A penny placed in a clear glass of water will appear to be not as deep if viewed from above because of refraction.

Many other examples could be given, but the main application to be made is to the refraction of light by lenses. Refraction is the process by which lenses focus light rays to form images of distant objects. If a light ray strikes the top part of a double convex lens such as that in Figure 19–4(a), it is bent downward upon entering the glass and again when it exits from the glass. This can be seen from the nature of refraction as described above. The ray is bent toward the perpendicular *a-a'* as it enters the glass, and away from the perpendicular *b-b'* as it leaves the glass. If the lens is symmetric, all rays are bent toward the centerline of the lens and focusing occurs. If the same observations are made for a double concave lens such as that in Figure 19–4(b), it is seen that all entering rays are reflected away from the centerline and the light is said to "diverge."

If parallel light rays strike an ideal convex lens, they will focus to a point at a fixed distance f beyond the lens. The distance f is called the *focal length* of the lens. Parallel rays passing through a concave lens will diverge such that they appear to be emanating from a single point some distance f behind the lens (Figure 19–4(b)). The length f is said to be the focal length of the concave lens, but it is given a negative sign. It is customary to express the focusing ability of a lens in terms of the reciprocal of the focal length, $1/f$. This parameter is usually called the "strength" of the lens; the strength S of the lens in *diopters* is defined as the reciprocal of the focal length

$$S_{(\text{diopters})} = \frac{1}{f(\text{meters})}$$

where f is expressed in meters. For example, a lens which focuses parallel light rays to a point 10 cm past the lens has a strength

$$S = \frac{1}{f} = \frac{1}{0.10 \text{ meter}} = 10 \text{ diopters.}$$

A lens which causes parallel light rays to diverge so that they appear to be coming from a

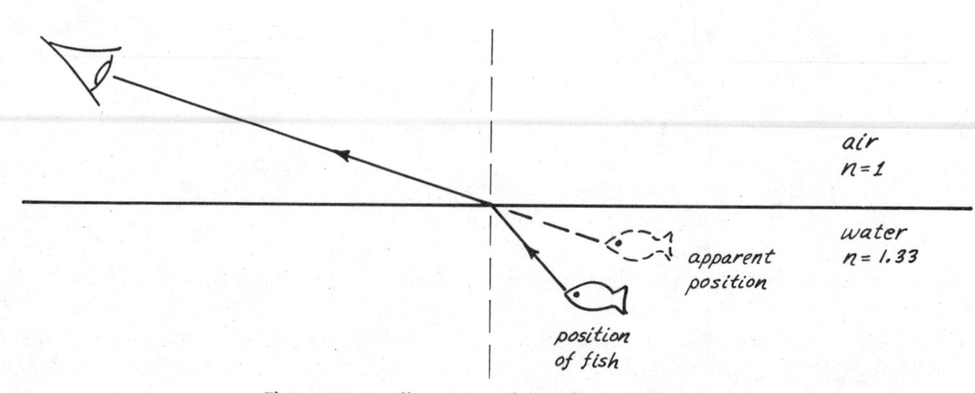

Figure 19–3 Illustration of the effect of refraction.

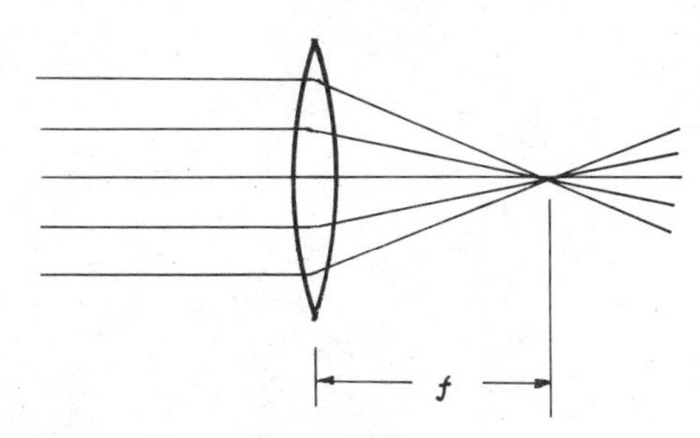

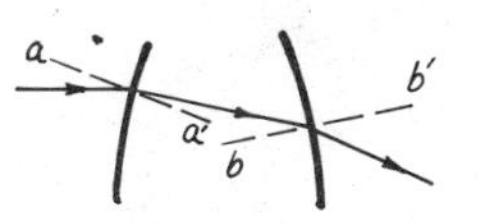

a. A double convex lens focuses light.

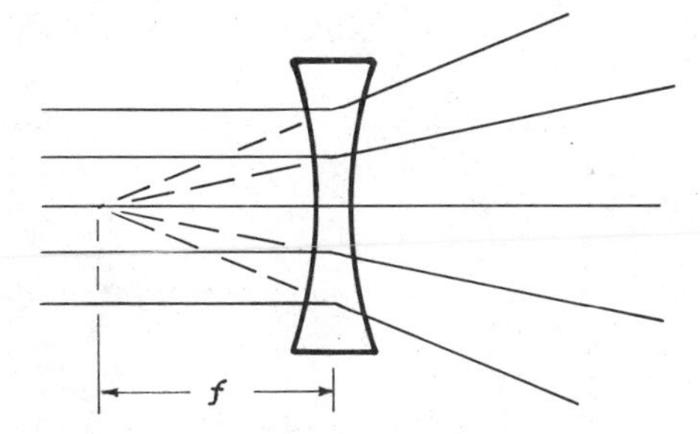

b. A double concave lens causes light to diverge.

Figure 19–4 Refraction in lenses.

point 20 cm behind the lens has a strength

$$S = \frac{1}{f} = \frac{1}{-0.20 \text{ meter}} = -5 \text{ diopters.}$$

With this convention in mind, converging lenses are often referred to as "positive" lenses, while diverging lenses are "negative" lenses.

Converging lenses can be used to form real images of an object, as illustrated in Figure 19–5. For example, the lens in a slide projector can be used to form an enlarged image of a slide on a distant screen. Note that the real image is inverted (Figure 19–5); this is why the slide must be placed in the projector upside-down in order to get a properly oriented picture on the screen. To locate the position of the image with a lens like that in Figure 19–5, the paths of the four light rays labeled a, b, c, and d can be traced. The rays a and d are parallel to the lens axis and will be bent to pass through the focal point at distance f from the lens. The rays b and c pass through the center of the lens and are undeflected. The point where rays a and b cross locates the image of the head of the arrow; all other rays from the head of the arrow which pass through the lens will also focus at that point. Similarly, the point where rays c and d cross locates the point at which the image of the tail of the arrow is projected.

The distance, O, from the object to the lens and the distance, I, from the lens to the projected image are related by the equation

$$\frac{1}{O} + \frac{1}{I} = \frac{1}{f} \qquad \textbf{19–1}$$

where f is the focal length of the lens. This equation is rigorously true only for an idealized thin lens (1), but it is good approximation for all the applications to be considered here. The magnification, M, achieved by the process is equal to the ratio of the image distance to the object distance:

$$\text{M} = \frac{\text{image size}}{\text{object size}} = \frac{-\text{image distance}}{\text{object distance}} = \frac{-I}{O} \qquad \textbf{19–2}$$

where the minus sign in the equation is a reminder that the image is inverted with respect to the object.

Example. An object of length 2 cm is placed 12 cm away from a lens which has a focal length f = 10 cm. Calculate the appropriate distance at which to place a screen for the projection of a focused image. How large will the image be?

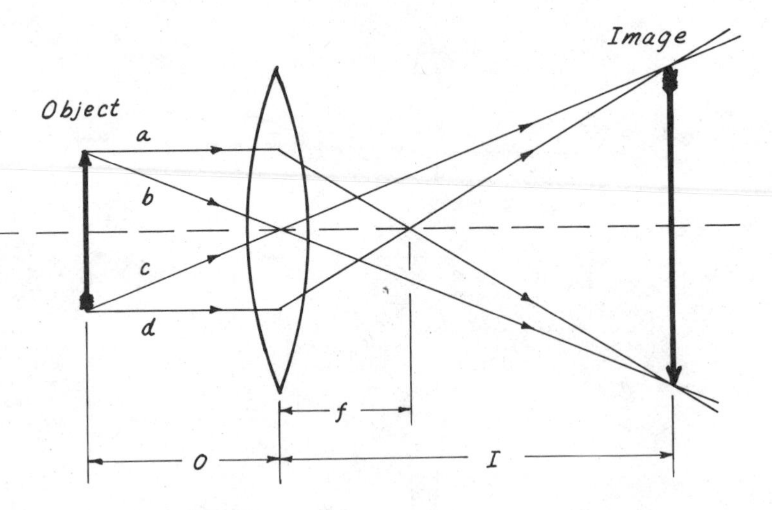

Figure 19–5 A convex lens forms a real, inverted image.

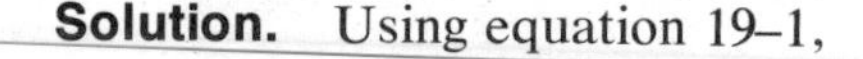

Solution. Using equation 19–1,

$$\frac{1}{12\text{ cm}}+\frac{1}{I}=\frac{1}{10\text{ cm}},$$

$$\frac{1}{I}=\frac{1}{10\text{ cm}}-\frac{1}{12\text{ cm}}=\frac{1}{60\text{ cm}},$$

$$I=60\text{ cm}.$$

The magnification is

$$M=\frac{-I}{O}=\frac{-60\text{ cm}}{12\text{ cm}}=-5.$$

The image is formed at a distance of 60 cm past the lens and is 5 times as large as the object, or 10 cm in length.

Example. If the same lens, $f=10$ cm, is used to form an image of a light which is 15 cm in diameter and located 5 meters from the lens, calculate the location and size of the image.

Solution. Using the lens equation, 19–1:

$$\frac{1}{I}=\frac{1}{10\text{ cm}}-\frac{1}{500\text{ cm}}=\frac{49}{500\text{ cm}},$$

$$I=10.2\text{ cm}.$$

The magnification is $M=-10.2$ cm/500 cm $=-0.02$; therefore, the image diameter is

$$(15\text{ cm})(0.02)=0.3\text{ cm}.$$

This illustrates the fact that the images of very distant objects are formed very near the focal point and that the image size is very small. If the object distance $O=\infty$, then the lens equation becomes

$$\frac{1}{\infty}+\frac{1}{I}=\frac{1}{f}$$

which reduces to $I=f$. If a magnifier is held in sunlight, a bright dot will be formed when it is held at a certain height. This bright dot is the focused image of the sun, formed at the focal length of the lens. This is a convenient method for measuring the focal length of converging lenses.

Diverging lenses do not form real images and therefore cannot be used to project an image on a screen. If you look at an object through a diverging lens, the object appears smaller and farther away. This "image" formed by the diverging lens is called a virtual image; it can be observed with the eye but cannot be projected onto a screen. Your eye, however, can focus the diverging rays to form an image on the retina.

IMAGE FORMATION BY THE EYE

When light from an object strikes the eye, it passes through the cornea, the crystalline lens, and the transparent vitreous humor to form an image on the retina. The nerve endings in the retina transmit electrical impulses to the brain via the optic nerve. The image formation is accomplished by refraction in the cornea and crystalline lens; a focused real image is formed on the retina as shown in Figure 19–6.

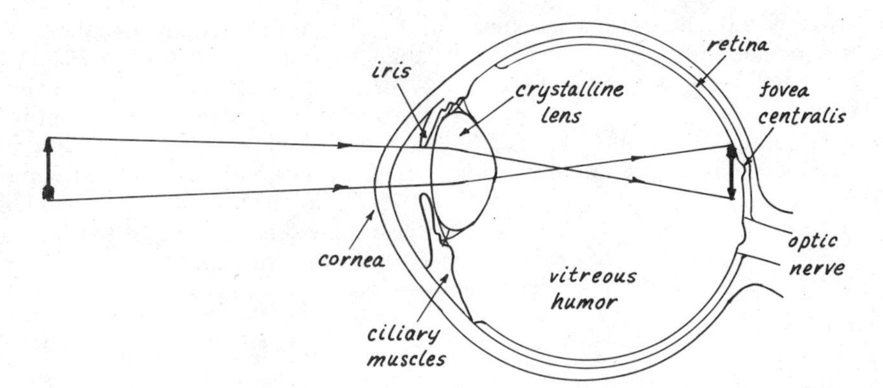

Figure 19–6 Image formation by the eye.

Most of this refraction occurs in the cornea, which has a fixed focal length (2). The crystalline lens is adjustable to provide for accommodation of the focus of the eye for objects at different distances. This accommodation is accomplished by changing the shape of the crystalline lens to alter its focal length. A ring of muscles called the ciliary muscles surrounds this crystalline lens; when these muscles are relaxed, the lens is held in a strained position by ciliary fibers as shown in Figure 19–7(a). In this position the normal eye is focused upon a distant object. To focus on a closer object, a "stronger" lens is required to form the image. The ciliary muscles contract, loosening the ciliary fibers and allowing the crystalline lens to take on a more rounded shape (Figure 19–7(b)). This shortens the focal length of the crystalline lens and increases its refracting capability.

Although the eye is a two-lens system (3,4), it is instructive to consider a simplified model in which the two lenses (cornea and crystalline lens) act together as a single lens to form images on the retina. The effective position of this single lens is about 2.2 cm from the retina in the average human eye (2). As seen from the examples in the previous section, when the object viewed is at a large distance from the eye, the image is formed near the focal length of the lens. If the object distance is infinite, then from equation 19–1 it can be seen that the image distance is exactly equal to the focal length. Therefore, the effective focal length of the eye when viewing distant objects must be about 2.2 cm, the distance to the retina. Therefore, the strength of the eye's lens in diopters is

$$S = \frac{1}{f} = \frac{1}{0.022 \text{ meter}} = 45 \text{ diopters.}$$

When the object viewed is closer to the eye, the image is no longer formed at the focal length of the lens, but at the position given by equation 19–1. Since the image distance is fixed, the focal length of the lens must change to produce a focused image on the retina.

Example. Suppose that the distance from the effective center of the eye's lens sys-

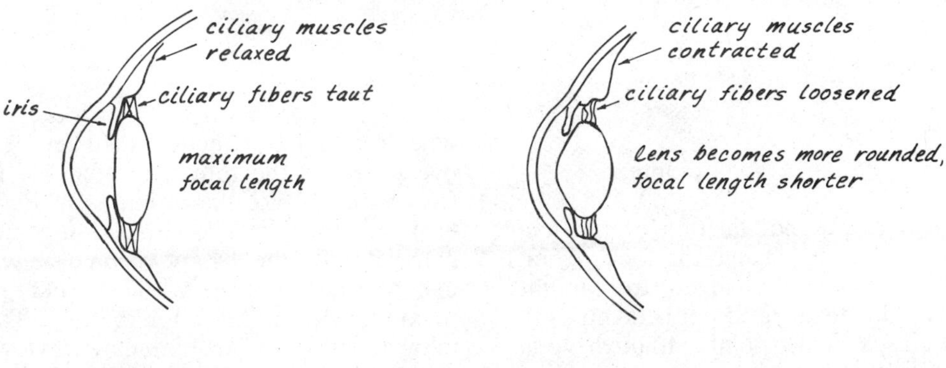

ciliary muscles relaxed
iris
ciliary fibers taut
maximum focal length

ciliary muscles contracted
ciliary fibers loosened
lens becomes more rounded, focal length shorter

a. Distant vision

b. Close vision

Figure 19–7 Accommodation of focus.

tem to the retina is 2.0 cm for a particular person. Find the focal length and lens strength required to focus on objects at ∞, 50 meters, 1 meter, and 25 cm.

Solution. Equation 19–1 may be used with $I = 2.0$ cm as a fixed parameter. For $O = \infty$:

$$\frac{1}{\infty} + \frac{1}{2\text{ cm}} = \frac{1}{f}.$$

Since $1/\infty = 0$, then

$f = 2$ cm = image distance

$$S = \frac{1}{0.02\text{ meter}} = 50\text{ diopters.}$$

This would apply to the viewing of a star; its distance is effectively infinite. This should be the relaxed focal length, since the normal eye should be completely relaxed when viewing a distant object. For an object at 50 meters:

$$\frac{1}{5000\text{ cm}} + \frac{1}{2\text{ cm}} = \frac{1}{f},$$

$$\frac{1}{f} = \frac{5002}{10000},$$

$$f = 1.9992\text{ cm},\ S = 50.02\text{ diopters.}$$

For most practical purposes, these results are the same as those obtained for an object at an infinite distance. Very little adjustment of the focal length is required to change the focus from very distant objects to objects as close as 10 to 20 feet. Most of the adjustment is required to view very close objects. To focus on an object at 1 meter:

$$\frac{1}{100\text{ cm}} + \frac{1}{2\text{ cm}} = \frac{1}{f},$$

$$f = 1.96\text{ cm},\ S = 51\text{ diopters,}$$

and for an object at 25 cm:

$$\frac{1}{25} + \frac{1}{2} = \frac{1}{f},$$

$$f = 1.85\text{ cm},\ S = 54\text{ diopters.}$$

Note that a larger modification of the focal length is required to change the focus from 1 meter to 25 cm than to change it from infinity to 1 meter. The distance 25 cm is taken as the standard closest-focus point, although young people with normal vision can usually focus somewhat closer than 25 centimeters.

The focal length of the lens system changes only about 6% to 8% in the entire range of focus of the eye. However, since the cornea remains fixed and the entire change is accomplished by the crystalline lens, this represents a large change in focal length for that part of the lens system. The focal length of the crystalline lens is thought to change by about twenty per cent (2).

COMMON VISION DEFECTS

If the focal length of the eye's lens is too short as shown in Figure 19–8(a), the light rays will focus before they reach the retina, resulting in a blurred image on the retina. This condition is referred to as myopia or "nearsightedness." It can be corrected by inserting a diverging or "negative" lens in front of the eye to cause the rays to diverge slightly before entering the eye (Figure 19–8(b)). If the focal length of the eye is too long (hyperopia or "farsightedness"), then the light would focus only at a distance greater than the distance to the retina. A converging lens (Figure 19–8(d)) will correct this defect. The focal length of the necessary correcting lens can be calculated approximately by the relationship

$$\frac{1}{f_{eye}} + \frac{1}{f_{lens}} = \frac{1}{f_{corrected}} \qquad \textbf{19–3(a)}$$

or equivalently

$$S_{eye} + S_{lens} = S_{corrected}. \qquad \textbf{19–3(b)}$$

This is a good approximation for thin lenses which are placed very close together, and is reasonably accurate for calculating prescriptions for eyeglasses. For viewing distant objects (object distance approaching ∞) the lens equation 19–1 yields

$$\frac{1}{f_{corrected}} = \frac{1}{d}$$

where d is the effective distance from the eye's lens to the retina, so that the focused rays would strike the retina as in Figure 19–8(b) and 19–8(d). The correction is normally calculated for the distant vision case when the eye's lens is completely relaxed and f_{eye} is the maximum focal length for the eye. This then allows the eye to accommodate to view closer objects by decreasing f_{eye} as described previously.

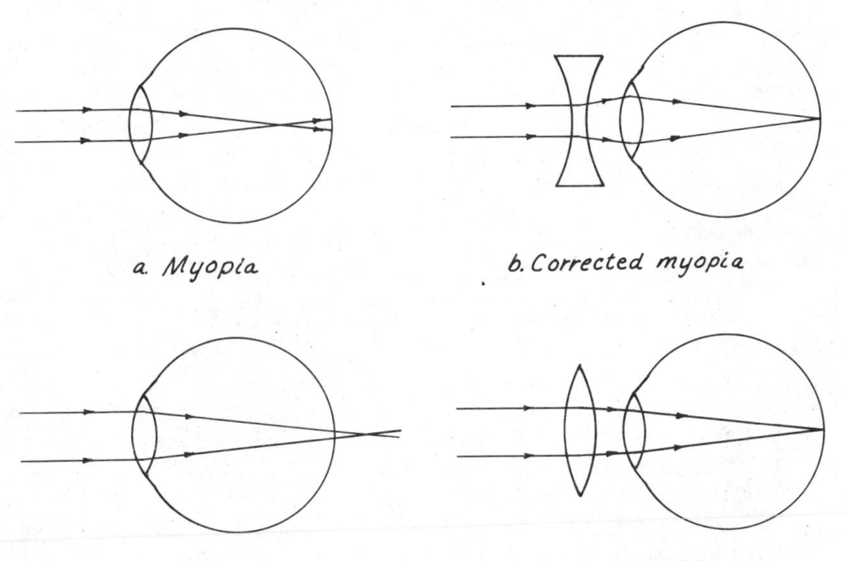

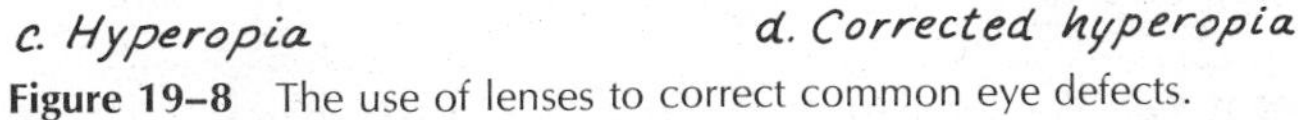

Figure 19–8 The use of lenses to correct common eye defects.

Example. If the distance to the retina is 2 cm and the relaxed focal length of the eye's lens is 1.96 cm, calculate the range of vision assuming that the focal length can change by 8%. What lens strength in diopters would be required to correct the vision?

Solution. With $I = 2$ cm and $f = 1.96$ cm, the maximum focus distance can be calculated from equation 19–1:

$$\frac{1}{O} + \frac{1}{2\text{ cm}} = \frac{1}{1.96\text{ cm}}$$

$$\frac{1}{O} = 0.51 - 0.50 = 0.01$$

$$O = 100\text{ cm} = 1\text{ meter.}$$

No object more distant than 1 meter can be seen clearly, because the relaxed focal length is the maximum focal length. Given an 8% focusing range, tightening the ciliary muscles can reduce the focal length to 1.80 cm. Therefore, the close focal point can be calculated:

$$\frac{1}{O} = \frac{1}{1.80\text{ cm}} - \frac{1}{2.0\text{ cm}}$$

$$O = 18\text{ cm.}$$

Therefore, this myopic eye can focus from 1 meter down to 18 cm, compared to a focus range from infinity down to 25 cm for the normal eye. Using equation 19–3(a) to calculate the correction required,

$$\frac{1}{1.96\text{ cm}} + \frac{1}{f_{\text{lens}}} = \frac{1}{2.0\text{ cm}},$$

$$f_{\text{lens}} = -1\text{ meter}, \; S_{\text{lens}} = -1\text{ diopter.}$$

The calculation yields a negative focal length, indicating that a diverging lens is needed. Since lens prescriptions are commonly stated in diopters, equation 19–3 (b) is often more convenient.

$$S_{\text{eye}} = \frac{1}{.0196\text{ meter}} = 51\text{ diopters}$$

$$S_{\text{corrected}} = \frac{1}{.02\text{ meter}} = 50\text{ diopters}$$

$$51\text{ diopters} + S_{\text{lens}} = 50\text{ diopters}$$

$$S_{\text{lens}} = -1\text{ diopter.}$$

Note that the corrective lens (-1 diopter) has a strength which is rather small compared to the approximately 50 diopter strength of the normal lens system.

Example. Repeat the calculation above for an eye lens system which has a focal length of 2.04 cm (hyperopia).

Solution. Using the parameters from the previous example, an 8% focus range would allow focal length variation from 2.04 cm to 1.88 cm. Note that the focal length must be reduced to 2.0 cm to focus on distant objects, so the useful range is from 2.0 cm to 1.88 cm. The results of the calculations are:

Distant focus point: infinity

Close focus point: 31 cm

Corrective lens strength to change relaxed focal length to 2.0 cm: +1 diopter.

It might seem that no corrective lens is indicated in this case, since the focus range from infinity to 31 cm may be adequate. But the ciliary muscles are under constant strain because they must tighten to reduce the focal length from 2.04 cm to 2.00 cm even for the viewing of distant objects. The normal eye would be completely relaxed when viewing distant objects. The constant strain and greater ciliary muscle tension for close objects tend to cause headaches and other symptoms.

Up to this point it has been assumed that the lenses considered were symmetric and with ideal optical properties. The common eye defect known as astigmatism occurs when the lens has different focal lengths for light rays striking it in different planes, as illustrated in Figure 19–9(a). Light coming from a distant bright object such as a star will therefore not focus to a point behind the lens because of this asymmetry. If a screen were placed at distance *f* in Figure 19–9(a), the image would be a short bright line in the horizontal plane rather than a bright dot, since the light in the vertical plane is focused at that distance but that in the horizontal plane is not. The astigmatism arising from asymmetry in the curvature of the eye's lens can be detected with the aid of a fan chart such as that in Figure 19–9(b). To a person with normal vision the lines of the chart will appear to have equal darkness. If astigmatism is present, some of the lines will appear darker than others because the eye is presumably focused on the lines in one plane, while the lines in another orientation will be slightly out of focus.

Other vision defects may become evident in the comparison of the clarity of vision in dim light and bright light. It has been assumed that all parallel rays striking a lens will focus at one point, the principal focus point. This is never precisely true for actual lenses. Light striking the outer part of the lens will focus at a slightly different point than the light striking the center of the lens, resulting in a slightly blurred image. This effect occurs even with perfectly spherical glass lenses and is referred to as *spherical aberration*. The clarity of the image can be improved by inserting a small aperture in front of the lens so that the light passes through only a small portion of the center of the lens. The iris of the eye forms such an aperture, and its action in response to changing light intensity is shown schematically in Figure 19–10. In dim light the iris opens to admit more light as shown in Figure 19–10(a). The effects of different focal lengths in different parts of the lens are greatly exaggerated to show that the light from a point on a distant bright object will form a small disc on the retina instead of a bright point. In bright light the aperture formed by the iris is smaller and the image is therefore sharper. The sharpness of the image is degraded by a wide iris opening, even for a perfectly symmetrical lens, and the effects of any lens defects are made more pronounced in dim light because

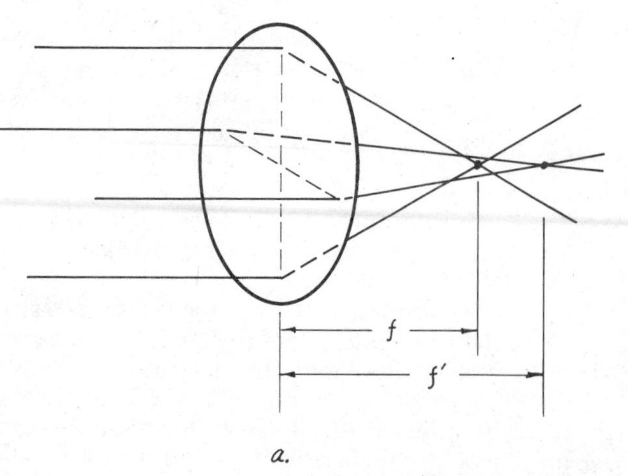

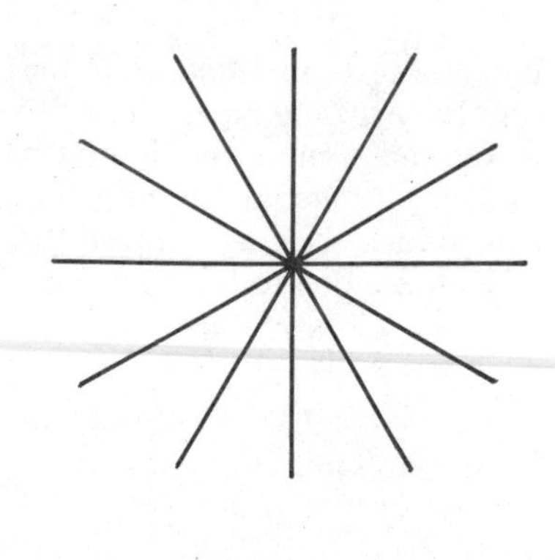

Figure 19–9 Astigmatism.

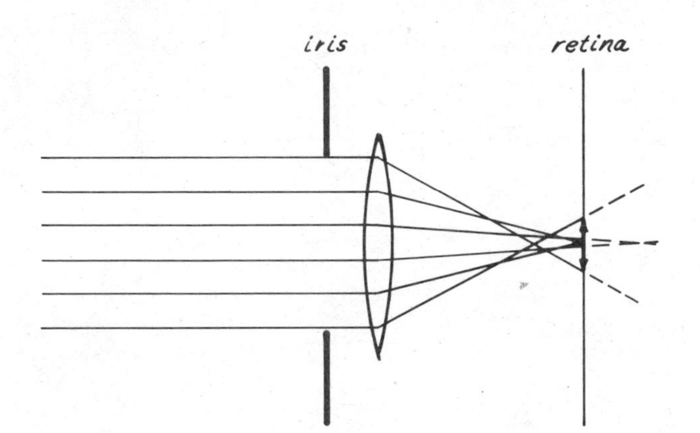

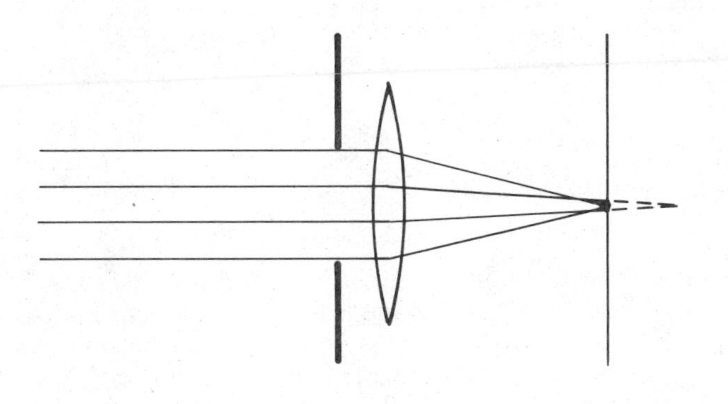

Figure 19–10 The effect of aperture size upon focusing.

more of the lens area is being used to form the image on the retina. The increased sharpness of image gained by using a small aperture is the reason for the practice of "squinting" by nearsighted persons.

SIMPLE OPTICAL INSTRUMENTS

An additional lens or collection of lenses can be used to aid the lens of the eye, either by providing a larger image of an object to be observed or by making possible the viewing of areas which would otherwise have been inaccessible. Some simple examples will be given here to illustrate the physical principles involved.

THE SIMPLE MAGNIFIER. A single converging lens can be used to produce a larger image on the retina than that which could be formed with the unaided eye. As noted above in the discussion of the lens system of the eye, the normal eye cannot focus clearly on an object which is closer than about 25 cm from the eye. The image formed with an object distance of 25 cm is therefore the largest image which can be formed on the retina, as illustrated in Figure 19–11(a). However, if a converging lens is held in front of the eye as shown in Figure 19–11(b), the eye can be moved closer to the object, resulting in a larger retinal image. To achieve such magnification the object must be placed at some distance from the magnifying lens which is less than its focal length. For example, suppose the magnifying lens has a focal length $f = 10$ cm and the object to be viewed is placed at an object distance $O = 7.1$ cm from the lens. When the lens equation, 19–1, is applied,

$$\frac{1}{7.1 \text{ cm}} + \frac{1}{I} = \frac{1}{f},$$

the image distance resulting is -25 cm. The minus sign indicates that the image is a *virtual image* formed on the same side of the lens as the object. The meaning of a virtual image in this case is that if one looks through the lens at the object, it will appear to be at a distance of 25 cm, as shown in Figure 19–11(b). The mag-

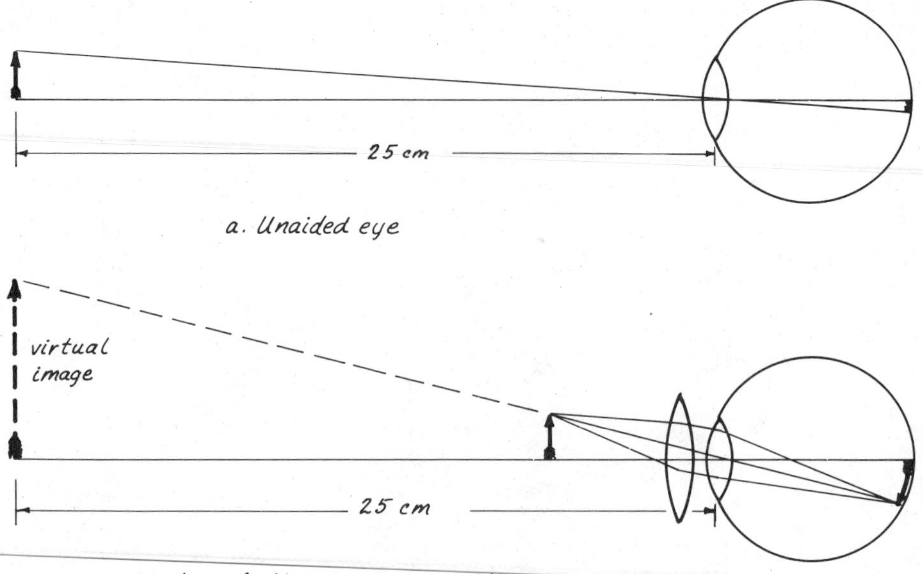

Figure 19–11 The simple magnifier.

nification as calculated from equation 19–2 will be

$$M = \frac{25 \text{ cm}}{7.1 \text{ cm}} = 3.5,$$

indicating that the virtual image seen at the apparent distance of 25 cm will be 3.5 times as large as the object if viewed by the unaided eye at that distance. This condition achieves the maximum magnification which can be obtained for this magnifier, but the ciliary muscles of the eye are under considerable strain since they are focused at 25 cm.

Suppose the object were placed at a distance of 10 cm from the lens, that is, exactly at its focal length. From equation 19–1,

$$\frac{1}{10 \text{ cm}} + \frac{1}{I} = \frac{1}{10 \text{ cm}},$$

it is seen that the image distance is now infinite. Therefore, if you look through the lens you see an enlarged image which is focused when your eye is relaxed, as if you were looking at a distance object. The magnification is not as great as in the previous case, but eye strain is reduced. The magnification when the virtual image is formed at infinity is (Reference 10)

$$M = \frac{25}{f}$$

which in this case is $M = 2.5$ compared to $M = 3.5$ when the image is formed at the close focal point (25 cm). This would usually be referred to as a 2.5× magnifier or a "2.5 power" magnifier, assuming that it would normally be used to form a distant image.

The simple magnifier has been discussed in detail because it is a constituent of many optical instruments, serving as the "eyepiece" lens for instruments such as microscopes, telescopes, and camera viewfinders. In the practical use of such instruments, the eyepiece can usually be adjusted to focus the image. For prolonged instrument use, it is important to focus the image so that the effective image distance is infinite, so that the eye (i.e., the ciliary muscles) can be relaxed while viewing the image.

THE COMPOUND MICROSCOPE. The magnifier discussed above is sometimes called the "simple microscope," in contrast to the two-lens arrangement shown in Figure 19–12, which is called the compound microscope. The far lens, which is called the *objective,* is a lens of very short focal length which is placed close to the object to be viewed. In the case of microscopes which are used to view microorganisms, the focal lengths of the objective lenses are so short that they are almost in contact with the sample. Sometimes the objective lens is actually immersed in oil or the liquid in which the microorganisms are suspended. The position of the objective lens is

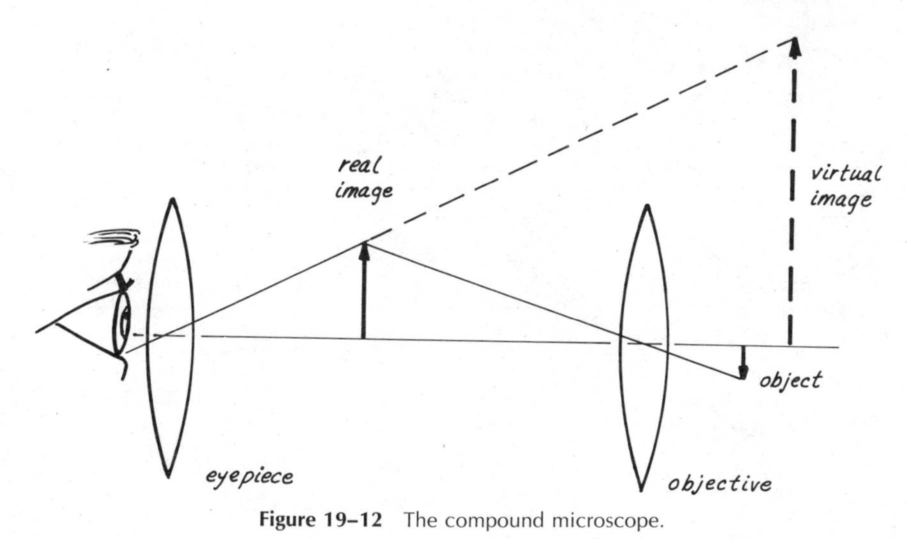

Figure 19–12 The compound microscope.

such that the object distance is slightly greater than the focal length of the objective. A magnified real image is then formed at some point inside the barrel of the microscope as shown in Figure 19–12. The eyepiece lens then acts as a simple magnifier with which to view this image. The eye views the enlarged virtual image which has been magnified by both lenses, as indicated by the enlarged image in Figure 19–12. The magnification of the compound microscope is the product of the magnifications of the objective and eyepiece,

$$M = M_{\text{objective}} \times M_{\text{eyepiece}}.$$

With laboratory microscopes, many different magnifications are obtainable by changing objectives and eyepieces. Often three objectives will be mounted on a turret for convenient magnification change; typical magnifications might be 10×, 20×, and 50× for the available objectives. Eyepieces may then be changed at will to change magnification. If a 10× eyepiece is used with the above objectives, the magnifications available are 100×, 200×, and 500×.

From what has been said thus far, it could be assumed that the magnification could be increased without limit by increasing the power of the microscope lenses. However, limits arise from two different considerations. First, the images formed by a lens system are never perfectly "sharp" because of imperfections in the lenses. These lens imperfections cause a slight blurring or overlap of the images of two adjacent points on the object, so that increasing the magnification beyond a certain power provides no new information about the object. The second consideration is the theoretical limit imposed by the wave nature of light. Optical images are formed by reflecting light from an object, but to be distinguished an object must be somewhat larger than the wavelength of the light which is reflected from it. In limiting cases, the resolution of a microscope can be improved by using monochromatic blue or violet light sources, which have the shortest wavelengths in the visible spectrum. If an object is small compared to the wavelength of light, however, it cannot be observed by a light microscope. Many viruses are so small that they cannot be resolved by a light microscope. The increased magnification of an electron microscope is made possible by the shorter effective wavelength achieved, as discussed in Chapter 20.

The Ophthalmoscope. This is an instrument for viewing the retina of the eye. In its simplest form it consists of a light source and a mirror with an aperture as shown in Figure 19–13. Light is reflected from the mirror into the subject's eye and illuminates the retina. If the subject's eye is normal and focused for the viewing of a distant object, then the focal length of the eye's lens will be at the retina. Therefore, light rays reflected from a point on the retina will be rendered parallel by the eye's lens when passing out of the subject's eye as shown in the figure, and these rays can be focused by the observer's eye to form a clear image of the retina. Note that the lens of the subject's eye is being used as a simple magnifier to provide an enlarged image of the retina for the observer. If the subject's eye is abnormal, correcting lenses may be added to bring the image of the retina into focus.

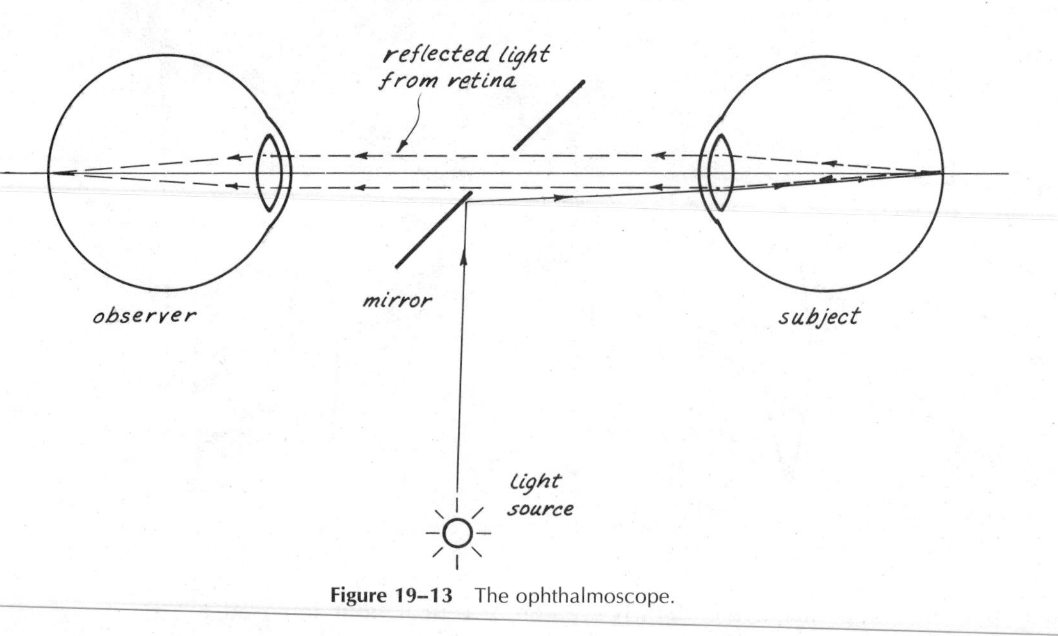

Figure 19–13 The ophthalmoscope.

FIBER OPTICS. Complicated optical systems have been developed for bronchoscopes, cystoscopes, and other instruments for viewing the internal tracts of the body. More recently, fiber optics probes have been developed which may supplant and improve upon such systems.

Under the proper conditions, light will be transmitted through a thin fiber with very little loss. In most cases, a light ray striking a boundary between two transparent media will be partially reflected and partially transmitted. If the path of the ray is perpendicular to the boundary, most of it will be transmitted, but if the direction is gradually changed so that it becomes more nearly parallel to the boundary, more and more of it will be reflected. For a given type of boundary there is a certain critical angle, beyond which all the light is reflected from the boundary. If light enters the end of a small fiber, it will pass along the fiber by means of multiple internal reflections since its angle with the boundary will always be less than the critical angle. A large bundle of such fibers constitutes a "light pipe" which will transmit light even if the bundle is bent into tight curves and complicated shapes. If an illuminated object is placed at one end of the bundle, an image of the object can be transmitted to the other end (Figure 19–14). Under conditions of total internal reflection, the light from adjacent fibers does not mix, and therefore each fiber transmits information about the luminosity of a localized area of the object. A "mosaic" image is formed by the light and dark ends of the fibers, analogous to the construction of newspaper photographs from dots of varying blackness.

A flexible bundle of fibers can easily follow the change of direction in an internal tract of the body. Light can be passed down the outer fibers to illuminate the tract ahead of the bundle. Light reflected back through the central core of the bundle can provide a detailed image of the tract if the fibers are small enough. Fiber-optic bundles are now often used in gastroscopes (11) and other medical optical instruments. The flexible bundles may be manufactured with lengths in excess of 10 ft, and some experimentation has been done with the transmission of high intensity laser light for cancer treatment and other therapeutic treatments in inaccessible areas.

COLOR VISION

When a beam of white light is passed through a prism as shown in Figure 19–15, it is separated into colored bands by refraction in the prism. This separation occurs because the index of refraction of the prism varies with the frequency of the light, and therefore different frequencies are bent through different angles. The prism serves to illustrate the fact that white light is composed of all the colors of the visible spectrum and the fact that color is associated with the frequency (or wavelength) of the light.

The human eye is sensitive to electromagnetic waves in a small range of frequencies, extending from about 3.9×10^{14} Hz to $7.9 \times$

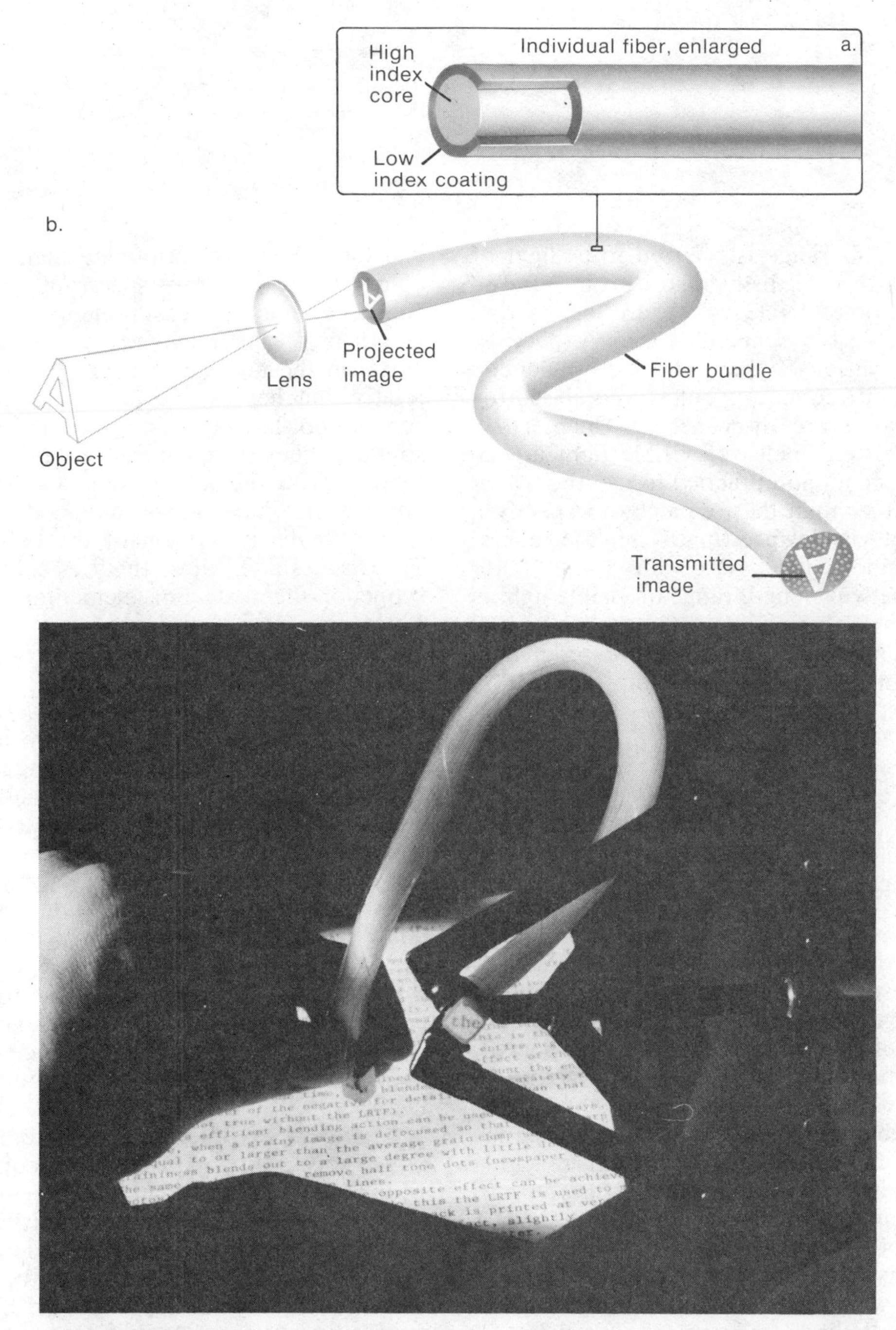

Figure 19–14 A bundle of very thin fibers, each acting as a light pipe, will transmit an image if the arrangement of the fibers is the same at both ends. (Photo by M. Velvick, Regina.) (From L. H. Greenberg, *Physics with Modern Applications,* W. B. Saunders, 1978.)

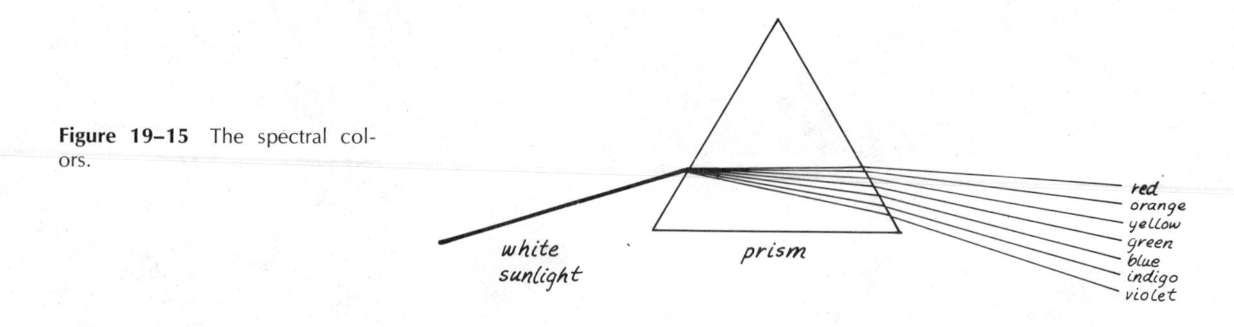

Figure 19–15 The spectral colors.

10^{14} Hz. This range is referred to as light or "visible light" to distinguish it from infrared and ultraviolet light, which are below and above the visible frequency range, respectively. The various types of electromagnetic waves with frequencies below and above the visible range are discussed in Chapter 20. Since the frequencies of visible light are so large, it is common practice to use the wavelength rather than the frequency to specify a part of the visible spectrum. Using the general wave relationship $v = f\lambda$ with $v = 3 \times 10^8$ m/sec, the wavelength range of visible light is found to be about 3.8×10^{-7} to 7.7×10^{-7} meters. This range is usually expressed in angstrom units, where one angstrom (Å) = 10^{-10} meters:

Visible light:

colors	red	to violet
frequencies	3.9×10^{14} Hz	to 7.9×10^{14} Hz
wavelengths	7700 Å	to 3800 Å

The range from red light (lowest frequency, longest wavelength) to violet light (highest frequency, shortest wavelength) is only about a factor of two in frequency, or one octave. This can be compared to the 10 octave audible frequency range for sound (20 to 20,000 Hz). This visible range, of course, corresponds closely with the range of wavelengths reaching the earth in greatest abundance from the sun. Sunlight is essentially white light, having a mixture of all wavelengths in the visible spectrum. This spectrum may be readily seen in a rainbow, which is formed by the refraction of sunlight in falling raindrops.

The retina of the eye contains two types of light-sensitive nerve ending commonly referred to as "rods" and "cones." The cones are responsible for the ability to discriminate between colors. For example, if light with wavelength 5000 Å strikes the eye, most persons can identify the color as green. The eye can identify and discriminate many different colors in the rainbow spectrum: this corresponds roughly to measurements of the wavelength of the light. No corresponding ability exists in the hearing of most persons. While relative pitches can be perceived, most persons do not have the ability to distinguish absolute pitches (frequencies), a feat analogous to identifying the color of a light source. Various theories have been advanced to explain the color discrimination of the cone vision. The discussion of these theories is outside the scope of the text. For elementary reviews, References 5 to 8 are suggested, and Reference 9 offers a more detailed development of the subject.

The rods have little color discriminating ability, if any. The rods are more numerous than the cones, typically numbering about 130 million compared to 7 million cones (6). The cones are more concentrated near a central spot on the retina called the fovea centralis (2), and the number of cones is very small in the extremities of the retina. For this reason, color sensitivity is greatest when looking directly at an object, because the light rays from the object fall near the fovea centralis. The color sensitivity associated with extreme peripheral vision is very slight, since the light falls on areas of the retina which are deficient in cones. This can be demonstrated by moving two objects of different color from a point directly in front of your eye to a point beside your head where they can just be seen. At this point the colors will be indistinguishable.

When the light intensity is low, vision occurs primarily by stimulation of the rods. As discussed in Chapter 16, a nerve cell requires a stimulus above a certain threshold to depolarize the cell and produce the electrical "action potential," which in this case would carry the visual information to the brain. The threshold for the rods is considerably lower than that for the cones. Since the rods are less concentrated near the direct vision part of the retina (fovea), the sensitivity of direct vision

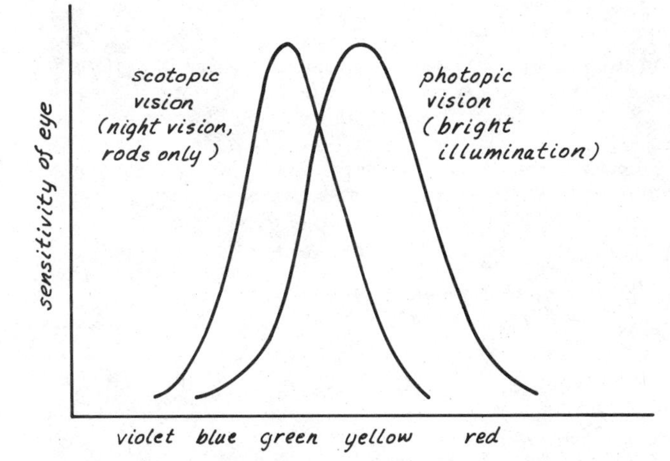

Figure 19–16 Approximate sensitivity curves for the eye.

in dim light is less than the sensitivity of peripheral vision. The sensitivity of the rod vision or "night vision" is significantly decreased by a deficiency in vitamin A.

Dark-adapted vision (almost entirely rod vision) is referred to in medical literature as "scotopic vision." It is most sensitive for light of wavelength near 5100 Å, which is in the green region. In bright light both rods and cones are active and the peak sensitivity occurs at about 5800 Å, which is in the yellow region of the spectrum and corresponds roughly to the wavelength which is most abundant in sunlight. This light-adapted vision is referred to as "photopic vision." Approximate sensitivity curves are shown in Figure 19–16. The difference in the wavelength associated with the peak sensitivity of the eye under bright and dim lighting conditions may be demonstrated by observing a red rose in twilight conditions. When the light is bright (photopic vision), the red petals will appear much brighter than the surrounding green leaves. But as the light level falls, the relative brightness of the green leaves will increase. Under very dim light (scotopic or rod vision) the green leaves will often appear brighter than the red petals.

SUMMARY

Refraction of light refers to the bending of a light ray when it crosses a boundary between two media in which its speeds are different. Refraction of light by lenses, including the lens of the eye, is what makes possible the formation of images of illuminated objects. Lenses may be characterized by the distance from the lens at which they would focus parallel light rays; this distance is called the focal length. The strength of a lens in diopters is the reciprocal of the focal length in meters. The distance I at which an image will be formed when an object is at distance O from a lens is given by the lens equation

$$\frac{1}{O}+\frac{1}{I}=\frac{1}{f}.$$

Since the image distance I in the human eye is fixed, the focal length f of the eye's lens must be changed to form clear retinal images of objects at different distances from the eye. This process is called accommodation.

If the focal length of the eye's lens is too short, the image of a distant object will be formed in front of the retina. This condition, myopia, may be corrected with a diverging lens of appropriate diopter strength. If the focal length of the lens of the eye is too long, the image would tend to be formed behind the retina. This hyperopic condition may be corrected with an appropriate converging lens. In the case of astigmatism, a correcting lens with different focal lengths along different axes must be used since the eye lens is not symmetric. The human eye can normally focus to within about 25 cm from the eye. Optical instruments such as the magnifier and compound microscope make possible the formation of enlarged images for the eye to view. Medical applications of devices such as the ophthalmoscope and fiber optics probes make possible the viewing of otherwise inaccessible areas.

The different colors of light represent different frequencies and consequently different wavelengths of light. The human eye is sensitive to a narrow range of frequencies, but within that range it can readily identify frequency or color differences. The retina contains two types of light sensitive receptors; rods and cones. The cones provide the color sensitivity, while the rods provide the greater overall light sensitivity needed for night vision.

REVIEW QUESTIONS

1. Why does an underwater object appear closer than it really is?
2. A shore fisherman must make allowances for refraction when spearing fish. Does a skin diver have to make similar adjustments when spearing fish? Explain.
3. Draw a sketch of the image formed by an object placed outside the focal length of a converging lens. Will the image be real or virtual? Erect or inverted? Magnified or diminished?.
4. What is the difference between a positive lens and a negative lens?
5. Describe what happens to the lens of the eye during "accommodation" to view a close object.
6. Considering the nature of accommodation, what changes may occur with age which may require the use of bifocal lenses for a normal range of vision?
7. Describe what happens to the light rays from a distant object in a myopic eye. How can this vision defect be corrected?
8. Describe what happens to the light rays from a distant object in a hyperopic eye. How can this vision defect be corrected?
9. Describe astigmatism and the methods of correcting it.
10. What is meant by spherical aberration?
11. What is the physical difference between red light and blue light?
12. Why are colors less distinct when viewed in dim light?

PROBLEMS

1. A magnifying lens is held in sunlight and is observed to focus the sun's rays to a bright point on a piece of paper when the lens is held 5 cm above the paper. What is the lens strength in diopters?
2. A common corrective lens for myopic vision might have a lens strength of −2 diopters. What is the focal length of this lens?
3. If you have a lens of focal length 3 cm and you wish to project an image on a white screen 12 cm away from the lens, where would you put the illuminated object? If the object is 3 cm high, how large will the image be?
4. If a bright object 20 cm from a lens projects a focused image 30 cm on the other side of the lens:

a. What is the focal length of the lens?
b. What is the lens strength in diopters?

5. An illuminated object 10 cm from a lens projects a focused image on a card 5 cm beyond the lens. What is the focal length of the lens?

6. An object which is 3 cm high is placed 5 cm away from a lens of focal length 4 cm.

   a. Where will an image of the object be projected?
   b. How large will the projected image be?
   c. What is the lens strength in diopters?

7. An object 6 cm from a lens of focal length 4 cm projects a real image.

   a. Find the distance from the lens to the image formed.
   b. If the object is 2 cm long, how long will the image be?

8. A lens with focal length 10 cm is used to project an image on a screen 50 cm away from the lens.

   a. How far must the object be placed from the lens for the image to be sharply focused?
   b. If the size of the object is 5 cm, what will be the size of the projected image?

9. A slide projector has a lens of focal length 10 cm and is used to project an image of the slide on a screen which is 3 meters away from the lens.

   a. How close to the slide projector lens must the slide be placed?
   b. If the maximum dimension of the slide is 3.5 cm, how large will the projected image be?

10. If an object is placed 25 cm from a lens of focal length $f = 5$ cm, where will the image be formed? What is the magnification?

11. A camera is focused by changing the distance from the lens to the film. If a camera lens with focal length $f=50$ mm is focused on an object at a distance of 2 meters, what is the distance from the lens to the film?

12. What is the focal length of a lens which forms an image 24 cm from the lens when the object is 40 cm away?

13. What magnification can be achieved by a magnifying lens with a focal length of 5 cm?

14. The process of forming an image of an object by reflecting waves from the object is such that the minimum resolvable distance is on the order of the wavelength of the wave. The dolphin emits sound frequencies as high as 250,000 Hz. If the speed of sound in water is 1460 m/sec what is the size of the smallest object a dolphin can "see" with his sonar imaging system?

15. A camera has a lens with focal length 5 cm. It focuses the images on the film by altering the distance between the lens and the film. If a picture is taken of a man 1.8 m tall at a distance of 5 m, what must be the lens-to-film distance? What is the lens-to-film distance if the man is 3 m away? What is the size of the image formed on the film in each case?

16. A camera lens of focal length 50 mm is used to photograph an object which is 1 meter from the lens.

    a. What is the distance from the lens to the film for a focused image?
    b. If the lens is 50 mm from the film when focused at infinity, how far must the lens be moved to change focus from infinity to 1 meter?
    c. If the size of the image on the film is 5 mm, how large was the object which was photographed?

17. A person has a camera with a 50 mm focal length lens and wishes to take a picture of an object such that the size of the image on the film is exactly equal to the actual size of the object, i.e., magnification = 1.

    a. How far away from the lens must the object be placed?
    b. The lens-to-film distance is normally 50 mm if focused on infinity. What length extension tube must be placed between the camera and the normal lens to get a magnification of 1?

18. If a book with 3 mm print is held at the closest focal point, 25 cm from the eye, what will be the height of the images of the letters on the retina? Assume the lens-to-retina distance to be 2 cm.

19. If the maximum distance for clearly focused vision for a person is 2 meters, what is the focal length of the lens of his eye? (Assume that the lens-to-retina distance is 2 cm.) What is the strength in diopters of a corrective lens which will allow him to focus on distant objects?

20. If a farsighted person has a closest focus point at 2 meters, what correcting lens should be used to allow him to see clearly at a distance of 25 cm? Assume 2 cm lens-to-retina distance.

21. The relaxed focal length of the lens of a nearsighted person's eye is 1.90 cm compared to a 2.0 cm lens to retina distance.

    a. What is the maximum distance for clear vision for this person?
    b. What lens strength in diopters would be required to correct the vision?

22. With advancing age, the range of accommodation of the eye is reduced, causing difficulty in close-focusing for reading, etc. For this problem, assume the lens to retina distance to be 2 cm.

    a. At age 6 a person had a range of accommodation from $f = 2.0$ cm to $f = 1.67$ cm for the lens of his eye. What is his closest focusing distance?
    b. At age 60 the same person could accommodate the focal length of his eye's lens only over the range $f = 2.0$ cm to $f = 1.92$ cm. What is the shortest distance at which he can focus clearly?
    c. What lens prescription (lens strength in diopters) would you give to him to use as reading glasses to focus clearly at 25 cm?

REFERENCES

1. Tippens, P. E. *Applied Physics,* 2nd ed. New York: McGraw-Hill, 1978.
2. Ackerman, E. *Biophysical Science.* Englewood Cliffs, N.J.: Prentice-Hall, 1979.
3. Glasser, O., ed. *Medical Physics.* Chicago: Year Book Publishers, Inc., 1944, Vol. 1.
    a. Luckiesh, M., and Moss, F. K. "Light, Vision, and Seeing," pp. 672–684.
    b. Sheard, C. "Optics: Ophthalmic, With Applications to Physiologic Optics" pp. 830–869.
4. Ogle, K. N. *Optics: An Introduction for Ophthalmologists,* 2nd ed. Springfield, Illinois: Charles C Thomas, 1976.
5. Blackwood, O. H., Kelly, W. C., and Bell, R. M. *General Physics,* 4th ed. New York: Wiley, 1973.
6. White, H. E. *Modern College Physics,* 6th ed. New York: D. Van Nostrand Co., 1972.
7. Land, E. H. *Experiments in Color Vision,* Scientific American, May 1959, p. 84.
8. Time-Life Books, *Light and Vision,* Life Science Library. New York: Time-Life Books, 1965.
9. Le Grand, Y. *Light, Colour and Vision,* 2nd ed. London: Chapman and Hall Ltd., 1968.
10. Jenkins, F. A., *Fundamentals of Optics,* 4th ed. New York: McGraw-Hill, 1975.
11. Hett, J. H. "Medical Optical Instruments," Chapter 9 in *Applied Optics and Optical Engineering, Volume V,* Kingslake, R., ed. New York: Academic Press, 1969.

CHAPTER TWENTY

Light and Modern Physics

INSTRUCTIONAL OBJECTIVES

After studying this chapter, the student should be able to:

1. State what is meant by "quantization" and name several atomic variables which are quantized.
2. List the common types of electromagnetic waves in order of frequency or wavelength.
3. Outline the experimental evidence supporting the "wave" nature of light and the "particle" nature of light.
4. Calculate the energy associated with a photon, given Planck's constant and the frequency.
5. Calculate the frequency of light emitted or absorbed in an atomic transition, given the energy change and Planck's constant.
6. Describe the process of identifying the elements present in substance by means of its emission or absorption spectrum.
7. Describe the physiological effects of radio frequency, infrared, ultraviolet and x-ray radiation.
8. Outline the atomic or molecular processes involved in the basic laser process.
9. State the characteristics of laser light which make it medically useful.

Most of the topics discussed thus far are in the realm known as "classical physics," as opposed to "modern physics." One of the characteristics of the classical concepts of mass, velocity, acceleration, and other quantities is that they have a continuous range of values. When phenomena on the atomic scale are considered, it is found that many of the physical relationships we have discussed do not apply directly and must be modified. The modified framework of physics necessary to explain such phenomena is often referred to as quantum mechanics or quantum physics. One of the characteristics of nature on the atomic scale is that many physical variables are "quantized." That is, they can take only certain discrete values, and all other values are forbidden. In this chapter some topics in modern physics will be discussed along with a further treatment of light and the broad class of electromagnetic waves.

THE ELECTROMAGNETIC SPECTRUM

Visible light, x-rays, microwaves, and the "radio waves" used for communication purposes are all basically the same type of phenomena. They are examples from the broad class of *electromagnetic waves*. They have some properties that are analogous to those of sound waves or waves on the surface of a lake; for example, they have characteristic frequencies, wavelengths, and amplitudes. Yet, unlike sound or water waves, they do not require a material medium through which to travel. Electromagnetic waves propagate freely through empty space. They are transverse waves that can be visualized as electric and magnetic fields distributed in space, as indicated schematically in Figure 20–1. All electromagnetic waves travel through empty space at the same enormous speed $c = 3 \times 10^8$ meters/sec (about 186,000 miles per second). When they travel through a material medium, their speed will, in general, be less than the free-space speed, but there is no apparent movement of the particles of the medium to mark their passage.

The electromagnetic spectrum is commonly divided into several bands as indicated in Figure 20–2, which lists them in order of increasing frequency. The carrier waves for AM radio signals start at a frequency of about 500,000 Hz. Using the basic relationship $c = v = f\lambda$, the wavelength associated with this frequency is 600 meters. The center of the visible light region corresponds to a wavelength of 6×10^{-7} meter or 6000 Å, and gamma rays may have wavelengths shorter than 1 Å (1 Å = 10^{-10} meter).

The following numerical examples illustrate the wavelength-frequency relation and show how the wave properties of some of these phenomena affect their uses.

Example: The wavelengths of light visible to the human eye are about 4000 Å to 7000 Å. What are the corresponding frequencies?

Solution: The basic relation $c = f\lambda$ is used with the values $c = 3 \times 10^8$ m/sec and 1 Å $= 10^{-10}$ meter.

$$\text{For } \lambda = 4000 \text{ Å}, f = \frac{3 \times 10^8 \text{ m/sec}}{(4000 \text{ Å})(10^{-10} \text{ m/Å})} = 7.5 \times 10^{14} \text{ Hz.}$$

$$\text{For } \lambda = 7000 \text{ Å}, f = \frac{3 \times 10^8 \text{ m/sec}}{7 \times 10^{-7} \text{ m}} = 4.3 \times 10^{14} \text{ Hz.}$$

Example: Signals for radio or television broadcasts are superimposed on electromagnetic "carrier waves." An electrical "image"

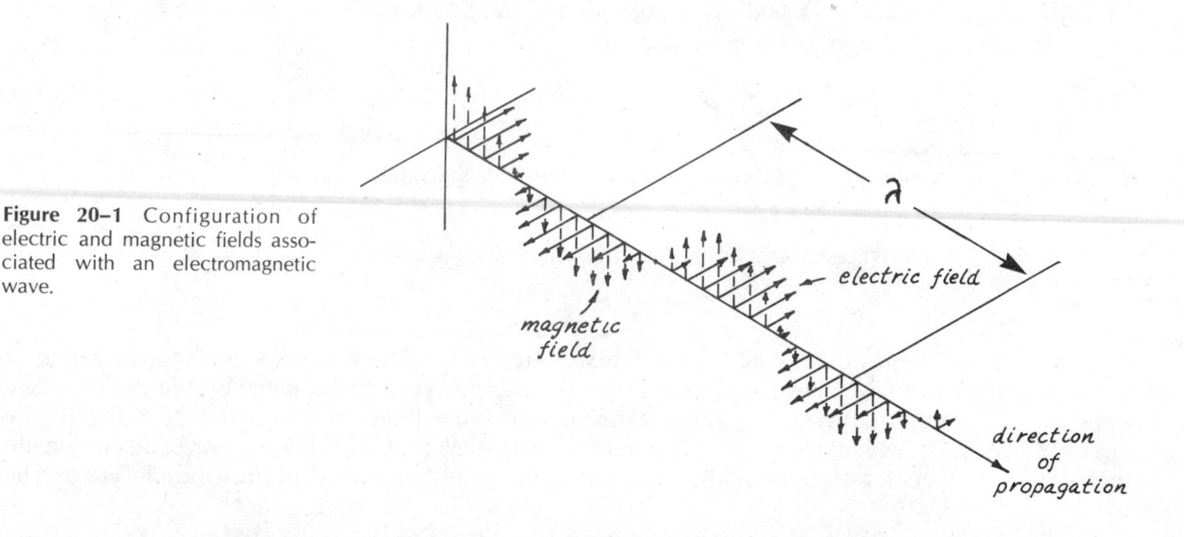

Figure 20–1 Configuration of electric and magnetic fields associated with an electromagnetic wave.

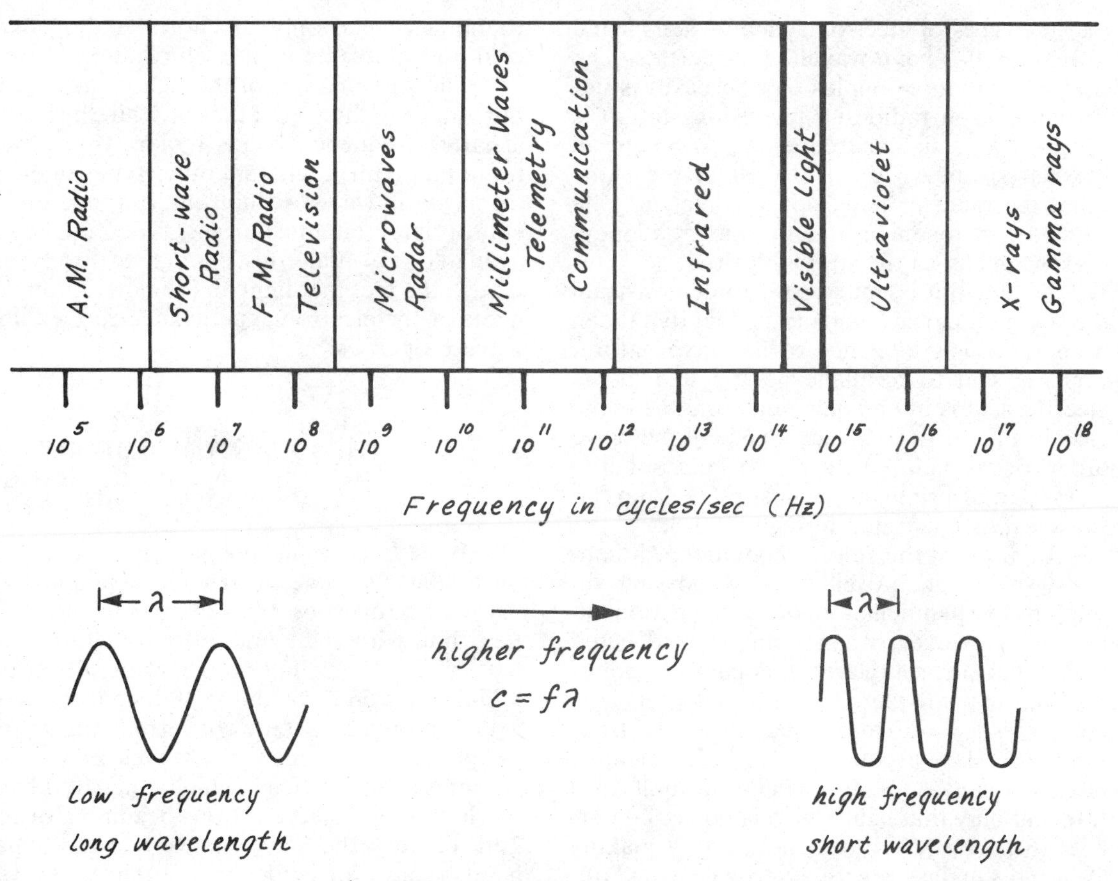

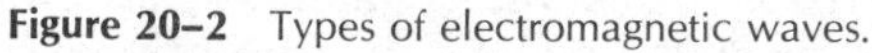

Figure 20–2 Types of electromagnetic waves.

of the sound or picture signal is used to alter or "modulate" the carrier wave by amplitude modulation (AM) or frequency modulation (FM) so that the carrier wave transports the information. The wave properties of the carrier waves determine the dimensions of antennas appropriate for receiving these transmitted waves. The antenna reception is optimized for a straight metallic antenna if it has a length equal to half the wavelength of the carrier wave (the standing wave condition discussed in Chapter 17). If the television Channel 2 has a carrier wave for the picture signal of frequency 55.25×10^6 Hz (55.25 MHz), how long should the antenna element for Channel 2 in your television antenna be?

Solution: The wavelength of the carrier wave is

$$\lambda = \frac{c}{f} = \frac{3 \times 10^8 \text{ m/sec}}{55.25 \times 10^6 \text{ Hz}} = 5.43 \text{ meters}$$

$$= 17.8 \text{ ft.}$$

Therefore the antenna element should be of length

$$L = 2.7 \text{ m} = 8.9 \text{ ft.}$$

Example: An unmanned satellite is to scan the surface of Venus to examine small detail. Visible light is unsatisfactory because of cloud cover, which is essentially opaque in the visible region of the spectrum. Transmission efficiency is found to increase as the frequency is lowered. If you must resolve surface detail as small as 1 cm, and the resolution is approximately equal to the wavelength, what minimum frequency could be used? In what portion of the electromagnetic spectrum does this fall?

Solution:

$$f = \frac{c}{\lambda} = \frac{3 \times 10^8 \text{ m/sec}}{.01 \text{ m}} = 3 \times 10^{10} \text{ Hz.}$$

This frequency lies in the microwave region, the region used by radar techniques. So this scanner could be referred to as a high resolution radar scanner.

These examples have been selected from the wide variety of phenomena which show that the types of electromagnetic waves listed in Figure 20–2 have wavelike properties. The communication examples can be easily studied since each radio and television station is assigned a particular frequency. Although all these frequencies are in the air at the same time, the radio or television receiver may be "tuned" to resonate at only one frequency, corresponding to the station desired.

Although it is not so easy to demonstrate the wave properties of light, the fact that it can be polarized is evidence of its wave nature. Light is said to be plane polarized if all its electric field is in one plane in space, as is the case in Figure 20–1. We make use of the wave properties of light when we use Polaroid sunglasses to discriminate against glare from flat surfaces, as illustrated in Figure 20–3.

Light from the sun is unpolarized; it contains waves which oscillate in all possible directions perpendicular to its propagation direction. However, when sunlight is reflected from a flat surface (glare), it is partially polarized in the plane parallel to the reflecting surface. This is a horizontal plane for most troublesome glare-producing surfaces (water, beach, roadways, etc.). Polaroid materials transmit only that light which is polarized parallel to their "pass" direction, so by making Polaroid sunglasses with a vertical "pass" direction, much of the troublesome glare can be eliminated. This is accomplished without excessive darkening of other parts of the surroundings since only the horizontally polarized portion of the light is eliminated.

The wave nature of the light is also demonstrated by the separation of white light into separate frequency bands (colors) by refraction in a prism. Light and other types of electromagnetic waves exhibit interference under appropriate circumstances. The examples available are numerous enough to firmly establish the fact that light and the other components of the electromagnetic spectrum exhibit wave properties.

THE QUANTUM THEORY OF LIGHT

With large scale phenomena like waves on a lake, we have no trouble distinguishing "wave" properties from "particle" properties, but with light and its interaction with matter, things are not so clear-cut. From the middle ages up through the 19th century there was a continuing debate concerning the nature of light: was it composed of waves, or was it a stream of tiny particles which emanated from the luminous object? At the beginning of the 20th century the wave theory had the upper hand because of evidence like that presented in the previous section. During the first few years of this century, new experimental evidence was found which seemed to dem-

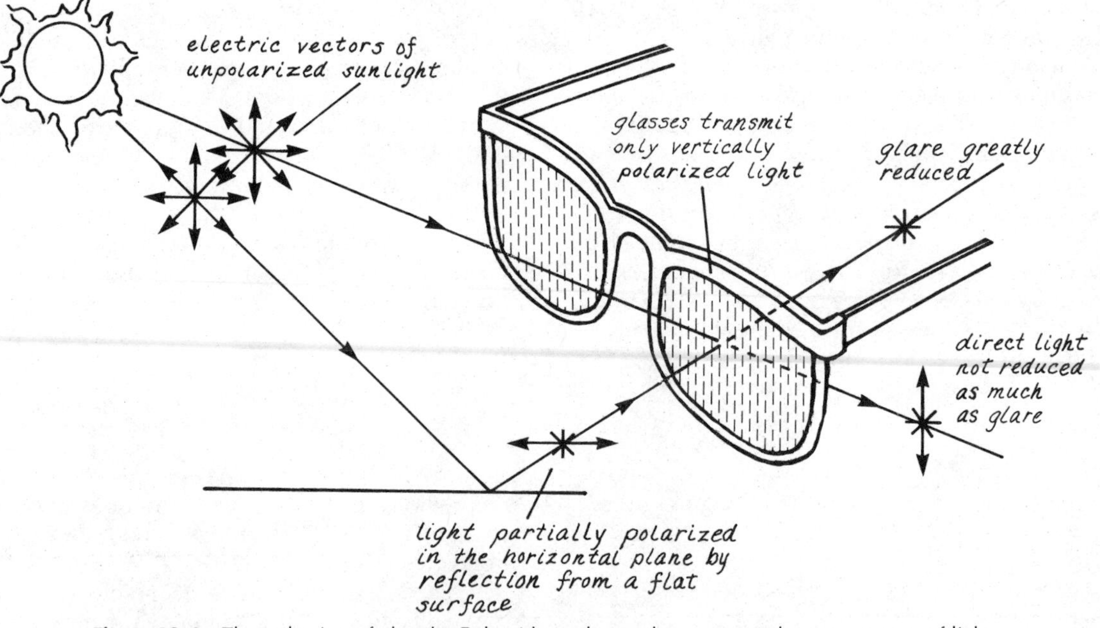

Figure 20–3 The reduction of glare by Polaroid sunglasses demonstrates the wave nature of light.

onstrate just as clearly that light exhibits particle properties in interactions on the atomic scale.

One of the most important discoveries was the photoelectric effect, illustrated schematically in Figure 20–4. It was found that when light falls on the surface of metals like sodium and potassium, electrons are ejected from the surface of the metal, indicating that the energy of the light is given to the electrons to enable them to escape the forces holding them in the metal. This is not so extraordinary in itself, and at first glance it might seem that the wave theory of light could explain it, since waves carry energy. But further analysis of the experimental details are indicated in Figure 20–4 leads to the conclusion that the light falling on the metal is exhibiting particle properties.

In the attempt to explain the experiment using the wave theory of light, several difficulties arise. First, waves would tend to distribute the energy uniformly in the surface layers of the metal and it would be extremely unlikely that any single electron would get enough energy to escape from the surface. Since it is observed that metals normally must get very hot before electrons escape, it would be expected that a considerable time lag would be observed before even localized heating would give any electron the energy necessary to escape the surface. Yet these electrons are ejected instantaneously. Second, it would be expected that increasing the intensity of the light would increase the speed of the escaping electrons, since more energy is being given to the metal. Experimentally the effect of increased light intensity is to increase the number of the electrons ejected, but the maximum velocity stays the same. This indicates that the light beam consists of a stream of particles with the same energy which collide with electrons and knock them out of the surface. A more intense beam implies more particles and therefore more collisions, but the same energy available in a single collision. Third, the frequency of the light would not be expected to have any effect on the experiment, since the energy of light waves should depend upon the intensity of the beam and not on its color. It is found that changing the color of the light toward the blue end of the visible spectrum (increasing its frequency) increases the maximum speed of the ejected electrons. This indicates not only that the light beam consists of particles but that the energy of the particles increases with the frequency of the light.

This experiment and others led to the formulation of the quantum theory of light. Light is postulated to exist in the form of quanta called photons, each of which has a definite amount of energy. This is *not* the equivalent of visualizing a light beam as millions of tiny billiard balls traveling through space. Light has no mass* associated with it and cannot be stopped and localized at a particular point. But the energy is quantized and the light beam can be visualized as tiny packets of energy traveling through space with the velocity of 3×10^8 m/sec. As indicated by the photoelectric effect experiment, the energy of each photon is determined by its frequency. The energy of a photon is found to obey the relationship:

$$E = hf \qquad \textbf{20–1}$$

*I.e., no rest mass.

Change in experiment	Change in number of electrons	Change in maximum velocity of electrons
Make light more intense (increase number of photons)	INCREASE	NO CHANGE
Increase frequency of light (increase radiant energy of photons)	NO CHANGE	INCREASE
Make light less intense but of higher frequency (decrease number, but increase energy of photons)	DECREASE	INCREASE

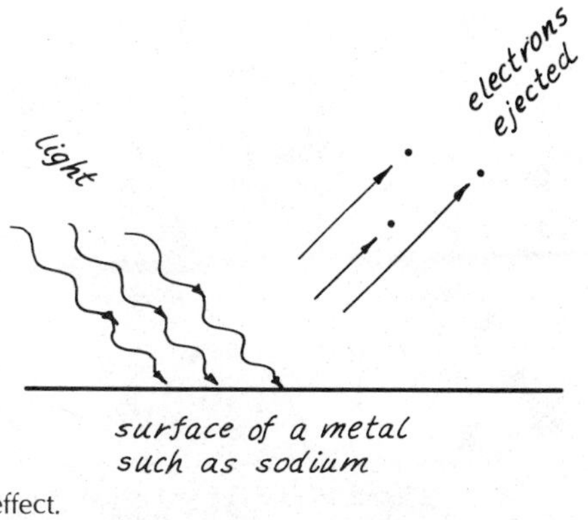

Figure 20–4 The photoelectric effect.

where E is the energy in joules, f is the frequency, and h is an apparently universal constant called Planck's constant and has the numerical value $h = 6.63 \times 10^{-34}$ joule-sec. This dependence of the energy upon the frequency is a crucially important concept in modern physics. It applies to all parts of the electromagnetic spectrum. From this relationship it is seen that a photon of blue light has more energy than a photon of red light. Light photons are more energetic than microwave or radio wave photons, and x-ray photons are more energetic than those of visible light. The innumerable experimental applications of the quantum theory of light have established it firmly as a part of our understanding of nature on the atomic and molecular scale.

This quantum theory of light explains all aspects of the photoelectric-effect experiment. The maximum kinetic energy given to the ejected electrons follows the relation

$$E_{\text{max}} = hf - w, \qquad \textbf{20–2}$$

where w is referred to as the "work function" of the surface and represents the minimum amount of work required to remove an electron from the metal surface. Other electrons are ejected with lower speeds, implying that they were deeper in the metal and lost more of their energy in the process of escaping.

It is no doubt evident at this point that two seemingly contradictory but well-documented positions have been established concerning the nature of light. In the early part of the century this problem was debated and highly publicized as the "wave-particle paradox." The answer to the question, "Which is correct?" must be that both are correct. The concept of light must be broad enough to encompass both the wave properties and the particle properties, because each property is exhibited under the appropriate experimental conditions. The wave/particle dual nature of light is an example of the different frameworks necessary to describe large-scale, ordinary phenomena which fit the framework of classical physics and the small-scale molecular, atomic, and nuclear phenomena which require the newer framework of quantum physics. There is no essential contradiction between quantum physics and classical physics; the laws of classical physics are seen as approximations which are appropriate only for large-scale phenomena. If the laws of quantum physics are extended to apply to large-scale phenomena, they produce agreement with the classical laws of physics, but they involve a great deal more mathematical complexity.

The wave/particle dual nature of light can be made somewhat more comprehensible by the "wave packet" model. A photon can be visualized as a packet of energy which has some wave characteristics but which is localized in space to the extent necessary to show particle-like properties under the appropriate circumstances. Figure 20–5 uses this model to depict the incoming light in the photoelectric effect. The localization of the photon makes it possible for the total energy of the photon to be transferred to a single electron in an interaction like a particle-particle collision. Yet the photon has a characteristic wavelength associated with it; when a large number of photons interact at the same time, the individual photon effects are not evident and the total effect is determined by this wavelength, giving the light an essentially wave-like nature.

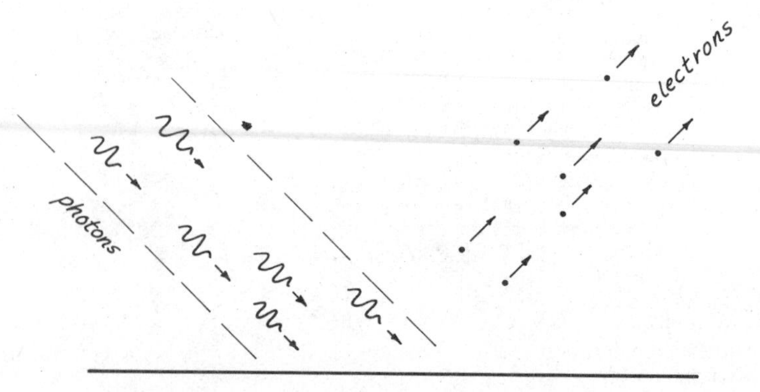

Figure 20–5 Incident photons and the photoelectric effect.

MATTER WAVES: THE ELECTRON MICROSCOPE

After the photoelectric effect had established the essentially dual nature of light, Louis de Broglie suggested in 1923 that the things which we had always regarded as particles might also have a dual nature and exhibit wave properties under the appropriate circumstances. If this duality is a fundamental part of nature, then the electrons which have been depicted as particles in Figure 20–5 might be observed to have wave properties in other experiments. It was predicted by de Broglie that a particle of mass m moving with a velocity v would demonstrate a wavelength

$$\lambda = \frac{h}{mv} \qquad \textbf{20–3}$$

where h is Planck's constant.

The proposed wave nature of the electron would imply that a beam of electrons could be refracted, reflected, and focused to form an image. The confirmation of the wave properties of electrons was made in 1927 by experimenters Davisson and Germer who were scattering electrons from metal crystals. They found that the electrons reflected very strongly at certain special angles and not at others. Upon anaylsis, this behavior could be explained only by invoking wave properties, and the angles were consistent with the wavelength proposed by de Broglie. Since then other particles such as protons and neutrons have been shown to exhibit such wave properties.

The wave properties of larger objects are not observable because the wavelengths are so small compared to the sizes of the objects. For example the wavelength associated with a 0.15 kg baseball piched at 40 m/sec would be

$$\lambda = \frac{h}{mv} = \frac{6.6 \times 10^{-34}\ \text{joule-sec}}{(0.15\ \text{kg})(40\ \text{m/sec})}$$
$$= 1.1 \times 10^{-34}\ \text{m},$$

much too small to be observable. For subatomic particles, however, the wave properties often have profound effects. If an electron is traveling at a speed of 6×10^6 m/sec, the speed it would attain by being accelerated by an electrical potential difference of about 100 volts, its wavelength would be

$$\lambda = \frac{h}{mv} = \frac{6.6 \times 10^{-34}\ \text{joule-sec}}{(9.11 \times 10^{-31}\ \text{kg})(6 \times 10^6\ \text{m/sec})}$$
$$= 1.2 \times 10^{-10}\ \text{m}.$$

This is 1.2 Å, compared to about 3900 Å for the shortest wavelength of visible light. Since the smallest resolvable object is comparable in size to the wavelength of the radiation used to view it, this example implies that electrons can be used to view much smaller objects. Since the electron wavelength decreases with increasing electron speed, the resolution can be further improved by using higher speed electrons.

The higher resolution inherent in the short electron wavelengths has proved to be of immense practical value in the *electron microscope*. This instrument focuses the electrons and uses them to form images in a way directly analogous to that of a projecting light microscope, as shown in Figure 20–6. A high energy electron beam is produced by an electron gun similar in principle to that discussed in Chapter 12 in connection with the oscilloscope. Magnetic fields are used to focus the electrons, as indicated schematically in Figure 20–6. The final image is formed on a fluorescent screen, which produces visible light when struck by the high speed electrons, or it is recorded on photographic film.

It was mentioned in Chapter 19 that there is a theoretical limit on the resolution of an optical microscope; objects significantly smaller than the wavelength of light cannot be observed. Even with special techniques (6) the limit of resolution is around 2000 angstroms. With the commonly used acceleration potential of 50,000 volts on the electron gun, the electrons in the electron microscope have a wavelength of 0.08 angstrom, but because of limitations on the magnetic "lenses" which focus the electrons, the actual limit of resolution is about 10 angstroms. The large improvement in resolution compared to that of the optical microscope makes the electron microscope an invaluable tool for biomedical applications. It can be used for the observation of small organisms such as viruses and for the observation of the structural details of cells which are too small for observation with the optical microscope.

As will be shown, the wave nature of the electron also plays an integral part in our understanding of the allowed "quantized" orbits which electrons may occupy in atoms, and is important in establishing the "rules" govern-

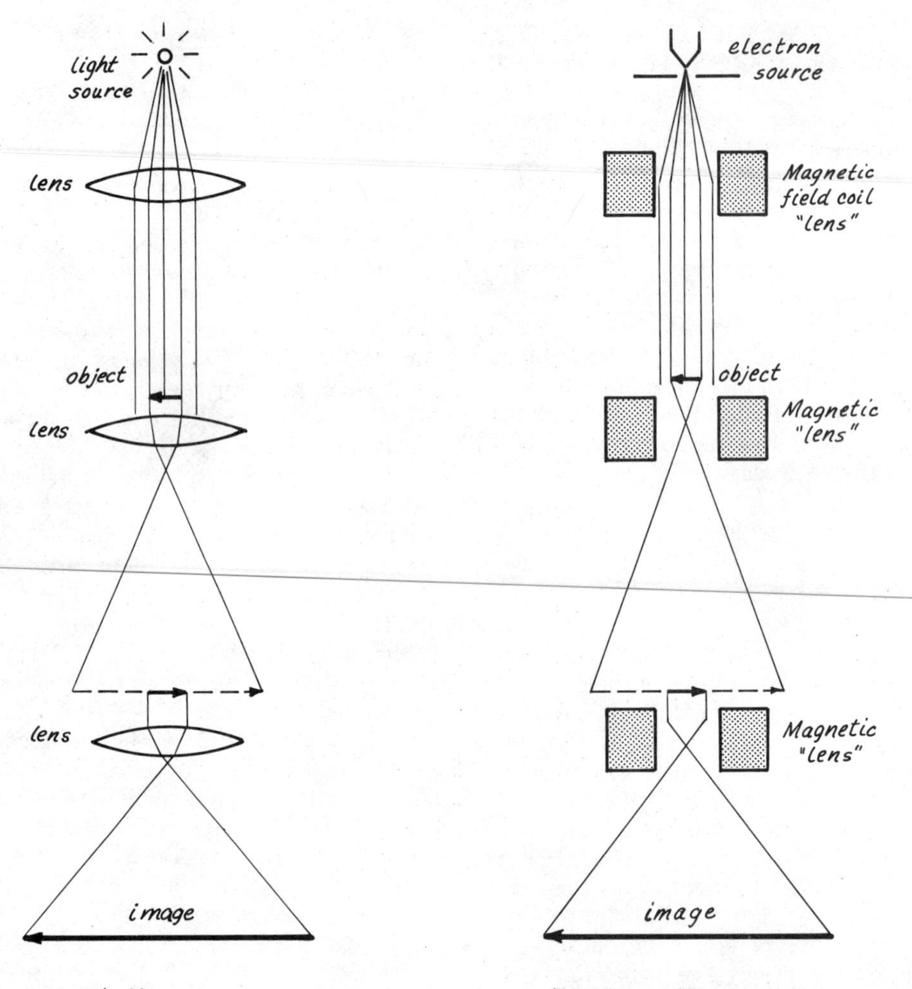

Figure 20–6 Comparison of electron and optical microscopes.

ing the electron configurations and chemical properties of the elements.

QUANTUM THEORY OF THE ATOM

As discussed briefly in Chapter 12, the atom is composed of a positively charged nucleus surrounded by negatively charged electrons in orbit around this nucleus. It was mentioned that each electron possesses one quantum of negative charge (1.6×10^{-19} coulombs) while each proton possesses one positive quantum and the neutron is uncharged. The very strong electrical attraction between unlike charges provides the force to collect a number of negatively charged electrons which is equal to the number of positive charges (protons) in the nucleus. This number of protons determines what chemical element the atom constitutes, and is the only nuclear property of interest at this point. The study of the electrons and the outer structure of the atom is termed "atomic physics" to distinguish it from "nuclear physics," the study of the nucleus itself. The discussion of nuclear properties will be deferred to Chapter 21.

Experiments with atoms yield the information that the electrons which surround them can occupy only certain preferred energy levels. These energy levels are very precise and reproducible, and the electrons are not observed at other energies; other energy values are said to be "forbidden." It is attractive to think of these as prescribed "orbits" in analogy with the paths of the planets around the sun, but the analogy is very limited and should not be pushed very far. The commonly used physical terminology to express this limitation of the electron energies to specific, discrete values is to say that the energy is "quantized." A brief description of this quantization is given here because it is the key to the basic structure of atoms and can give some insight into the chemical behavior of the elements.

Why are some energy levels allowed and other energies forbidden? In the previous section we saw that electrons often act like waves rather than like the miniature billiard balls of a particle model. If the electron travels around the nucleus as a wave, then it may undergo either constructive or destructive interference. If we make the analogy to traveling waves in a stretched string as discussed in Chapter 17, then we might expect the wave motion to be sustained only for certain "standing wave" or "resonant" modes where the interference is constructive. This is like expecting guitar strings to produce only certain pitches when plucked since the constructive interference of the waves of these particular wavelengths (allowed wavelengths) will sustain these pitches, and destructive interference will discriminate against other wavelengths (forbidden wavelengths). In the atomic orbit, the condition for constructive interference is that the circumference of the orbit is equal to an integer number of electron wavelengths. This is illustrated in Figure 20–7(a) for an orbit equal to four wavelengths. If the pathlength is not an integer number of wavelengths, it undergoes destructive interference, as illustrated in Figure 20–7(b) and the orbit cannot be maintained.

This process provides the first step toward building a model of atomic structure by establishing definite energy levels in which the electrons may be found. These levels are often referred to as "shells" and labeled by a "quantum number" n which may take positive integer values. The value $n = 1$ refers to the innermost orbit and corresponds to a pathlength of one electron wavelength. The orbit illustrated in Figure 20–7(a) would correspond to $n = 4$ in this model. Since other quantum numbers will be introduced, n will be distinguished by calling it the principal quantum number.

The energy associated with the orbit arises from the attraction between the electron and the positively charged nucleus. For the hydrogen atom it has the form

$$E_n = -\frac{R}{n^2} \qquad \textbf{20–4}$$

where R is the Rydberg constant and has the value 2.17×10^{-18} joule. Note that the energies are negative, with the innermost orbit ($n = 1$) having the largest negative value. A free electron at rest is the reference for zero energy in this case, and work would have to be done on a hydrogen electron to free it from the attractive force of the nucleus and put it in a zero energy state. Likewise, all electrons in "bound states" in atoms have negative energies, with the most tightly bound electrons having the largest negative energies. Since all physical systems seek the lowest possible energy, an electron will seek the most tightly bound state which is available to it. In the hydrogen atom the electron will normally be found in the $n = 1$ state, which is referred to as its "ground state." If it is raised to a higher state, an "excited state," it will tend to return to the ground state.

The energy obtained from equation 20–4 for the ground state of the hydrogen atom is

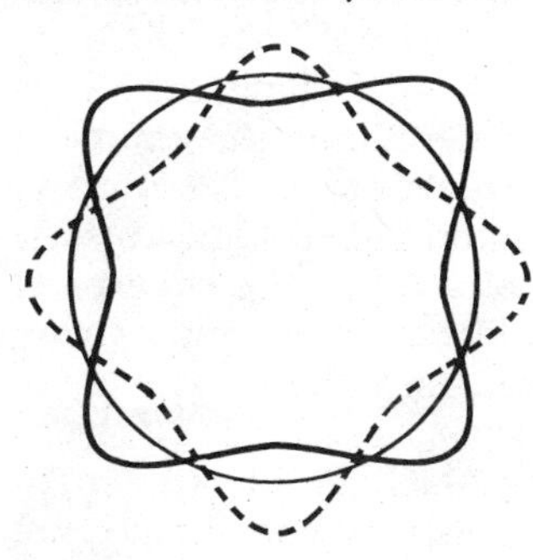

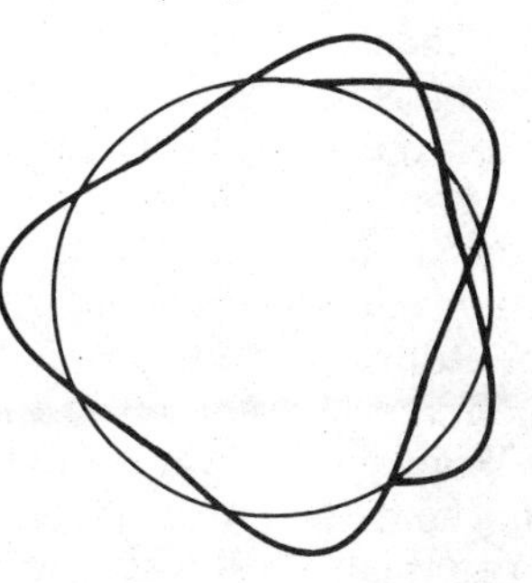

Figure 20–7 The allowed electron orbits correspond to an integer number of complete electron wavelengths.

$$E_1 = -\frac{(2.17 \times 10^{-18}\ \text{joule})}{(1)^2}$$
$$= -2.17 \times 10^{-18}\ \text{joule}.$$

Since the energies of single electrons are so small, another energy unit, the electron-volt (eV) is commonly used to describe atomic and nuclear phenomena. The electron-volt is the energy which an electron would attain by being accelerated by one volt of electric potential. The electron charge is 1.6×10^{-19} coulomb, so with the use of equation 12–5:

$$1\ eV = qV = (1.6 \times 10^{-19}\ \text{coul})(1\ \text{joule/coul})$$
$$= 1.6 \times 10^{-19}\ j.$$

The ground state hydrogen energy is then

$$E_1 = -\frac{(2.17 \times 10^{-18}\ \text{joule})}{(1.6 \times 10^{-19}\ \text{joule/eV})} = -13.6\ \text{eV}.$$

Since the constant R in equation 20–4 can be expressed in electron volts, the energies of the hydrogen levels can be calculated from

$$E_n = \frac{-13.6}{n^2}\ \text{eV}.$$

We shall refer to these energy levels in our subsequent discussion of the interaction of electromagnetic waves with hydrogen gas.

In addition to establishing quantized energy shells for atoms, it is necessary to account for the fact that a limited number of electrons can occupy each shell. This limited occupancy is experimentally shown by the chemical properties of the elements. The limitations are imposed by other physical properties of the electrons in orbit which are also quantized. While the energy of an electron is mainly determined by the shell in which it resides (i.e., by the principal quantum number n), three other properties are required to completely describe the configurations of the electrons. They are generally represented by the quantum numbers l, m and s. While a complete description of the physical meaning of these four parameters is beyond the scope of this text, a brief discussion seems in order because they can give a great deal of insight into the basic structure of atoms and the chemical behavior of the elements.

The quantum number l is called the *orbital quantum number* and specifies the angular momentum of the electron; for a planet's orbit this would be associated with the shape of the orbit, a highly elliptical orbit having a low angular momentum. The quantum number m is called the *magnetic quantum number* and is associated with the quantum number l, in that it specifies the orientation of the orbit. It is called the magnetic quantum number because the energy of the electron will change with orbit orientation (and thus with changing value of m) if the atom is placed in a magnetic field. The change in energy occurs because the electron in orbit acts like a small electromagnet and interacts with any external magnetic field. The final quantum number, s, is called the *electron spin quantum number*. Electrons act in some respects as if they were tiny bar magnets. Even though there is really no ordinary analog to this electron property, a crude model in which the electron is visualized as a spinning sphere of charge is somewhat helpful. The spinning charge would represent an electric current and would produce a magnetic field similar to the electromagnet in Figure 12–13. This "electron spin" aligns itself either along an external magnetic field or directly against it. These two possible quantum states of the electron spin are assigned quantum number values $s = +1/2$ and $s = -1/2$, respectively. The quantum numbers m and s are associated with the orientation of the electrons in space and illustrate the property called "space quantization." Whereas an ordinary object can take any space orientation desired, the electron orbit and electron spin can take only a small number of definite orientations. The number of possible values for the quantum numbers are discussed below.

Although at this point it may seem unlikely that such a set of four quantum numbers can clarify the basic structure of the atom, it will be seen that they lead to some profound results. One key to atomic structure lies in a fundamental principle called the *Pauli exclusion principle,* which states that no two electrons in an atom can exist with the same set of quantum numbers n, l, m, and s. It is as if each set of quantum numbers defines a particular slot which can be occupied by only one electron. As will be shown, the order of filling these slots is a major factor in determining the chemical properties of the elements.

When electrons enter bound states (orbits) around a nucleus, they enter the innermost shell ($n = 1$) first and begin to fill outer shells only when the inner shells are filled. The limits on the number of electrons which can occupy a given shell are determined by the range of values which the quantum numbers can take. The rules which establish the allowed values for the quantum numbers are summarized in

Table 20–1. For example, to find out how many electrons can occupy the first shell, we must determine how many distinct sets of quantum numbers *n, l, m,* and s there are with $n = 1$. From Table 20–1 it is seen that the orbital quantum number l can take only the value 0 and that m can takes only the value 0. The spin quantum number s can take the values $+\frac{1}{2}$ and $-\frac{1}{2}$. Therefore, there are only two distinct sets of quantum numbers for n= 1

n	l	m	s
1	0	0	$\frac{1}{2}$
1	0	0	$-\frac{1}{2}$

Therefore, only two electrons can occupy the first shell.

The shell configuration is very closely linked to the chemical behavior of the elements. A filled shell represents a very stable configuration, and all elements have a tendency either to add electrons to fill the outermost shell or, if the outer shell is less than half full, to get rid of the electrons in this shell to drop back to the next closed shell configuration. These tendencies can be seen with this first shell. The element hydrogen has only one proton in its nucleus and in its normal state will have one electron in shell $n = 1$. It is very active chemically because of the strong tendency toward an empty $n = 1$ shell. By contrast, helium with two electrons has a filled $n = 1$ shell and is therefore stable or chemically inert.

For the next shell, characterized by $n = 2$, the quantum numbers can take the following values:

	n	l	m	s
8 electrons	2	0	0	$\frac{1}{2}$
	2	0	0	$-\frac{1}{2}$
	2	1	-1	$\frac{1}{2}$
	2	1	-1	$-\frac{1}{2}$
	2	1	0	$\frac{1}{2}$
	2	1	0	$-\frac{1}{2}$
	2	1	1	$\frac{1}{2}$
	2	1	1	$-\frac{1}{2}$

TABLE 20–1 Allowed Values of Quantum Numbers.

Principal quantum number:	$n = 1, 2, 3, \ldots$
Orbital quantum number for each n:	$l = 0, 1, 2, \ldots$ up to $n - 1$
Magnetic quantum number for each l:	$m = -l, -(l-1), \ldots 0, \ldots, l-1, l$
Spin quantum number for each m:	$s = -\frac{1}{2}, +\frac{1}{2}$

Note from the guidelines in Table 20–1 that if $n = 2$, l can take the values 0 and 1, m the values -1, 0, and 1, and s the values $\frac{1}{2}$ and $-\frac{1}{2}$. The total of eight different sets of values for the quantum numbers, coupled with the exclusion principle, indicates that no more than eight electrons can occupy the second shell. The elements which make up the second row of the periodic table have one or more electrons in the second shell. The elements of the first two rows with their electron configurations are listed in Table 20–2. A schematic representation of the electrons which make up the first two shells is given in Figure 20–8(a). The quantum numbers *n, l,* and *m* are indicated on the diagram, and the electron spin orientation is indicated by the small arrows.

When there is more than one shell, the chemical behavior of the element is mainly determined by the outermost electron shell. In Figure 20–8(b) it is indicated that the element lithium has a closed first shell and only one electron in the second shell. It has a tendency to lose that electron and form a positive ion with one quantum of positive net charge. It is said to have a valence of +1. The other elements, such as sodium and potassium, which have closed shells with one electron in the next highest shell tend to have similar chemical properties and form the Group I elements on the periodic chart of the elements. By contrast, the element fluorine (Figure 20–8(c)) has seven electrons in the second shell and would like to pick up one more to form a closed shell. This would make it a negative ion with a charge of −1; it is therefore said to have a valence of −1. The expected valences for the other elements in the second row of the periodic table can be deduced by a similar process. The element neon has a complete second shell and is therefore chemically stable. The details become more complicated in the third shell because the third and fourth shells mix, but the periodic variation of the chemical properties associated with the filling of the outermost shell may still be noted.

The picture here is obviously far from complete, but the intention has been to show that the quantum or discrete nature of the electron parameters along with the Pauli exclusion principle provide some insight into the structure of the electrons surrounding the nuclei of atoms. The chemical and physical consequences of the shell structure of the atom are far-reaching. For more extensive elementary treatments of atomic structure, References 1 and 2 are recommended, with Reference 3 providing more advanced treatment.

TABLE 20–2 Partial List of the Chemical Elements

	GROUP: I	II	III	IV	V	VI	VII	VIII
Element	H							He
Shell 1	1							2
Element	Li	Be	B	C	N	O	F	Ne
Shell 1	2	2	2	2	2	2	2	2
Shell 2	1	2	3	4	5	6	7	8

The basic quantum nature of the atomic electrons is manifested not only in the chemical properties of the atoms but also in a wide variety of other phenomena. The fact that the electrons can exist only in certain allowed energy levels is the key to understanding the interaction of light and other forms of electromagnetic waves with matter.

THE INTERACTION OF ELECTROMAGNETIC WAVES WITH MATTER

It has been emphasized that the various components of the electromagnetic spectrum are essentially wavelike in nature, differing only in frequency and wavelength. However, many common phenomena illustrate the fact that their interactions with matter differ widely. For example, the electromagnetic waves transmitted by local radio and television stations pass through the walls of your house with little attenuation while the same walls are completely opaque to visible light. On the other end of the spectrum, x-rays will also pass through this wall with relative ease. The brick wall which so completely blocks out visible light is quite transparent to some waves above and below the visible frequency range.

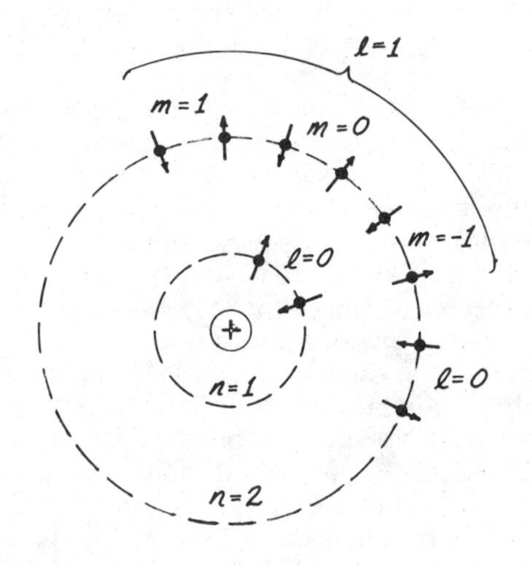

a. Representation of the electrons which can occupy the first two shells. (Not an indication of actual position.)

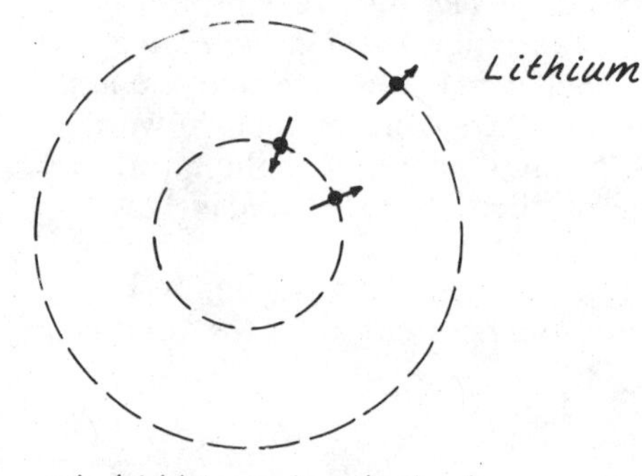

b. Lithium tends to lose one electron, leaving it with one net positive charge. (valence = +1).

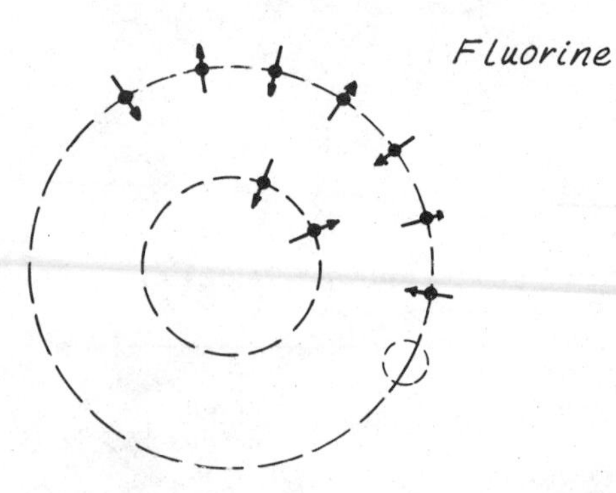

c. Fluorine tends to pick up one electron. (valence = -1).

Figure 20–8 Electron configurations in the first two atomic shells.

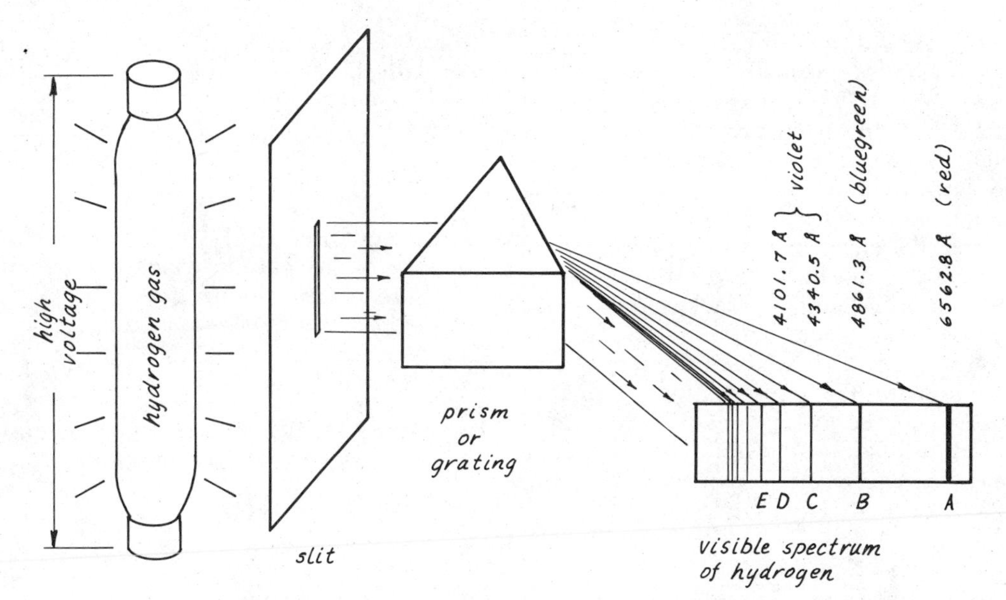

Figure 20–9 Example of an emission spectrum: atomic hydrogen gas.

The study of the selective absorption or emission of electromagnetic waves is known as spectroscopy. The emission spectrum of hydrogen is one of the most thoroughly studied examples. If hydrogen gas is placed in a sealed glass container and subjected to a high voltage, it will emit pink light along the electrical discharge path and usually emits blue light in other parts of the container. If this light is passed through a diffraction grating to separate the colors as indicated in Figure 20–9, the spectrum will appear as a series of bright lines of different colors. In contrast to the continuous distribution of colors obtained by passing white light through a prism, only certain discrete colors (frequencies) are emitted by the hydrogen. These frequencies are determined by the quantized energy levels which the hydrogen electron can occupy. The energy level diagram in Figure 20–10 shows the possible energy lev-

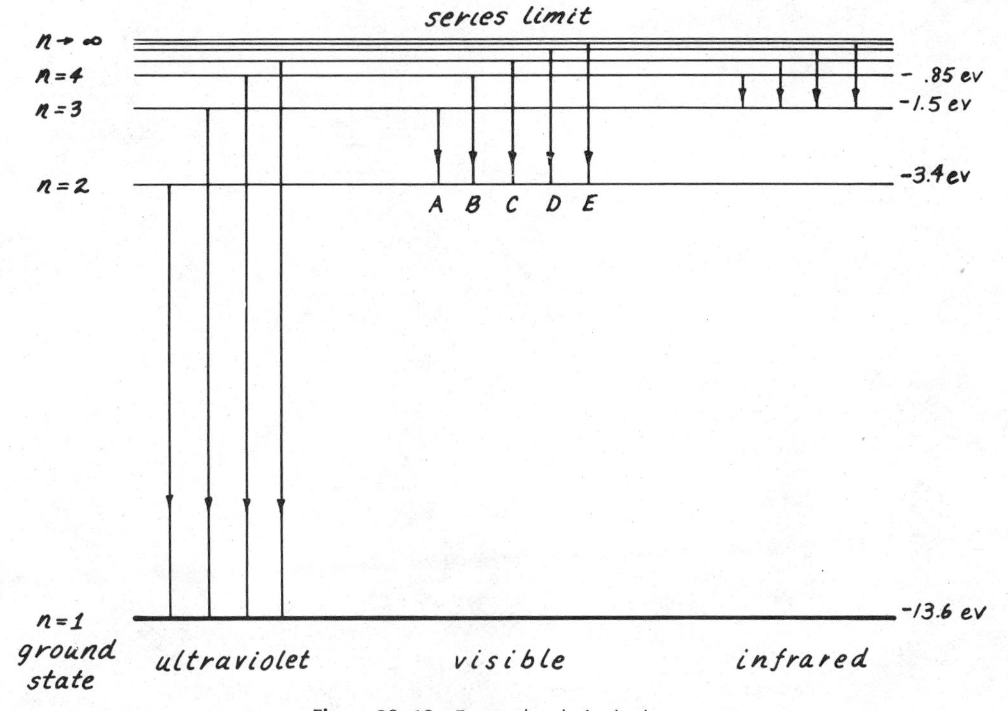

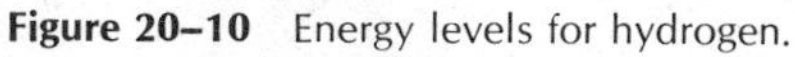

Figure 20–10 Energy levels for hydrogen.

els. As discussed in the previous section, the hydrogen electron normally occupies the first shell ($n = 1$). If it receives energy from an electrical discharge it can jump up to any one of the higher levels. The $n = 1$ level is referred to as the "ground state" for the electron and the higher states are called "excited states." After excitation, the electron has a strong tendency to drop down to one of the lower states, continuing the process until it reaches the lowest energy level, the ground state. In order to drop to a lower state, it must give off energy to be consistent with the conservation of energy principle. This energy is given off in the form of a photon, a quantum of light. The frequency (color) of light given off is determined by equation 20–1

$$\Delta E = E_{\text{initial}} - E_{\text{final}} = hf.$$

A larger difference between energy levels will produce radiation at a higher frequency. The energy level separation associated with transition A in Figures 20–9 and 20–10 is such that the emitted light is red. The wavelength associated with it can be calculated with the use of equations 20–1 and 20–4. Transition *A* is associated with the movement of an electron from the $n = 3$ to the $n = 2$ level. Since the spectra can be measured very accurately, it is necessary to use physical constants with an accuracy of about five significant digits to compare with the experiment. The change in energy associated with this transition is

$$\Delta E = E_3 - E_2 = -13.600 \left[\frac{1}{(3)^2} - \frac{1}{(2)^2}\right] eV$$
$$= 1.8889 \text{ eV}.$$

From equation 20–1 we can obtain the frequency of the photon which must be emitted to make this transition:

$$f = \frac{\Delta E}{h}$$
$$= \frac{(1.8889\ eV)(1.6021 \times 10^{-19}\ \text{joule}/eV)}{(6.6256 \times 10^{-34}\ \text{joule}\cdot\text{sec})}$$
$$= 4.5674 \times 10^{14}\ \text{Hz}.$$

Using the wave relationship $c = f\lambda$, (equation 17–2) the wavelength is

$$\lambda = \frac{c}{f} = \frac{2.9979 \times 10^{8}\ \text{m/sec}}{4.5674 \times 10^{14}\ \text{Hz}}$$
$$= 6.564 \times 10^{-7}\ \text{m} = 6564\ \text{Å}.$$

This is extremely close to the experimentally measured wavelength 6562.8 Å. The small deviation which remains may be attributable to small inaccuracies in the constants used, neglect of the motion of the nucleus, and other small effects.

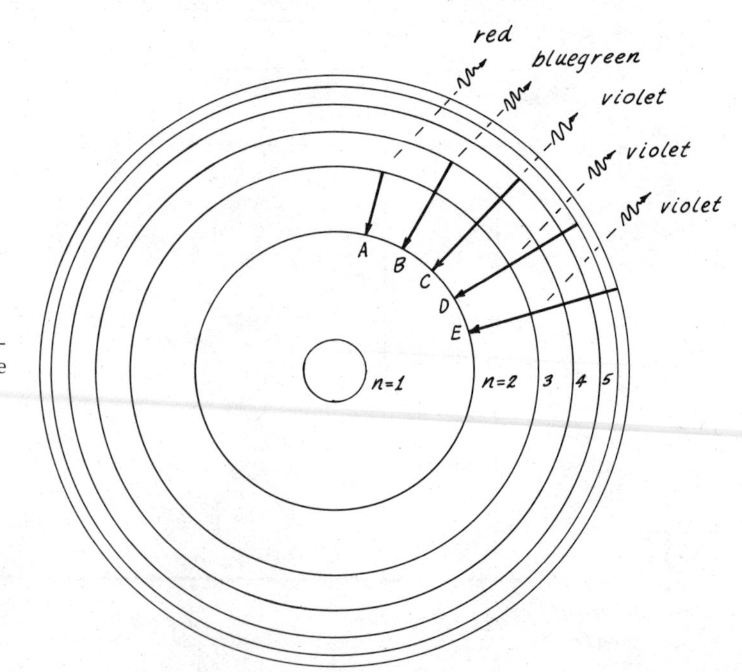

Figure 20–11 Schematic of hydrogen transitions emitting visible light.

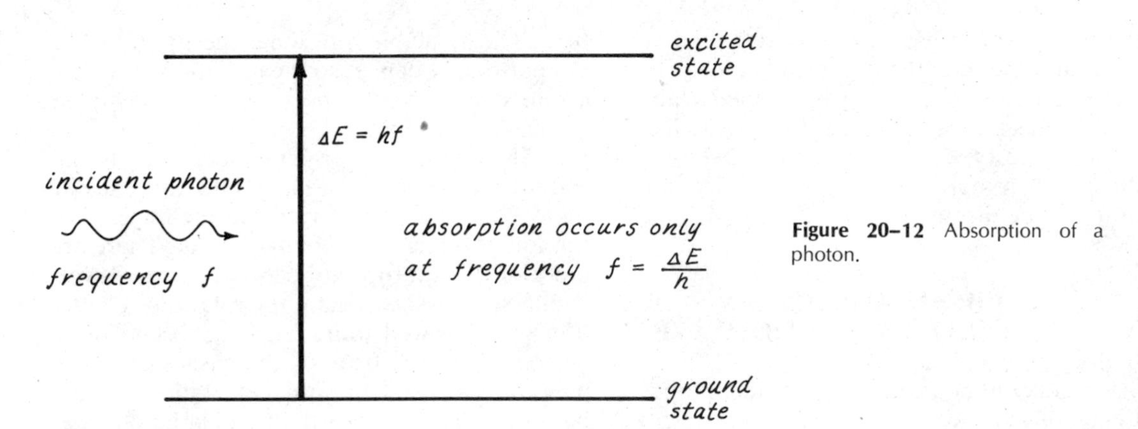

Figure 20–12 Absorption of a photon.

The transition B involves a greater energy change and produces bluegreen light, a higher frequency. The transitions producing visible light are illustrated in Figure 20–11. Other transitions produce ultraviolet or infrared light, outside the sensitive range of the human eye.

The bright line spectra of hydrogen and other gases can be explained in detail by the use of the quantum theory of light and the quantized energy levels of the atomic electrons. Historically, such spectra were strong influences leading to the development of the quantum theory. Since measurements of the frequencies associated with the spectral lines can be used to calculate the energy levels of the electrons involved, atomic spectroscopy is the basic tool for the study of atomic structure. The quantum theory of the atom discussed in the previous section is based upon such measurements.

The frequencies of the spectral lines for a given element are unique and precisely reproducible. Each element has a characteristic set of allowed energy levels for the electrons. The spectral lines can be used to identify the element emitting them, even when they are mixed with spectra from many other elements. For example, the discovery of a red spectral line at wavelength 6562.8 Å would indicate the presence of hydrogen. A check for the presence of the other known wavelengths in the hydrogen spectrum (Figure 20–9) would provide a positive demonstration of the presence of hydrogen. Such "fingerprints" from an emission spectrum may be used in spectrochemical analysis to check for the presence of trace quantities of toxic elements such as arsenic or lead.

Absorption spectra are also commonly used for the identification and study of elements. If white light is passed through cool hydrogen gas and then passed through a diffraction grating, the result will be dark bands superimposed upon the continuous rainbow spectrum characteristic of the white light. These dark bands are associated with absorption of light by the cool gas and will occur at the same wavelengths as the bright lines of the emission spectrum of hydrogen. This experiment is just the reverse of the one described above. Since the electrons can exist only in specific energy states, only those photons with energies equal to the separation between allowed energy states can be absorbed. The condition for absorption is $\Delta E = hf$, as indicated in Figure 20–12. An example of absorption spectroscopy is the identification of the chemical elements present in stars. If the light from the sun or another star is passed through a grating, the result is a continuous spectrum with superimposed dark bands corresponding to the absorption of light by the cooler gases surrounding the star. From wavelength measurements the chemical elements can be identified, even across the vast distances of space.

The atoms and molecules of solid objects interact with each other and produce many more allowed energy levels which merge into continuous bands of possible energy levels. The energy bands of a solid object may offer a continuum of levels such that any frequency in the visible range will be absorbed, causing that object to be opaque to visible light. It may, however, be transparent to x-rays or other frequencies for which there are no corresponding energy level separations in the material. The colors of solid objects are often the result of selective absorption. The nature of a green object is such that it absorbs all of the visible spectrum except green, which is reflected or re-emitted to provide the perceived color. The energy condition $\Delta E = hf$ may be applied in any part of the electromagnetic spectrum. Applications to other types of radiation will be introduced where appropriate to the clinical application to be discussed.

CLINICAL APPLICATIONS OF ELECTROMAGNETIC WAVES

A number of clinical treatment techniques and diagnostic processes make use of electromagnetic waves. Some representative applications are considered here, proceeding in the order of increasing frequency of the waves.

Recall that all types of electromagnetic waves are physically similar, differing only in frequency and wavelength. Yet different parts of the electromagnetic spectrum have very different physiological effects. Your body is transparent to radio waves, becomes opaque as the frequency rises to the visible light region, and becomes more transparent again as the frequency rises into the x-ray region. The key to understanding these different effects is the quantum theory, as discussed in the previous section. Basically, the physiological effects are different because the quantum energies ($E = hf$) are different for the different frequencies of electromagnetic waves, and they produce different types of physical interactions. The nature of the physiological effect also depends upon how strongly the radiation is absorbed, and this too depends upon the quantum theory for explanation.

As indicated by the sketch in Figure 20–12, the requirement for absorption is that a pair of energy levels of just the right separation exists in the material so that the photon can raise the system from the lower to the higher state. If a substance has many such pairs of energy states corresponding to a given radiation frequency, then that radiation will be strongly absorbed. If there are no pairs of energy states in the material such that the energy difference $\Delta E = hf$ for that frequency, then the material will be transparent to that frequency.

The interactions of the different types of radiation are summarized in Figure 20–13. There are very few mechanisms by which tis sue can absorb the lower radio frequencies, so they pass through almost unattenuated. In the microwave region the quantum energies of the photons are large enough to cause molecular rotation and torsion, which is experienced as heat. The available rotational energy states are not extremely dense, so the radiation may penetrate deeply and a portion of it may pass through the tissue.

The quantum energies associated with infrared radiation are appropriate for causing molecular vibration: periodic stretching and/or torsion of internal molecular bonds. There are more of the appropriate pairs of energy levels available in tissue than for the rotational excitations, so infrared radiation is absorbed more strongly than microwaves and has less depth of penetration. The vibrational excitations are experienced as heat, but the associated changes in internal energy are close to the surface rather than the volume changes in internal energy associated with microwave radiation. In the visible and ultraviolet ranges the quantum energies are large enough to excite electrons to higher orbits. Since there are vast numbers of available electron energy levels, these types of radiation are absorbed strongly and generally do not penetrate past the skin. The available energy levels for ultraviolet photons are so dense that all incident ultraviolet radiation is usually absorbed in a very thin outside layer of the skin.

The amounts of energy absorbed in transitions of electrons from one orbit to another are the largest amounts of energy which an atom or molecule can absorb and remain intact. When the radiation frequency is increased into the upper ultraviolet or into the x-ray or gamma ray ranges, the quantum energy hf of the photons is so large that it can only be absorbed by disrupting the atom or molecule (ionization). Further details about these interactions will be given in the following paragraphs.

Radio Frequency and Microwave Radiation

Medical applications of the lower frequencies in the electromagnetic spectrum have usually involved frequencies in the range from 500 kilohertz to 2500 megahertz. The lower part of this range is classified as the radio frequency region, and the upper part is in the microwave region. From the quantum viewpoint, the photon energies in this part of the electromagnetic spectrum are very low. As discussed above, the effect of such radiation on a biological specimen is to produce molecular agitation; that is, it heats the specimen. The quantum energy as determined by the frequency, $E = hf$, is much too low to produce ioniza-

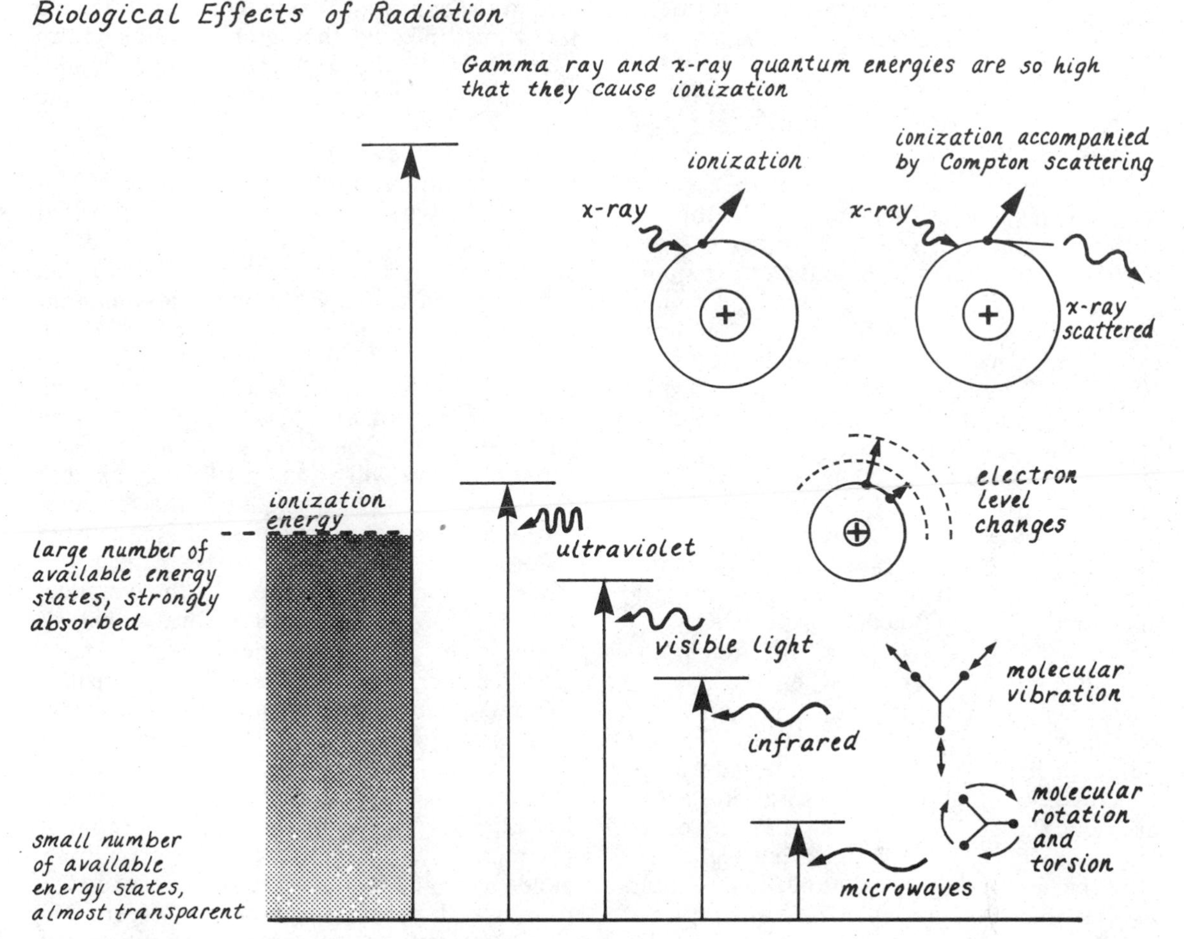

Figure 20–13 Interactions of the different types of electromagnetic radiation.

tion, molecular dissociation, or other direct chemical effects. It is a common misconception that microwave radiation can produce radiation damage similar to that produced by x-rays and nuclear radiation. As will be seen, the specific effects of x-rays are associated with their large quantum energies; no comparable effects can be produced by microwave and radio frequency radiation. This is not to say that microwaves present no hazards; exposure to high intensities can produce internal and external burns.

The therapeutic effectiveness of radio frequency and microwave radiation is based upon the fact that they penetrate the body and raise the temperature throughout the portion of the body which is exposed to the radiation. The use of such radiation for heat therapy is referred to as diathermy. It can be used for relief of muscular pain, inflammation of the skeleton, and other conditions in which warming of deep tissues is beneficial.

Although such radiation does not produce instantaneous chemical changes in the body as in the case of x-rays, it may produce internal burns with excessive exposure and may produce surface burns if the coils or transmitting elements are in contact with the skin. Microwave radiation will often interfere with the operation of an implanted cardiac pacemaker. Other problems with diathermy occur because it operates in the frequency range used by radio and television communication and radar. Diathermy is now closely regulated in frequency and power output because of potential interference with communication networks. Microwave ovens operate on the same principle as diathermy. They are now regulated by the Federal Communications Commission and limited to the two frequencies 900 megahertz and 2560 megahertz to prevent interference with radar networks and other communications.

Infrared Radiation

The term "infrared" refers to a broad range of frequencies, beginning at the top end of those frequencies used for communications purposes and extending up to the low frequency end (the red end) of the visible light region. The wavelength range is from about 1 millimeter (10^{-3} m) down to 7500 angstrom units (7.5×10^{-7} m). The low frequency end of the infrared range is called the "far infrared" range and the upper end is called the "near infrared" range, the words "near" and "far" referring to its separation from the visible light region. Most medical applications of infrared radiation make use of the near infrared range, extending below the visible range from wavelengths of 7500 angstroms to about 30,000 angstroms.

Any hot object gives off infrared radiation; the radiant heat felt when the hand is placed near a red hot object is mostly infrared radiation. As the temperature of any object is raised, it has more available internal energy which can be released in the form of electromagnetic waves. From the quantum viewpoint, more available energy implies that higher energy (higher frequency) photons can be emitted. This fact is shown when the temperature of a piece of metal rises. At a certain temperature it will begin to glow with a dull red color. This implies that the quantum energy has increased to the point where the associated frequency lies in the visible light region. At a higher temperature it will finally reach the condition known as "white hot" in which all the wavelengths of the visible spectrum are being emitted.

Infrared lamps are sometimes used for heat therapy, but the characteristics of such warming are quite different from the diathermy warming discussed above. Infrared radiation tends to cause molecules to vibrate, and the energy transfer involved causes the radiated energy to be absorbed very quickly in solid and liquid materials. Therefore, infrared waves do not penetrate very far into the body and are not suitable when the therapeutic effect desired is the warming of deep tissues. On the other hand, they do penetrate more deeply than visible light, and this fact has been used to advantage in infrared photography for diagnostic purposes. Since infrared light penetrates the skin, it can be used to photograph veins and other structures beneath the skin which are invisible to the eye. Infrared photography has also been used to study the pattern of healing under certain scabs and for diagnostic studies of the eye.

Infrared thermography is another diagnostic use of infrared radiation. It is similar to infrared photography except that no infrared illumination is used. The technique involves a pictorial representation of the infrared radiation given off by the body. The amount of radiation given off is a function of the temperature, so the thermograph is essentially a display of the temperature variations of the skin. One use of the thermograph is for the early detection of breast cancer. The temperature of a tumor will often be 1°C to 2°C higher than that of normal tissue, and skin temperature elevations of as much as 5°C have been recorded as a result of breast cancer (4). Thermographs are also used for the examination of burns and frostbite, and for the analysis of the vitality of various types of skin grafts.

The thermograph receiver is a sensitive infrared detector which scans over successive small strips of area and converts the infrared intensity into an electrical signal which is amplified and displayed on an oscilloscope screen. Recalling the discussion of the oscilloscope in Chapter 12, the brightness of the display depends upon the number and speed of the electrons which strike the phosphor screen. The signal from the thermograph receiver modulates this electron beam so that the brightness of the oscilloscope screen follows the intensity of the infrared radiation emitted by the patient. As successive strips of area are scanned, a composite picture of the desired area of skin is produced on the screen in a manner analogous to the successive lines which make up the picture on a television screen. This display may then be photographed for a permanent record.

Ultraviolet Radiation

Ultraviolet radiation ranges upward in frequency and quantum energy from the high frequency end of the visible light range (the violet end) to the low frequency end of the x-ray range. The wavelength range is from about 4000 angstroms at the end of the visible spectrum down to less than 400 angstroms at the x-ray region. Most of the ultraviolet phenomena appropriate for discussion here are in the wavelength range from 2500 to 4000 angstroms, which might be called the "near-ultraviolet" range.

The quantum energies of ultraviolet radiation are such that it is absorbed strongly by most forms of matter, even air. The separations between allowed electron energy levels in many atoms and molecules correspond to photon energies in the ultraviolet range. Therefore, photons with those energies are very strongly absorbed. The high energy ultraviolet photons may excite electrons to higher levels, eject electrons from atoms or molecules to form ions, dissociate molecules into their constituent atoms, or produce other instantaneous chemical changes. Even though the sun is a strong source of ultraviolet radiation, very little radiation below 3000 angstroms wavelength reaches the earth. The atmospheric gases, chiefly oxygen and ozone, absorb the radiation before it reaches the earth. Therefore, depletion of the ozone layer by atmospheric contaminants could have very serious consequences. This absorption is fortunate, because the shorter wavelengths would be injurious to living tissue. Sunburn is largely attributable to the ultraviolet rays which do penetrate the atmosphere. Since the higher frequency ultraviolet rays can cause ionization, ultraviolet radiation has been implicated as a cause of skin cancer (see Chapter 21).

Ultraviolet radiation normally does not penetrate into tissue deeper than about 1 mm. The therapeutic uses of such radiation are for the treatment of skin conditions such as psoriasis and acne and for certain cosmetic effects. The radiation is also effective for killing fungi and bacteria on the skin. Ultraviolet radiation has been used for the sterilization of the air in operating rooms and for instrument sterilization. The eye is the part of the body most susceptible to damage by ultraviolet light. Welders must wear protective eye shields because the ultraviolet content of the light from a welding arc can produce acute inflammatory conditions of the eye. A similar inflammation occurs in the case of "snow-blindness." The normal terrain absorbs almost all of the ultraviolet light incident upon it, but snow reflects a large amount of it to the eyes.

X-Ray Radiation

X-rays are electromagnetic waves with wavelengths in the approximate range from 100 to 0.1 angstrom units. They are essentially the same as gamma rays, except that the term "gamma ray" refers to radiation which originates in the nucleus of the atom and which generally has higher energies. X-rays are produced by accelerating electrons through high voltages and allowing them to strike a metal target (Figure 20–14). The accelerating voltages used are usually in the range from 20,000 to 200,000 volts. When the high speed electron

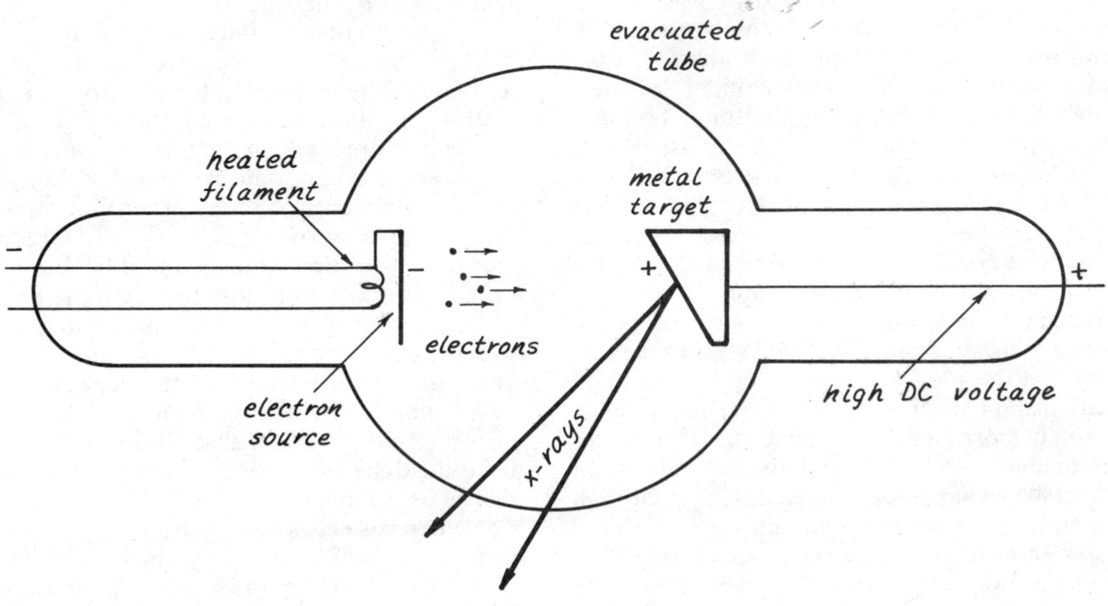

Figure 20–14 X-ray production.

approaches a metal atom, it is strongly repelled and decelerated by the electron cloud of the atom, thereby losing kinetic energy. Most of this energy goes into raising the temperature of the metal target, but about 1% of it is given off in the form of x-rays. Because different electrons may be decelerated at different rates, x-rays can be produced with a wide spread of wavelengths. This process is essentially the reverse of the photoelectric effect discussed earlier.

The quantum energies of x-ray photons are so great that they are not strongly absorbed by ordinary tissue. They can be absorbed only by disrupting atoms or molecules, and the probability for such violent events is lower than for other absorption processes. If the entire quantum energy of an x-ray photon is absorbed, it will generally eject an electron in the photoelectric process discussed earlier. The electron will be ejected with a kinetic energy equal to the quantum energy, hf, of the photon minus the work required to tear the electron out of its molecular environment (equation 20–2). Usually this ejected electron will have enough energy to ionize many other atoms. For example, a typical x-ray photon in a diagnostic medical x-ray might have an energy of 50,000 electron volts (50 keV). By comparison, it would take only 13.6 eV to strip the electron from a hydrogen atom, leaving the ejected electron with enough surplus energy to cause the ionization of many other atoms.

A more common interaction in the energy range used for diagnostic x-rays is *Compton scattering*. In this case the incident x-ray photon gives only a fraction of its energy to the ejected electron, with the remainder "scattering" off as a lower energy photon. This scattered photon would have a lower frequency and longer wavelength, as illustrated in Figure 20–13. Even though the ejected electron has only a fraction of the original photon energy, it will have enough energy to ionize many atoms in the vicinity of the original event. Other x-ray and gamma ray interaction mechanisms occur at very high energies (5), but Compton scattering and the photoelectric effect are the main attenuation mechanisms in the energy range used in most medical applications. Neither interaction has a high probability, so most of an x-ray beam would be expected to pass through normal tissue without interacting.

The ionizing capability sets x-rays and gamma rays apart from the lower frequency electromagnetic waves. The disruptive ionization events, though relatively small in number, can cause biological damage. The discussion of the radiation hazards will be deferred to Chapter 21 for discussion along with the biological effects of nuclear radiation. While the benefits to be derived from diagnostic x-rays generally far outweigh the risks, they should nevertheless be used conservatively. It is to be hoped that ultrasonic techniques and other nondestructive tests can be found to supplant x-rays for diagnostic purposes.

The use of x-rays for the examination of bones is a conceptually simple matter. X-radiation can be used to expose photographic film just as visible light is used. The fact that the soft tissues of the body are more transparent to x-rays than the bony structures can be used to form what is essentially a shadow pattern when the body is placed between the x-ray source and the film. The same technique can be used for the x-ray examination of organs, but the difference in x-ray transmission by different types of soft tissue is quite small and it is more difficult to get sufficient contrast on the film. One way to increase the contrast is to introduce air or another gas into a cavity normally filled by fluid. The x-ray absorption of the air is lower than that of the tissue providing the needed contrast. In other areas it is easier to increase the contrast artificially by inserting materials which are more nearly opaque to x-rays. Barium sulfate is sometimes administered in aqueous suspension for examination of the upper and lower gastrointestinal tract. The barium strongly absorbs x-rays and provides a light outline of the tract of interest.

A general understanding of the differences in x-ray absorption by various materials may be obtained by examining the electron energy levels of the atoms involved. The energy level separations are small in the lighter atoms such as carbon, oxygen, and hydrogen, which are major constituents of tissue. Therefore, these atoms cannot absorb x-ray photons directly and the bulk materials made up of light atoms are nearly transparent to x-rays. On the other hand, heavier elements such as barium and lead have energy levels in their inner shells which have separations comparable to the x-ray photon energies and therefore readily absorb x-rays. This absorption efficiency makes lead the most efficient common material for use as a radiation shield.

When it is necessary to observe the motion of internal parts of the body, an x-ray fluoroscope is used. The patient is placed between the x-ray source and a fluorescent screen. The nature of a fluorescent material is such that it absorbs radiation of one wavelength and emits radiation of another wavelength. When a material like cadmium tungstate is used on a screen, it will absorb x-rays and emit visible light. Therefore, the x-rays striking the screen will produce a luminous visible image via the fluorescent process. The movements of a joint can be directly observed, or the insertion of a cardiac catheter may be visually guided. Internal organs or the gastrointestinal tract may be examined with the aid of an opaque tracer material. Photographic records or video tape recordings can be made of the fluoroscopic images. The advantages of such techniques are obvious, but they of course expose the patient to more x-radiation than does the single exposure x-ray photograph.

X-RAY COMPUTED TOMOGRAPHY

Even with the methods for enhancing contrast, the conventional x-ray photograph is essentially a shadow picture, superimposing the absorption patterns from many layers of tissue and bone on top of each other, making accurate diagnosis difficult. Major advances in x-ray diagnosis have been made with computer-assisted scanning methods which produce a cross-sectional view of part of the body. Such techniques are generally called computed tomography or CT scans (8). The basic idea behind such scans is that, if a thin pencil-beam of x-rays is passed through a section of tissue a large number of times from different directions so that all the beam exposures have a common crossing point, the results of all the exposures can give a detailed evaluation of the x-ray absorption at that crossing point. With multiple x-ray beams and multiple detectors, modern scanners can collect such information about thousands of points within a few seconds of exposure time, storing the information in a computer memory for the purpose of constructing a two-dimensional view of the x-ray absorption of a section of the body.

Computed tomography scans are an impressive application of computer capability in medicine. Besides storing the data, the computer is used to process it to eliminate distortions of the image which arise from the nature of the technique, and to produce a finished display on a television screen or a photograph. Access to computer processing makes possible the assigning of different display colors to different x-ray absorption for a color-coded two-dimensional map of x-ray absorption. With x-ray exposures no greater than those of conventional x-ray photographs, the CT scans can detect changes in x-ray absorption on the order of a hundred times smaller than the minimum detectable changes on conventional x-ray photographs.

The success of x-ray tomography has spurred research with other tomography techniques. The ultrasonic scans discussed in Chapter 17 are also tomographs (9). (Tomograph means an image of a slice or cross-section.) The research on x-ray tomographs will undoubtedly lead to improvements in ultrasonic techniques. Another sectional scan technique uses magnetic resonance (NMR or nuclear magnetic resonance) to map the concentration of hydrogen atoms in tissue, with changes in signals with different molecular environments for the hydrogen atoms (10). This developing process, called zeugmatography, holds the promise of two-dimensional or three-dimensional imaging without the use of ionizing radiation (only magnetic fields and radio frequency waves are used). Zeugmatography is apparently the most sensitive technique for detecting small amounts of fluid in the lungs and holds promise for tumor detection and for studies of the heart.

THE LASER AND ITS APPLICATIONS

Few scientific discoveries have made as great an impact in a short time as the laser. The word LASER is an acronym representing Light Amplification by Stimulated Emission of Radiation. The property of laser light which has made it valuable for medical applications is the collimation of the light in a very narrow beam which can be further focused to an almost microscopic point, yielding enormous energy densities in the area of focus. The basic physical mechanisms involved in the laser and some of the medical applications will be discussed.

As discussed earlier, the emission of light by a material is associated with the transition of an atomic electron from a high energy state to a lower one as indicated in Figure 20–15(a). An electron will tend to move to the lowest available energy level, and this tendency leads to the spontaneous emission of a photon whose energy, hf, is equal to the energy lost by the electron in transition, $E = E_2 - E_1$. If a photon of energy E interacts with the atom while the electron is in the lower level E_1, the photon may be absorbed, raising the electron to energy E_2

a. Emission of a photon

b. Absorption and upward transition

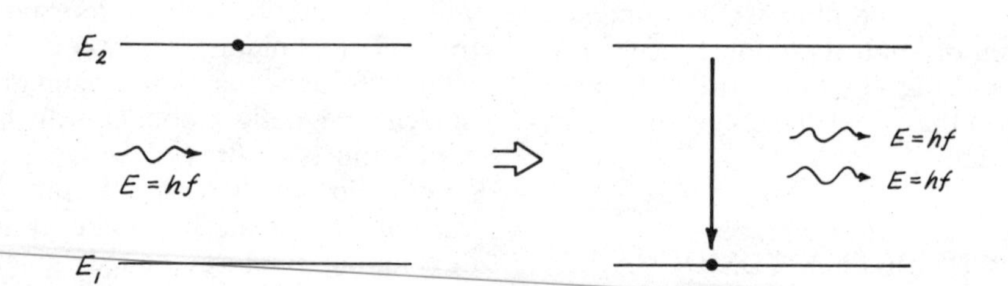

c. Stimulated emission of radiation

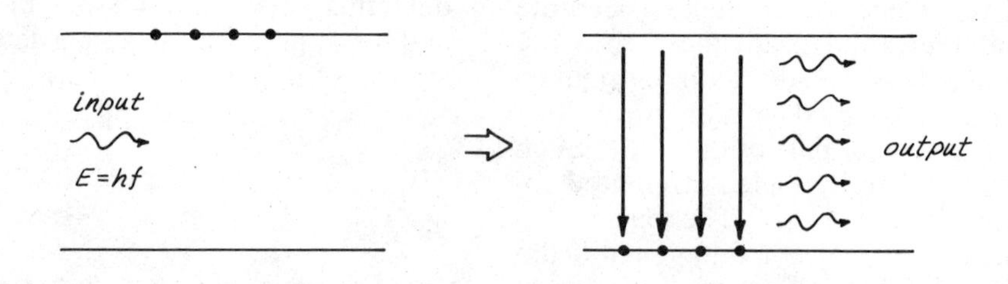

d. Light amplification by stimulated emission of radiation (laser).

Figure 20–15 The stimulated emission process involved in laser action.

(Figure 20–15(b)). If the photon interacts with the atom when the electron is in the higher state E_2, it may cause it to make the downward transition sooner than it would have by the spontaneous process. This is called *stimulated emission,* and the probability for its occurrence is exactly equal to the probability for absorption of a photon (Figure 20–15(c)). Therefore, if a large number of photons of this energy are incident upon the material, both absorption and stimulated emission occur. For normal materials, more electrons will be in the lower state, so absorption would occur more frequently. However, if a *population inversion* can be produced by some means so that the upper states have more electrons than the lower states, then *light amplification* can be achieved (Figure 20–15(d)). A chain reaction can occur, with one photon triggering the emission of another photon and proceeding to produce two more, until a large number of photons is produced. This is the basic process by which light amplification may be achieved by stimulated emission of radiation.

The basic requirement for producing a laser is to find two atomic or molecular energy levels for which a sizeable population inversion can be produced. With most energy levels, spontaneous emission will occur so rapidly that it is difficult to keep electrons in the upper state. Some energy states, called metastable states, have much longer lifetimes before spontaneous emission occurs. With most excited states of electrons, spontaneous emission will occur in about 10^{-8} second, but metastable states may have lifetimes from 10^{-6} sec up to

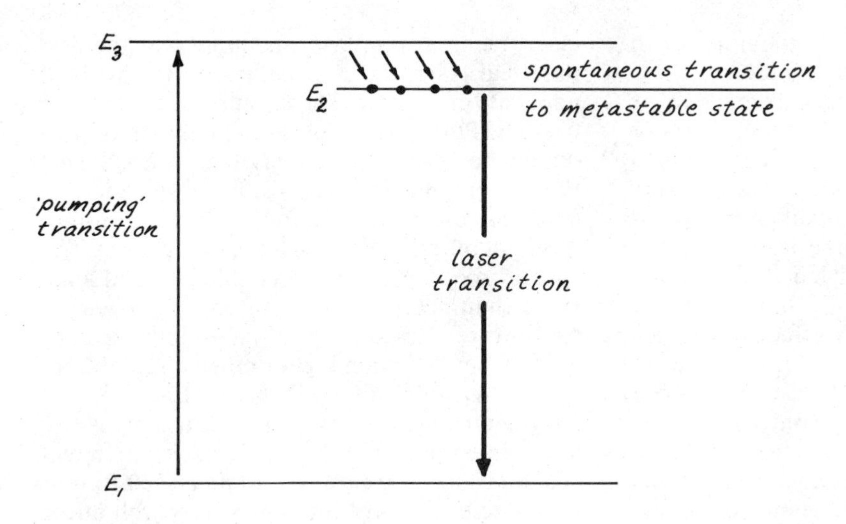

Figure 20–16 An example of the "pumping" process used to obtain the population inversion necessary for laser action.

several minutes before the electron drops to a lower energy level. However, the same physical principles which prevent rapid emission also tend to prevent absorption to elevate electrons to that level. Most lasers operate with three energy levels as shown in Figure 20–16. The electrons are first "pumped" into a higher energy level, E_3, by supplying energy in the form of light, electric discharge, or high energy molecular collisions. The elevated electrons then drop quickly into the metastable state E_2 by spontaneous emission or loss of energy through collisions. A large population inversion can be obtained because the electrons will remain in state E_2 for a considerable length of time. When a photon of appropriate energy interacts with one of these "trapped" electrons, it stimulates the electron to emit a photon and make the transition to the ground state E_1.

To improve the efficiency of the laser it is important to have the light pass through the laser medium several times, since the light is amplified each time it passes through. This is made possible by placing parallel mirrors at the ends of the laser medium, as shown in Figure 20–17. One mirror is nearly totally reflective and the other is usually made partially reflective to let some light out upon each reflection. Very precisely aligned mirrors are required to get efficient laser action; this is the most critical mechanical factor involved in constructing the laser. Besides providing greater light amplification, this precise mirror alignment produces a very narrow, collimated beam of light through the partially reflective mirror which doesn't spread out like ordinary light sources. The spreading of the light beams from most light sources causes the available light intensity to drop off rapidly with distance. The rays of a laser beam are so nearly parallel that the diameter of the beam of a laboratory laser may be less than three feet at a distance of 10 miles from the source. This property of a laser beam made it possible to determine the earth-moon distance by laser ranging. A pulsed laser beam was reflected from a mirror system placed on the moon in 1969 by astronauts Armstrong, Aldrin, and Collins. The detection of the reflected laser light at the earth's surface provided a

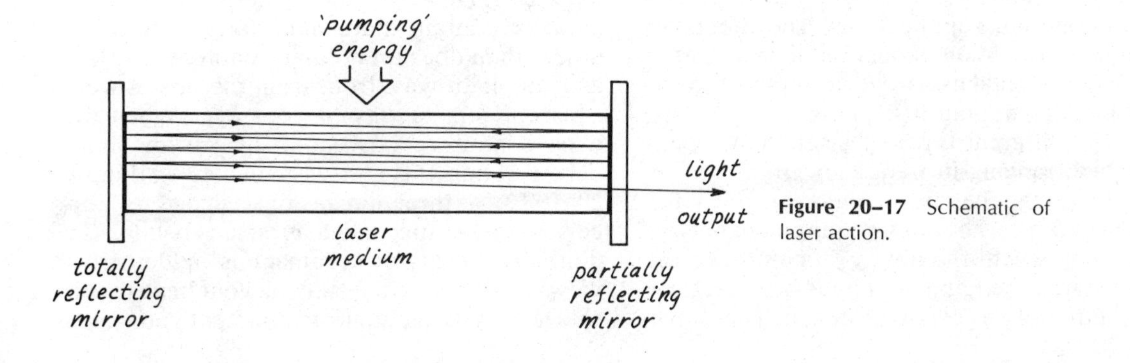

Figure 20–17 Schematic of laser action.

determination of the distance to the moon with an uncertainty of only about 130 feet.

The facts that the laser beam is made up of very nearly parallel rays and that it consists of only one wavelength of light make it possible to focus the beam to an extremely tiny spot. This is the feature which is most significant in present medical applications of the laser. The widest medical use of the laser has been in ophthalmology. A laser, emitting light in short bursts, has been proven very effective in photocoagulation of the retina (7). Hemorrhages in the retina may be treated by focusing intense light through the lens of the eye onto the affected part of the retina. Before the advent of the laser, a xenon-arc flash lamp was used for this purpose. A pulse of laser light can accomplish this task in less than 1 millisecond, whereas a full second was sometimes required for the xenon lamp. Much less heating of the ocular media results from the laser treatment, making possible more doses per treatment session. A further advantage of the laser treatment is the fact that the beam can be focused on an area having a diameter of 50 microns compared to 500 to 1000 microns for the xenon lamp. This is crucially important if the treatment area includes the fovea. As described in Chapter 19, the fovea contains a large fraction of the color receptors (cones) and is essential for color vision. This spot is only about 1000 microns in diameter. Therefore, the small focus area of the laser is essential if photocoagulation is to be attempted in this region. The laser is also the preferred method for treating retinal tears and detachments and some other retinal conditions.

The enormous energy density and precise focusing of the laser beam may lead to wider applications in delicate surgery. In addition to precision, the beam has a cauterizing effect and offers the possibility of bloodless surgery in specific applications. There are competing methods such as high frequency electrosurgery which must be considered in surgical procedures. Experiments are continuing with numerous other laser applications. The effects of high intensity light on pigment molecules have led to experimental use of the laser for removal of birthmarks and tattoos.

Many different types of lasers have been developed, ranging in frequency from the far infrared through the visible region and into the ultraviolet range. The basic requirement of two energy states with the ability to pump excess electrons into the upper state has been met by many different types of systems. The ruby laser, using a solid laser medium, was the first practical laser and is widely used for medical applications where short pulses of light are needed. Pulsed ruby lasers are capable of producing powers on the order of 10^9 watts during their short pulses. Gaseous laser systems are generally more efficient for producing continuous light output with a high average power. The carbon dioxide laser is capable of producing more than 5000 watts of continuous power in the infrared region while maintaining a very narrow beam. Much higher powers are obtainable with a more divergent beam. Liquid lasers have been developed which make use of energy levels in complicated dye molecules which will sustain laser action at a number of different frequencies. A large amount of research effort is being spent on the further development of semiconductor lasers which consist of two types of semiconductor crystals joined together and which emit light at the junction between the crystals. These semiconductor lasers have enormous potential in communications because both the intensity and frequency of the laser light can be modulated, making it possible to transmit information via laser beam.

Holography: Three-Dimensional Images

One of the most interesting applications of the laser is the production of three-dimensional images by a process known as holography. Holography is a photographic process, but instead of forming a focused image on the film with the use of a lens, the film is directly exposed to the reflected light from the object as indicated in Figure 20–18. Part of the light from the laser is reflected from a mirror and directed toward the film where it mixes with the light reflected from the object. The film records the interference pattern between these two beams of light and produces a photographic pattern, which appears as irregular fringes with no resemblance to the object when viewed in ordinary light. However, these interference fringes actually contain more information about the object than does an ordinary photograph. It is as if the light wavefront from the object were "frozen" on its way to your eye. When the wavefront is reconstructed by passing laser light through the film as indicated in Figure 20–18(b), a three-dimensional image is perceived to be suspended in space behind the film. If one part of the image is hidden from view, it can be seen by moving your head to one side in exactly the same manner that you would

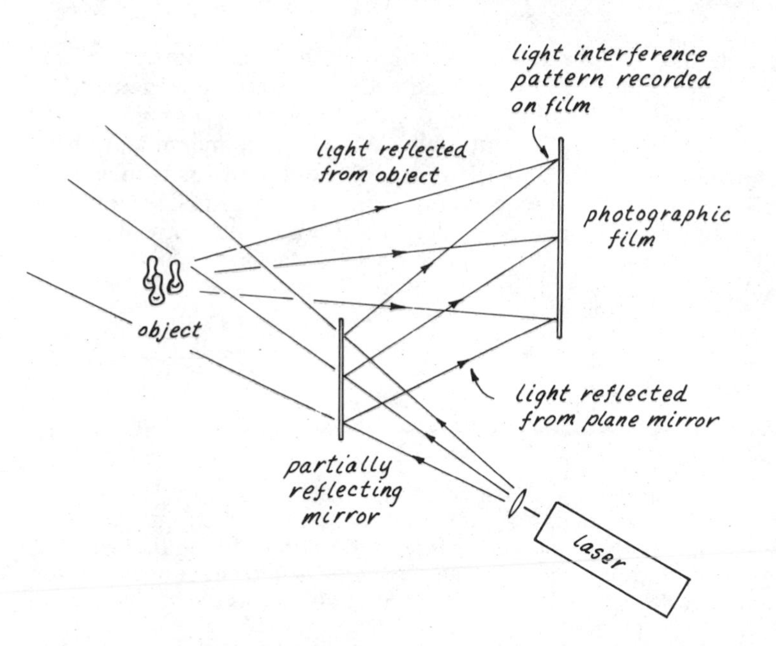

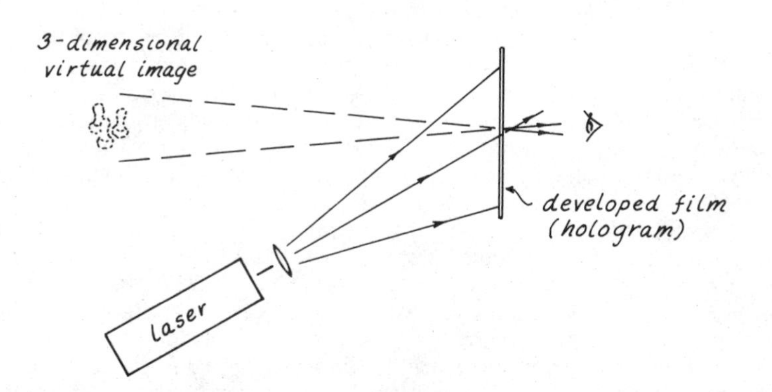

Figure 20–18 The hologram.

move to see all parts of the real object. In fact, a full 360 degree hologram can be made to allow you to rotate the hologram and see all sides of the object.

A hologram cannot be made with ordinary light because such light is "incoherent"; that is, it occurs in random bursts from different atoms which have no definite time or space relationship to each other. The coherence of laser light is one of its most important properties. Since the light produced is a cascade of photons produced by stimulated emission (Figure 20–15(c)), the waves have a definite space relationship to each other and can interfere with each other constructively or destructively in a manner similar to the adding and subtracting of waves on a water surface. It is this interference which allows the information to be recorded on the film.

Some of the first practical applications of holography were in the area of microscopy (7). Besides the obvious advantages of a permanent three-dimensional recording of microscopic events, the hologram provides improved resolution. The theoretical limits of resolution for microscopy are set by the wavelength of the light and the size of the lens aperture. Since a hologram c n be made using a lensless imaging process, it can provide better resolution than a standard microscope of comparable magnifying power. The hologram has the potential to overcome the wavelength limitation of the light microscope, since a hologram can be made with one wavelength of light and reconstructed with another. For example, an enormous increase in magnification could be achieved if the hologram magnification achieved would be approximately equal to the ratio of the wavelengths, so that a hologram made with a coherent x-ray source with wavelength 5 Å and reconstructed with a laser with a wavelength of about 5000 Å would represent a magnification of about 1000. With further magnification by standard optics, three-dimensional images of objects approaching molecular size could be made visible. These possibilities await the development of an x-ray laser.

SUMMARY

Many of the properties of matter on the atomic scale are found to be quantized. Properties such as mass, charge, energy, and others can have only certain discrete, allowed values. A study of the quantum energy levels for electrons associated with atoms reveals patterns of atomic structure which explain many of the features of the periodic table of the elements and the chemical nature of the elements.

Experiments such as the photoelectric effect have shown that light is quantized, with the quanta having energies equal to Planck's constant times the frequency of the light. Light quanta are called photons, and under the appropriate circumstances they exhibit particle-like properties. In large scale phenomena, the individual quantum effects are not observed and light behaves like a traveling wave. This wave-particle duality is a part of the basic nature of light. Visible light constitutes a small segment of the broad spectrum of electromagnetic waves. In order of increasing frequency, some of the types of electromagnetic waves are radio waves, microwaves, millimeter waves, infrared, visible light, ultraviolet, x-rays, and gamma rays. These types of radiation have various clinical applications which depend upon their energy, penetration, and specific physiological effects.

When electromagnetic waves interact with matter, they can be absorbed by the matter only if the photon energies are equal to the separation of two allowed energy states for the atoms or molecules of the matter. A gas may absorb one frequency of light very strongly but be transparent to a nearby frequency. On the other hand, the gas will emit only certain frequencies corresponding to the dropping of an electron from a high energy state to a lower one. The discrete nature of the frequencies emitted and absorbed forms the basis of spectroscopy, by which chemical elements and compounds can be identified by the frequencies emitted or absorbed. The single frequency of light emission associated with a given electron transition makes possible the intense monochromatic light from the laser. Electrons are pumped to an upper energy state and stimulated to drop down in response to an incident light photon of appropriate energy. The light is amplified and collimated into a narrow beam which is useful for surgical and other applications. Holography, a type of three-dimensional photography possible with the laser, has many potential applications in medicine and elsewhere.

REVIEW QUESTIONS

1. What does "quantization" mean with regard to physical properties? What physical properties of atoms are quantized?

2. What is the difference between light and x-rays? What is the difference between x-rays and radio waves?

3. What features of the photoelectric effect experiment could not be explained by a classical wave theory of light?

4. From the point of view of the quantum theory of light, what explanation could you offer for the fact that x-rays produce damaging radiation effects in the body, while radio waves do not?

5. Describe the effect of a barium sulfate solution taken before a stomach x-ray.

6. What precautions should be taken when administering an ultraviolet treatment?

7. What type of radiation is primarily responsible for sunburn?

8. How can the light emitted by a substance be used to identify the chemical elements present in the substance?

9. What are the properties of the laser which make it medically useful?

10. What would be the advantages of an acoustic hologram over x-rays as a tool for the study of internal organs and bones?

PROBLEMS

1. How long did it take for a radio signal to reach the Apollo astronauts on the moon's surface, 240,000 miles from the earth?

2. What is the wavelength of the radio waves from an AM radio station which is broadcasting at 750 kHz?

3. Radar makes use of short microwave pulses which travel outward from the transmitter at the speed of light and produce an "echo" by reflecting back to the antenna. If an echo pulse is received from an aircraft 6 microseconds after the initial phase, what is the distance from the antenna to the aircraft?

4. If the television picture carrier wave frequencies for TV channels 5 and 11 are 77.25 MHz and 199.25 MHz respectively, what length antenna elements would be needed to make half-wavelength antennas for these channels? (The actual optimum lengths are somewhat shorter because of "end effects" at the ends of straight antenna segments.)

5. If your vision process made use of microwaves of frequency 1×10^9 Hz instead of visible light, what would be the smallest detail your eyes could resolve? Assume that the limit is roughly equal to the wavelength of the electromagnetic waves used.

Worked Example: A molecule will usually have a large number of quantized energy levels of different separations corresponding to different phenomena within the molecule. Calculate the frequency of the photon required to cause the molecule to make the following energy level transitions, and identify the type of electromagnetic radiation absorbed to cause the transition.

(a) 2.0 electron volts, transition between electron orbital energy levels.
(b) 0.1 eV, transition between molecular vibration states.
(c) 0.0001 eV, transition between rotational energy states for the molecule.

Solution: (a) An electron volt = 1.6×10^{-19} joule. From the photon energy relationship $E = hf$:

$$2 \times 1.6 \times 10^{-19} \text{ joule} = (6.62 \times 10^{-34} \text{ joule-sec})(f)$$
$$f = 4.8 \times 10^{14} \text{ Hz.}$$

From Figure 20–3, this is seen to be in the visible light region. The corresponding wavelength is about 6200 Å, which corresponds to orange light.

(b) The energy 0.1 eV corresponds to a frequency

$$f = \frac{0.16 \times 10^{-19} \text{ joule}}{6.62 \times 10^{-34} \text{ joule-sec}} = 2.4 \times 10^{13} \text{ Hz}$$

which is in the infrared range. Molecular vibrations and infrared radiation are associated with heat energy.

(c) The energy 0.0001 eV corresponds to a frequency

$$f = \frac{1.6 \times 10^{-23} \text{ joule}}{6.62 \times 10^{-34} \text{ joule-sec}} = 2.4 \times 10^{10} \text{ Hz}$$

which is in the microwave frequency range.

6. The frequency associated with a certain diagnostic x-ray photon is 1.7×10^{19} Hz.

   a. What is its energy in joules?
   b. What is its energy in electron volts?
   c. If it takes 13.6 eV to ionize hydrogen, how many hydrogen atoms could theoretically be ionized by one such x-ray photon?

7. If a microwave oven operates at a frequency of 2560 MHz, what is the quantum energy in electron volts associated with the microwave photons? What multiple of this microwave quantum energy would be required to ionize one hydrogen atom (13.6 eV)?

8. Calculate the frequency and the wavelength in angstroms for the photon emitted when an electron makes a transition from the $n = 2$ to the $n = 1$ level of hydrogen.

9. Calculate the frequency and wavelength in angstroms associated with the transition B in Figure 20–11 if it corresponds to a transition of an electron from the $n = 4$ level of hydrogen to the $n = 2$ level.

10. The high intensity yellow street lights make use of the emission of light from sodium vapor. If the principal wavelength for this emission is 5890 angstroms, what is the frequency and the energy in electron volts?

11. An electron is accelerated by a voltage of 1000 volts.

    a. What speed will the electron have if it starts from rest?
    b. What wavelength in angstroms will be associated with the wave properties of the electron?

REFERENCES

1. Krauskopf, K. B., and Beiser, A. *The Physical Universe,* 3rd ed. New York: McGraw-Hill, 1973.
2. Freeman, I. M. *Physics: Principles and Insights,* 2nd ed. New York: McGraw-Hill, 1973.
3. Beiser, A. *Concepts of Modern Physics,* 2nd ed. New York: McGraw-Hill, 1973.
4. Dummer, G. W. A., and Robertson, J. M., eds. *Medical Electronic Equipment 1969–70,* Volume I, Clinical, Diagnostic and Therapeutic Equipment. New York: Pergamon Press, 1970.

5. Stanton, L. *Basic Medical Radiation Physics*. New York: Appleton-Century-Crofts, 1969.
6. Burns, D. M., and MacDonald, S. G. G. *Physics for Biology and Pre-Medical Students,* 2nd ed. Reading, Massachusetts: Addison-Wesley, 1975.
7. Goldman, L., and Rockwell, R. J. *Lasers in Medicine*. New York: Gordon and Breach, 1971.
8. Swindell, W. and Barrett, H. H. *Computerized Tomography: Taking Sectional X-rays*. Physics Today, Vol. 30, p. 32, December 1977.
9. Devey, G. B., and Wells, P. N. T. *Ultrasound in Medical Diagnosis*. Scientific American, V238, p. 98, May 1978.
10. Lubkin, G. B. *NMR Imaging Technique Provides High Resolution*. Physics Today, Vol. 31, p. 17, May 1978.

CHAPTER TWENTY-ONE

Nuclear Radiation

INSTRUCTIONAL OBJECTIVES

After studying this chapter, the student should be able to:

1. Name the basic constituents of the nucleus and define the following terms: atomic number, mass number, isotope, radioactivity.

2. Name and describe the physical nature of the three basic types of radioactivity.

3. Define half-life and calculate the amount of a radioactive substance remaining after a stated time interval, given the original amount and its half-life.

4. Define the "biologic half-life" and calculate the effective half-life, given the physical and biologic half-lives.

5. Describe the biological effects of radiation with respect to (a) resulting physiological damage and (b) possible therapeutic effects.

6. Compare the medical usefulness of alpha, beta, and gamma radiation.

7. Name the structures in the body which are most sensitive to radiation.

8. Describe the basis for the "therapeutic ratio" for cancer therapy.

9. State the physical variables which determine the radiation dose received by a person who is exposed to a radioactive material.

10. State the characteristics of radioactive nuclei which make possible the tagging of chemical compounds for diagnostic studies.

11. Define nuclear fission and nuclear fusion and describe their actual and potential usefulness as energy sources.

Nuclear radiation refers to those particles or waves which emanate from the atomic nucleus. The energy associated with such radiation is quite high, and nuclear radiation is often classified with x-rays as "ionizing radiation" because the energies are large enough to strip electrons from atoms or otherwise alter the structures of atoms and molecules. The nature of the interactions between ionizing radiation and biological materials has led to widespread use of x-rays and nuclear radiation for diagnostic and therapeutic purposes. Some of the basic properties of nuclei will be examined here with emphasis on radioactivity, radioisotopes, and their clinical applications.

A SCALE MODEL OF THE ATOM

By 1900, a considerable amount of information about electrons had been established, but little was known about the other constituents of the atoms. In particular, the role of the electron in electric currents was established and it was known that atoms contained electrons. From measurements of the charge and mass of electrons it could be implied that the remainder of the atom had a net positive charge and that it contained most of the mass of the atom. It was popularly assumed that the negatively charged electrons were distributed in a uniform positively charged mass, a model which was sometimes called the "raisin pudding" model of the atom.

A major development in physics occurred in 1911 when Sir Ernest Rutherford of England bombarded thin metal foils with alpha particles, which are helium atoms stripped of their two electrons. From the scattering of these charged particles by the metal atoms Rutherford was able to establish that the mass and positive charge were not evenly distribtuted but concentrated in a volume which was very small compared to the volume of the atom. This was the beginning of the nuclear model of the atom. By detailed analysis of his experimental data, Rutherford was able to show that the radius of the entire atom must be at least 10,000 times as large as the radius of the nucleus. The nucleus is then visualized as a tiny "particle" surrounded by the electrons in a large amount of empty space. This incredible revision of the concept of the atom was not readily accepted and Rutherford's reputation suffered for a time, but when later experiments supported his results he became justifiably famous.

Some insight into the nature of nuclear and atomic phenomena can be gained by constructing a scale model of an atom and comparing it to a solar system model. It must be strongly emphasized that such a model is a conceptual aid only, and not an accurate representation of the atom. As pointed out in the discussion of the quantum theory of the atom, electrons and other atomic constituents do not behave like tiny billiard balls and do not obey the ordinary laws of classical physics. Atomic electrons show wavelike properties and must be viewed as distributed in space, somewhat like a negatively charged "cloud" around the nucleus. Nevertheless, having pointed out that the model contains some swindles, we proceed to construct it.

The element gold has been chosen for a model calculation, although any element could be used with similar results. The data necessary for the calculation are listed in Table 21–1. A rough calculation of the radius of a gold atom can be made using the density and atom-

TABLE 21–1 Data for Scale Models of a Gold Atom and the Solar System.

A. Gold Atom.
Nuclear density* $\simeq 2 \times 10^{17}$ kg/m^3
Density of solid gold = 19.32 gm/cm^3
Atomic mass = 196.97 amu (1 mole = 196.97 grams)
1 amu = 1.66×10^{-27} kg
Avogadro's number = 6.02×10^{23} atoms/mole
Calculated atomic radius = 1.3×10^{-10} m
Calculated nuclear radius = 7.3×10^{-15} m

B. Solar System.
Radius of sun = 432,000 miles
Radius of earth = 3963 miles
Sun-earth distance = 93×10^6 miles
Sun-Pluto distance = 3666×10^6 miles

*See Reference 1.

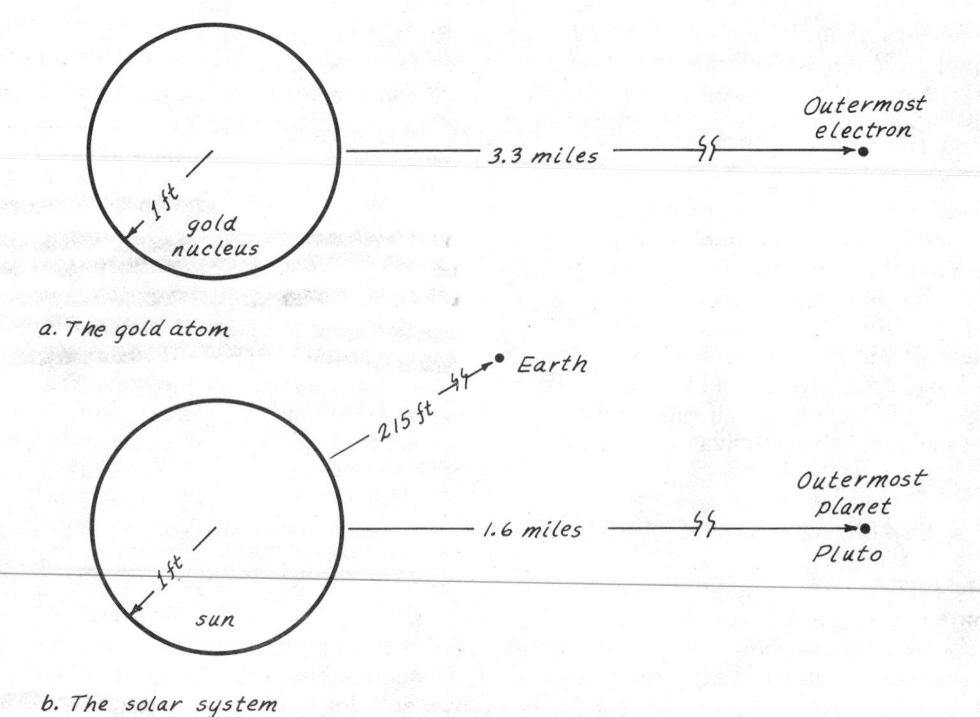

Figure 21–1 Scale models of a gold atom and the solar system.

ic mass of gold, assuming a cubic volume for each atom. This yields a radius $R = 1.3 \times 10^{-10}$ meters. Nuclei are observed to have a density of about 2×10^{17} kg/m^3. This enormous density amounts to more than 3 billion tons per cubic inch of nuclear material (1). The use of this density and the nuclear mass yields a nuclear radius of about 7.3×10^{-15} meters. The calculated atomic radius is nearly 18,000 times the nuclear radius. If the scale is expanded until our nuclear model is a sphere of radius 1 foot, as shown in Figure 21–1, the outermost electron will be a small pea-sized object about 3.3 miles away. Even with 79 such electrons associated with the gold nucleus, this crude model illustrates the fact that the atom is mostly empty space.

When the corresponding model of the solar system is calculated by reducing the sun in scale until it has a radius of 1 foot, the earth is about 215 feet away and the most distant planet, Pluto, is about 1.6 miles away from the sun. We tend to think of the solar system as isolated planets in the void of space and to think of a gold ring as a very solid object. Some insight into the nuclear structure of the atom may be obtained from the realization that the relative amounts of empty space in the two models are comparable.

THE NATURE OF THE NUCLEUS

Our understanding of the nucleus, with its extremely small size and enormous density, is far from complete. Though many mysteries remain, a large amount of information about nuclei has been gathered. Nuclear physics is a comparatively young field, with much of our present understanding of the nucleus having been developed since 1930. Although the internal structure of the nucleus is not well understood, it is known to be composed of protons and neutrons. The proton has one positive quantum of charge and a mass of 1836 times the mass of an electron, and the neutron is an uncharged particle with 1839 times the mass of an electron. The nuclei of light atoms tend to have about an equal number of neutrons and protons. In such cases the nucleus contains about 99.97% of the mass of the atom, while occupying an extremely tiny part of the volume of the atom.

The number of protons in the nucleus is called the *atomic number* and is represented by the symbol Z. The atomic number determines the chemical species or chemical element to which the nucleus belongs. The *neutron number* is represented by N: the total number of neutrons plus protons is called the *mass number* and is represented by the letter A. Nuclei with the same number of protons but a different number of neutrons are nuclei of the same chemical element but are said to be different isotopes of that element. The standard nomenclature for an isotope of a chemical element is:

$$^{A}_{Z}X$$

where

X represents the chemical symbol for the element
Z = the atomic number = number of protons
N = neutron number
A = mass number = $Z + N$.

For example, the isotopes of carbon may be represented as follows:

$$^{12}_{6}C$$

$$^{13}_{6}C$$

$$^{14}_{6}C \text{ (radioactive).}$$

The subscript 6 is redundant since this is implied by the chemical symbol C for carbon, so the isotopes are often written ^{12}C, ^{13}C, and ^{14}C. These isotopes differ only in the number of neutrons in the nucleus; they have identical chemical properties.

For each number of protons in the nucleus (i.e., for each chemical element) there is an optimum number of neutrons for maximum stability of the nucleus. If the neutron number is too small *or* too large, particles and sometimes electromagnetic waves are emitted from the nucleus until it reaches a stable configuration. This emission process is often referred to as radioactivity or radioactive decay. Very few radioactive isotopes are found in nature because the process of radioactive decay converts them into other elements as described below. Several elements have more than one stable isotope. About 99% of all carbon consists of the isotope ^{12}C, with about 1% of the isotope ^{13}C which has one additional neutron. A very small quantity of the radioactive isotope ^{14}C exists in the earth's atmosphere because it is continually produced in the upper atmosphere by cosmic rays. Some elements such as phosphorus and fluorine have only one naturally occurring stable isotope, and some heavier elements have several isotopes. The element tin has ten stable isotopes, the largest number of naturally occurring isotopes. Radium has been the most prominent naturally occurring radioactive element for medical application, but most present medical applications make use of artificially produced radioisotopes, as discussed below.

Radioactivity is the result of nuclear instability. An examination of the forces in the nucleus will reveal something about the cause of this instability. From the models presented, it is clear that the collection of protons and neutrons which makes up the nucelus is extremely dense. Recalling the nature of the electrical force as described in Chapter 12, this implies that the positively charged protons in the nucleus repel each other with enormous forces. The magnitude of this electric repulsion force may be more fully appreciated by comparing it with the gravitational force. For example, the hydrogen atom consists of an electron orbiting around a single proton. The electron and proton are attracted toward each other by both a gravitational force and an electrical attraction. In this case the electrical force is about 10^{39} times stronger than the gravitational force — illustrating the fact that gravitational forces are totally negligible in determining atomic orbits. The electric force of repulsion between two protons in a nucleus is on the order of 100 million times as strong as this attractive force, which keeps the electrons in orbit. Nuclear instability would then be expected, and it is remarkable that any nuclei are stable. The stability of many nuclei demonstrates the existence of a third force, a powerful attractive nuclear force which holds the protons and neutrons together despite the electric repulsion forces. This nuclear force is capable of maintaining stability only if the appropriate number of neutrons is present to moderate the repulsive forces; if there are too many *or* too few neutrons the nucleus becomes unstable (radioactive). In view of the enormity of the forces inside the nucleus, it is perhaps not surprising that particles are emitted with very large energies when the nucleus is unstable.

THE THREE BASIC TYPES OF RADIOACTIVITY

Early experimenters with nuclear radiation found that a narrow beam of radiation could be produced by placing a mixture of radioactive materials in a narrow hole in a lead block, since the lead absorbed the radiation in all other directions. It was discovered that if this beam was passed through a strong electric

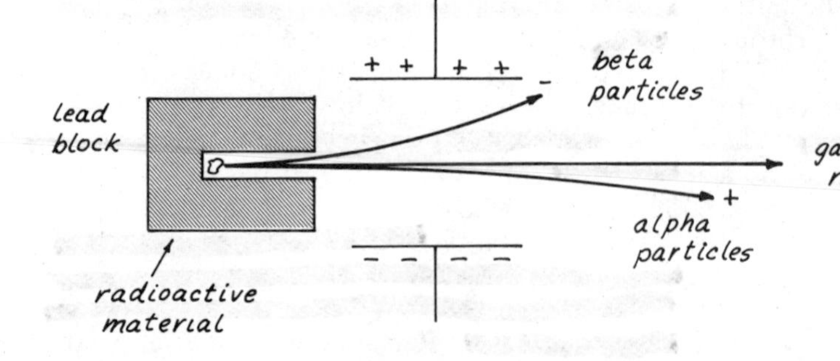

Figure 21–2 The three common types of nuclear radiation, from a mixture of different radioactive materials.

or magnetic field, it split into three beams as indicated in Figure 21–2. A single radioisotope will not emit all three types of radiation, but a sample containing several radioactive species may produce the three beams. The three types of radiation were labeled alpha, beta, and gamma rays. In the experiment represented in Figure 21–2, one beam is deflected slightly toward the negative plate, demonstrating that it is composed of positively charged particles (alpha particles). One beam is deflected strongly toward the positive plate, showing that it consists of negatively charged particles (beta particles). The third beam is undeflected and very penetrating, the uncharged gamma ray. The nature of these three types of radiation is now well understood and their properties are summarized below.

Type of Radiation	**Physical Nature and Description of Effects**
Alpha	Composed of two protons and two neutrons, it is a nucleus of the element helium. Because of its very large mass (more than 7000 times the mass of the beta particle), it has a very short range. Not suitable for radiation therapy since its range is less than a millimeter inside the body. Main radiation hazard comes when it is ingested into the body; it has great destructive power within its short range.
Beta	An electron. It has a greater range of penetration than the heavier alpha particle, but is much less penetrating than gamma rays. Its radiation hazard is greatest if it is ingested.
Gamma	Electromagnetic ray. Distinguished from x-rays only by the fact that it comes from the nucleus. Somewhat higher in energy than x-rays, generally. Very penetrating. It is the most useful type of radiation for therapy, but at the same time it is the most hazardous because of its ability to penetrate large thicknesses of material.

A nucleus which emits a particle is transformed into a nucleus of a different chemical element in the process. For example, a nucleus which emits an alpha particle loses two protons and two neutrons from the nucleus. The uranium isotope ^{238}U is transformed into a thorium nucleus by alpha emission. This radioactive decay process can be represented by the following:

$$^{238}_{92}U \rightarrow {}^{4}_{2}\alpha + {}^{234}_{90}Th.$$

The sum of the subscripts is the number of positive charges, which is the same before and after the process in order to conserve charge. The sum of the superscripts is the total mass number, which must always remain the same in the decay process in order to conserve mass. The emission of a beta particle is also accompanied by the transmutation of the nucleus into another chemical element. The thorium nucleus produced in the above process is radioactive and is transformed into the element protactinium by the emission of a beta particle:

$$^{234}_{90}Th \rightarrow {}^{0}_{-1}\beta + {}^{234}_{91}Pa. + \bar{\nu}$$

The superscript 0 on the β indicates that it doesn't change the nuclear mass number, and the subscript -1 indicates the negative charge of the β (an electron) and denotes the fact that the positive charge of the nucleus must be increased by one quantum in order to maintain the same total charge. Whereas the alpha emission subtracted two protons from the nucleus, the beta emission process results in an addition of one proton and the subtraction of one neutron in an essentially instantaneous transformation process. Another particle called an antineutrino ($\bar{\nu}$) is also emitted during the beta decay process, but its properties will not be discussed here (2). Since the gamma ray is an electromagnetic ray with no charge or rest mass, gamma radiation does not change the nuclear species.

A characteristic common to α, β, and γ radiation is a very large amount of energy. This raises the question "How can a nucleus which is just sitting there suddenly throw out a particle or ray with such enormous energy for penetration or radiation damage?" The answer lies in the fact that a small amount of mass is converted to other forms of energy in the radiation process. A complete statement of the conservation of energy principle includes the fact that mass must be considered to be a form of energy, and that "mass energy" can be converted to other forms of energy according to the Einstein relation

$$E = mc^2, \qquad \textbf{21–1}$$

where E is the energy yield, m is the amount of mass converted, and c is the speed of light ($c = 3 \times 10^8$ m/sec). The enormous energy yield from mass conversion can be seen from the fact that c is a very large number, and it is squared in the Einstein equation so that the conversion of one kilogram of mass yields 9×10^{16} joules of energy! The production of energy by nuclear processes is discussed further in the final section of this chapter, but the point of interest here is the fact that a very small amount of mass conversion can give a very high kinetic energy to an α or β particle or a very high frequency to a γ ray (γ ray energy $E = hf$). For example, when radioactive uranium-238 is transformed, the sum of the masses of the resulting α particle and thorium atom is slightly less than the mass of the original uranium atom. The lost mass is transformed into kinetic energy, giving a very high velocity to the α particle and a smaller recoil velocity to the thorium atom.

Another type of particle, the positron, is emitted from some artificially produced radioisotopes. It is identical to the electron except that it has a positive charge. The nitrogen radioisotope ^{13}N is transmuted into carbon by the emission of a positron in the transformation

$$^{13}_{7}N \rightarrow {}^{0}_{+1}\beta + {}^{13}_{6}C + \nu,$$

where the beta symbol with a +1 subscript is used to represent the positron. The positron is the anti-particle of the electron; when a positron encounters an electron, they annihilate with the production of two gamma ray photons. This production of energy in the form of gamma rays is a conversion of mass to energy in other forms, as described above. While there are some potential medical applications of positrons and other types of nuclear radiation (2, 3), our subsequent discussion will be limited to the three most common types of nuclear radiation.

RADIOACTIVE DECAY AND HALF-LIFE

The more unstable the nuclear species, the larger is the percentage of the nuclei which will emit radiation in a given time period, and thus the more radioactive is the species. When a nucleus emits a particle or gamma ray it is said to "decay" into another species of nucleus in the case of particle emission or to a more stable configuration of the same nucleus in the case of gamma emission. In either case, the number of nuclei of the original radioactive species decreases with time. A useful parameter for classification of radioactivity is the *half-life*, the time for one half of the nuclei to decay, regardless of the original number. The fact that the radioactive isotope iodine-131 may be said to have a half-life of eight days is a result of our observation that, regardless of how many ^{131}I nuclei you have originally, you will have half as many eight days later.

The radioactive decay process may be visualized as a "decay curve," as in Figure 21–3. The concept of the half-life is based upon the experimental evidence that each nucleus has an equal probability of emitting radiation in a given time interval, independent of the number of other nuclei present or any other physical conditions. No one can predict when a given nucleus will decay, but we can statistically predict that if there are twice as many nuclei, twice as many will decay in a given time period. This statistical treatment produces the curve in Figure 21–3, which has been experimentally verified to a high degree of precision. It should be noted that the precision of any statistical calculation increases with the number of objects you are dealing with. If you have a few grams of a radioactive substance, you have a number of nuclei on the order of Avogadro's number (6×10^{23}). Therefore, statistical methods could be expected to be very precise.

The radioactive transformation processes in which particle emission (α, β) is involved can be easily studied because of the fact that what is left after the particle is emitted is a

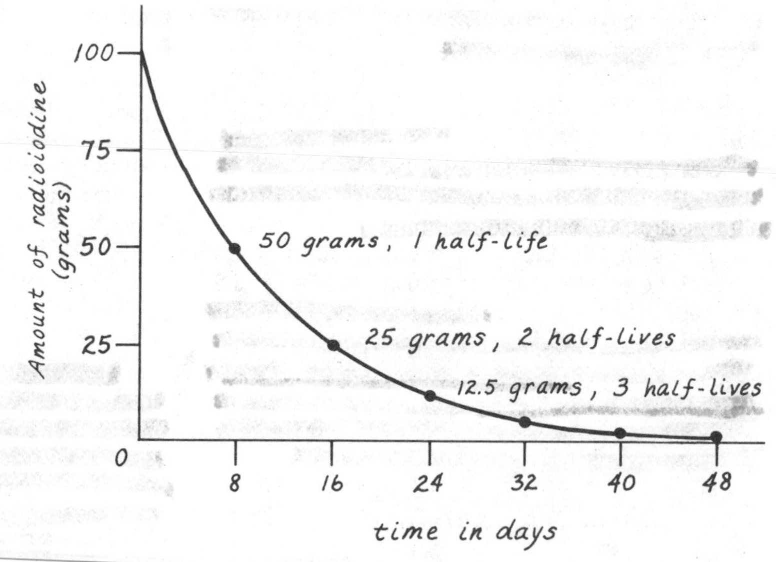

Figure 21–3 Radioactive decay curve for iodine-131.

different chemical element which can be separated or detected by using its different chemical properties. For example, consider an element *A* which emits a particle and is transformed into element *B* in a process which has a half-life of 10 years. If there is a 100 gram mass of element *A* and none of element *B* at some initial time, then after 10 years there will be 50 grams of *A* and 50 grams of *B*, after 20 years there will be 25 grams of *A* and 75 grams of *B*, and so forth, as summarized in Table 21–2. If a sample were found which contained 99 grams of *B* and one gram of *A*, it could be surmised that the sample had been there just under 70 years, if one were justified in making the assumption that the material was originally 100% *A*. This kind of reasoning is the basis for radioactive dating processes. In geological dating calculations, several radioactive decay processes must be used to give enough data to calculate the original amounts of each element present, since the assumption of 100% purity for any given element is seldom justified.

TABLE 21–2 Radioactive Decay of a Substance with a Ten Year Half-Life.

Time in Years	Amount of Element *A* in Grams	Amount of Element *B* in Grams
0	100	0
10	50	50
20	25	75
30	12.5	87.5
40	6.3	93.7
50	3.1	96.9
60	1.6	98.4
70	0.8	99.2
80	0.4	99.6
90	0.2	99.8
100	0.1	99.9

An interesting application of the half-life concept is the carbon-14 dating process. The radioactive isotope ^{14}C emits beta particles and has a half-life of 5570 years. Although it is not found in minerals, it is continually produced at a more-or-less constant rate in the upper atmosphere when nitrogen atoms are bombarded by cosmic-ray neutrons. Living organisms ingest a small amount of ^{14}C from the atmosphere and therefore emit a measureable amount of beta radiation. This level of radioactivity remains nearly constant while the organism lives because of continuous ingestion, but decreases when the organism dies. If a buried plant or animal specimen is found, the level of beta radiation can be measured and compared to the normal radiation intensity for a living organism. If the radiation intensity is one half that of the living organism, it can be deduced that the organism lived about 5570 years ago. One fourth of the living radiation level would imply an age of 11,140 years, and so on. However, the experimental accuracy diminishes with the level of radioactivity, and the determination of ages greater than about 20,000 years is not feasible using ^{14}C because the radiation level is too low.

In the case of medical radioisotopes the half-life must be known for accurate dose calculations. Radioisotope doses are usually measured in terms of the curie, the basic unit of radioactive quantity, which is equivalent in activity to one gram of radium. More convenient units for medical use are the millicurie (mCi) and the microcurie (μCi). For example, a typical dose of a radioiodine-containing compound might be 5 microcuries. Since the half-life of the isotope ^{131}I is 8 days, the radioactive quantity will be only 2.5 microcuries after 8 days. The radioactive quantity will normally be specified on the day of delivery, and must be calculated at later times from the half-life of the isotope.

Example. A radioiodine compound has 5 microcuries of radioactivity on a given date. How much radioactivity remains after 40 days?

Solution. Since the quantity decreases by half in each successive half-life, the quantity present after 40 days may be found.

Time	Quantity
0	5 μCi
8 days	2.5 μCi
16 days	1.25 μCi
24 days	0.63 μCi
32 days	0.31 μCi
40 days	0.16 μCi

After 40 days (5 half-lives) the radioactive quantity is only 0.16 microcurie. Radioiodine compounds for diagnostic use are typically discarded after about 20 days because of the decrease in radioactive quantity.

In the case of radioisotopes which have been introduced into the body, the radioactive quantity decreases with time as a result of natural biologic elimination processes as well as through radioactive decay. It is sometimes convenient to define a "biologic half-life" as the time required to reduce the quantity of radioisotope to one-half by elimination processes (3). For a given radioisotope dose in microcuries, the maximum total radiation in the body will occur if both the physical and biologic half-lives are long. If either the physical or biologic half-life is short, the radiation level in the body will drop rapidly. Though the physical half-life can be considered to be a precise parameter, the biologic half-life is subject to considerable variation. The process of clearing the radioactive chemical from an organ does not always follow a half-life type reduction. With the assumption that the biologic half-life is a valid concept, an "effective half-life" may be calculated from the relationship

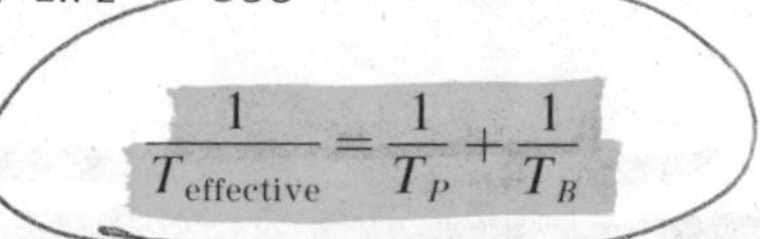

$$\frac{1}{T_{\text{effective}}} = \frac{1}{T_P} + \frac{1}{T_B}$$

where T_P and T_B are the physical and biologic half-lives, respectively. Note that this is exactly the same form as equation 13–3 for calculating the effective electrical resistance for two resistances in parallel. The physical and biologic half-lives may be thought of as representing two parallel pathways for reducing the amount of radioactive material. It is usually more convenient to express the effective half-life in the equivalent form

$$T_{\text{effective}} = \frac{T_P T_B}{T_P + T_B}. \qquad \textbf{21–2}$$

This effective half-life may be used to calculate the reduction of radioactive quantity in the body as a result of both radioactive decay and natural elimination.

Example. A 5 microcurie dose of radioiodine is administered to a patient for diagnostic purposes. If the physical half-life of the isotope is 8 days and the biologic half-life is 2 days, what radioactive quantity will remain after 8 days?

Solution. From equation 21–2 the effective half-life can be calculated.

$$\text{effective half-life} = \frac{T_P T_B}{T_P + T_B}$$

$$= \frac{(8 \text{ days})(2 \text{ days})}{(8 \text{ days} + 2 \text{ days})}$$

$$= 1.6 \text{ days}.$$

The decrease during successive half-lives can then be calculated.

Time	Quantity
0	5 μCi
1.6 days	2.5 μCi
3.2 days	1.25 μCi
4.8 days	0.62 μCi
6.4 days	0.31 μCi
8.0 days	0.16 μCi

The remaining radioactive quantity in the body after eight days is 0.16 microcurie, compared to 2.5 microcuries which would have remained if the reduction had been due to radioactive decay alone. The remaining 2.34 microcuries existing after the 8 day physical half-life have been excreted from the body.

In addition to the type of radiation and its half-life, the energy of the radiation must be known for the full description of a radioactive decay process. The basic metric energy unit, the joule, is much too large for the description of a process involving a single nucleus. The basic unit used for nuclear processes is the *electron volt*, abbreviated eV, which is defined as the energy attained by an electron when it is accelerated by 1 volt of electric potential (see Chapter 20):

$$1 \text{ eV} = 1.6 \times 10^{-19} \text{ joules.}$$

Though this is a very small energy unit, it is very convenient for the description of many atomic and nuclear phenomena. To provide a context for the appreciation of nuclear radiation energies, Table 21–3 contains some typical energies in electron volts. Note that the nuclear radiation energies are much higher than the energies of the atomic and molecular processes. It is convenient to express these energies in millions of electron volts (MeV for megaelectronvolt).

Since many isotopes decay by more than one process or with the emission of more than one type of radiation, it is often convenient to represent the radioactive decay by an energy level diagram as shown in Figure 21–4. The phosphorus radioisotope ^{32}P decays by beta emission to form the stable sulfur nucleus ^{32}S (Figure 21–4(a)). The energy 1.7 MeV is the maximum energy for the emitted electron; the energy is shared between the beta particle and the anti-neutrino which is emitted at the same time, so the energy of the electron can take any value between zero and 1.7 MeV. The anti-neutrino accompanying it passes through most matter without interacting, so it has no clinical significance. Radium decays by alpha emission to form the radioactive gas radon, ^{222}Rn (Figure 21–4(b)). An energy of 4.79 MeV must be given off in the process. This energy usually goes to a single alpha particle, but about 1.2 per cent of radium nuclei decay by successive emission of an alpha particle and a gamma ray. Cobalt-60 decays by successive emission of a beta particle and two gamma rays to form a stable isotope of nickel, ^{60}Ni.

MEDICAL RADIOISOTOPES

Medical radioisotopes which are to be administered internally are chosen on the basis of the type and energy of the radiation emitted, the half-life, and the rapidity and completeness of their excretion from the body. The naturally occurring radioisotopes which are medically useful are radium and its radioactive decay products. Radium has been used in hospitals since about 1901, but has been supplanted in many applications by the artificially produced radioisotopes that have become available since 1946.

Radium decays by alpha emission with a half-life of 1622 years. This alpha emission is not medically useful since it has a range of less than 1 mm inside the body; a sheet of paper would stop essentially all of the alpha particles. The usefulness of a radium sample comes from the fact that the alpha decay process produces the radioactive gas radon, and successive decays produce other radioisotopes with short half-lives. Radium itself is actually one stage of a long radioactive series which begins with the uranium isotope ^{238}U and ends with the stable lead isotope ^{206}Pb. Because of its long half-life, radium can be isolated from uranium ore and used as a source for its shorter-lived products, which emit medically useful radiation. Since over three tons of uranium must be processed to obtain 1 gram of radium (3), it is quite costly.

TABLE 21–3 Some Typical Energies in Electron Volts.

Room temperature thermal energy of a module	0.04 eV
Visible light photons	1.5–3.5 eV
Energy for the dissociation of an NaCl molecule into Na^+ and Cl^- ions	4.2 eV
Ionization energy for atomic hydrogen	13.6 eV
Approximate energy of an electron striking a color television screen	20,000 eV
High energy medical x-ray photon	200,000 eV (0.2 MeV)
Typical energies from nuclear decay:	
(1) gamma	0–3 MeV
(2) beta	0–3 MeV
(3) alpha	2–10 MeV

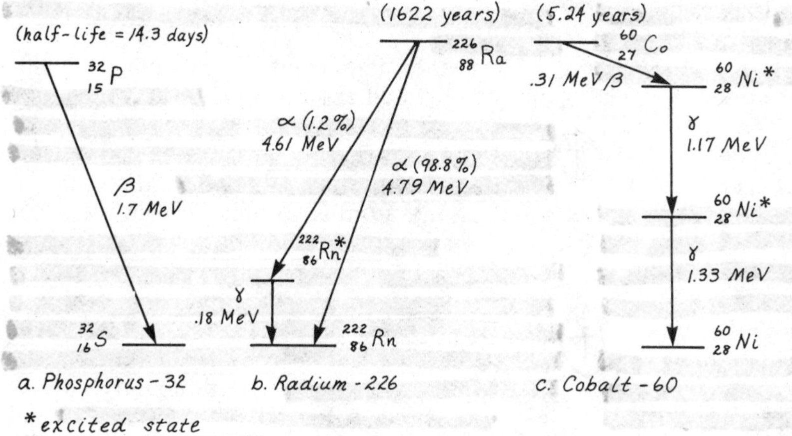

Figure 21–4 Radioactive decay diagrams.

The radioactive series, starting with radium, is listed in Table 21–4 with the types and energies of the emitted radiation. Note that radium produces eight successive "daughter" isotopes before reaching the stable lead isotope at the end of the series. Since the second member of the series is a gas, radon, the radium must be sealed in a capsule in order to make use of the entire series. It is usually encapsulated in gold or platinum alloy containers. Besides containing the radon, the capsule stops all the alpha particles and most of the beta particles, so that the radium source is essentially a gamma emitter. The high energy gamma rays come from the isotopes of lead ^{214}Pb and bismuth ^{214}Bi, which are the fourth and fifth members of the radium series (Table 21–4).

After a radium source is sealed into a capsule, it takes about 30 days for the series of successive nuclear decays to come to equilibrium with each other. After that, it is an essentially constant intensity gamma source, since the radioactive quantity is determined by the radium, with which its 1622 year half-life decreases in activity only 1% in 25 years. This constant activity is desirable for some applications, but makes it unsuitable for permanent implantation in the body — the radiation level must decrease rapidly in the case of permanent implants. The nature of the radium series demands care in the handling of the radium capsule source. If the capsule is broken, the radon gas escapes, rendering the capsule medically useless because of the absence of the gamma

TABLE 21–4 The Radium Series.

			Radiation Energy in MeV		
Element	**Isotope**	**Half-life**	*Alpha*	*Beta*	*Gamma*
radium	^{226}Ra	1622 yr	4.79(98.8%)	—	—
			4.61(1.2%)	—	0.18(1.2%)
radon	^{222}Rn	3.83 days	5.49	—	—
polonium	^{218}Po	3.05 min	6.00	—	—
lead	^{214}Pb	26.8 min	—	0.65	0.241(11.5%)
					0.350(45%)
					0.294(25.8%)
bismuth	^{214}Bi	19.7 min	—	3.17	0.607(65.8%)
					0.766(6.5%)
					0.933(6.7%)
					1.120(20.6%)
					1.238(6.3%)
					1.379(6.4%)
					1.761(25.8%)
					2.198(7.4%)
polonium	^{214}Po	164 μsec	7.68	—	—
lead	^{210}Pb	19.4 yr	—	0.017	—
bismuth	^{210}Bi	5.0 days	—	1.17	—
polonium	^{210}Po	138.4 days	5.30	—	—
lead	^{206}Pb	STABLE			

emitters produced by the radon. The radon gas is quite damaging inside the body because of the ionizing power of its high energy alpha emission. The properly sealed radium source is a valuable high intensity gamma source for therapeutic applications and will be discussed further in the radiation therapy section below.

Artificially produced isotopes are used in diagnostic applications because of their relatively short half-lives and the ease of introduction into chemical compounds which are utilized by the body. These radioisotopes are produced by bombarding normal atoms with energetic particles from nuclear reactors, cyclotrons, or other high energy particle accelerators. Some of the commonly used isotopes are listed in Table 21–5. Note that most of the half-lives are on the order of a few days, compared to the 1622 year half-life for radium. This property is essential for tracer isotopes which are injected for circulation studies or otherwise allowed to move freely inside the body. Since a portion of the material will remain in the body, it is important for its radioactivity to decrease rapidly and reach a safe level in a matter of a few days.

Another important feature of the artificially produced radioisotopes is the fact that they are isotopes of ordinary chemical elements which are constituents of many molecules found in bone and tissue. Since the nucleus of any atom is so remote from the outer electrons, and since the outer electrons are generally the only ones involved in ordinary chemical processes, the radioactive decay processes of nuclei are independent of the chemical state or the chemical compound in which they are found. The radioactive decay rates are independent of the temperature and independent of whether the compounds are in solid, liquid, or gaseous form. Thus, the radioactive nuclei can be introduced into whatever chemical compound is desired for the most convenient physiological applications. Radioactive isotopes are often put into human albumin (iodine-131), vitamin B_{12} (cobalt-60), and other compounds which are generally or selectively assimilated by the body.

TABLE 21–5 Artificially Produced Radioisotopes for Medical Use.

Element	Isotope	Physical Half-Life	Beta Particle Energies in MeV	Gamma Ray Energies in MeV
sodium	^{22}Na	2.58 yr	0.542	1.28
phosphorus	^{32}P	14.3 days	1.707	–
sulfur	^{35}S	88 days	0.167	–
chromium	^{51}Cr	27.8 days	–	0.321
iron	^{59}Fe	45 days	0.271(46%) 0.462(54%)	1.289(43%) 1.098(57%) 0.191(2.5%)
cobalt	^{57}Co	270 days	–	0.123(93%) 0.137(7%)
	^{60}Co	5.24 yr	.31	1.17 1.33
rubidium	^{86}Rb	18.8 days	1.77(91%) 0.68(9%)	– 1.08(9%)
strontium	^{85}Sr	64 days	–	0.513
technetium	^{99m}Tc	5.996 hr	–	0.002(98.6%) 0.140(98.6%) 0.142(1.4%)
iodine	^{131}I	8.08 days	0.608(87%) 0.335(9%) 0.250(3%)	0.722(3%) 0.637(9%) 0.364(81%) 0.284(6%) 0.080(2%)
cesium	^{131}Cs	9.69 days	–	0.030 (x-rays)*
xenon	^{133}Xe	5.27 days	0.345	0.081
gold	^{198}Au	2.7 days	0.959(99%) 0.283(1%)	0.412(95%)
mercury	^{197}Hg	2.66 days	–	0.077(99%)* 0.279(1.2%)
mercury	^{203}Hg	46.5 days	0.21	0.279

*These nuclei decay by capturing an electron from the $n = 1$ shell and subsequently emitting x-rays and/or gamma rays. This process is called electron capture and designated by E.C.

Thermoluminescent Dosimetry Crystals.

THE DETECTION OF RADIATION

The methods for the detection of x-rays and nuclear radiation are based upon the fact that the high energy radiation ionizes or otherwise directly alters atoms and molecules in materials through which it passes. The early experimenters used photographic film to detect and measure radiation. The effects of ionizing radiation upon film are similar to the effects of visible light, and the relative blackening of the film is proportional to the amount of radiation received. In addition to the familiar use of film for medical and dental x-rays, the film badge is the most widely used method for monitoring the amount of radiation received by personnel who work with radiation sources. A small square of film is covered with a material which is opaque to light and is kept on the person at all times. At regular intervals the film is developed; any blackening of the film will be caused by ionizing radiation.

If more accurate personnel monitoring is needed, small crystal radiation detectors are used. These crystals are called thermoluminescent dosimetry (TLD) crystals. The absorption of radiation stores energy in the crystals by causing reversible electron energy transitions. Later, to measure the dose received, the crystals are heated and the electrons make the reverse transitions to lower energy states, emitting light in the process (thermoluminescence). The amount of light emitted is proportional to the radiation dose received. The average radiation dosage can be measured from the film or crystal monitors and compared with the established safety standards to be discussed later.

Another widely used portable radiation detector is the radiation dosimeter, which is based on the electroscope principle discussed in Chapter 12. Instead of two metal foils which repel each other when charged (Figure 12–2), these dosimeters make use of a quartz fiber which deflects when charged. After the fiber is charged by a battery, it is sensitive to radiation because ionizing radiation acts to remove charge from the fiber and alter its deflected position. These dosimeters are often about the size of a pencil and have a magnifying eyepiece which is marked with a radiation dose scale. The position of the fiber is measured to get an indication of the radiation dose received since charging. Such dosimeters can be carried in a pocket or placed in an irradiated area. Although the accuracy is limited, the dosimeter is valuable for an immediate indication of the total dose received in a given time interval.

The widely used Geiger-Müller counter makes use of a sealed glass tube containing a gas such as argon at a pressure of about 1/10 atmosphere. Inside this tube are positive and negative electrodes with a voltage of some 800 to 2000 volts maintained between them. This voltage is almost enough to ionize the gas and cause a current to flow between the electrodes. When ionizing radiation passes through the walls into the tube and violently removes electrons from gas atoms, the resulting charged ions are accelerated by the electrode voltage and undergo successive collisions, causing an avalanche of ionizations. The resulting electrons are attracted to the positive electrode and can be measured as an electric current. With amplification, the radiation level can be indicated on a meter and/or converted into the familiar audible clicks popularly associated with the Geiger-Müller counter. Because of its high sensitivity, ruggedness, and portability, the Geiger-Müller counter is a popular instrument for area monitoring.

There is a wide variety of detection instruments known as ionization chambers (3). The basic feature of the ionization chamber is a sealed tube containing an ionizable gas and electrodes with an applied voltage on them. The ionization produced by penetrating radiation yields electrical pulses which may be counted or used to operate a display device. The usual terminology divides this type of detector into three classes: ionization chambers, proportional counters, and Geiger-Muller counters. All three are basically the same type of detector and differ mainly in terms of the voltage applied to the electrodes in the ionizable gas.

The term "ionization chamber" usually refers to a chamber with a fairly low voltage on the electrodes such that the ions produced by the radiation do not themselves gain enough energy from the electrode voltage to produce further ions by collision. The electrical pulses produced at the electrodes are records of the ions produced directly by the radiation and by the high energy electrons ejected by the radiation. Such counters can be used to identify the type of radiation, since an alpha particle directly produces many more ions than a beta particle, for example. The number of ions is also proportional to the energy of the radiation, so an indication of the radiation energy is obtained. Such instruments are accurate and highly reliable, but their sensitivity is very low.

If the electrode voltage is raised, there is a narrow voltage range in which a number of secondary ions are produced by the collisions of primary ions, but the size of the electrical signal produced is nevertheless proportional to the energy of the original radiation. The tube under these conditions is called a proportional counter; it produces electrical pulses up to 100,000 times the output of the simple ionization chamber while retaining the ability to discriminate between types and energies of radiation.

If the electrode voltage is further increased to a point near the threshold for direct electrical ionization, ions and photoelectrons may be produced in the entire chamber by a single ionizing event, and a pulse up to one hundred million (10^8) times the ionization chamber pulse can be produced. In this type of operation it is called a Geiger-Müller tube, as described above. The disadvantage of the G-M tube is that an alpha particle which directly produces thousands of ions will give the same output pulse as a particle which produces only one or two direct ionizations, so all discriminating ability is lost.

The scintillation counter is another type of radiation detector; it is rapidly becoming the primary detection instrument for medical radioactive tracer work. This instrument is based on the fact that the absorption of an x-ray, gamma ray, or particle by some types of crystals is followed by the emission of a flash of light. These flashes of light can be counted to indicate the number of absorption events and therefore the intensity of the radiation. In addition, the light intensity for a given event is proportional to the energy of the absorbed radiation. This capability is made more attractive by the fact that the sensitivity of such scintillation counters can be made quite high, much higher than that of the basic ionization chamber. The attainment of this sensitivity requires a sophisticated application of physics and electronics (3) and involves the use of photomultiplier tubes which receive the light flashes and convert them to electric impulses, which can be amplified by factors up to 10,000,000. While scintillation counters may be quite sensitive, they cannot easily be made portable. The electronic hardware for such detectors is largely responsible for the large size and great expense of scanning machines for radioactive tracer studies.

BIOLOGICAL EFFECTS OF IONIZING RADIATION

At the atomic and molecular level the effects of radiation are well defined. Since the energies of nuclear radiation and x-rays are much too large to be absorbed by the molecules and leave them intact, the radiation either passes through without interacting, or interacts by ejecting an electron from the molecule or by breaking a molecular bond. Either interaction leaves behind a charged species (ion), hence "ionizing radiation" is used as a general term to include x-rays and alpha, beta, gamma, and other types of nuclear radiation.

When x-rays or gamma rays enter tissue, they may give all their quantum energy to an electron in ejecting it from a molecule (photoionization) or they give only a fraction of their energy to the electron and then "scatter" the remainder off in the form of a lower energy, lower frequency quantum of radiation (Compton scattering, see Chapter 20). In either case the ejected electrons have very high speeds and enough energy to ionize many other atoms. They may act directly on cellular molecules or may interact with water molecules to dissociate H_2O into chemically active H and OH free radicals. In the case of high energy particle radiation such as alpha and beta radiation, the particles may act directly on vulnerable cellular components to break chromosomes, deactivate enzymes, or alter the DNA synthesis process (2). The ionization resulting from all these types of radiation produces very active chemical species which may disable cellular components or produce toxins.

At the cellular level, radiation damage is not quite so clear-cut. The molecular damage discussed above would produce observable chemical changes if there were a sufficient number of events, but observable biological effects occur at radiation levels far too low to produce large scale chemical effects. It has been shown that radiation can break chromosome chains, and it can clearly disrupt essential molecular constituents of the cells by direct interaction. It appears that a relatively small number of such violent events can "throw a monkey wrench" into the delicate machinery of a cell.

The most directly observable effect of radiation at the cellular level is that the cell will fail to reproduce. Since the damage is at the molecular level, the cell may continue to function in an apparently normal fashion, so the damage may be "latent" and not be apparent until the time the cell would normally reproduce. While radiation damage is incompletely understood, this cellular "reproductive death" model seems to be in agreement with the observed facts. In particular, it agrees well with the observed "latent" nature of large scale radiation damage. Exposure to ionizing radiation produces no pain or other sensory response, but produces damage which shows up at a later date. Since cells are being continuously reproduced in most tissue, other cells may make up for the lost reproductive capacity of a few damaged cells, but clearly if enough cells are prevented from reproducing, the organism will die. Again, this model agrees with the most commonly observed facts, that exposure to a small amount of ionizing radiation produces no measurable effects, but exposure to intense radiation produces serious illness or death (2).

In a much smaller number of cases the radiation damage may not prevent reproduction but may disrupt the reproductive machinery so that an altered or "mutated" cell results. Most such mutated cells would be destroyed by the body's defense mechanisms. Occasionally a mutation occurs which is close enough to normality to escape the body's defense mechanisms, yet regressive enough to be undifferentiated, to take in all the nourishment it can, and to reproduce rapidly without recognizing bounds to its growth as do normal cells (cancer). There is adequate experimental evidence from laboratory animals and from the victims of Hiroshima and Nagasaki to demonstrate that cancers as well as genetic defects can result from large radiation doses.

The latency of radiation effects, plus the fact that a person can be exposed to ionizing radiation without feeling anything, is responsible for the extreme caution in the use of sources of ionizing radiation. However, with reliable radiation detectors such as those discussed in the previous section, radiation alarm systems, and well-defined radiation exposure standards, even the professional radiation therapist can expect to work for a lifetime without measurable radiation effects. As will be seen, the properties of ionizing radiation which make it hazardous are also the properties which make it a tool of great value for the treatment of cancer.

MEASUREMENT OF RADIATION EXPOSURE

Nuclear radiation and x-rays produce molecular ionization effects which can be accurately and reproducibly measured. While some of the steps leading from these specific molecular effects to the final biological results are unclear, those results must be proportional to the microscopic effects. Accordingly, radiation exposure standards are stated in terms of the measurable ionization effects and the amount of energy transferred to the tissue. Since all of us are exposed to some ionizing radiation from cosmic rays and naturally occurring radioactivity, to which we add diagnostic x-rays and possibly other sources of radiation, it is important to establish standards for maximum permissible exposure. The units for exposure measurement are presented here, followed by some comments about present radiation exposure standards.

Since several types of units are used in the measurement of radiation, it is probably best to pick one unit for concentration to develop your understanding of radiation hazards, normal background radiation levels, etc. For reasons discussed below, the *rem* is the unit which will be used wherever possible in the subsequent discussion.

The *curie* (Ci) is the unit used for stating the "strength" or "activity" of a given radioactive sample. A source of activity of 1 curie is equivalent to one gram of radium. As the unit of radioactive quantity, it is formally defined as the amount of material which will produce 3.7×10^{10} nuclear decays per second. The curie is not a satisfactory unit for radiation exposure, since the energy and type of radiation strongly affect its ionizing power. For example, if a gamma ray produced a single ionization in traveling a certain distance in air, a beta particle traveling the same distance might produce 100 ions and an alpha particle might produce 10,000 ions in the same distance (4).

The *Becquerel* (Bq) is the unit for source activity in the new Système Internationale (SI) unit system, and it is defined as one nuclear decay per second, so that one curie is equal to 3.7×10^{10} Becquerels.

The *roentgen* (R) is a measure of radiation intensity for x-rays or gamma rays. The formal definition of one roentgen is the radiation intensity required to produce an ionization charge of 0.000258 coulombs per kilogram of air. The roentgen is one of the standard units for radiation dosimetry, but it is not applicable to alpha, beta, and other particle radiation and does not accurately predict the tissue effects of gamma rays of extremely high energies. The roentgen is used for the calibration of the output of x-ray machines.

The *rad* (rd) is a unit of absorbed radiation dose in terms of the energy actually deposited in the tissue. The rad is defined as an absorbed dose of 0.01 joules of energy per kilogram of tissue. The rad is the basic dose unit for clinical applications. For example, a typical radiation treatment for cancer might be stated as a 200 rad treatment to a specific area. However, it has been observed that the same doses of different types of radiation have different degrees of biological effectiveness. This relative biological effectiveness (RBE) is found to depend upon the amount of energy transferred to the tissues per unit length along the path of the radiation (i.e., it is range dependent). For example, the linear energy transfer (LET) for alpha particles is much higher than that for gamma rays. Therefore, a smaller dose in rads of alpha particles is required to inhibit cell division in a biological specimen.

The *Gray* (Gy) is the new SI unit for absorbed radiation dose and is defined as 1 joule of absorbed energy per kilogram of tissue. Therefore 1 Gray is equivalent to 100 rads.

The *rem* is a unit designed to measure the radiation dose in terms of its biological effectiveness in man, and the unit name is an acronym for "rad-equivalent-man." The dose in rems is defined as the dose in rads multiplied by a "quality factor," which is an assessment of the biological effectiveness of that particular type and energy of radiation. For alpha particles, the quality factor (2) may be as high as 20, so that one rad is equivalent to 20 rems. However, for x-rays and gamma rays, the most commonly used types of radiation in medicine, the quality factor is usually taken as 1, so that the roentgen, rad, and rem are essentially equivalent. That is, for x-rays and low energy gamma rays, incident radiation of intensity 1 roentgen on human tissue would result in an absorbed dose of one rad, which in this case is equivalent to one rem. The rem is the unit used for the establishment of radiation safety standards, since it measures the biologically effective dose.

The most radiosensitive structures in the human body are considered to be the blood-forming organs, the gonads, and the lens of the eye. On the level of individual cells, the most radiosensitive cells in the body are those of the bone marrow, lymphoid, and epithelial tissues. Blood changes, such as a reduction in white blood cell count and blood platelet count, and the destruction of lymphocytes are among the earliest observed effects following the absorption of radiation. Standards for maximum permissible doses for persons in radiation-related occupations were recommended by the International Commission on Radiological Protection (ICRP) in 1965 (Reference 5). The limit proposed for the above organs was 5 rem/yr average after age 18 with no more than 2.5 rem in any three month period. The suggested limits for other less sensitive organs were: skin, bone, thyroid, 30 rem/yr; hands, feet, forearms, 75 rems/yr; and other organs, 15 rem/yr. Maximum permissible concentrations of radioactive materials have been established for air and water which are low enough that a 50 year exposure would not give more than the above maximum permissible doses to any organ.

The Environmental Protection Agency (EPA) has recommended that the radiation exposure from sources other than natural or medical radiation sources be limited to 0.17 rem/year, a much more stringent standard than that discussed above. This annual radiation figure arises from the fact that when the 5 rem/year maximum occupational exposure was set, it was recommended that non-occupational exposure be kept to less than one tenth of that level for individuals (0.5 rem/year). It was then further recommended that the average population exposure be kept below one third of the individual maximum, hence 0.17 rem/year was the recommended maximum population dose.

To help put all of these numbers in perspective, the radiation limits are displayed in bar-graph form in Figure 21–5 along with some average exposures. The units used there are millirem (mrem). In some mountain villages of the Espirito Santo state of Brazil the natural radiation levels are as high as 1200 millirem/year, yet the effects on mortality are low enough to be undetectable in a number of studies (14). There are significant differences in natural background radiation in the United States, with a range of about 60 to 200 millirem/yr, but no naturally occurring levels approaching the levels in Brazil and some parts of India. The maximum permissible radiation from a nuclear power plant is 5 mrem/yr. On the high side, radiation therapy for cancer may involve exposures of over 5000 rems in a few weeks to localized body areas. The average U.S. radiation exposures are summarized in Table 21–6. These sources of radiation may be compared with the known effects of large radiation doses summarized in Table 21–7.

While the molecular effects of nuclear radiation and x-rays are well understood, and the acute biological effects of large radiation doses are well understood, the long-term effects of low level radiation present a much more difficult problem (13). Given the accuracy of the measurement of radiation dose and the years of study of radiation effects, it is probably true

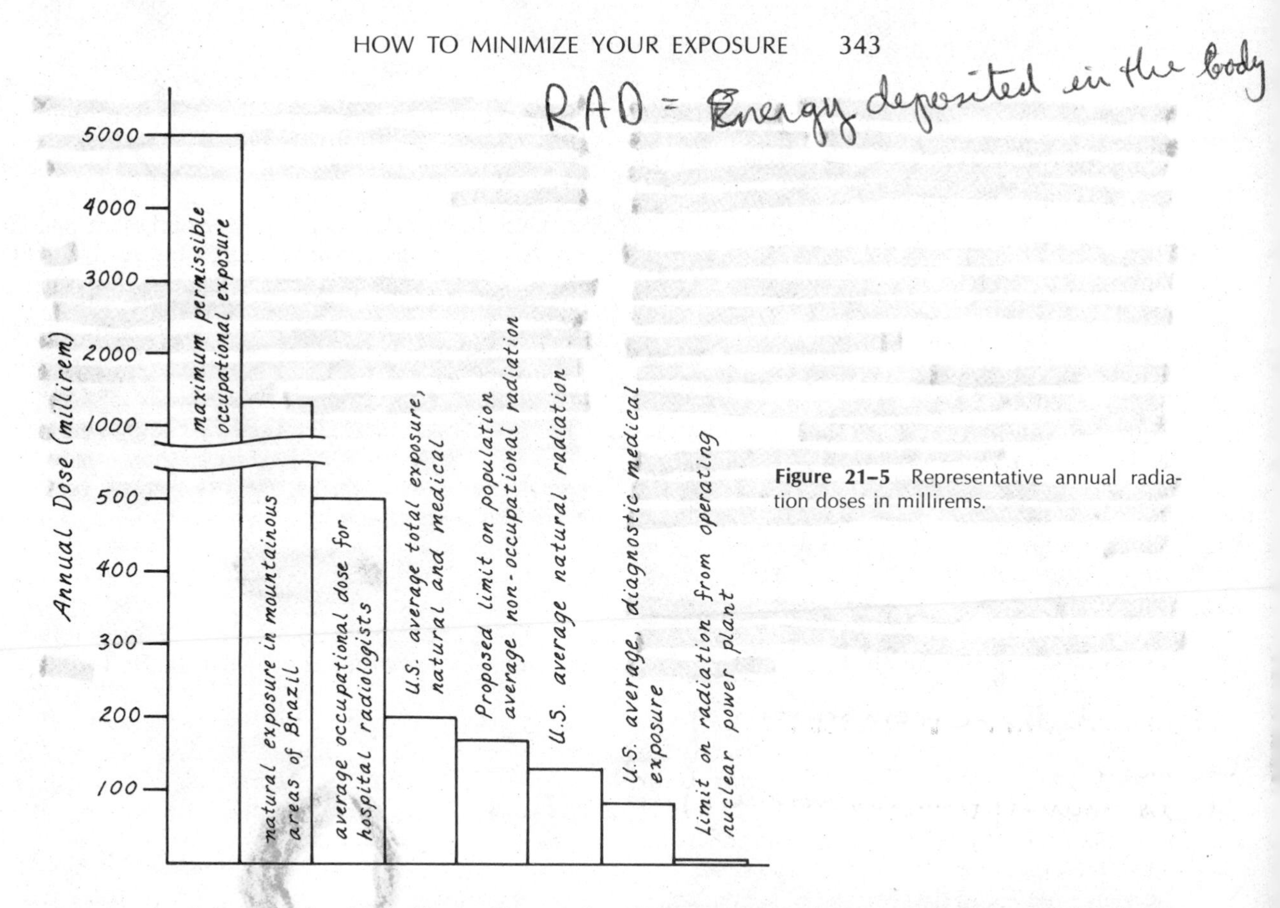

Figure 21–5 Representative annual radiation doses in millirems.

that we understand the biological effects of ionizing radiation far better than we understand the effects of food additives and other chemicals which we take into our bodies. Yet, as long as there are even small risks of cancer or genetic defects showing up years after exposure, it is best to be conservative with radiation exposure. One of the pertinent questions about radiation exposure is whether or not there is a "threshold" dose, below which no harm is done. The existence of such a threshold has not been clearly demonstrated for long-term, low level doses, so it is best to assume that any radiation dose is potentially harmful. While the guidelines for occupational exposure probably represent safe limits, it is a good policy to avoid *any* unnecessary radiation exposure.

HOW TO MINIMIZE YOUR EXPOSURE

The steps to be taken to minimize your radiation exposure from a radioactive source are similar in principle to the steps you would take to minimize the light reaching you from a light bulb in the room. That is, you can increase

TABLE 21–6 Average U.S. Radiation Exposures(13).

Source	Exposure (in millirem/yr)
Cosmic rays	45
External radiation from radioactive ores etc.	60
Internal exposure from radioactive material ingested into the body.	25
Diagnostic X-rays	70
Total	200

TABLE 21–7 Effects of Large, Whole-body Radiation Doses.

Dose (rems)	Effect
0–25	No observable effect
25–100	Slight blood changes
100–200	Significant reduction in blood platelets and white blood cells (temporary).
200–500	Severe blood damage, nausea, hair loss, hemorrhage, death in many cases.
>600	Death in less than two months for over 80%

your *distance* from the source, decrease the *time* you spend near the source, and place some kind of *shielding* between yourself and the source to keep the radiation from reaching you (see Figure 21–6). In occupational exposures, a film badge or some other type of personnel monitor should be worn to measure your radiation exposure in rems for a permanent record. The piece of film in the film badge is in a light-tight envelope, so that any fogging of the film is attributable to the ionizing radiation which penetrates the envelope. Film badges can determine the dose of x-ray or gamma ray exposure to a minimum limit of about 10 millirem above background with optimum conditions. Practical conditions may degrade the sensitivity to 50 millirems above background, but the film badge is the most convenient and most widely used form of personal monitor. For more accurate monitoring, tiny thermoluminescent crystal (TLD) monitors can achieve a sensitivity of about one millirem above background radiation levels from the environment.

While the effects of shielding and minimizing your exposure time are straightforward, the effects of your distance from the source upon your exposure merit some further discussion. If the radiation source is small enough to be considered a point source, it emits its radiation uniformly in all directions. The radiated energy can be considered to be spread over a sphere of area $4\pi r^2$ at any distance r from the source. Since the radiation intensity is equal to the energy per unit area

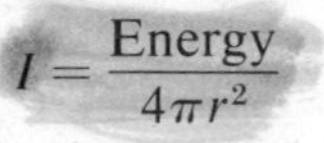

$$I = \frac{\text{Energy}}{4\pi r^2}$$

it follows the *inverse square law* described in Chapter 17 for other types of radiation. For

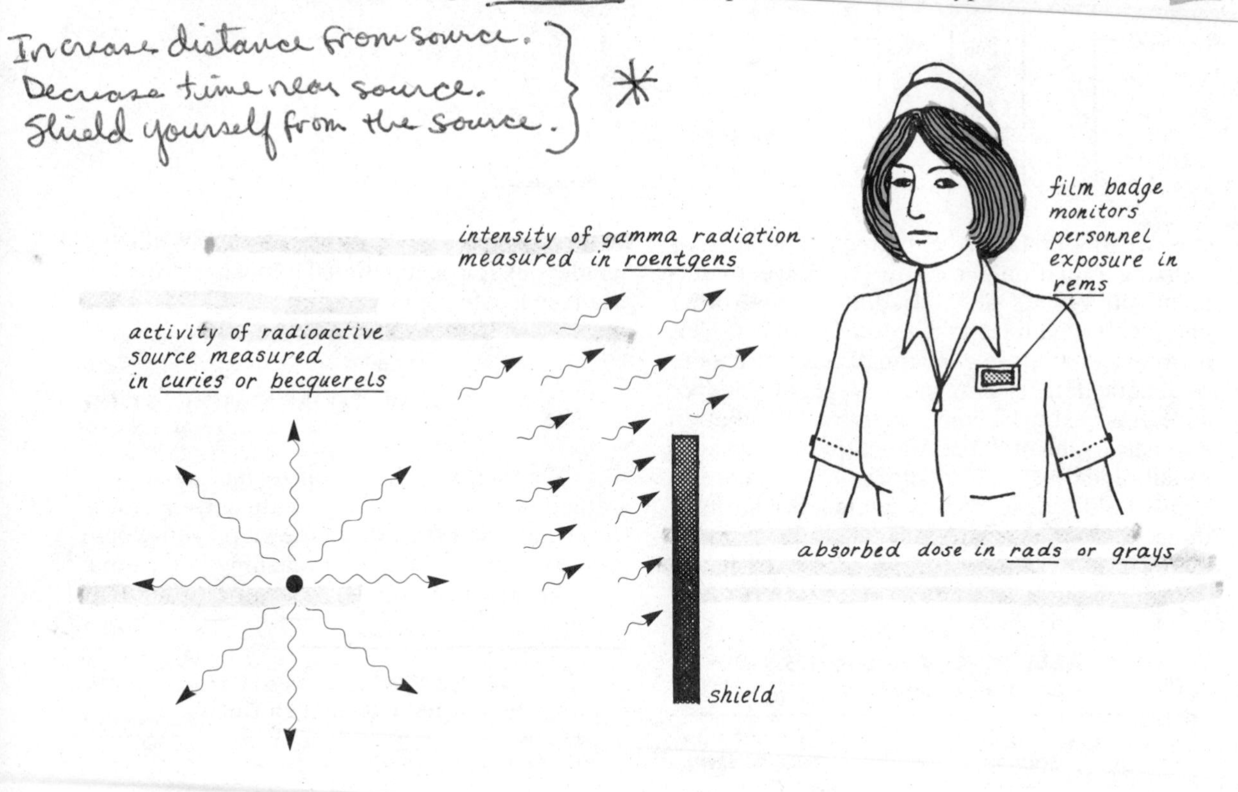

Safety measures:

1. Maximize distance between you and the source.
2. Minimize the time you spend near the source.
3. Use shielding whenever possible.
4. Use a film badge or other monitoring device to record the radiation dose you have received.

Figure 21–6 Minimizing the radiation exposure of personnel.

example, if you are twice as far away from the radiation source, you get only one fourth as much radiation. If you are 10 times as far away, you get only one hundredth as much radiation, and so on. If the radiation intensity is I_0 at a distance r_0 from the source, the intensity at I at any other distance r will be given by

$$I = I_0\left(\frac{r_0^2}{r^2}\right).$$

Since the radiation dose is proportional to the intensity, the dose D can be calculated by the same type of relationship

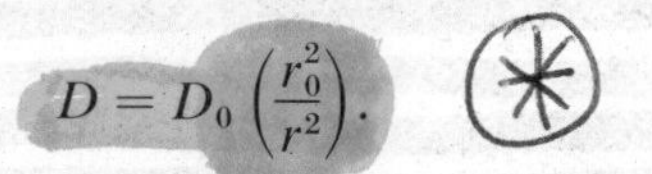

$$D = D_0\left(\frac{r_0^2}{r^2}\right).$$

Example. If a nurse received a radiation dose of 100 millirem (0.1 rem) by spending one hour at a distance of 2 ft from a radioactive implant in a patient, how much would she have received at a distance of 4 ft? at a distance of 20 ft?

Solution. The dose for the same time spent at 4 ft would be

$$D = (100 \text{ mrem})\frac{(2 \text{ ft})^2}{(4 \text{ ft})^2} = \frac{100 \text{ mrem}}{4} = 25 \text{ mrem.}$$

At 20 feet the dose would be

$$D = (100 \text{ mrem})\left(\frac{2 \text{ ft}}{20 \text{ ft}}\right)^2$$

$$= \frac{100 \text{ mrem}}{100} = 1 \text{ mrem.}$$

The rapid decrease in radiation dose with distance makes distance a major ally in minimizing your radiation exposure.

The 5 rem per year occupational maximum which has been used as a guideline for health care personnel since 1965 translates to roughly 100 millirem per week. Though the level of exposure should be kept at a much lower level than that, this maximum along with the application of the inverse square law can be used to calculate maximum exposure times for hospital personnel, as shown in the following example.

Example. If a patient has received 100 millicuries of cobalt-60 as a therapeutic measure, the dose received by an attendant at a distance of one foot from the patient might be on the order of 200 millirems per hour (9). How long could the attendant spend at this distance in a week's time? How long could he spend at 2 ft? At 10 ft?

Solution. Using the maximum of 100 mrem/week, at a distance of one foot the maximum exposure time is

$$\frac{100 \text{ mrem/week}}{200 \text{ mrem/hour}} = 0.5 \text{ hour per week maximum}$$

If the distance is 2 ft, the dose rate is

$$\frac{D}{t} = (200 \text{ mrem/hour})\left(\frac{1 \text{ ft}}{2 \text{ ft}}\right)^2 = 50 \text{ mrem/hr}$$

and the maximum exposure time is

$$\frac{100 \text{ mrem/week}}{50 \text{ mrem/hour}} = 2 \text{ hours per week}$$

At a distance of 10 ft the dose rate is

$$\frac{D}{t} = (200 \text{ mrem/hour})\left(\frac{1 \text{ ft}}{10 \text{ ft}}\right)^2 = 2 \text{ mrem/hr}$$

and the maximum exposure time is

$$\frac{100 \text{ mrem/week}}{2 \text{ mrem/hr}} = 100 \text{ hours per week.}$$

The same method could be used to calculate maximum exposure times using whatever maximum weekly exposure dose is adopted as a standard. The exposure rate at some reference distance should be measured by the doctor administering therapeutic radiation doses or by the hospital's health physicist. Guidelines can then be set for personnel exposure times.

It must be noted that the inverse square law applies only to point sources of radiation. When in the vicinity of x-ray machines or x-ray therapy units, the main personnel dose is from radiation scattered from the walls and objects in the room. Since these are not point sources, the radiation from them does not drop off so rapidly with distance, and personnel must rely on shielding and minimizing the time of exposure to reduce the dose. Generally, the scattered radiation is of low intensity, around 0.1% of the primary beam intensity (13), so the problem is not as severe as in the treatment of radiation implant patients.

Having emphasized the risks of radiation exposure, it must be noted that there are many circumstances in which the benefits to be derived from radiation outweigh the risks. While it is wise to be conservative with radiation, diagnostic x-rays, diagnostic nuclear medicine procedures, and radiation therapy procedures have made major positive contributions to health care. Unnecessary diagnostic x-rays should be avoided, and any radiation procedures should be performed by well-trained personnel with properly calibrated and maintained equipment, but within those guidelines the risks are apparently quite small. Recent estimates of the long-term risks of radiaton exposure are (15):

Cancer risk (lifetime)	1.5 per 10,000 per rem of exposure
Genetic defect (in 5 generations)	1.3 per 10,000 per rem of exposure

Since the number of cancers and genetic defects attributable to low level radiation is quite small, and the elimination of other possible contributing effects is difficult, such statistical calculations of risk are highly unreliable. Such risks will remain the subject of lively debate for a long time. Accurate measurement and documentation of radiation exposure can help to clarify the risks.

RADIATION THERAPY

Ionizing radiation, particularly x-rays and gamma rays, has proved to be a useful therapeutic tool for the treatment of cancer. This may seem paradoxical after discussing radiation hazards and the possibility of a radiation-*caused* cancer, but recall that the risk of a cancer-causing mutation is quite small and that the most probable result of the interaction of radiation with a cell is that the cell will fail to reproduce. Since the basic problem with a cancer is that the cells reproduce uncontrollably, then irradiation would appear to be an excellent tool to use against it. The drawback is that there is no way to irradiate the cancer without also damaging the normal tissue surrounding it. The value of radiation therapy hinges upon our ability to find ways to damage the cancer *more* than the normal tissue, or at least to limit the damage to normal tissue so that there is no unacceptable loss of function. The ratio of cancer cells killed to normal cells killed is called the "therapeutic ratio." The effectiveness of the therapy depends upon making the therapeutic ratio as high as possible. Some of the factors which increase the therapeutic ratio will be discussed here, followed by some comments about the radiation sources used. A more detailed description of the factors affecting the therapeutic ratio may be found in reference 8.

It has been observed that undifferentiated cells and rapidly dividing cells are more sensitive to radiation than well-differentiated and slowly reproducing cells. Accordingly, radiation is observed to have a greater effect upon rapidly dividing cancer cells than upon the surrounding normal cells. However, a number of types of normal cells also rapidly reproduce and must be carefully monitored during radiation therapy; examples are the white blood cells and blood platelets. Also, a common side effect of abdominal radiation therapy is nausea or "radiation sickness." To explain this to the suffering patient you say, "Well, the intestinal membrane would cover an area larger than a football field and it completely reproduces itself every two weeks." While the scale may not be accurate, the point is made that the rapidly reproducing membrane cells are quite susceptible to radiation damage.

The therapeutic ratio may be further increased by the fact that normal tissue apparently recovers faster and more completely from radiation effects than does cancerous tissue. The recovery of normal tissue is seen in the fact that a considerably greater total radiation dose is required to produce a given level of damage in normal tissue if it is given in several small doses separated by a significant time interval than if it is given in one large dose (6). Properly spaced treatments can reduce the damage to normal tissue associated with a given level of damage to the cancerous tissue. A typical radiation therapy course of treatment may involve a total dose on the order of 5000 rads to a selected area given in single daily treatments of 100 to 200 rads over a period of several weeks. This is an *enormous* dose compared to the occupational exposure limits discussed above. Though some damage to normal cells is inevitable, the "therapeutic ratio" often makes it possible to destroy a tumor while preserving enough of the surrounding normal tissue for patient survival with reasonable function.

The "oxygen effect" is a significant factor related to the radiation sensitivity of tissue. As mentioned earlier, the specific chemical effects of radiation result from the formation of "free radicals," chemically unstable fragments of molecules. The presence of oxygen greatly increases the production of the most active free radicals. The partial pressure of oxygen in tissue strongly influences the biological response to x-rays, gamma rays, and beta particles. Some tumor cells are deficient in oxygen (hypoxic), while surrounding normal cells may be essentially saturated with oxygen. This, of course, decreases the therapeutic ratio and may make it impractical to destroy the tumor cells because of excessive damage to normal cells. A number of experiments have been performed in which the oxygen supply to tumor cells was increased (7). Since the surrounding normal cells were already nearly saturated with oxygen, there was little increase in the radiosensitivity of normal cells, but an increase of more than a factor of two was reported for the radiosensitivity of the tumor cells.

High energy gamma rays from radioisotopes such as cobalt-60 and high energy x-rays are commonly used for therapy. Such sources tend to irradiate a considerable amount of normal tissue. The cobalt-60 gamma radiation is emitted uniformly in all directions and can be controlled only by shielding the source in all directions except for a small aperture directed toward the tumor. While x-rays cannot be focused by a lens like visible light, the sources can be contoured to concentrate the x-ray beam on the selected target. There is a present trend toward high energy electron accelerators which produce higher energy x-rays than were practical with older x-ray sources. The x-rays from these accelerators provide a more controllable dose and decrease the irradiation of surrounding tissue.

Another option for radiation therapy is the implantation of a radioactive source within a cancerous tumor. Implantation, of course, provides the maximum radiation intensity for a given source strength. It also has the advantage of decreasing the radiation to normal tissue. An external beam usually has to go through normal tissue to get to the tumor, while the radiation from a small implant may be considered to drop off according to the inverse square law, giving outside normal tissue less radiation. For example, if an implanted source delivered a radiation dose of 100 rads at the extremity of a tumor which was 1 cm from the source, the dose in an equal volume of normal tissue 2 cm away from the source would be

$$D = 100 \text{ rad} \left(\frac{1 \text{ cm}}{2 \text{ cm}}\right)^2 = 25 \text{ rad}.$$

At a distance of 10 cm the dose would be

$$D = 100 \text{ rad} \left(\frac{1 \text{ cm}}{10 \text{ cm}}\right)^2 = 1 \text{ rad}.$$

This rapid decrease in radiation dose with distance ensures minimum exposure to normal tissue.

Among the radiation sources used for cancer therapy are radium, cobalt-60, and gold-198. The characteristics of radium were described in the previous section on medical radioisotopes. Small encapsulated radium sources referred to as "seeds" and "needles" have been commonly used for implantation but are now being replaced by artificially produced radioisotopes with shorter half-lives. Cobalt-60 is often used in the same form. Both types of sources must be removed, since they have long half-lives. Gold-198 is used in a colloidal suspension for treatments in body cavities such as the pleural cavity. Since gold-198 has a short half-life (2.7 days), its activity decreases rapidly and it need not be removed. Other sources, such as iridium-192 and tantalum-182, are employed for specific purposes. No attempt will be made here to survey all of the available sources.

Radioiodine, ^{131}I, is unique as a therapeutic agent because it is selectively taken up by the thyroid gland. In addition to its use for the treatment of thyroid malignancies, it is used for the treatment of hyperthyroidism, toxic nodular goiter, and certain cardiac conditions (9). In hyperthyroidism and in toxic nodular goiter, the aim of therapy is to destroy part of the thyroid gland, which in these conditions is producing too much thyroid hormone. In certain cardiac conditions radioiodine is used to destroy a small amount of thyroid tissue even when it is normal, thus reducing the metabolic rate and lessening the work load on the heart.

Radiation therapy with heavy particles such as protons has proved to have some advantages over x-ray and gamma ray therapy. One advantage comes from the fact that when protons enter tissue, they deposit most of their energy near the end of their range, and this range is dependent upon the particle energy. This implies that by controlling the particle energy, most of the dose can be deposited in an internal tumor while minimizing the dose in the surrounding tissue. Another advantage is that the damage to a tumor is not as dependent upon its oxygen content as in the case of x-ray and gamma irradiation. The oxygen effect discussed above is much less evident since the protons tend to act directly upon vulnerable cellular components rather than indirectly through free radical formation. Since it is not uncommon for tumors to outgrow their oxygen supply, heavy particle radiation is often useful.

Another way to increase the therapeutic ratio for particle radiation therapy is to inject boron into a tumor and irradiate with low energy neutrons. Since boron interacts with neutrons much more strongly than ordinary tissue, this results in a much higher dose to the tumor. These are examples of the many avenues being explored in radiation therapy.

Regardless of the method or source used, the radiation doses are extremely high, on the order of one thousand times the maximum permissible annual occupational dose for the health care personnel in attendance, so extreme care must be taken to avoid excessive exposure. With beam or external radioisotope exposure, all health care personnel are out of the room while the patient is being irradiated. With implanted radiation sources, however, interaction of the patient during irradiation is necessary and careful monitoring of dose is necessary, as described in the previous section. One final point, to remove a popular misconception: a person who is subjected to an intense x-ray, gamma ray, or particle beam for therapy is not made radioactive or in any way dangerous by that exposure. As soon as the beam is turned off, it is quite safe to care for the patient normally.

DIAGNOSTIC USE OF RADIOISOTOPES

The development of sensitive radiation detectors such as the scintillation counters discussed in a previous section has made it possible to employ very small amounts of radioisotopes as tracers for diagnostic studies. The fact that the radioactive decay process is totally independent of the chemical compound in which the isotope resides makes it possible to "tag" or "label" many convenient molecules with radioisotopes. Whole body studies, such as circulation studies, may be done by injecting a radioactive species into the venous system. Organ scans may be made by tagging a chemical species which is selectively absorbed by a specific organ. For example, liver scans can be made by labeling a dye, rose bengal, with radioiodine or by injecting a dilute colloidal suspension of gold-198; both of these are quickly taken up by the liver. The plasma volume may be determined by the intravenous injection of a few microcuries of radioiodinated (^{131}I) serum albumin. Measurement of the dilution of the tracer in a subsequent blood sample makes possible the calculation of the total plasma volume. Cobalt-57 is used to label vitamin B_{12}, and the labeled vitamin forms the basis for the Schilling test for pernicious anemia.

The body retention of certain elements may be an index of physiologic activity (for example, calcium-47 retention in evaluating bone metabolism, iron-59 retention for the evaluation of anemia, etc.). The metabolic pathways of elements like copper, cobalt, zinc, and manganese may be studied by the use of radioactive isotopes of these elements. With readily available radioisotopes of many common elements, the diagnostic and research potentials of tracer techniques are enormous. Details of the many diagnostic uses of radioactive tracers may be found in the references.

The brain scan will be considered as a straightforward example of a diagnostic application of a radioisotope. Iodine-131 in the form of radioiodinated serum albumin (human) has been used for this purpose. A small amount of the radioactive material is injected intravenously and allowed to distribute itself throughout the circulatory system. A scintillation counter is then used to scan back and forth across the head to detect the gamma radiation from the blood which reaches the brain. The radiation received may be indicated in the form of marks transferred to a sheet of recording paper, as indicated in Figure 21–7. Since the amount of blood supply to the brain is rather small, a relatively small number of radiation counts is recorded for the normal brain (Figure 21–7(a)). However, if any pooling of blood occurs as a result of some lesion, it shows up as an area of higher gamma activity, as indicated in Figure 21–7(b). This information is difficult to obtain by any other nonsurgical means, and the radioactive quantity used is on the order of microcuries so that it is a non-destructive diagnostic method.

Although caution should be used with any radioactive material, the radioactive quantities used in diagnostic tests are usually so small that they present no radiation hazard in normal use. For example, a 5 microcurie dose of ^{131}I used for a brain scan would probably produce a radiation level on the order of 0.001 millirem/hour at a distance of 1 foot from

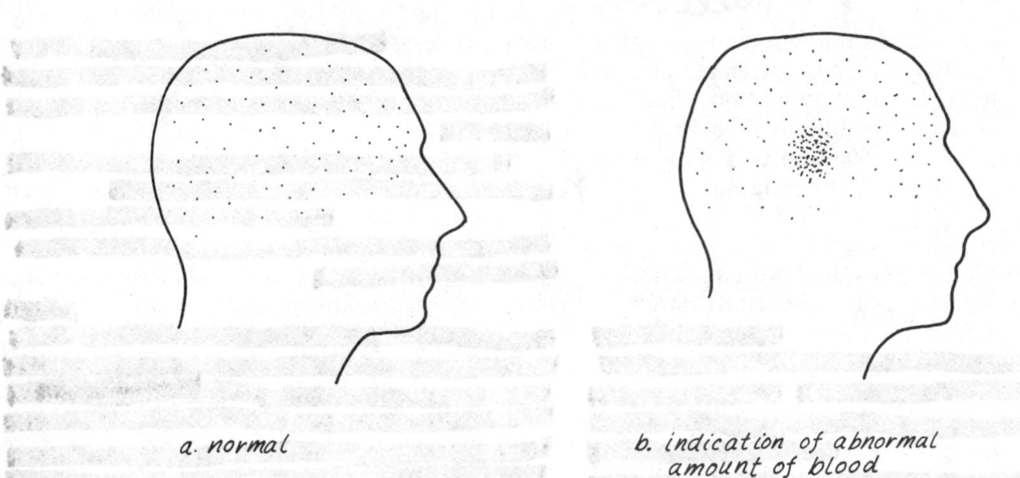

Figure 21–7 Brain-scan presentation.

the patient. Continuous exposure at this level would deliver only about a sixth of the allowed occupational radiation level. This is coupled with the fact that the ^{131}I activity decreases rapidly because of its 8 day half-life and the fact that it is gradually eliminated from the body. Precautions should be taken in administering tracer doses to avoid spills and to shield a person who must administer repeated doses to several patients.

NUCLEAR ENERGY

Since most of the necessary nuclear concepts have been developed, it seems appropriate to consider briefly the use of nuclear processes for the generation of useful energy. Such processes will become more and more important for large scale energy production as the supplies of petroleum and other fossil fuels are exhausted. Nuclear processes may also be applied for the long term generation of power for implanted devices such as cardiac pacemakers, as mentioned in Chapter 16.

The key to the enormous amounts of energy available from nuclear processes lies in the fact that changes in nuclear structure involve changes in the total mass of the particles involved. When mass is lost, it is transformed into energy according to Einstein's famous energy equation, $E = mc^2$. For example, consider the process of building an alpha particle (a helium nucleus) from two protons and two neutrons. The proton has a mass of 1835.7 times the mass of an electron, and the neutron has a mass of 1839 times the mass of an electron. It would be expected that the mass of an alpha particle would be the sum of the masses of two protons and two neutrons, but that is not the case:

neutron masses $= 2 \times 1839\ m_e =$	$3678\ m_e$
proton masses $= 2 \times 1835.7\ m_e =$	$3671.4\ m_e$
total	$7349.4\ m_e$
mass of alpha particle	$7294\ m_e$
mass discrepancy	$55.4\ m_e$

The mass of the alpha particle is less than the sum of the masses of its constituent parts by $55.4\ m_e$, where m_e is the mass of an electron. This mass is converted into energy in the process of "fusing" the neutrons and protons together to form the alpha particle. To obtain more convenient units and to provide a clearer picture of the energy-producing capability of the "nuclear fusion" process, let us consider the above process in gram quantities rather than in multiples of the electron mass. If 3678 grams of neutrons and 3671.4 grams of protons were combined to form alpha particles:

mass of original particles (fuel)	7349.4 grams
mass of resulting alpha particles	7294 grams
mass converted into energy	55.4 grams

Using Einstein's equation, $E = mc^2$, where $c = 3 \times 10^8$ m/sec, the energy yield would be

$$E = (0.0554 \text{ kilograms})\ (3 \times 10^8 \text{ m/sec})^2$$

$$= 5 \times 10^{15} \text{ joules.}$$

This enormous energy yield can be more fully appreciated by comparing it with the United States energy consumption for 1970, which was 6.8×10^{19} joules (12). The above energy, 5×10^{15} joules, would have supplied the entire United States with energy for nearly 40 minutes at the 1970 level, with the conversion of about 2 ounces of mass into energy and with a total of about 16 pounds of original fuel. In another context, this energy is more than the explosive yield of one billion tons of TNT, compared to about 20,000 tons of TNT energy yield for the Hiroshima and Nagasaki nuclear fission bombs.

Although the process described above is not the most feasible one for future production of energy, it illustrates the enormous potential of *nuclear fusion* processes. Nuclear fusion simply refers to the fusing of light nuclei together to form heavier nuclei. Any decrease in mass achieved in such processes is released as energy. It also is possible to release energy by *nuclear fission*, the splitting of the nuclei of heavy elements such as uranium into two or more lighter nuclei. As illustrated in Figure 21–8, nuclei of intermediate masses have the greatest binding energy per nuclear particle, which is equivalent to having the minimum mass per nuclear particle. Therefore, if heavy nuclei can be split to form two intermediate nuclei, the total mass is reduced and energy is released (nuclear fission). Alternatively, if light nuclei are combined to form intermediate nuclei, the total mass is reduced and energy is released (nuclear fusion). All present nuclear power generation is done by the nuclear fission process. The fission of uranium-235 produces large amounts of energy, but it is scarce and expensive and produces radioactive waste materials.

The nuclear fusion process produces more energy for a given mass of fuel, as seen from Figure 21–8. It also makes use of cheap and abundant fuel and produces no radioactive waste products. It offers the greatest hope for an abundant future supply of energy, but it has not yet been possible to initiate and control the fusion process to the degree necessary to produce useful energy. The main reason is that it requires temperatures of several million degrees Fahrenheit, about the temperature of the center of the sun (which produces its energy by the nuclear fusion process). Besides the problems of generating and sustaining such temperatures, no material container can withstand such temperatures. Laboratories around the world are working on means for containing and controlling the nuclear fusion process. It seems likely that the controlled nuclear fusion reactions will take place in a superhot "plasma" of charged particles held in a "magnetic bottle" composed of strong magnetic fields which contain the plasma and keep it from touching any solid matter.

The present nuclear power plants make

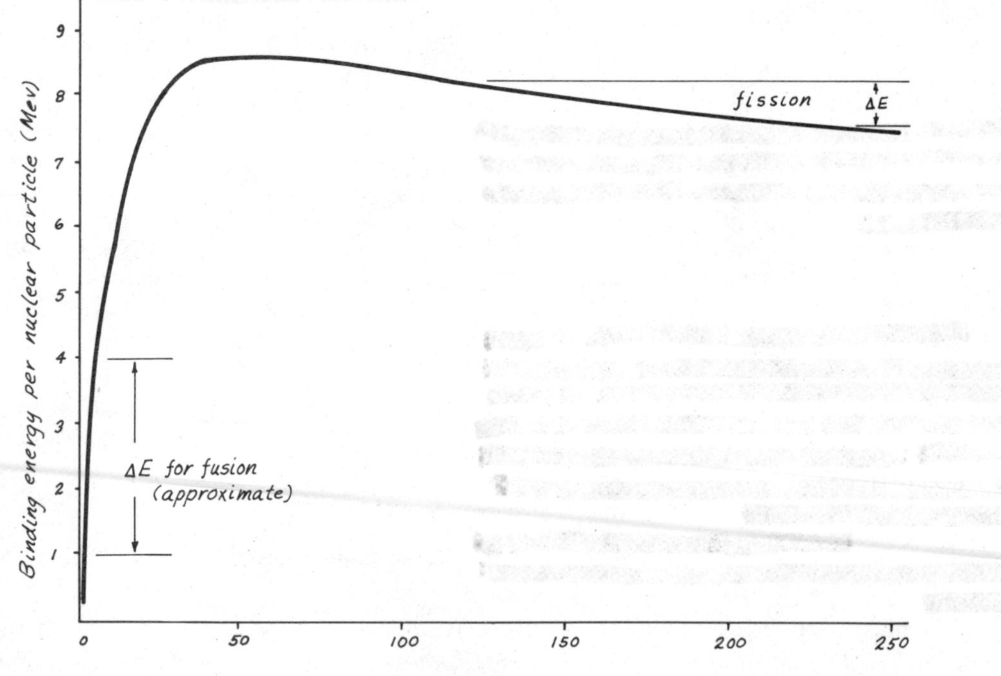

Figure 21–8 The binding energy changes of nuclei make possible the release of energy by either nuclear fission or nuclear fusion.

use of the energy released by nuclear fission to heat water to produce steam. This steam is used to turn turbines which generate electricity. The nuclear power sources which power pacemakers and other small devices make use of neither fission nor fusion, but simply utilize the energy of emitted alpha particles or other particle radiation to produce heat. This energy raises the temperature of the junctions of a large number of thermocouples, which produce an electric voltage. Although the energy supplied is small, it is sufficient to power small devices. The main advantage of such power sources is the fact that they can operate steadily for many years if powered by a radioisotope which has a long half-life.

SUMMARY

Most of the mass of the atom consists of positively charged protons and neutral neutrons which are concentrated in a tiny fraction of the atom's volume. In the case of stable nuclei the enormous coulomb repulsive force between the protons is overcome by the stronger nuclear attractive force. Other nuclei are unstable or "radioactive" and emit radiation to transmute them into more stable nuclei. The most common types of radiation from the nucleus are gamma rays (electromagnetic waves), beta particles (electrons) and alpha particles (helium nuclei). Many elements have more than one stable isotope and some have a large number of known radioactive isotopes, including those produced artificially in reactors and accelerators. Radioisotopes vary widely in the rapidity of their radioactive decay. The most commonly used parameter for classifying radioactive decay rates is the half-life, the time required for one half of the nuclei in any given sample to decay. For medical radioisotopes the biological half-life, the time for the removal of half of a sample from the body by biological means, must also be considered. A composite effective half-life can be determined by consideration of both half-lives.

Radioisotopes which emit gamma rays have been used for therapeutic and diagnostic purposes. Alpha and beta radiation have found very limited application because of their short ranges. Radium has been the most useful of the naturally occurring radioisotopes because of the gamma rays emitted by several of its decay products. In contrast to the long half-life of radium (1622 years), most of the artificial radioisotopes have half-lives on the order of a few days. Radiation affects tissue by ionizing or otherwise altering molecules, usually producing some degree of "reproductive death" or disabling of the cell reproduction mechanism. A "therapeutic ratio" exists when cancerous tumors are irradiated because the rapidly reproducing tumor cells are more susceptible to radiation damage and show a smaller degree of recovery from such damage than the surrounding normal tissue. For standardizing radioactive doses, the physical units used are the curie, roentgen, rad, and rem. The curie is a measure of the number of radioactive disintegrations per second, and the roentgen, rad, and rem are related to the ionizing effectiveness of the radiation. Specific radiation dose limits are enforced for health care personnel who are exposed to radiation. The distance from the radioactive source should be given particular attention in radiation safety, since the radiation intensity follows the inverse square law and therefore drops off rapidly with distance.

Many diagnostic techniques make use of the fact that the radioactive decay rate of an isotope is independent of its physical state or the chemical compound in which it resides. Many chemicals can be tagged with radioisotopes for whole body scans or for mapping of specific organs. Diagnostic radioisotopes do not normally present a radiation hazard to personnel.

Because of the enormous amounts of energy produced when mass is converted to energy, as calculated by Einstein's equation $E = mc^2$, nuclear processes seem to hold the key to future energy sources for the earth. Energy can be produced by either nuclear fission or nuclear fusion, with fission accounting for all present peacetime energy sources. Fusion is preferable because of abundant fuel and low radioactive wastes, but its development as an energy source is not yet complete.

REVIEW QUESTIONS

1. Define the terms atomic number, neutron number, and mass number.

2. Explain the meaning of the term isotope.

3. What are the three most common types of nuclear radiation, and what is the physical makeup of each?

4. Why are there only a few naturally occurring isotopes for each chemical element?

5. What is meant by radioactive half-life? By biologic half-life?

6. When radium is found in nature, uranium and lead are also found. Explain why.

7. If radium were stored in a sealed container, could the decomposition of the radium be stopped? Explain.

8. Explain why a radon seed could be permanently implanted for radiation therapy but a radium seed could not.

9. What is meant by "reproductive death" in radiation damage? Why are radiation damage effects not evident until a considerable time after exposure?

10. Explain the origin of the "therapeutic ratio" in the treatment of cancer by radiation; i.e., why does radiation damage the cancer more than the surrounding normal cells? What effect does the presence or absence of oxygen have on the therapeutic ratio?

11. What is meant by "tagging" an albumin molecule with radioiodine? What effect does this have on the radioactive decay rate of the radioisotope iodine-131?

12. From the scale model of the atom and the nature of chemical interactions, can you suggest why nuclear decay rates are found to be independent of all external physical and chemical conditions?

13. What other physical phenomena besides nuclear radiation intensity drop off with the distance from the source according to the inverse square law?

PROBLEMS

1. A cube of matter which is 0.5 mm on each side would be approximately the size of the period at the end of this sentence. If this matter had the density of the nucleus, what would be its mass? What would be its weight in pounds?

2 A cobalt-60 gamma ray has an energy of 123 keV. If it takes 13.6 eV to ionize a hydrogen atom, how many hydrogen atoms could theoretically be ionized by this gamma ray photon?

3. Determine the type of radiation emitted in the following radioactive decays.

   a. Radium-226 decays to radon gas.

$$^{226}_{88}\text{Ra} \rightarrow \times + ^{222}_{86}\text{Rn}$$

b. Iodine-131 decays to xenon

$$^{131}_{53}I \rightarrow \times + ^{131}_{54}Xe$$

4. To determine a patient's blood volume, a small volume of solution containing 200 microcuries (μCi) of iodine-131 is injected into a vein. After 10 minutes to allow circulation of the radioactive material, a 1 cm^3 volume of blood is taken from the patient and found to have 0.025 μCi of the radioiodine. What total blood volume does this indicate, assuming uniform distribution of the tracer? Would you have to worry about loss of the radioiodine by nuclear decay during the 10 minute test?

Worked Example: A patient who is given a 100 millicurie therapeutic dose of radiogold (gold-198, half-life 2.7 days) will subject a person at a distance of 2 ft from the bed to a radiation dose of about 15 mrem/hr. How long would it take for the dosage level to come down to the maximum permissible level of 2.5 mrem/hr for continuous exposure during a 40 hr work week? Calculate the time by assuming that no biological elimination of the radioisotope occurs, and then calculate it assuming a 5 day biological half-life.

Solution: With the physical half-life, 2.7 days, the radiation level at 2 ft would be reduced as follows:

Number of half-lives	Time in days	Radiation dose level
0	0	15 mrem/hr
1	2.7	7.5 mrem/hr
2	5.4	3.75 mrem/hr
3	8.1	1.88 mrem/hr

After 8 days, continuous exposure at a distance of 2 ft from the bed would provide less than the maximum permissible dosage. [For an exact calculation of the time required to drop to 2.5 mrem/hr, a logarithmic relationship must be used. Time = (2.7 days)(log (D_0/D)/log 2) = 6.97 days. Calculation to the nearest half-life is usually sufficient.]

If the radioisotope were eliminated from the body with a biological half-life of 5 days, in addition to the physical decay, the effective half-life would be

$$T_{\text{effective}} = \frac{T_P T_B}{T_P + T_B} = \frac{(2.7)(5)}{2.7 + 5} = 1.75 \text{ days.}$$

After three half-lives, or 5.3 days, the level would be down to 1.9 mrem/hr at a distance of 2 ft from the bed. (This calculation is not necessarily realistic in terms of the dose rate and biological half-life, and it is best to be quite conservative about the amount of radiation received. The instructions of the attending physician and the radioisotope supplier should be followed carefully.)

5. If 100 millicuries of the radioisotope gold-198 are administered for cancer therapy, how much of it will remain two weeks later if none of it is eliminated from the body by biological means?

6. An isotope A is known to undergo radioactive decay with a half-life of 300 days, forming the stable element B. A sample is examined and found to contain 2 grams of A and 398 grams of B. If it could be assumed that none of element B was present initially, how long had the sample been in its present state?

7. The half-life of radioiodine is about 8 days. If you started with 1 gram of it, how much would be left after 24 days?

8. The radioisotope iodine-131 (half-life = 8 days) is received from a medical supplier in a solution such that 0.5 cm^3 of the solution has enough activity for injection for a certain organ scan at the time of supply.

   a. If the solution is kept for 24 days, how many cm^3 must be used for the same scan?
   b. If it is impractical to use more than 8 cm^3 of the solution per scan, what is the useful life of the solution?

9. A radioisotope used for an organ scan has a physical half-life of 8 days. However, it is eliminated from the organ by biological processes at a rate characterized by a biological half-life of 3 days. What is its effective half-life in that organ? If the original dose in that organ was 100 microcuries, what dose will remain after 11 days?

10. Tritium (^3H), the heavy isotope of hydrogen, is sometimes used as a radioactive tracer for whole body scans. If it has a physical half-life of 12.3 years and a biologic half-life of 19 days, what is its effective half-life in the body?

11. The following radioisotopes are found to have the following biologic half-lives in the indicated areas of the body. Calculate the effective half-life in each case.

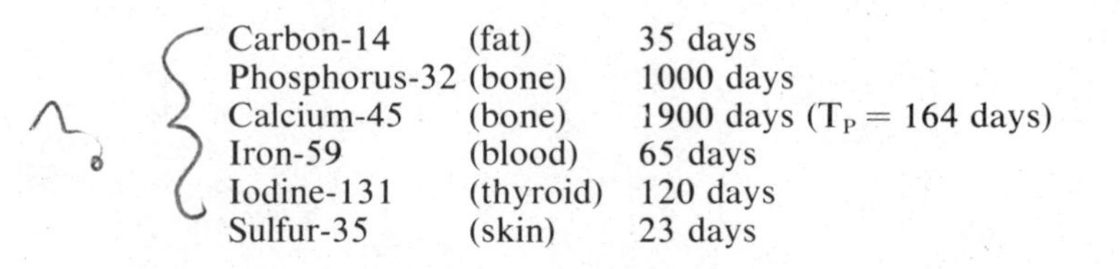

Carbon-14	(fat)	35 days
Phosphorus-32	(bone)	1000 days
Calcium-45	(bone)	1900 days (T_P = 164 days)
Iron-59	(blood)	65 days
Iodine-131	(thyroid)	120 days
Sulfur-35	(skin)	23 days

12. If you could safely spend 1 hour at a distance of one foot from a patient with a radium implant, how much time could you spend at a distance of 10 feet without a radiation hazard?

13. Radiogold, with a half-life of approximately 3 days, is given to a patient for cancer therapy. A nurse at a distance of 1 foot from the patient receives a radiation dose of 10 mrem in one hour when the radioisotope is first given.

   a. How much radiation dose would the nurse receive in an hour if she were 20 feet away?
   b. How much radiation would be received in one hour at a distance of one foot three weeks later (assuming only the 3 day physical half-life acts to remove the radioisotope)?

14. A dose of 50 mrem/hr is received at a distance of 2 ft from a given source.

   a. How long could you stay there in a work week and meet the minimum standards for radiation safety?
   b. How long could you stay at a distance of 20 feet during a work week?

15. How much mass would have had to be converted to energy in nuclear processes to supply the United States energy needs in 1970 (6.8×10^{19} joules)?

REFERENCES

1. Beiser, A. *Concepts of Modern Physics*, 2nd ed. New York: McGraw-Hill, 1973.
2. Hurst, G. S., and Turner, J. E. *Elementary Radiation Physics*. New York: Wiley, 1970.
3. Stanton, L. *Basic Medical Radiation Physics*. New York: Appleton-Century-Crofts, 1969.
4. White, H. E. *Modern College Physics*, 6th ed. New York: D. Van Nostrand Co., 1972.
5. International Commission of Radiological Protection. *Recommendations of the International Commission on Radiological Protection*. ICRP Publication No. 9. London: Pergamon Press, 1966.
6. Puck, T. T., and Marcus, P. I. *Action of X-rays on Single Mammalian Cells*. J. Exp. Med. Vol. 103, p. 653, 1956.
7. Gray, L. H., Conger, A. D., Ebert, M., Hornsey, S., and Scot, O.C. *The Concentration of Oxygen Dissolved in the Tissues at the Time of Irradiation as a Factor in Radiotherapy*. Brit. J. Radiol., 26, 1953, p. 638.
8. Schwartz, E. E., ed. *The Biological Basis of Radiation Therapy*. Philadelphia: J. B. Lippincott Co., 1966.
9. Abbott Laboratories. *Radioisotopes in Medicine*. Radio-Pharmaceutical Products Division, North Chicago, Illinois 60064, 1970.
10. Deeley, T. J., Hart, J., Clarke, E., Chaters, J. M., and McCarthy, M. *A Guide to Radiotherapy Nursing*. London: E. & S. Livingstone, 1970.
11. Howl, B. *Simplified Radiotherapy for Technicians*. Springfield, Illinois: Charles C Thomas, 1972.
12. Fisher, J. C. *Energy Crises in Perspective*. Physics Today, Dec. 1973, p.40.
13. Bushong, S. C. *Radiation Exposure in Our Daily Lives*. The Physics Teacher. Vol. 15, p. 135, March 1977.
14. Freire-Maia, A., and Krieger, H. *Human Genetic Studies in Areas of High Natural Radiation-IX. Effects on Mortality, Morbidity and Sex Ratio*. Health Physics. Vol. 34, p. 61, 1978.
15. Cohen, B. L. *The Disposal of Radioactive Wastes from Fission Reactors*. Scientific American. Vol. 236, p. 21, June 1977.

Appendix A

Tables of Units and Physical Data

Abbreviations Used

Å	angstrom unit
AC	alternating current
amp	ampere
atm	atmosphere
BTU	British thermal unit
C	degree centigrade or Celsius
c	velocity of light
Cal	food calorie (kilocalorie)
cal	calorie
CGS	centimeter-gram-second (system of units)
cm	centimeter
coul	coulomb
d	density
$\mathbf{d}_w$	weight density
D	diopter
db	decibel
DC	direct current
deg	degree
E	energy; electric field
emf	electromotive force
eV	electron volt
F	force, degree Fahrenheit
$\mathscr{F}$	volume flow rate
f	frequency
ft	foot
gm	gram
g	acceleration due to gravity
h	Planck's constant
HP	horsepower
hr	hour
Hz	Hertz (=cycle per second)
I	electric current
in	inch
j	joule
K	degree Kelvin
kcal	kilocalorie(= Cal)
kg	kilogram
kHz	kilohertz (1000 cycles/sec)
km	kilometer
kW	kilowatt
kWh	kilowatt-hour
l	liter
λ	wavelength
lb	pound
m	meter
mEq	milliequivalent
MeV	million (or mega-) electron volts
mg	milligram
min	minute
MKS	meter-kilogram-second
ml	milliliter
mph	miles per hour
nt	newton
oz	ounce
π	pi = 3.14
Q	energy gained or released
sq	square
T	temperature
t	time
τ	torque
v	voltage
W	watt
$\mathscr{W}$	work
yr	year

TABLE T–1 The Three Basic Unit Systems

QUANTITY	MKS	CGS	BRITISH
Length	meter	centimeter	foot
Time	second	second	second
Mass	kilogram	gram	slug
Velocity	m/sec	cm/sec	ft/sec
Acceleration	m/sec^2	cm/sec^2	ft/sec^2
Force	newton (kg-m/sec^2)	dyne (gm-cm/sec^2)	pound (slug-ft/sec^2)
Work, energy	joule (nt-m)	erg (dyne-cm)	ft-lb
Power	watt (joule/sec)	erg/sec	ft-lb/sec
Torque	nt-m	dyne-cm	lb-ft
Pressure	nt/m^2	dyne/cm^2	lb/ft^2

TABLE T–2 Conversion Factors

A convenient way to accomplish unit conversions is to multiply by an appropriate numerical factor such that the old units are canceled and the desired units remain. For example, since there are 2.54 cm in one inch, a length of 10 inches can be converted to cm by the multiplication:

$$10 \text{ in} = (10 \cancel{\text{in}}) \left(2.54 \frac{\text{cm}}{\cancel{\text{in}}}\right) = 25.4 \text{ cm}.$$

When there is more than one unit to be converted, a series of factors may be used so that all the old units are eliminated. For example:

$$60 \frac{\text{miles}}{\text{hour}}$$

$$= \left(60 \frac{\cancel{\text{miles}}}{\cancel{\text{hour}}}\right) \left(5280 \frac{\text{ft}}{\cancel{\text{mile}}}\right) \left(\frac{1}{3600} \frac{\cancel{\text{hour}}}{\text{second}}\right)$$

$$= 88 \frac{\text{ft}}{\text{sec}}.$$

TABLE T–2 Conversion Factors *(continued)*

To Convert:			Multiply by:
inches	to	cm	$2.54 \frac{cm}{in}$
feet	to	cm	$30.48 \frac{cm}{ft}$
meters	to	feet	$3.28 \frac{ft}{m}$
miles	to	feet	$5280 \frac{ft}{mi}$
miles	to	meters	$1609 \frac{m}{mi}$
miles	to	kilometers	$1.609 \frac{km}{mi}$
angstroms(Å)	to	cm	$10^{-8} \frac{cm}{angstrom}$
angstroms(Å)	to	meters	$10^{-10} \frac{meters}{angstrom}$
miles/hr	to	ft/sec	$1.467 \frac{ft/sec}{mi/hr}$
m/sec	to	mi/hr	$2.24 \frac{mi/hr}{m/sec}$
pounds	to	newtons	$4.45 \frac{nt}{lb}$
dynes	to	newtons	$10^{-5} \frac{nt}{dyne}$
pounds	to	dynes	$4.45 \times 10^{5} \frac{dyne}{lb}$
slugs	to	kilograms	$14.59 \frac{kg}{slug}$
lb/ft^2	to	$newtons/m^2$	$47.88 \frac{nt/m^2}{lb/ft^2}$
lb/in^2	to	$newtons/m^2$	$6895 \frac{nt/m^2}{lb/in^2}$
atm	to	$newtons/m^2$	$1.013 \times 10^{5} \frac{nt/m^2}{atm}$
atm	to	lb/in^2	$14.7 \frac{lb/in^2}{atm}$
lb/in^2	to	mm Hg	$51.7 \frac{mm\ Hg}{lb/in^2}$
mm Hg	to	mm H_2O	$13.6 \frac{mm\ H_2O}{mm\ Hg}$
cm H_2O	to	$dyne/cm^2$	$980 \frac{dyne/cm^2}{cm\ H_2O}$
mm Hg	to	$dyne/cm^2$	$1333 \frac{dyne/cm^2}{mm\ Hg}$
$dyne/cm^2$	to	$newtons/m^2$	$0.1 \frac{nt/m^2}{dyne/cm^2}$
mm Hg	to	$newtons/m^2$	$133.3 \frac{nt/m^2}{mm\ Hg}$
cm H_2O	to	$newtons/m^2$	$98 \frac{nt/m^2}{cm\ H_2O}$
cm^3	to	m^3	$10^{-6} \frac{m^3}{cm^3}$
in^3	to	m^3	$1.639 \times 10^{-5} \frac{m^3}{in^3}$
in^3	to	cm^3	$16.39 \frac{cm^3}{in^3}$
ft^3	to	m^3	$2.832 \times 10^{-2} \frac{m^3}{ft^3}$
liters	to	m^3	$.001 \frac{m^3}{l}$
gallons	to	m^3	$3.785 \times 10^{-3} \frac{m^3}{gal}$
gallons	to	cm^3	$3785 \frac{cm^3}{gal}$
gallons	to	in^3	$231 \frac{in^3}{gal}$
ft-lb	to	joules	$1.356 \frac{joule}{ft\text{-}lb}$
ergs	to	joules	$10^{-7} \frac{joule}{erg}$
calories	to	joules	$4.186 \frac{joule}{cal}$
kilocalories	to	joules	$4186 \frac{joule}{kcal}$
kilowatt-hours	to	joules	$3.6 \times 10^{6} \frac{joules}{kWh}$
electron-volts	to	joules	$1.602 \times 10^{-19} \frac{joule}{eV}$
ft-lb/sec	to	watts	$1.356 \frac{watts}{ft\text{-}lb/sec}$
horsepower	to	watts	$746 \frac{watts}{HP}$
erg/sec	to	watts	$10^{-7} \frac{watts}{erg/sec}$

At sea level, where $g = 980.665\ cm/sec^2 = 32.174\ ft/sec^2$:

1 kg weighs 2.2 lb
1 slug weighs 32.174 lb
453.6 gm weighs 1.0 lb

The conversion factors are listed with units so that the indicated conversion can be accomplished by multiplication. The reverse conversion can be accomplished by dividing by the numerical factor.

TABLE T-3 Physical Constants

Acceleration of gravity (standard)	$g = 980.665$ cm/sec$^2 = 9.80665$ m/sec^2 $= 32.174$ ft/sec^2
Universal gravitation constant	$G = 6.67 \times 10^{-11}$ m^3/kg-sec^2
Speed of light	$c = 2.9979 \times 10^8$ m/sec $\approx 3 \times 10^8$ m/sec $= 186{,}000$ mi/sec
Gas constant	$R = 8.32$ joules/mole-°K
Avogadro's number	$n = 6.02 \times 10^{23}$ molecules/mole
Planck's constant	$h = 6.6256 \times 10^{-34}$ joule-sec
Electronic charge	$e = 1.6021 \times 10^{-19}$ coul
Mass of electron	$m_e = 9.11 \times 10^{-31}$ kg
Mass of proton	$m_p = 1.6726 \times 10^{-27}$ kg $= 1836 m_e$
Mass of neutron	$m_n = 1.6749 \times 10^{-27}$ kg $= 1839 m_e$
Atomic mass unit	$u = 1.66 \times 10^{-27}$ kg
Coulomb's constant	$K = 9 \times 10^9$ nt-m^2/coul^2

TABLE T-4 Densities of Selected Materials

MATERIAL	MASS DENSITY GM/CM^3	WEIGHT DENSITY LB/FT^3
A. Liquids		
Water at 4°C	1.000	62.6
Water at 20°C	0.998	62.4
Gasoline	0.7	44
Mercury	13.6	850
Milk	1.03	64
B. Solids		
Magnesium	1.7	106
Aluminum	2.7	169
Copper	8.3–9.0	520–560
Gold	19.3	1200
Iron	7.8	490
Lead	11.3	708
Platinum	21.4	1334
Uranium	18.7	1167
Osmium	22.5	1409
Ice (0°C)	0.92	58
C. Gases at 0°C and 1 atmosphere pressure		
Air	1.293×10^{-3}	0.0807
Carbon dioxide	1.977×10^{-3}	0.1234
Carbon monoxide	1.250×10^{-3}	0.0781
Hydrogen	0.090×10^{-3}	0.0056
Helium	0.178×10^{-3}	0.0111
Nitrogen	1.251×10^{-3}	0.0781
Nitrous oxide	1.978×10^{-3}	0.1235
Oxygen	1.429×10^{-3}	0.0892

TABLE T-5 Thermal Expansion Coefficients

MATERIAL	FRACTIONAL EXPANSION PER DEGREE AT 20°C (68°F)	
A. Linear expansion coefficients	α(1/°C)	α(1/°F)
Glass, ordinary	9×10^{-6}	5×10^{-6}
Glass, pyrex	4×10^{-6}	2.2×10^{-6}
Quartz, fused	0.59×10^{-6}	0.33×10^{-6}
Aluminum	24×10^{-6}	13×10^{-6}
Brass	19×10^{-6}	11×10^{-6}
Copper	17×10^{-6}	9.4×10^{-6}
Iron	12×10^{-6}	6.7×10^{-6}
Steel	13×10^{-6}	7.2×10^{-6}
Platinum	9×10^{-6}	5×10^{-6}
Tungsten	4.3×10^{-6}	2.4×10^{-6}
Gold	14×10^{-6}	7.8×10^{-6}
Silver	18×10^{-6}	10×10^{-6}
B. Volume expansion coefficients	β(1/°C)	β(1/°F)
Alcohol (ethyl)	10×10^{-4}	5.5×10^{-4}
Glycerine	5×10^{-4}	3×10^{-4}
Mercury	1.8×10^{-4}	1.0×10^{-4}
Water	2.1×10^{-4}	1.2×10^{-4}
Ether	16.6×10^{-4}	9.2×10^{-4}
Acetic acid	10.7×10^{-4}	5.9×10^{-4}

TABLE T–6 Specific Heats of Common Substances

MATERIAL	SPECIFIC HEAT (CAL/GM-°C OR BTU/LB-°F)
Water at 15°C	1.00
Ice at 0°C	0.51
Steam (at constant pressure)	0.48
Aluminum	0.217
Brass	0.090
Copper	0.092
Gold	0.031
Iron	0.11
Lead	0.030
Silver	0.056
Glass (ordinary)	0.16
Alcohol (ethyl)	0.60
Glycerine	0.60
Mercury	0.033
Wood	0.4
Porcelain	0.26
Human body (avg.)	0.8

TABLE T–7 Thermal Conductivities of Common Substances in (cal/sec)/(cm^2 × °C/cm)

MATERIAL	THERMAL CONDUCTIVITY (k)
Silver	1.01
Copper	0.99
Aluminum	0.50
Iron	0.163
Lead	0.083
Ice	0.005
Glass, ordinary	0.0025
Concrete	0.002
Water at 20°C	0.0014
Asbestos	0.0004
Hydrogen at 0°C	0.0004
Helium at 0°C	0.0003
Snow (dry)	0.00026
Fiberglass	0.00015
Cork board	0.00011
Wool felt	0.0001
Air at 0°C	0.000057

Appendix B

Arithmetic Operations Using Powers of Ten

The use of the powers of ten notation offers many advantages in the handling of very large or very small numbers. The arithmetic operations normally encountered can be handled with the use of a few rules. In *addition* or *subtraction,* all numbers should be expressed to the same power of ten.

Example. To perform the addition

$$1.43 \times 10^4 + 5.6 \times 10^2 + 2.1 \times 10^3,$$

the numbers can be rewritten as multiples of 10^4:

$$1.43 \times 10^4 + .056 \times 10^4 + .21 \times 10^4 = 1.69\underline{6} \times 10^4.$$

Since there are only three significant digits in the largest term, it would be more appropriate to write the result as 1.70×10^4.

Example. The subtraction operation

$$6.58 \times 10^8 - 3.4 \times 10^3 - 2.3 \times 10^7$$

can be performed by rewriting the numbers

$$6.58 \times 10^8 - .000034 \times 10^8 - .23 \times 10^8 = 6.35 \times 10^8.$$

Note that this process can tell you when one factor can be reasonably neglected. The number 3.4×10^3 above is well beyond the range of significant digits of the larger numbers and does not contribute meaningfully to the result.

In *multiplication* of two powers of ten the exponents are added and in *division* the power of ten in the denominator is subtracted from that in the numerator.

Example. The product $(3 \times 10^4) \times (5 \times 10^3)$ is accomplished by the operation

$$(3 \times 10^4) \times (5 \times 10^3) = 3 \times 5 \times 10^{4+3} = 15 \times 10^7 \quad \text{or } 1.5 \times 10^8.$$

Example. If the exponent is negative, the negative sign must be maintained in this process.

$$(2 \times 10^6) \times (4 \times 10^{-4}) = 8 \times 10^{6-4} = 8 \times 10^2$$

$$(4 \times 10^{-3}) \times (5 \times 10^{-6}) = 20 \times 10^{-3-6} = 20 \times 10^{-9}$$

$$\text{or} \quad 2.0 \times 10^{-8}$$

Example. In division the numbers multiplying the powers of ten are divided, and the power of ten in the denominator is subtracted from that in the numerator.

$$\frac{9 \times 10^8}{2 \times 10^3} = \frac{9}{2} \times 10^{8-3} = 4.5 \times 10^5$$

$$\frac{6 \times 10^9}{2 \times 10^{-2}} = \frac{6}{2} \times 10^{9-(-2)} = 3 \times 10^{11}$$

If a power of ten is moved from the numerator to the denominator of a fraction, or vice versa, its sign is changed.

Examples.

$$\frac{6}{3 \times 10^3} = \frac{6 \times 10^{-3}}{3} = 2 \times 10^{-3}$$

$$\frac{2 \times 10^{-4}}{3} = \frac{2}{3 \times 10^4}$$

$$\frac{1}{10^{-5}} = 10^5$$

If a power of ten is raised to a power, the two exponents are multiplied.

Examples.

$$(3 \times 10^3)^2 = 3^2 \times 10^{3\times 2} = 9 \times 10^6$$

$$(2 \times 10^6)^3 = 2^3 \times 10^{6\times 3} = 8 \times 10^{18}$$

$$\sqrt{4 \times 10^6} = (4 \times 10^6)^{1/2} = 4^{1/2} \times 10^{6\times 1/2} = 2 \times 10^3$$

Note that the number multiplying the power of ten is separately raised to the desired power. Then the exponent of ten is multiplied by the power to which the number is to be raised. Since taking the square root corresponds to raising a number to the power 1/2, it can be handled in the same way.

Answers to Odd-Numbered Problems

Chapter 1: **1.** 16 cm^2 **3.** 2.92 ± .04 cm **5.** 6.2 ± .6 m/sec, 6.15 ± .07 m/sec

Chapter 2: **1.** 76 cm, 91.4 cm **3.** 144 in^2, 10.76 ft^2 **5.** 68 mi/hr **7.** 10 m/sec, 22.4 mi/hr **9.** 80.7 ft/sec, 48 ft **11.** 134 × 10^6 mi, 2.16 × 10^{11} m **13.** .54 m **15.** (a) 10 m/sec^2, (b) 275 m **17.** (a) −25 ft/sec^2, (b) 50 ft **19.** 11 ft/sec **21.** (a) 0.5 sec, (b) −80 cm/sec^2 **23.** (a) −9.4 m/sec, (b) 0.4 m/sec **25.** 25.9 m **27.** 3.0 sec **29.** 36.4 sec, 1456 m

Chapter 3: **1.** 16 nt **3.** 6 nt **5.** (a) 5 m/sec^2, (b) 40 m **7.** (a) 20 lb down, (b) 20 lb up, (c) 32 lb down, (d) 32 lb up, (e) 12 lb, (f) 0 **9.** 20 lb **11.** 196 nt, 7 slugs **13.** (a) 39.3 ft/sec^2, (b) 2361 ft/sec **15.** 20 m **17.** (a) 10 ft/sec^2, (b) 0.8 sec **19.** 172 lb **21.** (a) 200 lb, (b) 120 lb, (c) 0 **23.** 124 nt **25.** 2246 gm **27.** 30 cm from pivot, at 20-cm mark **29.** 70 lb, 100 lb **31.** 235 nt, 500 nt **33.** (a) 96 lb, (b) 216 lb **35.** .33 ft **37.** 6 ft from lighter end **39.** (a) 16 nt, (b) 2.43 × 10^{20} nt **41.** (a) 774 lb, (b) .24, (c) 56.6 ft/sec **43.** (a) 40 m/sec^2, (b) 3486 nt **45.** 176 ft braking distance, both cars **47.** friction coefficient .06

Chapter 4: **1.** (a) 1000 joules, (b) 200 watts, (c) to overcome resistance to motion **3.** (a) 20 m/sec, (b) 40 m, (c) 1000 joules, (d) 250 watts, (e) 1000 joules **5.** 7840 joules, 19.8 m/sec **7.** 480,000 nt, 108,000 lb **9.** 500 watts **11.** 1.2 horsepower **13.** 40,000 watts **15.** (a) 98 nt, (b) 196 joules, (c) 196 joules, (d) 19.6 nt **17.** 392 joules by friction, zero by gravity **19.** 406,880 ft-lb, 102 ft high **21.** 50 lb, M.A. = .2 **23.** (a) 40 lb, (b) 4, (c) 120 lb, (d) 80 ft-lb

Chapter 5: **1.** 1374 lb **3.** 2.3 × 10^6 lb/ft^2 **5.** 0.52 cm Hg, 7 cm H_2O **7.** 5.2 lb/in^2 **9.** 100 lb, output work 100 ft-lb, 5 ft input distance **11.** 1.67 gm/cm^3 **13.** 2.2 gm/cm^3 density, specific gravity 2.2 **15.** (a)

400 cm^3/min, (b) 100 cm^3/min, (c) 3200 cm^3/min, (d) 50 cm^3/min **17.** 41 cm^3/min **19.** 3.5 cm^3/min

Chapter 6: **1.** (a) 32.9 lb/in^2, (b) 0.7 liter **3.** 150 ft^3 **5.** 95.2 cm high **7.** 289 lb **9.** 20 liters **11.** 2569 lb/in^2 gauge pressure **13.** 150°K **15.** (a) 1.33 liters, (b) 24 liters, (c) 4 liters **17.** 33,000 lb **19.** 2 cm^3 **21.** 15.8 ft^3 **23.** 34.3 lb/in^2 (gauge) **25.** 13.4 ft maximum depth

Chapter 7: **1.** 480 mm Hg **3.** (a) 66 cm^3/min, 152 mm Hg pressure to restore normal flow, (b) 24 cm^3/min, 416 mm Hg pressure required to restore normal flow **5.** 260 cm^3/min **7.** 50 cm/sec where area is 2 cm^2, 200 cm/sec where area is 0.5 cm^2, volume flow rate 100 cm^3/sec at all points **9.** 3.77×10^{-9} cm^3/sec per capillary, 2.1×10^{10} capillaries required **11.** (a) 60 cm^3/sec, (b) 60 cm^3/sec, (c) 0.03 cm/sec **13.** 8.5×10^4 dyne/cm diastolic, 19.2×10^4 dyne/cm systolic **15.** 120 lb/in^2 gauge **17.** 8.8 mm Hg, 4.7 in H_2O **19.** 27 mm Hg drop **21.** 292 mm Hg at heart **23.** 1.11 watts

Chapter 8: **1.** 19,600 erg/cm^3, 19,600 erg/cm^3 **3.** (a) 955 dyne-sec/cm^5 for 60 cm tube, (b) 1274 dyne-sec/cm^5 for 5 cm tube **5.** 3.57 cm H_2O at end of 40 cm **7.** 1.3 cm H_2O drop

Chapter 9: **1.** 13.55% **3.** 6.2%

Chapter 10: **1.** 20°C, −17.8°C, 37.8°C **3.** 40°C **5.** .62 inches **7.** 10,560 calories **9.** 37,208 cal **11.** 25°C **13.** 5,000 cal **15.** 60°C **17.** 392 sec **19.** 0.26°C **21.** 103.1°F **23.** 404 kcal **25.** 1.75 kcal/gm

Chapter 11: **1.** 20°C **3.** (a) 480 sec, (b) 600 sec, (c) 3240 sec **5.** 15.65°C **7.** 36,000 cal **9.** 86.2°C **11.** 69%, dew point 57°F **13.** (a) 20% at 68°F, (b) 7.7% at 98.6°F **15.** 0.36 gm/min water supplied, 512 gm water exhaled per day, 297 kcal heat loss **17.** 6980 watts **19.** 3448 gm **21.** 5.7 watts for skin temp. 28°C, 13.4 watts for skin temperature 36°C **23.** 297 gm/hr

Chapter 12: **1.** 270 newtons, attractive **3.** 1.04×10^{-8} coulombs, fraction 2.7×10^{-14} **5.** 100 beats/min, 0.5 sec alarm threshold

Chapter 13: **1.** 6.25×10^{18} electrons/sec **3.** 2.4 ohms **5.** heater, 15 ohms, 960 watts; lamp, 60 ohms, 240 watts **7.** (a) 5 amps, (b) 24 ohms **9.** 1650 watts **11.** (a) 75 volts, (b) 5 amps **13.** (a) 15 ohms, (b) 0.6 watts **15.** (a) series, 1.5 amps, 180 watts, (b) parallel, 8 amps, 960 watts **17.** 10 amps, 12 ohms **19.** (a) 119.6 volts, (b) 117 volts **21.** (a) 0.5 ohms, (b) 118.8 volts at appliance, (c) 240 amps **23.** (a) 30 amps, (b) 2 amps **25.** (a) 200 watts, (b) 2203 watts, (c) 401,601 and 1202 watts with R_1, R_2, and R_3, (d) 1001 watts (R_1,R_2), 1602 watts (R_1,R_3), and 1803 watts (R_2,R_3) **27.** 9.6 kWh, 29¢

Chapter 14: **1.** (a) through intact skin, 2000 volts, (b) through direct path to heart, 1.0 volt **3.** 4000 volts

Chapter 17: **1.** 2.5 ft, **3.** 8.0×10^{-5} watts/cm^2 at 1 meter, 2.0×10^{-5} watts/cm^2 at 2 meters, 0.88×10^{-5} watts/cm^2 at 3 meters **5.** 178 Hz, 110 Hz, 68 Hz **7.** 722 m **9.** 7.5×10^{-5} meters

Chapter 18: **1.** 50 db **3.** 30 db **5.** 8.0×10^{-6} watts/cm^2, round to 10^{-5}w/cm^2, 110 db **7.** (a) I_A/100, (b) I_B = 60 db, (c) 1000 ft **9.** 0.56 ft for 2000 Hz, 0.22 ft for 5000 Hz **11.** 70 db **13.** 10,000

Chapter 19: **1.** 5 cm, 20 diopters **3.** object distance 4 cm, magnification 3, image size 9 cm **5.** 3.33 cm **7.** (a) 12 cm, (b) magnification 2, image size 4 cm **9.** (a) 10.34 cm, (b) magnification 29, image size 101.5 cm **11.** 51.3 mm **13.** 5 **15.** for object at 5 m, image distance 5.05 cm, image size 1.8 cm; for object at 3 m, image distance 5.08 cm, image size 3.1 cm **17.** (a) 10 cm object distance, (b) 5 cm extension tube **19.** −0.5 diopter correction **21.** (a) 38 cm, (b) −2.6 diopter correction

Chapter 20: **1.** 1.29 seconds **3.** 900 m **5.** 30 cm **7.** 1.06×10^{-5} eV, 1.28×10^{6} microwave photons **9.** 2.55 eV, 6.166×10^{14} Hz, 4862 angstroms **11.** (a) 1.87×10^{7} m/sec, (b) 0.39 angstroms

Chapter 21: **1.** mass 2.5×10^{7} kg, weight 55×10^{6} pounds **3.** (a) alpha particle, (b) beta particle **5.** 3 mCi, approximately **7.** 0.125 gram **9.** 2.2 day effective half life, approx. 3 microcuries remain **11.** 35 days for carbon-14, 14.1 days for phosphorus-32, 151 days for calcium-45, 26.6 days for iron-59, 7.6 days for iodine-131, and 18.2 days for sulfur-35 **13.** (a) 0.025 mrem/hr, (b) 0.078 mrem/hr **15.** 756 kg

INDEX

Entries followed by (t) indicate tables.

(F) (s)
Work = Force • Distance (Nt·meter (Joule) / Dyne·cm (Erg) / ft·#)

Power = Work / Time [(Nt·meter/sec (Joule/sec) (Watt)] / Dyne·cm/sec, Erg/sec / Ft·#/sec, 1 HP = 550 ft·#/sec

Power = Force • Velocity

Work = ΔKE + ΔPE / W = ΔKE

ΔKE = ½mv² / ΔPE = mGΔh

$F_B = D_L V_o$

Tension = Wt − F_B

F_B = Wt. (*)

Pressure absolute = P_{gauge} + 1 atm

Pressure = Force / area

Pressure = ρGΔh

Pressure = $D_{wt.}$ Δht.

Flow Rate = Velocity • area or F rate = Volume/Time

Mass Flowrate = $\dot{\rho}$ = F • ρ - mass Density

$V = S/T \quad S = \bar{V}T$

$S = V_oT + \frac{1}{2}aT^2$

$V_f^2 = V_o^2 + 2as$

Bernoulli's = $\Delta P = \frac{1}{2}\rho[(V_B^2 - V_A^2)]$

Poiseuille's = $\frac{8\eta L}{\pi r^4}$

$V_1A_1 = V_2A_2$